Fungi of Temperate Europe

VOLUME 1

Fungi of Temperate Europe

VOLUME 1

Thomas Læssøe
& Jens H. Petersen

Princeton University Press
Princeton and Oxford

Contents

Volume 1 (chanterelles, agarics and boletes)

Preface

"I am lost – I cannot fathom nature anymore . . ." exclaimed one of the authors to the other after having experienced parts of a forest completely devoid of fungal fruitbodies bordering other parts with thousands. The longer you are in the game, the more you realize how little you understand. We have now been engaged in mycology and fungal identification for more than 40 years and a synthesis of our combined experience would seem desirable. It would have been easier if mycological knowledge had stabilized and many of the problems had been settled. Instead, we experience a period with constant change and much confusion. Our ignorance still appears to be monumental. . . .

We have both worked on the dissemination of mycology for our entire mycological careers: from teaching evening classes, through courses organized by, *e.g.*, the Danish Mycological Society, to courses at university level in Denmark and abroad. At all levels we have felt that well-written and well-illustrated teaching material was lacking and we have, to our best ability, tried to provide such over the years. We have published basic fungal identification guides, advanced mycological textbooks, coffee table books, fungal keys and manuals. At the end of the nineties we realized that we had to go one step further.

Advanced fungal identification is often hampered by the difficulties of finding the right genus, so we initiated the MycoKey project: digital keys to fungal genera. Digital keys can be constructed as synoptic keys, whereby the user only enters characters that are absolutely certain. The computer analyzes these data and displays matching genera. When carefully constructed, such keys can be very accurate and highly useful. Unfortunately, however, a major part of the mycological community still preferred books, with their relative easy overview, to digital flexibility. So, again, we had to rethink.

An ambitious 5-volume publication dealing with all the Danish fruitbody-forming *Basidiomycota* was planned by Jan Vesterholt, Jacob Heilmann-Clausen, Thomas Læssøe and Jens H. Petersen in connection with the ´Atlas of Danish fungi´ (2009–13). However, partly due to the premature death of Jan Vesterholt in 2011, this plan was abandoned. As a result, the large image library, collated in part for that publication, was archived. In 2014, Axel Kielland (Gyldendal Publishing) approached Jens hoping he would revise Jan Vesterholt's out of print book *Danmarks Svampe* (The Fungi of Denmark). This hefty book, which covers 1,000 species, stands as a very important and lasting contribution to Danish mycology and Jens felt it would be wrong to rewrite it. Instead, Jens came up with the idea of publishing a completely new work that had a larger geographical coverage. The concept was to base the layout and communication ideas on MycoKey. Some of the graphics had been tested at a small scale in other books by Jens and collaborators, but the idea was to present and key out more or less the whole fungal kingdom as a series of so-called 'fungal wheels'. Thomas agreed to join the project and the result is this two-volume publication with more than 80 fungal wheels, about 2,800 species accounts and almost 10,000 illustrations and photographs.

The fungal kingdom is, at present, being 'split' at an unprecedented rate, largely as a result of molecular studies. Indeed, even while this publication was on its way to the printer it is likely that new genera will have been introduced. Many species are being split based solely on molecular characters and can no longer be identified using standard keys. Nevertheless, we hope that the concept applied in this publication will help the enthusiast to better understand the complexities and not feel completely at a loss when faced with an unfamiliar fungus.

Thomas Læssøe and Jens H. Petersen

Acknowledgements

First and foremost, we owe our gratitude to the many photographers that have allowed us to use their images. Pia Boisen Hansen/Jan Vesterholt (†) and Jens Maarbjerg have given us free access to their extensive archives. Many other mycologists with large image archives have allowed us to use images without any compensation other than the credits given under each image. We are honoured and grateful to be able to present their splendid pictures. Especially generous contributions were made by Enrique Rubio Domínguez, Jean-Marc Moingeon, Lucie Zíbarová, Michal Mikšík, Thomas Stjernegaard Jeppesen and Thomas Kehlet. Without these contributions, the result would have been a much less visually appealing publication. Credits are given under each image and a list of al contributing photographers is given on page 1714.

Our colleagues from the Danish fungal atlas, Jacob Heilmann-Clausen, Tobias Frøslev and Jan Vesterholt are thanked for allowing us to use their atlas texts for individual species written during the atlas project period (2009–2013) as templates for this work. In return, all the edited texts are now available within the online atlas project.

The illustrations of fungal structures, notably the microscopical ones, were drawn by Jens H. Petersen using the following main sources as templates: *Funga Nordica* (agarics and boletes), Ryvarden & Gilbertson (polypores), Eriksson *et al.* (corticioids), Hans-Otto Baral (inoperculate discomycetes), www.ascofrance.com (various other groups of *Ascomycota*), and various specialized papers and photographs from the internet.

We owe a huge thanks to Kirsten Bjørnsson and Mogens Holm for proof reading the Danish text and to Andy and Gill Swash for copy editing the English version.

Finally, we want to thank 15. Juni Fonden for financing work with the Danish text.

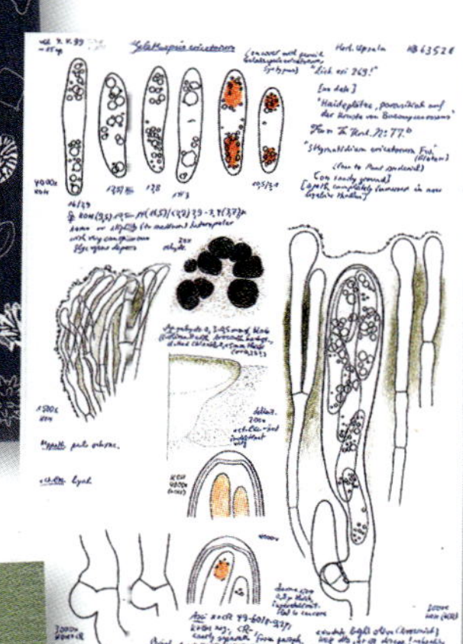

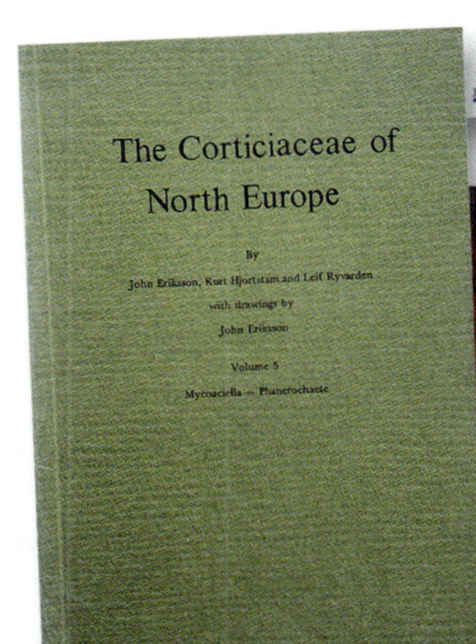

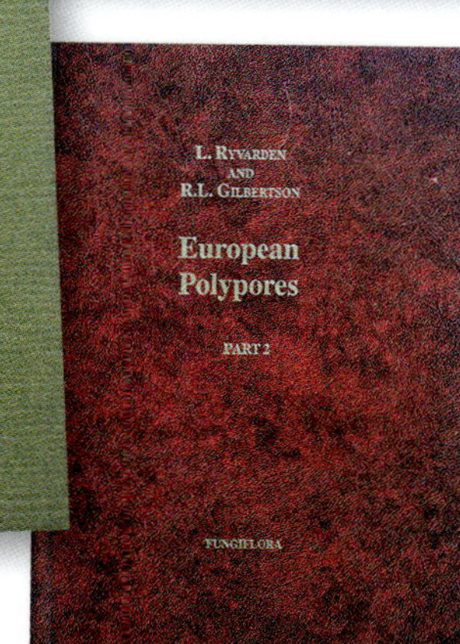

From the top Jan Vesterholt, Jens Maarbjerg, important source books, a plate from Hans Otto Baral's online archive and a photo-troll by Rolf Lidberg dedicated to Jens H. Petersen.

Scope and design

In this two-volume publication we have attempted to give a broad and deep overview of the fungi in the temperate zone of Europe. More than 2,850 fungal species from about 1,000 genera are depicted with approximately 10,000 illustrations (photographs of fruitbodies and other structures and drawings of microscopical characters). More than 1,500 additional species are discussed as potential look alikes. These figures may seem high but it is important to realize that the real fungal diversity in temperate Europe is probably in the range 15,000–20,000 species (see the discussion under the fungal kingdom p. 10). This publication is therefore nowhere near complete in terms of species coverage, and for accurate identification it may be necessary to use more specialized literature or perform DNA-sequencing of difficult material.

We have defined temperate Europe as extending from northern Norway down to the Alps, northern France and northern Spain (see the map, p. 20). We have not covered the fungi from arctic-alpine habitats that occur in parts of Scandinavia and in the Alps or Pyrenees. Towards the east we cover the Baltic states and Poland, but only in part the steppe regions of Hungary and Ukraine.

The fungi covered in this publication have been organized according to fungal phyla. The major groups are the fruitbody-forming *Basidiomycota* (spores on basidia) and *Ascomycota* (spores in asci), p. 12, 24. and 42. At the end of Volume 2, fungi without fruitbodies (p. 1622, 1626) and non-fungal groups traditionally studied by mycologists are outlined (p. 1646).

Most importantly, however, we have divided the fungi into manageable 'form groups' (see p. 42), *e.g.* gilled fungi, polypores, corticioids, pyrenomycetous fungi. Within the gilled fungi we have attempted to cover almost all the genera known from temperate Europe, while in some ascomycote groups only examples of characteristic, spectacular species are given (see more on coverage in the box below).

Individual species are treated with a short description of the most important characters, including spore measurements and ecological details, a section covering confusion species and, at the end, a section summarizing distribution and frequency within temperate Europe. Species mentioned in the discussion are marked with arrows or page references indicating where they are treated

Examples of the coverage in various groups

Agarics (gilled fungi)
Temperate Europe > 2,500; FTE 1,285 species. Nearly all genera so far described are included. For many small genera the coverage is 100%, but for large genera, such as *Cortinarius* with 125 species included, the coverage is about 20%.

Boletes
Temperate Europe > 100; FTE 74 species. Nearly all genera are covered and species coverage is high.

Polypores
Temperate Europe > 400, FTE 192 species. Most genera are covered and species coverage is fairly high.

Clavarioid fungi
Temperate Europe > 200; FTE 104 species. Almost all genera are covered and species coverage for small genera is high, but rather poor in, *e.g.*, the large genus *Ramaria*.

Corticioid fungi and jelly fungi
Temperate Europe > 800; FTE 217 species. All very common and easily recognizable genera and species are included. Species coverage is moderate.

Operculate cup-fungi
Temperate Europe > 650; FTE 136 species. The majority of the genera are covered, but species coverage is rather poor.

Inoperculate cup-fungi
Temperate Europe > 1,000; FTE 257 species. Very common or spectacular genera and species are included. Species coverage is rather poor.

Pyrenomycetous fungi (true and false flask fungi)
Temperate Europe > 1,000; FTE 209 species. A selection of common and spectacular species are included; coverage is poor.

Rusts, smuts, lichens, powdery mildews, *Laboulbeniales*, asexual fungi (hyphomycetes, coelomycetes) and slime moulds (Mycetozoa)
These groups are only covered by examples and identification of such fungi from this publication should not be attempted.

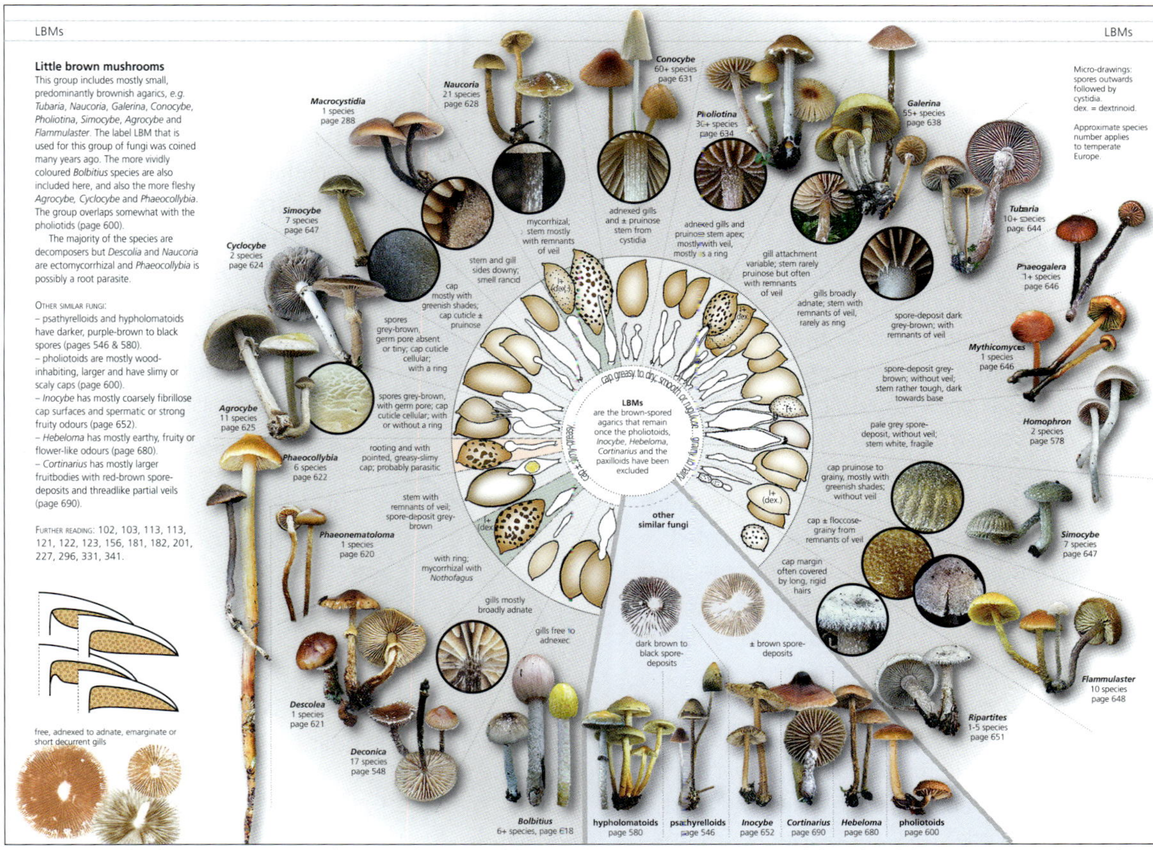

An example of a fungal wheel, one of the key features of this publication. The outer part of the wheel shows cut-out images of fruitbodies and enlarged details or tiny fruitbodies in inset circles. The inner part shows spores at approximately the same scale relative to each other. Green and red backgrounds indicate whether the groups are mutualists (mycorrhizals or lichens) or parasites.

or with a ⋉, indicating that they are not included elsewhere in the publication. Recommended edibles have a specific symbol attached, as do highly poisonous fungi (see box below).

We have not included detailed descriptions of the morphology of each species, partly because such descriptions are hard to follow and take up considerable space, and partly because images convey the message better than words. This has allowed room for more pictures showing the details that we consider to be diagnostic.

All the chapters in the systematic part of this publication start with one or more identification wheel introducing the group in question. Contrary to dichotomous, analytical keys, these wheels are not very rigid but should, by means of pictures and small texts, point the user in the right direction. We recommend starting at the bottom left of the wheel and working clockwise. The user can, depending on individual inclination, follow the pictures, or try to follow the internal logic of the wheel. This will hopefully provide a good lead that will make an identification possible when turning to the relevant species page. Several leads may have to be followed before an identification is confirmed.

The individual species accounts are presented in approximately the same order as in the wheels, and the page headings can be used as a guide to the contents of the page spreads.

In the process of writing and designing this publication, Jens H. Petersen has constructed the wheels, designed the entire work, photographed and processed the majority of the images and written some of the texts. Thomas Læssøe has written the majority of the texts and provided some of the images.

Edibility symbols – only important edible and seriously/deadly poisonous species are marked.

- ● acceptable edible after proper cooking
- ●● high quality edible after proper cooking
- †? presumed seriously poisonous
- † seriously poisonous
- †† deadly poisonous

References to presentation of the species

- △ = found above on the same page
- ▽ = found below on the same page
- ◁ & ▷ = found on opposite page
- ◁◁ = earlier in the chapter
- ▷▷ = later in the chapter
- ⋉ = not treated further

Why study fungi?

In nature, fungi constitute a vast group of organisms. Currently, each species of plant is expected to be matched by 6–10 species of fungi (see page 12). The only other group of eukaryotes (organisms with a nucleus in the cells) that surpass the fungi in diversity are the insects. For this reason alone, the study of fungi is of enormous importance if we want to attempt to understand nature.

Fungi are, nevertheless, a poorly known group of organisms. This is partly caused by their hidden life within various substrates, and partly caused by the relatively low number of professional mycologists compared to, for eample, the number of botanists or ornithologists. Typically, fungi are only noted when they produce fruitbodies, and in some species this is an infrequent event, even decades may pass between the fruitings of some individual fungi; this presents a huge challenge in their study.

Fungi play a very important role in ecosystems. They are responsible for a major part of the turnover of organic material in nature. They can, via enzymes, degrade plant material so it again becomes part of the nutrient cycle. They are also indispensable in plant growth (see p. 16). A comprehensive description of an ecosystem should always include the role that fungi play.

Fungi break down organic compounds, including cellulose and lignin, into nutrients that can be utilized in various ways by humans. Compared to the meat industry, fungi carry out this function without producing substantial emissions of greenhouse gasses. Thus, fungal protein helps feed an ever-growing human population in a sustainable way, and it also has vital amino acids that can be difficult to obtain via a plant-based diet.

Fungi are also important in the pharmaceutical industry. The antibiotic, penicillin, is, for example, produced, by a fungus in the genus *Penicillium* and many other compounds derived from fungi are in use. Not least in China, many fungal fruitbodies are used as medicines, including *Ophiocordyceps sinensis*, *Grifola frondosa* and species of *Ganoderma*. They are thought to contain antioxidants and other secondary metabolites that can boost the immune system and protect against life-threatening diseases. Other fungi are, through genetic engineering, used as vehicles in the production of pharmaceuticals.

Last but not least, fungi constitute a charming and beautiful group of organisms that constantly fascinate and surprise naturalists and mycologists. To learn more about these fascinating organisms, read the book *The Kingdom of Fungi* by Jens H. Petersen, also published by Princeton University Press.

The fungal kingdom

The fungal kingdom can crudely be defined as organisms that are made up of microscopic, thread-like cells (hyphae) and that propagate by tiny, airborne propagules (spores). From the hyphae most species form a complex network, a mycelium, which is able to take up nutrients from its surroundings. The hyphae are generally divided into compartments by transverse walls (septa) which have pores that allow the passage of cytoplasm, organelles and, in some cases, also nuclei.

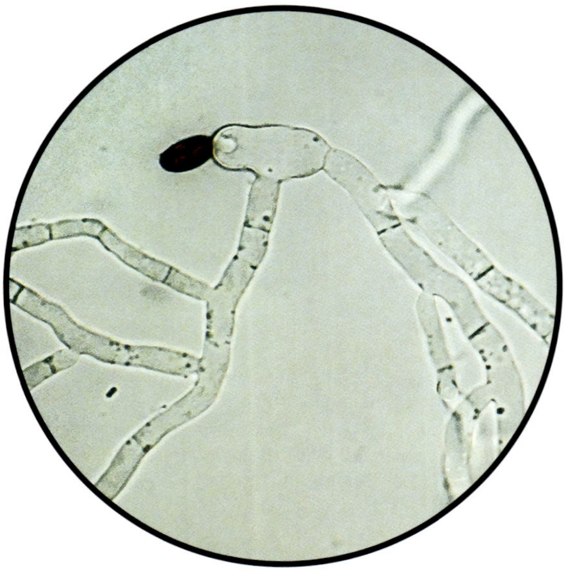

A *Coprinopsis* spore that has germinated to produce a branched mycelium with septa.

A mycelium can thus be viewed as made up of many cells or as one large, branched cell.

The majority of the species included in this publication also produce fruitbodies. These are small to large, more or less fleshy structures that are formed in connection with the sexual processes. Certain fungal groups do not form hyphae but only rounded, tiny cells that propagate asexually by budding. Such fungi are known as yeasts. Yet other species can produce both stages: a hyphal stage under certain conditions and a yeast stage under others. This is true for *Candida albicans* that occurs on our mucous membranes and in the gut. Yeasts are well known for their ability to ferment organic compounds, an ability that is put to good use in the brewing industry.

Fungal life-cycles have endless variations, but the majority have two main characteristics: the nuclei are, in the main part of the cycle, haploid, and there is a time gap between the fusion of cells and the fusion of nuclei (see illustration opposite page).

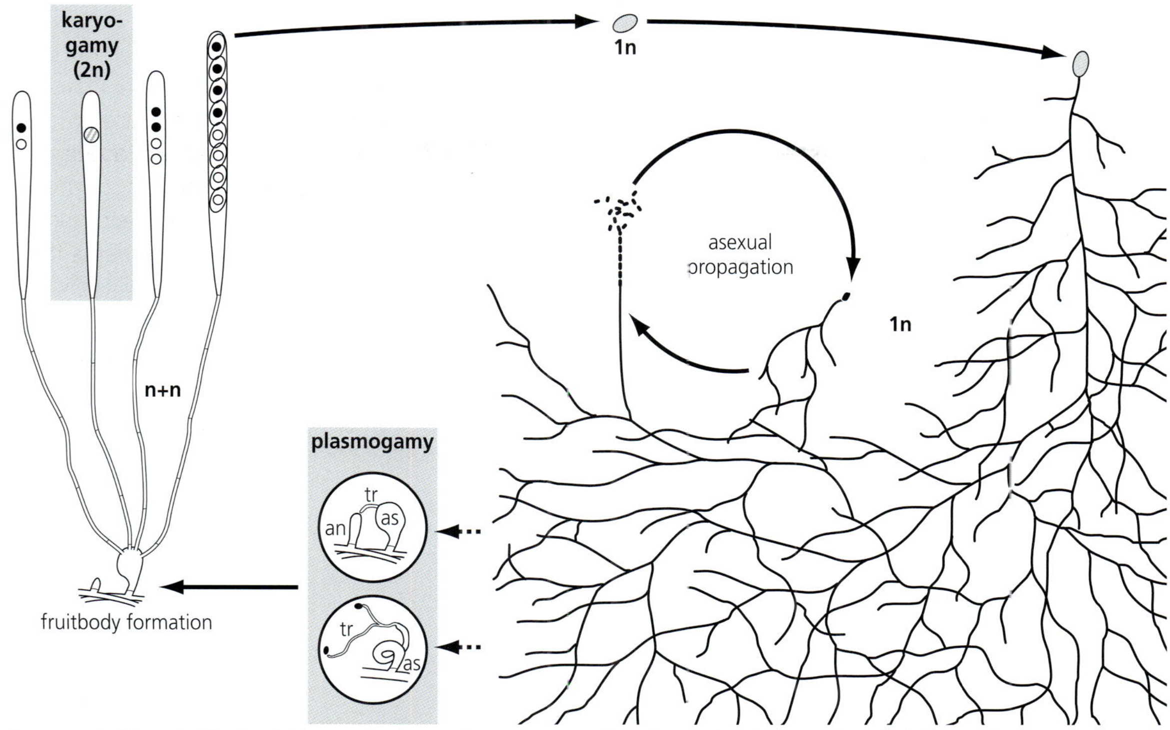

Ascomycota life-cycle. The haploid spore (meiospore) germinates and produces a haploid mycelium (1n). From this, asexual propagules (conidia) can be produced. The sexual propagation is initiated by the formation of an ascogonium (as) from where a hyphoid trichogyne appears. The trichogyne is responsible for plasmogamy with either a spermatium or an antheridium (an) and may, in this way, acquire genetically different haploid nuclei. The fruitbody is produced from two kinds of hyphae: haploid from the mycelium, and dikaryotic (n+n) from the ascogonium. The dikaryotic cells eventually form asci in a hymenium. Within the asci the karyogamy takes place with a resulting diploid (2n) nucleus that is followed by meiosis and then a mitosis, where the 8 haploid nuclei normally end up as nuclei within 8 spores (meiospores/ascospores).

Technical terms
haploid (1n): one nucleus with one set of chromosomes.
diploid (2n): one nucleus with two sets of chromosomes.
dikaryotic (n+n): each cell has two haploid nuclei.
meiosis: division of a diploid nucleus to 4 haploid nuclei.
mitosis: division retaining the number of chromosomes.

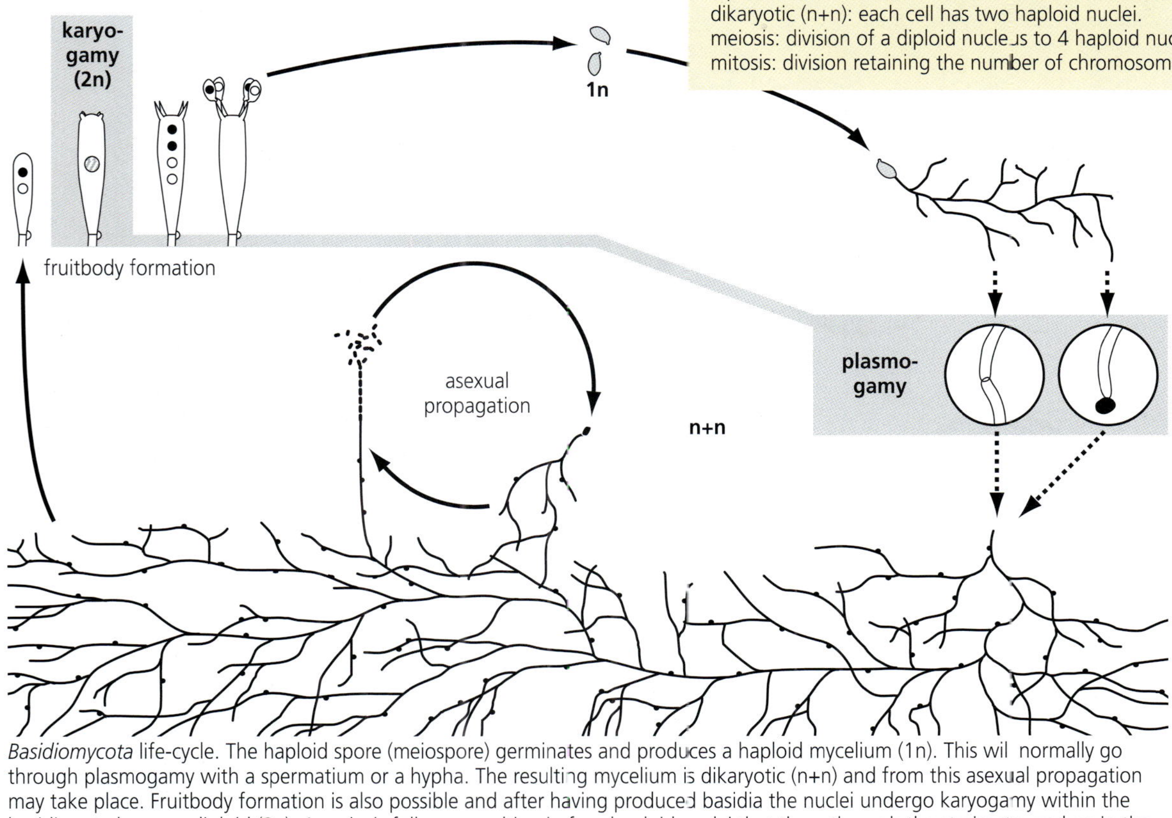

Basidiomycota life-cycle. The haploid spore (meiospore) germinates and produces a haploid mycelium (1n). This wil normally go through plasmogamy with a spermatium or a hypha. The resulting mycelium is dikaryotic (n+n) and from this asexual propagation may take place. Fruitbody formation is also possible and after having produced basidia the nuclei undergo karyogamy within the basidium to become diploid (2n). A meiosis follows, resulting in four haploid nuclei that then, through the sterimata, end up in the external spores (meiospores/basidiospores).

Many species belonging to *Ascomycota*, and the majority of *Basidiomycota*, produce fruitbodies. Within the remaining groups – *Chytridiomycota*, *Glomeromycota*, *Zygomycota* s.l. and *Blastocladiomycota* (see below) – no true fruitbodies are formed. *Neocallimastigomycota* and *Microsporidia* are here included within the fungal kingdom but may be closer to the protozoans.

The main biological diversity on planet Earth lies within the prokaryotes – organisms without nuclei – the *Bacteria* and *Archaebacteria*. Within the eucaryotes – those with nuclei – the fungal kingdom stands as number two in relation to diversity, only surpassed by the insects in the animal kingdom. The global fungal diversity is so immense that no consensus on the actual number of species has been reached. However, modern estimates have figures of between 1.5 and 5 million species, but even this is claimed to be conservative and much higher estimates have been published.

At present, more than 100,000 species of fungi are officially recognized, which means that for each known species there are presumably 15–50 unknown species. Many of these are cryptic and possibly invisible to the naked eye throughout their life-cycle. They live within plant tissues, in the soil, and in other hidden places and many never produce fruitbodies. Even amongst the more visible species,

Classification of the fungal kingdom

At present the fungal kingdom is commonly divided into eight major groups (phyla): *Microsporidia*, *Neocallimastigomycota*, *Chytridiomycota*, *Glomeromycota* (p. 1267), *Zygomycota* (p. 1267 & p. 1639–1644), *Ascomycota* (p. 1268–1625) and *Basidiomycota* (p. 46–1267). It is disputed whether or not the first two phyla belong in the kingdom or are closer to the *Protozoa*. The phylum *Zygomycota* should probably be split and all or some of the classes should be raised to phylum status. True fruitbodies are only found in the *Ascomycota*, the *Basidiomycota* and in a few in the *Zygomycota* (*Endogone*, p. 1267).

The wheel includes the phyla and most important classes of fungi. The figures give the estimated global number of species in each group.

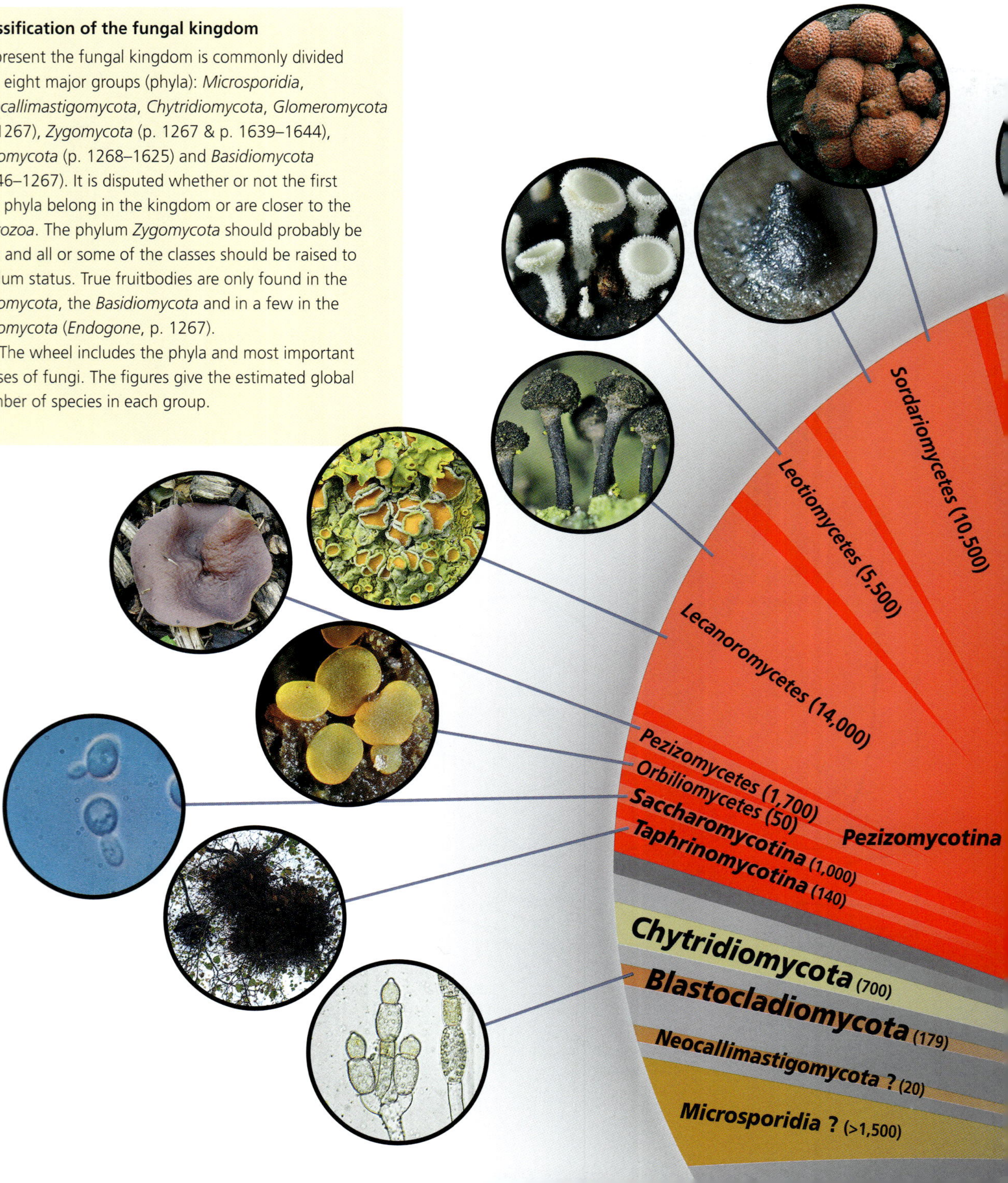

many new ones are discovered and described each year, including in Europe. Even in relatively well-surveyed Denmark, a five-year atlas project added more than 250 new species to the country's checklist, and in a plot-based biodiversity project that followed (130 plots of 40 × 40 m) more than 100 further species were added based on three visits to each plot.

In Denmark, the number of species of fungi currently known is about 8,000, and in the UK the figure is almost double that. For temperate Europe we estimate that the total number of species is higher than 20,000 and the number of fruitbody-forming fungi is at least 6,000–9,000 (including approximately about 2,000 species of lichens). As a rule, a given region should have 6–10 times as many fungal species as flowering plants.

Until near the end of the 20th century it was only possible to study fruitbody-forming fungi or fungi that produced asexual spores. With the new DNA-based techniques, fungi can be detected from any medium, including wood and soil samples. This means that our knowledge of the true occurrence of fungi in nature is increasing exponentially. It also means that many old dogmas stand to fall with regard to the nature of fungal biology and classification.

From the top: hyphae on a growth-medium in a petri-dish; brown rot; brown rot next to white rot; wood stained by *Chlorociboria aeruginascens*; wood stained by *Phanerochaete sanguinea*.

Fungal nutrition

Fungi are heterotrophic: they cannot use the energy from the sun as plants do in photosynthesis. They are thus dependent on breaking down organic substrates or gaining nutrients from living hosts or partners.

Decomposers (saprotrophs)

The majority of fungi are believed to live as decomposers, a lifestyle often termed saprotrophy. This process is carried out by penetrating the substrate with hyphae. Enzymes are released from the hyphae that break down the molecules into smaller units, which can then diffuse into the hypha through the hyphal wall. Many different enzymes have evolved in different fungi in order to decompose a range of different substrates.

The mode of nutrition has proved to be a conservative trait. Fungal genera with different nutritional modes represented have typically been found to be artificial when molecular phylogenies have been constructed. This has led to the description of many new genera. Enzyme systems are, for example, highly conservative, and white-rotting fungi with their enzymes are not closely related to brown rot fungi with another set of enzymes.

Wood is a very important fungal substrate, the primary organic compounds being cellulose and lignin. Some fungal species can only break down the fairly easily degradable cellulose and the brown lignin is left in the decayed substrate that splits in cubes; this is termed brown rot. Typical 'brown-rotters' include *Fomitopsis*, *Postia*, *Antrodia*, *Coniophora*, *Serpula* and *Tapinella*. Other species can degrade the lignin as well as the cellulose and the resulting white rot is whitish and fibrous. Typical 'white-rotters' include *Fomes*, *Tyromyces*, *Heterobasidion*, *Pleurotus*, most of the corticioids and most lignicolous agarics. There are some further types of wood degradation. *Xylobolus frustulatus* forms a rot full of small cavities, while species of *Physisporinus* form a very water-soaked soft rot recalling the inner part of a grapefruit. Some species stain the wood in a characteristic colour during decomposition: *Chlorociboria* stain wood green or turquoise; the corticioid *Phanerochaete sanguinea* stains wood orange-red; and the pyrenomycetous fungus *Camarops lutea* stains wood yellow. Some fungi, *e.g.* many pyrenomycetous fungi in the *Xylariaceae* and *Diatrypaceae*, can dry out wood and still survive with a decreased decomposition rate, but free of many competitors.

Some species are specialized decomposers of more exotic types of substrates. Members of the *Onygenaceae* are keratin-specialists. They can degrade feathers, hair, hooves *etc*. Others have even more special tastes, and can break down oils and plastic material. One species thrives on aviation fuel.

Many saprotrophic fungi are also highly specialized and are only found on certain hosts, such as the stems of just one plant genus or on the cones from certain conifers. If you pick up a fallen *Pinus* needle you may very often see transverse black lines that delimit individuals of the ascomycote *Lophodermium pinastri*. Each needle can house many individuals – and one can only imagine how common such a specialized fungus must be.

At the other end of the scale there are species that can decompose almost any plant material, *e.g.* species such as *Lepista nuda* and *Leucopaxillus giganteus*. Both can occur in woodlands and in more open habitats. Their mycelia penetrate large volumes of soil and litter, and form – like many similar fungi – fairy rings that expand outwards while decomposing all available organic material. Some of these fairy ring formers can even reach ages of many decades, and the rings can be several hundred metres in diameter. A species of *Armillaria* takes the prize for being the oldest and biggest: an individual in North America covers an area of 9.6 km^2 and has an estimated age of between 1,900 and 8,650 years.

While the generalist saprotrophs can be very old, another group of saprotrophs has a super-fast life-cycle. The coprophilous fungi, those that decompose faeces, have to colonize the substrate quickly, and sporulate quickly, before competitors (insects and other fungi) take away the nutrient source. One strategy that some species have evolved is to produce spores that can survive being eaten. The spores pass through the gut after having been ingested by a grazing animal and end up in the middle of the substrate. In just two weeks, species of *Pilobolus* can produce new sporangia and shoot them off ready to be eaten by another grazing animal. In the weeks that follow, the same dung is colonized by, *e.g.*, species of *Peziza*, *Ascobolus*, pyrenomycetous fungi and agarics. Many coprophilous fungi also use chemical warfare against the competition and, as with the wood degraders, some species can thrive in almost completely dry dung. Other fungi produce resting organs (sclerotia) in the dung, only later using the stored energy to produce fruitbodies.

From the top: *Onygens equina* on a ram's horns; mycelia on the surface coating of a photographic film; many individuals of *Lophodermiun pinastri* on a *Pinus* needle; *Phaeolepiota aurea* in a huge fairy ring; a fruiting fungal community on sheeps' dung.

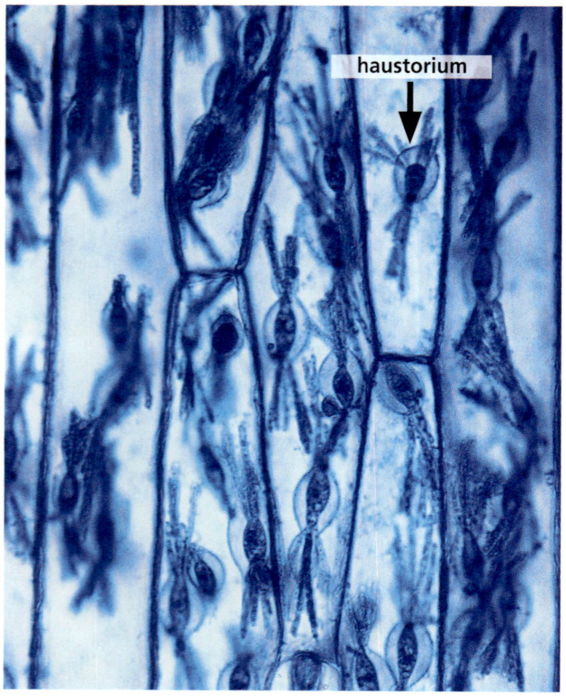

From the top: the rust fungus *Puccinia urticata* on *Urtica dioica*; the powdery mildew *Erysiphe alphitoides* on *Quercus* leaves; fruitbodies of *E. alphitoides*; haustoria within grass cells from the powdery mildew *Blumeria graminis* (photo: Hans Thordal-Christensen).

Biotrophs

Fungi that depend on other living cells to complete their life-cycle are termed biotrophs. This lifestyle is also termed a symbiosis. Plant-associated biotrophic fungi get their nutrients directly from living cells, and in some cases the transport goes both ways. Biotrophs can be divided into two groups: biotrophic parasites and mutualists.

Biotrophic parasites

Biotrophic parasites on plants exploit living plant cells without offering anything in return. Good examples are species of rust fungi (*Puccciniomycetes*) and the powdery mildews (*Erysiphales*) that mostly occur on living leaves and stems. The spores or conidia germinate on the surface and penetrate the cuticle and cell wall with special hyphae. Once inside, these special, often branched hyphae (haustoria) are used to steal the sugars produced through the photosynthesis of the plant. Attacks by such fungi can significantly diminish the harvest in plant crops.

Rust fungi and powdery mildews are very diverse groups with hundreds of species. There are other examples of less prolific parasites, *e.g.* some of the operculate discomycetes in the genera *Lamprospora* and *Octospora*. Species in these genera parasitize bryophytes and hepatics. Smut fungi form another important parasitic group, with many species attacking flowers and producing powdery smut spores inside seeds.

It is not only plants that are attacked by parasitic fungi. A number of fungal species cause symptoms on mammals, including *Homo sapiens*. Candidiasis (a vaginal yeast infection caused by *Candida albicans*) and ringworm (tinea) are fungal diseases on humans. Species of *Aspergillus* and *Schizophyllum commune* can infect lungs, typically in immunocompromized individuals.

A large number of fungi live on other fungi. They include gilled fungi such as *Volvariella surrecta* and species of *Asterophora*, many jelly fungi and various *Ascomycota*, *e.g.* many species in the *Hypocreales*.

Mutualists

A mutualistic symbiosis is an intimate, long-lasting relationship that benefits both partners. The two dominating types of mutualism are collaborations between plants and fungi via the roots (mycorrhiza), and partnership between fungi and green algae and/or cyanobacteria (lichens). In the mycorrhizal association, the two partners interact via the finer plant roots to allow photosynthesis products to reach the fungus, and nutrients and water obtained by the fungus to be transported into the plant tissues. The fine fungal mycelium can reach much larger volumes of soil than the coarser plant roots. Most plants depend on having mycorrhizal partners in order to thrive.

There are many types of mycorrhiza. One of the most important is called A-mycorrhiza (AM). Here the partner fungus penetrates the plant cell wall but not the cell membranes. It forms highly branched hyphae inside the plant cells, termed arbuscles, through which the exchange takes place. With the naked eye the rootlets appear uninfected – only microscopical analysis will reveal the mycorrhiza. All AM-fungi belong to the phylum *Glomeromycota* (p. 12–13 and 1267). This phylum is not very speciose, and no sexual fruitbodies are produced, so these fungi generally go about their business unnoticed. About 80% of all plants have A-mycorrhiza.

The most important type of mycorrhiza in relation to fruitbody-forming fungi is ectomycorrhiza (EM). With this type of mycorrhiza the fungal hyphae do not penetrate the cell walls of the partner plant, but position themselves between the cells and also form a thick mantle around the rootlets, making it easy to detect the association with the naked eye. The root tips can have a range of colours, and the root may form a characteristic branching pattern depending on the species of fungus and plant involved. The infected roots are easily found in the upper soil layers and also in very rotten wood and thick moss layers.

A large number of *Basidiomycota*, *Ascomycota* and *Zygomycota* are involved in EM. On the plant side, most of the partners involved are woody species (see table p. 18).

There are a few other types of mycorrhiza. Some species of *Morchella* have a mycorrhizal association with *e.g. Ulmus*. The relationship of some *Entoloma* species and members of the *Rosaceae* has been regarded as mycorrhizal, but now tends to be viewed as parasitic.

Orchids also depend on fungi but the evidence suggests that the orchids are the sole

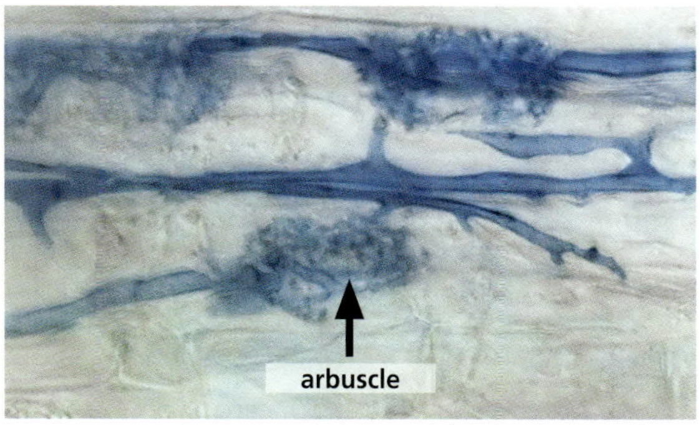

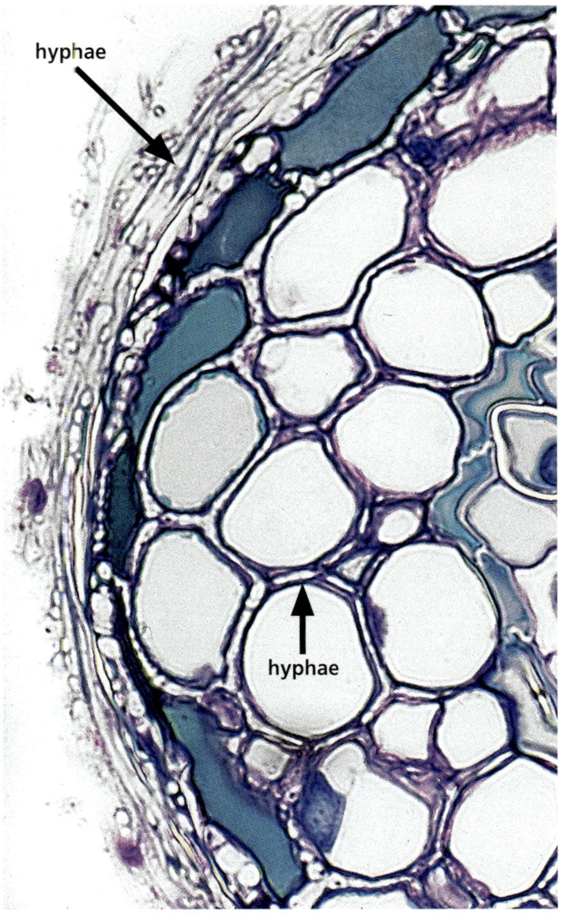

From the top: arbuscles in a root; transect of ectomycorrhizal root tip (photo Mikako Sasa); ectomycorrhizal roots with visible external mycelium.

Examples of ectomycorrhizal fungal genera:

Albatrellus, *Alpova*, *Amanita*, *Amphinema*, *Arcangeliella*, *Astraeus*, *Aureoboletus*, *Balsamia*, *Bankera*, *Boletopsis*, *Boletus*, *Byssocorticium*, *Cantharellus*, *Catathelasma*, *Chalciporus*, *Chamonixia*, *Choiromyces*, *Clavariadelphus*, *Clavulicium*, *Clavulina*, *Coltricia*, *Cortinarius*, *Craterellus*, *Descolea*, *Elaphomyces*, *Elasmomyces*, *Endogone*, *Entoloma*, *Gautieria*, *Geopora*, *Gomphus*, *Gyrodon*, *Gyroporus*, *Hebeloma*, *Helvella*, *Humaria*, *Hydnangium*, *Hydnellum*, *Hydnobolites*, *Hydnocystis*, *Hydnotrya*, *Hydnum*, *Hygrophorus*, *Hymenogaster*, *Inocybe*, *Laccaria*, *Lactarius*, *Leccinum*, *Leucogaster*, *Leucophleps*, *Lyophyllum*, *Melanogaster*, *Naucoria*, *Octaviania*, *Otidea*, *Pachyphloeus*, *Paxillus*, *Peziza* p.p., *Phellodon*, *Phylloporus*, *Picoa*, *Piloderma*, *Pisolithus*, *Plicariella*, *Porphyrellus*, *Pseudotomentella*, *Ramaria* p.p., *Rhizopogon*, *Russula*, *Sarcodon*, *Sarcosphaera*, *Scleroderma*, *Scutiger*, *Sebacina*, *Sphaerosporella*, *Stephanospora*, *Strobilomyces*, *Suillus*, *Tarzetta*, *Thelephora*, *Tomentella*, *Tomentellopsis*, *Tricholoma*, *Trichophaea*, *Tuber*, *Tylopilus*, *Tylospora*, *Wilcoxina*, *Wynnella*.

Examples of ectomycorrhizal plant genera:

Trees
Abies, *Alnus*, *Betula*, *Carpinus*, *Castanea*, *Cedrus*, *Corylus*, *Eucalyptus*, *Fagus*, *Larix*, *Nothofagus*, *Picea*, *Pinus*, *Populus*, *Pseudotsuga*, *Quercus*, *Salix*, *Tilia*.

Herbs
Helianthemum, *Bistorta*, *Dryas*.

beneficiary of the relationship, and can thus be viewed as parasites rather than mycorrhizals. Nevertheless, the literature is full of references to "orchid mycorrhiza".

We clearly have much more to learn about mycorrhizal associations in nature. A suspicion that many grassland fungi, *e.g.* waxcaps (*Hygrocybe*, *Cuphophyllus etc.*), some clavarioids (*Clavariaceae*), earthtongues (*Geoglossaceae*) and certain entolomas, have an unknown biotrophic lifestyle has long been discussed, and carbon isotope data support this assumption. Recently, it was shown that the waxcap *Hygrocybe coccinea* can be isolated from living tissues of *Plantago*, including the seeds. The details of this association have not been fully elucidated but it is assumed to be mutualistic.

Formation of a lichen is another kind of mutualism where a fungus produces a tissue (a thallus) within which green algae and/or cyanobacteria (called the photobiont) live and photosynthesize. The fungus provides a home in

Damp woodland on mull soil with a fungal community of waxcaps, clavarioids and earthtongues that is most likely to live mutualistically with plants.

environments that normally would be too hostile for an alga. In return, the photobiont produces sugars that keep the fungus alive. The photobiont may also acquire nitrogen from the air.

In the lichenization process, the fungal partner has the upper hand and shapes the thalli we recognize as *e.g.* reindeer lichens (*Cladonia*/*Cladina*). Members of the *Ascomycota* are by far the most dominant fungal partners in lichen associations, and most of them belong in the order *Lecanorales,* while the dominant algae are green algae, especially species in the genus *Trebouxia*. A few members of the *Basidiomycota* are also partners in lichens. In temperate Europe two genera are involved, the gilled *Lichenomphalia* and the clavarioid *Multiclavula*. Lichens are traditionally studied by specialized mycologists, known as lichenologists, who have developed special techniques for studying lichens and produce highly specialized literature; the subject is only touched upon in this publication (p. 1484).

From the top: thallus of *Xanthoria parietina* with orange apothecia in the centre; a section photographed through a microscope with the hymenium with asci above, and farther down a layer with cells of green algae; *Lichenomphalia hudsoniana* with a fruitbody and the blue-green, foliose thallus on the soil surface – also a thallus from a *Cladonia* to the right.

Fungal biogeography and habitats

The distribution of fungi in temperate Europe is governed by two climatic gradients, *i.e.* falling temperatures from south to north and decreasing precipitation from west to east. The fungal response to these gradients can in many cases be linked to the plant/fungus response. But there are also good examples of where the fungi themselves seem to be limited by these gradients.

Temperate climate

The temperate climatic zone is characterized by precipitation throughout the year. Different regions have precipitation maxima at different times, but on average there are no pronounced dry periods. The life-cycles of fungi are therefore to a large extent governed by the temperature variation caused by the height of the sun. The main season for fleshy fungi commences during the summer and continues until the first hard frosts. In northern parts, the season typically falls in July–August, whereas it can fall as late as December in France. In climate zones with a more pronounced dry period, the main season for fleshy fungi typically falls a few weeks into the rainy season.

Temperature

The temperature gradient from south to north plays an important role in the distribution of fungi in temperate Europe. Paris has a median August temperature of about 20°C, while Luleå in northern Sweden has a median of 14°C. Temperate Europe can be divided into a boreal zone towards the north dominated by *Picea*, *Pinus* and *Betula*. South of this a hemiboreal zone with *Quercus* can be found, and farther south again a nemoral zone with naturally occurring *Fagus* and other deciduous trees.

Some species of fungi occur throughout temperate Europe where suitable vegetation is present. *Lactarius torminosus* and *Leccinum versipelle* can, for example, be found with *Betula* in all the zones. Other species are more demanding with regard to temperature. A group of large, fleshy boletes is, towards the north, only present in thermophilic hotspots on south-facing slopes. *Rubroboletus satanas*, mycorrhizal with *Fagus*, *Quercus* and *Tilia*, occurs in warm places in southern Denmark and in Sweden on Öland and Gotland. It is absent from Norway and ± from mainland Sweden, although *Quercus* and *Tilia* occur north of both Oslo and Stockholm.

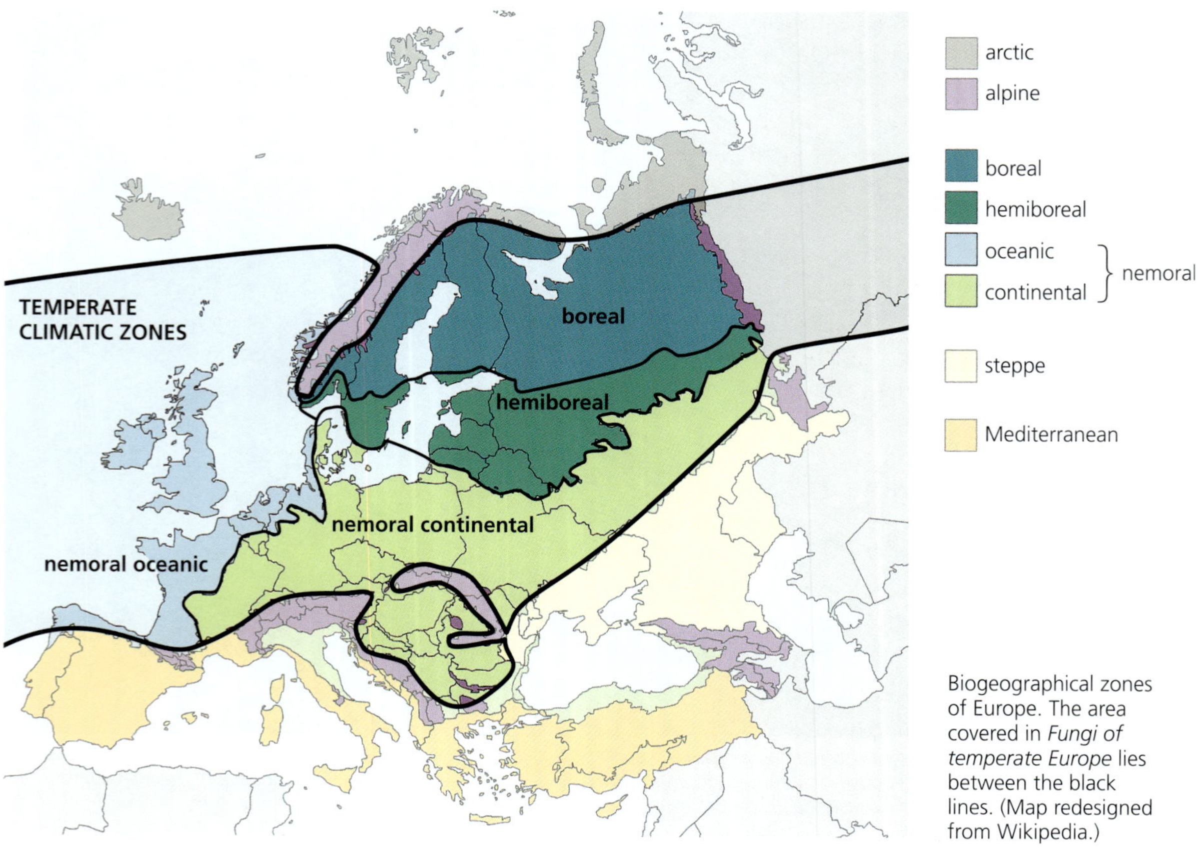

Biogeographical zones of Europe. The area covered in *Fungi of temperate Europe* lies between the black lines. (Map redesigned from Wikipedia.)

Precipitation

Western Europe with coasts along the Atlantic ocean has an oceanic climate, with wet, cool summers and mild, wet winters. Western Ireland has, for example, an annual precipitation of more than 1,400 mm, while certain places along the western coast of Norway exceed 3,500 mm (in part, also due to the influence of the steep Norwegian mountains). Some species, such as *Terana caerulea*, require a stable and wet climate, as found in westernmost Norway and western Britain and Ireland. Farther south, it can only survive on mountain slopes with high precipitation; it is common on the highest mountains in the tropics but considered extinct in Denmark and Sweden. Similarly, the polypore *Picipes tubaeformis* seems to favour extremely wet environments, also on a local scale.

Eastern Europe has, on the other hand, a continental climate, with cold winters, warm summers and less precipitation. In Kiev, the capital of Ukraine, the annual rainfall is 650 mm. A warm, dry summer climate is favourable to many gasteroid fungi, *e.g.* species of *Disciseda* and *Chlorophyllum agaricoides*. Species of the agaric genus *Floccularia* also thrive in a continental climate but have not been recorded under oceanic conditions.

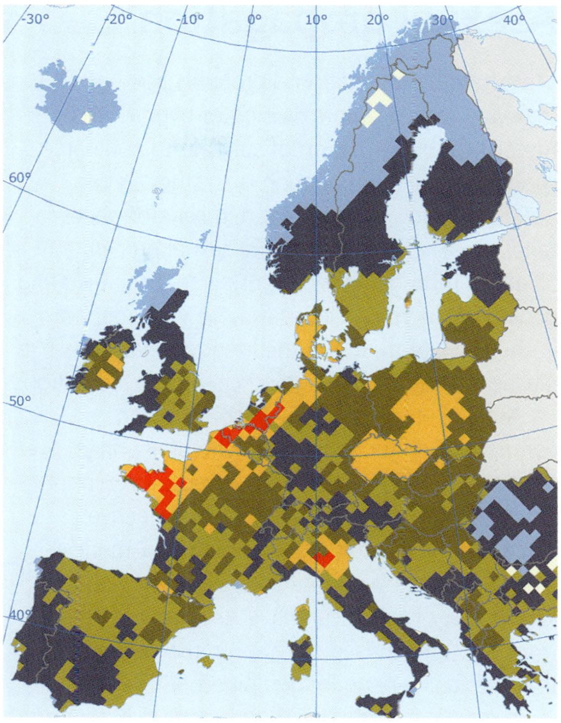

Human-induced nitrogen deposits in Europe. Blue is low, red high. Map produced by the European Environment Agency, 2010.

Human influences

Various human activities have resulted in changes in fungal distributions and frequencies. At a large scale, air pollution with sulphuric acid and nitrogen (*e.g.* ammonia) has had a very negative effect on many fungi, not least lichenized and mycorrhizal species. In the Netherlands, *Cantharellus cibarius* was previously common enough to harvest in large quantities but in 1989 was added to the red list. A similar decline was seen in other ectomycorrhizal fungi, but after a reduction in both types of air pollution an increase has been noted. The vast number of domesticated animals, coupled with intensive agricultural systems, has resulted in previously nutrient-poor (oligotrophic) ecosystems becoming rich in nutrients, leading to increased plant growth and a decline in many less vigorous plants, with similar negative effects on many fungal species. Nutrient-poor woodlands and heathlands with reindeer lichens are converted into moss- and grass-rich habitats, with a huge loss in lichen cover, and another set of fungi takes over. Lichens take up all their nutrients from the air, which explains why they are more sensitive than most other fungi. For this reason, indicator systems based on lichens have been designed to monitor pollution levels.

Artificial fertilizers and slurry are not only an air pollution problem. Just one application of a commercial fertilizer to an ancient, unimproved grassland can wipe out a whole fungal community. Waxcaps are a key component of such communities and the number of species present is often used to indicate the quality of a given grassland; they also provide a good proxy for other fungi, many of which have shown a similar decline.

Another important factor in human management of nature is the balance between disturbance and continuity. Natural ecosystems have both elements: forests and trees can reach impressive ages, but the inner dynamics cause islands of disturbance – for example, where a big tree has fallen, where a large animal has died or where the soil has moved on a bank. This typically coincides with a high biodiversity. On the other hand, human management often involves clear-felling and the planting of seedlings, thus creating an even-aged plantation, often with a different tree species from the one that was removed, resulting in a huge loss of microhabitats. Such breaks in continuity will significantly reduce fungal diversity, not least in the ectomycorrhizal fungal community which can take hundreds of years to recover.

Asexual propagation

Most fungal propagation is probably asexual. This may vary from a simple division of hyphae and mycelia to the formation of specialized asexual spores. Asexual propagation is a rather simple process where no, or a limited, mixing of genes takes place, as opposed to the sexual process that is all about mixing genes. The risk is that the resulting genetically identical clones may be well adapted in the short term, but in the long term lack the plasticity to adapt to ongoing changes in conditions.

Asexual propagation by simple division is commonplace. It can be seen when fairy rings, over the years, split into separate, curved parts without any connection. In all likelihood, many active mycelia are the result of such splitting. American studies have shown that hectares, or even square kilometres, of forest can have *Armillaria* mycelia from the same individual, but it must be assumed that the 'bits' are no longer joined.

In most cases, this splitting is a passive, random process, but some wood-degrading fungi have a strategy of producing hyphal cords that travel through the litter until a woody substrate is found and colonized. They act as scavengers, and in the process get split up. Lichens that produce thalli on the soil, such as reindeer lichens, also depend on being broken into bits, with each fragment acting as a propagule. The corticioid fungus *Amylostereum* has a very advanced way of propagating itself: woodwasps in the genus *Sirex* move the fungus

A system of hyphal cords by which a fungus may spread from one fallen branch to the next through the litter layer.

Hypholoma fasciculare spreads to adjacent stumps in the forest via hyphal cords.

from tree to tree and the larvae feed on the fungus-infected wood. In the same way, elm bark-beetles (*Scolytus* spp.) transport the parasitic fungus *Ophiostoma novo-ulmi* from elm to elm, spreading Dutch elm disease.

The most common means of asexual propagation is through the production of asexual spores (conidia) from ± specialized cells (conidiophores) formed on the mycelium. Such asexual states are called moulds (see page 1626).

Some species produce asexual spores in connection with fruitbody formation. In the agaric *Asterophora lycoperdoides*, for example, the entire cap flesh is transformed into dusty conidia, and the same is seen in the polypore *Postia ptychogaster*, where large, rounded, dusty structures are formed, sometimes with the sexual part – a small number of tubes with basidia – underneath. In some pyrenomycetous fungi a fleshy structure is formed, with the surface covered in conidiophores, but perithecia later develop underneath, and in the final stages take over from the asexual state; examples are found in the genera *Xylaria* and *Hypoxylon*.

In classic nomenclature, the different life-cycle stages bore different scientific names. However, a recent change in nomenclature rules now only allows one name for a given fungus – in principle the oldest available. For example, this has meant that what we used to call *Hypocrea pulvinata* is now *Trichoderma pulvinata*. In the species texts we have, in some cases, also given the older name, since the literature is still full of such synonymous names.

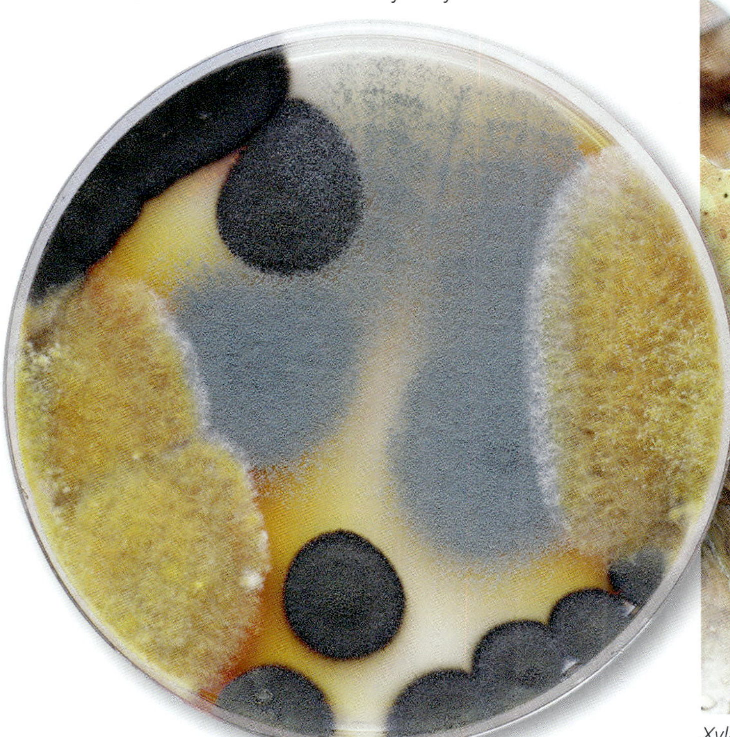

Petri-dish with sporulating moulds

Xylaria longipes initially forms a club-shaped structure (stroma) with conidia production from the surface (top image) but later develops fruitbodies (perithecia) underneath the black surface (bottom image).

From the top: gills from *Lacrymaria lacrymabunda* with spores in groups of four; fruitbody (apothecium) of an operculate discomycete (*Helvella*) during spore release (puffing).

Fruitbodies

The purpose of producing fruitbodies is sexual reproduction and the subsequent dispersal of propagules. The sexual process can run in various ways, but typically involves the fusion of two haploid nuclei into a diploid one, with a following meiosis that leads to the creation of four recombined haploid nuclei (see p. 11). This process takes place in special reproductive cells – basidia in the *Basidiomycota* and asci in the *Ascomycota* (see p. 34). The reproductive cells are typically organized in a palisade structure – a hymenium (see p. 31). To understand the structure and function of a fruitbody fully, it is of paramount importance to know the position of the hymenium.

Within *Basidiomycota* the spores are formed on the outside of the basidia and the release force of the spores is rather limited, only allowing for distances in the range 0.1–1 mm. In contrast, species of *Ascomycota* develop the spores inside the asci where a high pressure can be built up, allowing for much longer shooting distances of easily 5–10 cm away from the hymenium. This also means that *Basidiomycota* almost always produce basidia on the underside or on vertical structures (*e.g.* gill sides), allowing for gravity to take over after spore release. Within the *Ascomycota*, the hymenium is commonly found on the upper surface, enabling the spores to be shot into the more turbulent air above the fruitbodies. These differences may help to determine the type of fungus you are faced with.

To locate the hymenium, first look for a ± smooth surface. The hymenium can be completely exposed, as in corticioid fungi or in many cup fungi, or can be found on folds, hanging spines, gills or tubes that serve to increase the area from which spores can be released. If the hymenium is pointed upwards, the fungus is in all likelihood an ascomycote; if facing downwards or situated on a vertical structure it is likely to be a basidiomycote. If the hymenium is on an folded hymenophore (e.g. wrinkles, pores, gills or spines) the fungus is also most likely a basidiomycote because the short shooting distance of these prevents that the spores are shot into any opposing tissues. Folded hymenophores are not usefull in the ascomycotes which will normally have a longer shooting distance when the spores are released.

If no obvious hymenium can be located, this could mean that spore production is internal. Within the *Basidiomycota*, this is the case with the so-called gasteroid fungi (in earlier classifications *Gasteromycetes*), which include false truffles, puffballs, stinkhorns, *etc.*; these all lack a means of active spore dispersal.

Within the *Ascomycota*, several large groups have internal spore production and, at the same time, active spore release, *e.g.* pyrenomycetous fungi (true and false flask fungi). Here the asci are formed within small, globose to flask-shaped fruitbodies (perithecia), where the mature asci, one at a time, stretch to the tiny opening to release the spores (see p. 30). Other ascomycotes with internal spore production have passive spore dispersal, *e.g.* truffles and *Eurotiomycetes*. Such fungi are, like the gasteroid basidiomycotes, either dependent upon outside forces (*e.g.* the force from heavy raindrops or disturbance by animals that are attracted by their smells), or they simply rot away to expose the spores.

Macroscopy

Macroscopic features of fungi can be seen with the naked eye or with a hand lens. These typically include colours, surface structures and shapes.

Colours

When describing colours, the best approach is to refer to a standard colour chart, although most such charts are either rare, out of print or very expensive. In this publication we mostly follow the colour system published by the Danish Mycological Society in 1996 (Jens H. Petersen: *Foreningen til Svampekundskabens Fremmes farvekort*). Although this system includes 60 colours, it is still not sufficient to describe all the colours of a fruitbody with the small nuances of age and drought variations, but for a general description it is perfectly adequate.

In some species the colours change markedly with changes in humidity. If this colour change is reversible (typically starts to dry out at the centre) it is called hygrophanous. A good example is *Kuehneromyces mutabilis* (p. 606).

Surface structures

The surfaces of caps, gills and stems have many characteristics that are important for identification, although these are mostly microscopical. For example, when a surface is fibrillose-striate the hyphae will be arranged in a parallel pattern. When drying out, such a surface typically splits into fibres. Completely smooth surfaces are often constructed in a palisade (also called a cellular or hymeniform structure), where aggregated hyphae are orientated towards the surface. Such surfaces often split into random cracks. If the palisade structure is looser, this can

Some useful colours from the Danish Mycological Society colour chart.

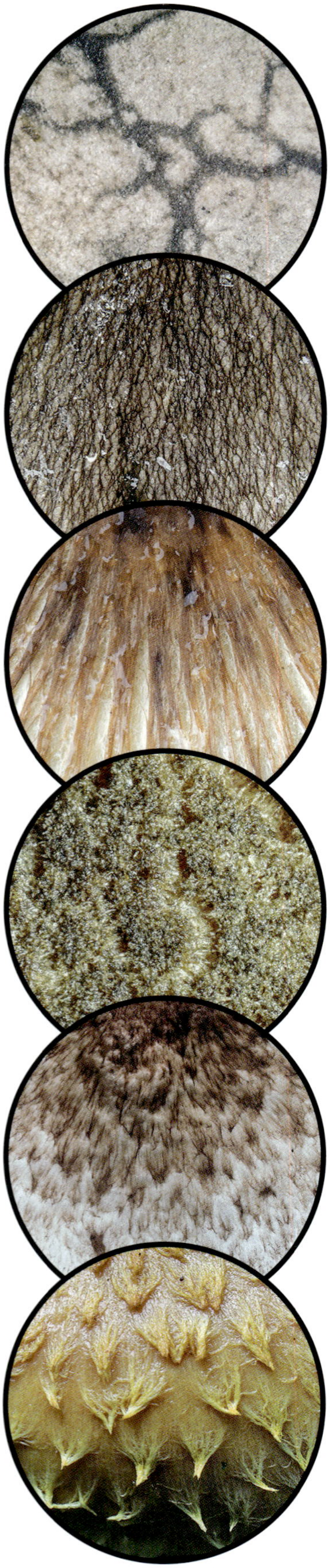

From the top: cap surface without structure (cellular/hymeniform); radially fibrillose; with divergent fibres (rimose); felty; appresed scaly; and with erect scales.

result in a downy, felty or hairy surface.

In some cases hairs are joined to form small scales. Coarse scales can be caused by the surface breaking up. As well as the underlying hyphal structure, the surface can be adorned by loose cells, *e.g.* from a universal veil (see below).

In principle, all hymenia are palisade structures of basidia (p. 33), basidioles (p. 31) and cystidia (p. 36) (*Basidiomycota*), or asci (p. 33) and paraphyses (p. 36) (*Ascomycota*), (see figs. p. 31) and thus lack a fibrous element. Only in special cases can hymenia be downy, caused by projecting, very long cystidia, special paraphyses or setae.

Shapes

Leafing through the species accounts quickly reveals the immense richness in morphologies within the fungal kingdom. Here, just a few of the terms relating to *Basidiomycota* and *Ascomycota* will be mentioned; other special terms are introduced under the appropriate groups.

Basidiomycotes with a cap and a stem

Within the group of fungi with a cap and a stem the tissue holding the basidia (the hymenium) is situated on an area called the hymenophore, which sits, protected, below the cap. The stem brings the hymenophore and cap free of the substrate, enhancing spore dispersal by wind. Cap-forming fungi without stems typically fruit above the soil level, on attached branches *etc.*, resulting in improved wind dispersal.

The hymenophore can be smooth as in *Craterellus cornucopioides*, veined as in species of *Cantharellus*, gilled as in agarics, spiny as in hydnoids, and tube-like as in boletes and polypores. The more folded it is, the more surface is created for the hymenium that holds the basidia and spores.

The entire young fruitbody (primordium) can be enveloped in a protective universal veil that gradually breaks during the growth of the fruitbody. This universal veil can be skin-like (membranous), thread-like (a cortina) or slimy, and remnants can be seen on the cap surface and/or on the stem, especially at the base. When membranous, it can be left as a bag-like structure at the stem base (a volva), *e.g.* as in *Amanita phalloides*. When thread-like, it can be left as threads on the surfaces, *e.g.* as in *Cortinarius*. When slime-like, it can be left as a slime on the surfaces, *e.g.* as in *Gomphidius*.

Agarics can also have a partial veil that protects the young gills. This connects the cap margin with the stem which, as with the universal

Mature fruitbody of *Amanita phalloides* (above) and a primordium in longitudinal section section (below).

Types of hymenophores. From the top: smooth (*Craterellus cornucopioides*); veined (*Cantharellus*); gilled (*Lepista*); spiny (*Hydnum*); and tubed (*Xerocomellus* and *Polyporus*).

veil, breaks up as the fruitbody enlarges. It may later be seen as a ring or ring zone on the stem, or as threads or teeth at the cap margin and, as with the universal veil, may be membranous, thread-like or slimy.

Most caps are initially ± convex but can, during expansion, become flat (applanate) or depressed. They may display a broad umbo at the centre or a small, acute papilla. They can also have a navel-like structure at the centre. Most gilled mushrooms have both full-length gills and shorter gills that reach only part-way towards the stem. If the long gills are without connection to the stem, they are called 'free' gills. If attached, they can be adnexed, adnate or decurrent, or sometimes decurrent with a narrow tooth. A small group belonging to a section of *Marasmius* differs in having gills that join to form a wheel-like structure around the stem (a collarium).

Gills can be narrow or broad, thick or thin and they can be well spaced (distant) or crowded together, with a full range of intermediates. The gill edge holds many characters that are useful for identification: it can, *e.g.,* be smooth, serrulate or a different colour from that of the gill surface.

The stem can be more or less barrel-shaped, club-shaped (clavate), cylindrical or attenuated. The base can be bulbous and/or rooting. The stem can be central on the cap, off-centre (excentric) or marginal (lateral), as typically seen in pleurotoid agarics. The stem may be covered in veil remnants, including a ring (see above) and can be solid or hollow.

From the top: depressed cap with a central umbo (*Infundibulicybe*); closely spaced (crowded) and well-spaced (distant) gills (*Rhodocollybia* and *Mucidula*); fimbriate and serrulate gill edges (*Bolbitius* and *Lentinellus*); and, at the bottom, coloured gill edges (*Mycena*).

Gills attached to a collarium free of the stem (*Marasmius rotula*). Photo Thomas Læssøe.

Free gills in *Parasola* and *Pluteus*.

Emarginate gills in an *Entoloma* and a *Cortinarius*.

Adnexed gills in a species of *Entoloma* and broadly adnate gills in a *Tubaria*.

Deeply decurrent gills in *Cantharellula* and in a species of *Mycena*.

Spore colours

The colour of the spore-deposit is a crucial character that governs the ordering of agarics in this publication (see p. 56–59). An agaric spore-deposit may be white, cream, yellow, brownish-rose, pale brown, rust-brown, grey-brown, chocolate-brown, black or, rarely, green or lilac.

If a fungus has coloured spores, it is often possible to see a spore-deposit directly on the fruitbody, *e.g.* on a stem ring, in the remnants of a veil or on a cap situated below the gills of another cap. With a hand lens you may also be able to see lumps of fallen spores on the sides of the gills (see below). However, if the spores are pale, they become more difficult to see and you may have to prepare a spore-deposit on paper or on a microscope slide, where they can be scraped together to form a thicker deposit.

Lumps of brownish-rose and rust-brown spores seen directly on the gills with a hand lens. To the right: spore-deposits.

Ascomycote morphology

Within the *Ascomycota*, somewhat technical terms are used for the various types of fruitbody: apothecia, perithecia and cleistothecia.

In apothecia, the hymenium is fully exposed at maturity, allowing asci to shoot spores simultaneously. Examples include cup fungi, and more unusual types are found in morels and earthtongues.

Perithecia are characterized by being almost completely closed but with a tiny opening (the ostiole) through which the spores are dispersed. In order to shoot and disperse the spores each mature ascus has to stretch to the opening. This limits the size that perithecia can attain and they are always less than 2 mm in diameter and mostly much smaller. Perithecial fruitbodies are found in several, not closely related groups (see overview on p. 1510); in those with two functional walls in the asci, the fruitbodies are often called pseudothecia, which can be further divided into thyriothecia (flattened, see p. 1604) and hysterothecia (laterally compressed, see p. 1614). Perithecia can be immersed in, or be superficial on, sterile, fluffy, fleshy or carbonous tissues – so-called stromata (sing. stroma), see also pages 1512 and 1526.

Cleistothecia are closed, mostly tiny fruitbodies (see p. 1500). At maturity the outer part of the fruitbody typically disintegrates so that the spores can escape without being actively released. Within the powdery mildews (*Erysiphales*), the cleistothecia-like fruitbodies will, however, crack open during the spring, allowing the spores to be ejected forcibly. These special fruitbodies are also called chasmothecia, and act as resting organs that can survive the winter.

Atypically large cleistothecia are found in the truffle genus *Elaphomyces* (p. 1276), the spores of which are dispersed by digging animals.

Apothecia with an exposed hymenium (blue).

Perithecia with asci inside (blue) and a narrow opening.

Several perithecia in a stroma.

Cleistothecia with internal asci (blue) and no opening.

Microscopy

Both living and dead (conserved) fungal material can be subjected to microscopic examination. Living cells display different characters from dead cells and for this reason working with living material is always preferable (also see p. 39). Fungal identification with the aid of a microscope often focuses on spore characters and the cells that produce them (basidia and asci, p. 33), typically those found in palisade-like hymenia. Other important elements, such as cystidia, paraphyses and setae, all p. 36, can also be present in the hymenium. The types of hyphae present in the flesh is important, for example, in polypore identification.

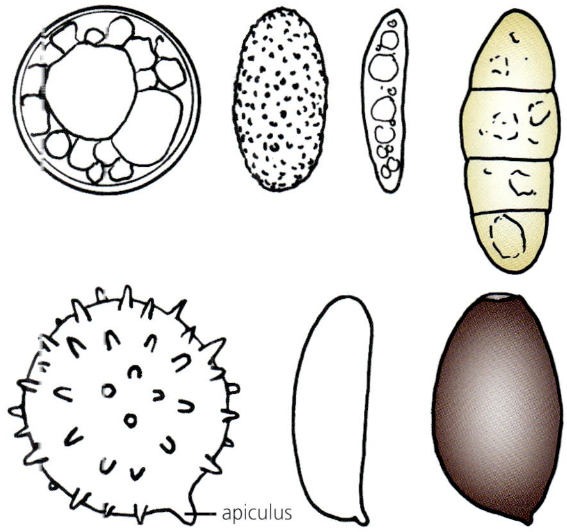

Ascomycote spores without apiculus (top) and basidiomycote spores with an oblique apiculus (bottom).

Spores

The key microscopic character used in fungal identification is spore morphology. There are two

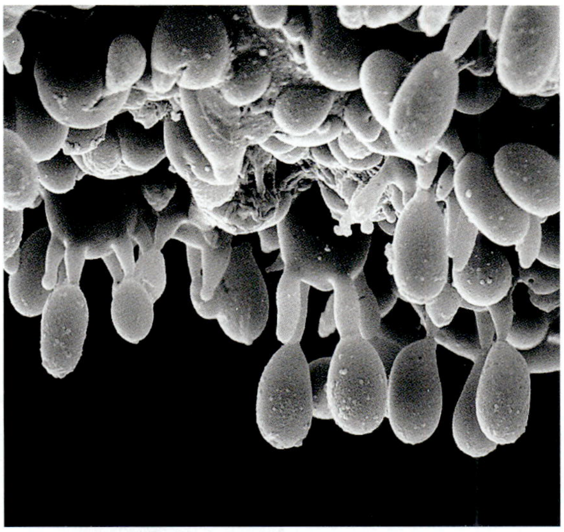

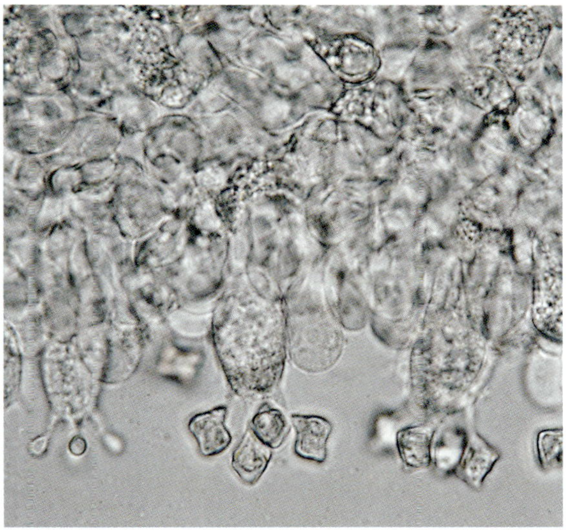

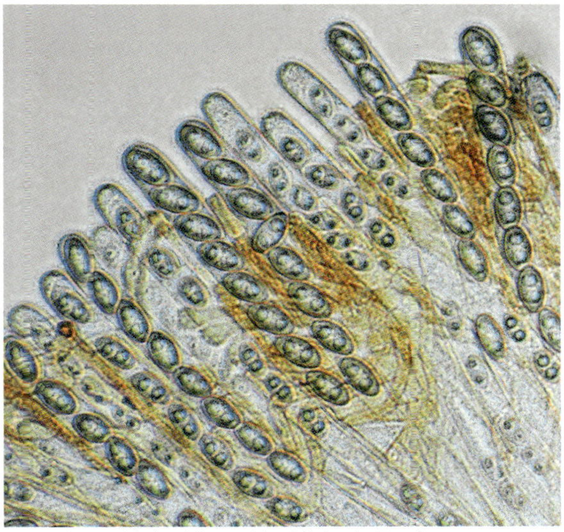

The hymenium is a palisade structure of parallel cells. In Basidiomyota (top) it consists of basidia with spores intermixed with basidioles; in Ascomycota (bottom) it consists of asci and paraphyses (SEM-pictures).

Hymenium with basidia and basidioles in *Entoloma conferendum* (top); hymenium with asci and paraphyses with orange contents in *Aleuria aurantia* (bottom) (compound microscope pictures).

Basis spore shapes in *Ascomycota* and *Basidiomycota*. From the top, left to right: (a) globose/spherical; (b) ellipsoidal; (c) amygdaliform/citriform (almond- or lemon-shaped); (d) fusiform; (e) drop/tear-shaped; (f) bread-shaped; (g) cylindrical; (h) spurred (bullet-shaped); (i) allantoid/sausage-shaped; (j) cucumber-shaped; (k) vermiform/worm-shaped; (l) coiled; (m) filiform/thread-like; (n) triangular; (o) heart-shaped; (p) discoid; (q) angular; and (r) star-shaped.

main types of spores: sexual spores (resulting from a meiosis – also known as meiospores) and asexual spores formed in connection with mitosis (conidia, mitospores). The shape of meiospores, hereafter just referred to as spores, is often a good indication of whether a fungus is a basidiomycote or an ascomycote.

Nearly all basidiospores are asymmetrical with a small peg at one end (the apiculus), where the spore was attached to the sterigma on the basidium (see p. 34). This feature is part of the active dispersal by the drop mechanism (see p. 34). If, on the other hand, the spores are symmetrical along the length axis they normally originate from an ascomycote (see fig. p. 31). Basidiomycote spores are mostly single-celled, while septate spores are rather common among some groups of ascomycotes. Spores evolve through maturation, and it is of paramount importance to ensure that the spores studied are mature. This can be achieved by allowing the fungus to produce a spore-deposit – best done on a microscope slide or coverslip. Alternatively, spores deposited on the fruitbody away from the hymenium can be examined – for example this is easily done with *Cortinarius* species where the cortina on the stem 'catches' numerous deposited spores. Many perennial fruitbodies are sterile when collected, and can therefore be very difficult to identify. Sometimes rewetting a fruitbody and allowing it to sit in a cool place overnight may induce a spore-deposit.

There is an enormous variation in spore shapes – *e.g.* globose (spherical), ellipsoidal, lemon-shaped, bread-shaped, cylindrical, fusiform, drop-shaped, sausage-shaped, cucumber-shaped, worm-shaped, thread-like, twisted, triangular,

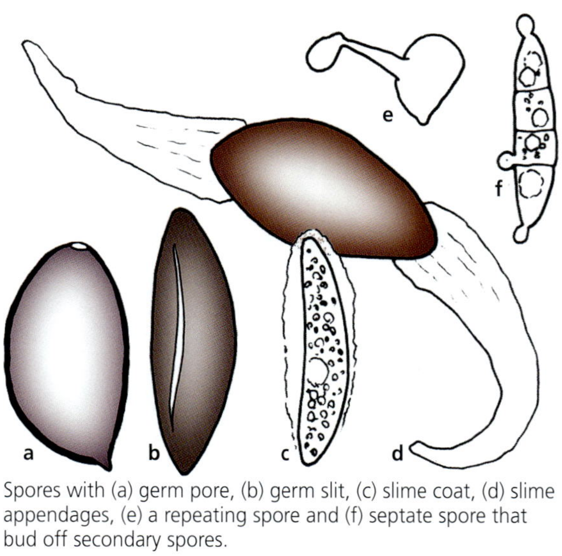

Spores with (a) germ pore, (b) germ slit, (c) slime coat, (d) slime appendages, (e) a repeating spore and (f) septate spore that bud off secondary spores.

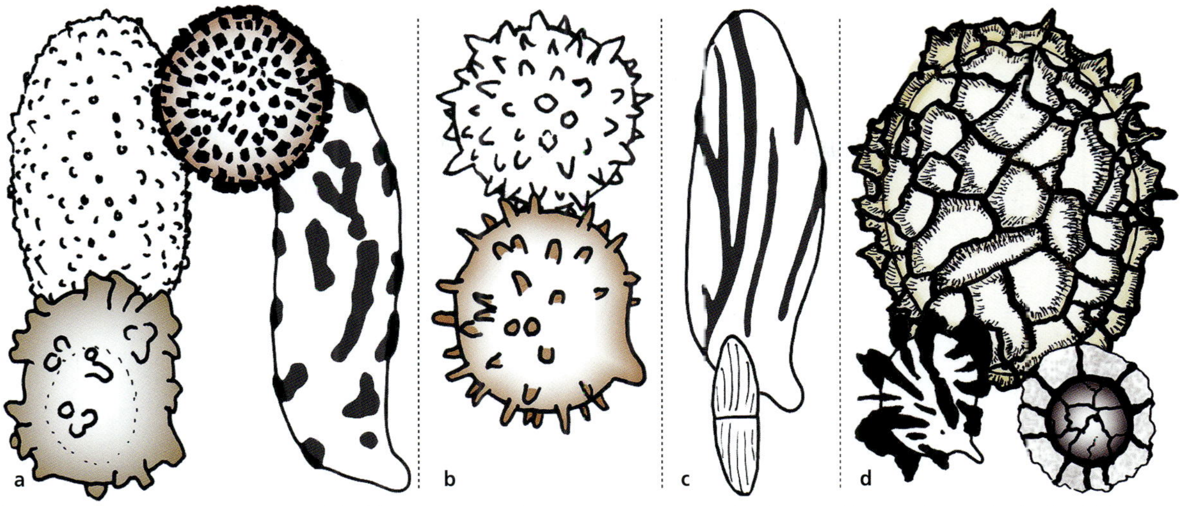

Examples of spore ornaments. From left: (a) warty spores with round, angular or twisted warts; (b) spiny; (c) longitudinally striate; (d) winged-crested or reticulate ornaments.

heart-shaped, disc-shaped and star-shaped, and many other bizarre shapes. Spores can be single-celled or 2- to many-celled. Septa can divide transversely but also longitudinally or diagonally. In some species, spores at maturity break up into part-spores, as seen in the genera *Trichoderma* and *Tolypocladium*. Some spores germinate with secondary spores rather than with hyphae (*e.g.* in *Ascocoryne*). The spores of many jelly fungi can produce a new, ± identical spore from the original; they are said to be 'repeating'.

The most simple spore surface is, of course, glabrous, but many spores have intricate ornaments adorning their surface, *e.g.* warty, spiny, striate, angled, winged, veined or reticulate. In some, the outer wall loosens – often best seen in a solution of KOH. Others have a slime coat surrounding the entire spore or only part of it, sometimes extended to form long appendages. Dark, pigmented spores often have one or several germ pores or a germ slit (seen as a pale line).

Spore colours play a crucial role in fungal identification. Within the agarics both the colour of the spore-deposit and the colour of the spores, as seen in water through a compound microscope, are key features (see spore-deposits on p. 29 and the fungal wheels for the gilled fungi). Within the *Ascomycota* the exact spore-deposit colour plays a minor role, but the colour of spores as seen in the microscope is important.

Some spores take up stains or react with chemical reagents (see p. 39). If the spores turn a bluish colour in Melzer's reagent or Lugol's reagent they are amyloid (I+). If they stain reddish-brown in Melzer's reagent they are dextrinoid. Correspondingly, they are called cyanophilic if they stain blue in cotton blue in lactic acid. In some cases the amyloid reaction is so pronounced that it is clearly visible macroscopically. A thick spore-deposit from a *Russula* or *Lactarius* will stain blue-black with a drop of Melzer's reagent.

One may wonder about the evolutionary explanation for all this variation in spore morphology. Some explanations are ± straightforward. For example, coprophilous fungi produce thick-walled, dark spores (often with slime coats), both because they are formed in direct sunlight (UV protection), and because they are adapted to pass through the entire gut of, *e.g.,* a cow and still germinate in the dung. They also tend to produce very large spores or spores that can cluster within a slime coat, which means they can be shot farther away from the dung. As grazers often avoid the grass surrounding their faeces (probably to avoid parasites), this increases the chance of spores being eaten by their associated herbivore. Fungi living in very wet environments, including directly in water, often have long, thin appendages on the spores, allowing them to float better. It has been shown that ectomycorrhizal fungi often have ornamented spores that may be an adaption related to the ability of small arthropods to transport spores down through the soil layers and closer to the root tips of the partner tree. A surprisingly high number of wood-decaying fungi have allantoid (sausage-shaped) spores, but quite why is anyone's guess!

Basidia and asci

The *Basidiomycota* produce their spores on basidia. Basidia can be single-celled or divided into up to four compartments, either with length-wise septa or with transverse septa (see also jelly fungus wheels for atypical types, p. 1162).

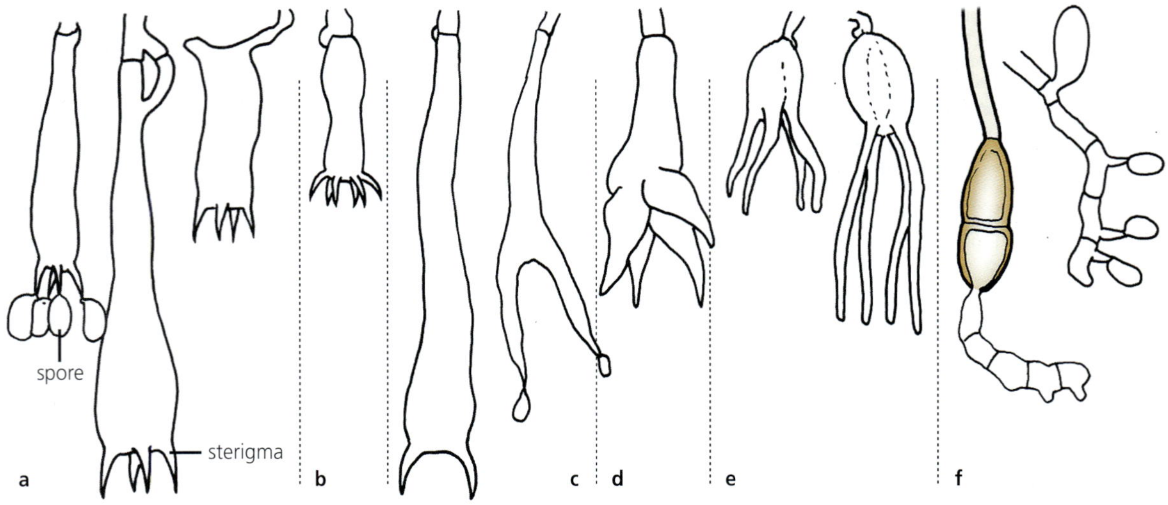

Examples of basidia. From the left: single-celled basidia with (a) 4, (b) 6 or (c) 2 sterigmata; (d) basidium with swollen sterigmata; (e) two basidia with longitudinal septa; and (f) basidia with transverse septa, one formed from a 2-celled spore (teliospore) and one with a probasidial sac.

Species with active spore dispersal produce basidia with ± curved sterigmata, normally situated on the top. As a result of the meiosis within the young basidium, four haploid nuclei are formed and this normally leads to the creation of four external spores on top of the sterigmata. At maturity, a small force created by the fusion of two droplets propels the spore a short distance away from the hymenium, *e.g.* into the space between opposing gills or halfway into the cavity of a tube. Species with passive spore dispersal typically have reduced or straight sterigmata, and the droplet fusion mechanism has been lost.

Within the *Ascomycota*, the spores are formed inside asci. At maturity a high internal pressure is created via osmosis and the spores are ejected with great force; they can easily be propelled up to 10 cm away from the hymenium. Asci are often ± cylindrical, but in species with passive spore dispersal, *e.g.* truffles, they tend to be sac-like or completely globose. The meiosis that takes place in the young ascus is normally followed by a mitosis, leading to the creation of eight nuclei which typically end up in eight spores. In rarer cases, the ascus only contains one, two or four spores, but sometimes even higher numbers (16, 32, 64, 128, *etc.*).

Some asci have a special mechanism at or in the tip that helps to control the release of pressure when spores are ejected (the apical apparatus). The asci of the operculate discomycetes have a lid (an operculum) that

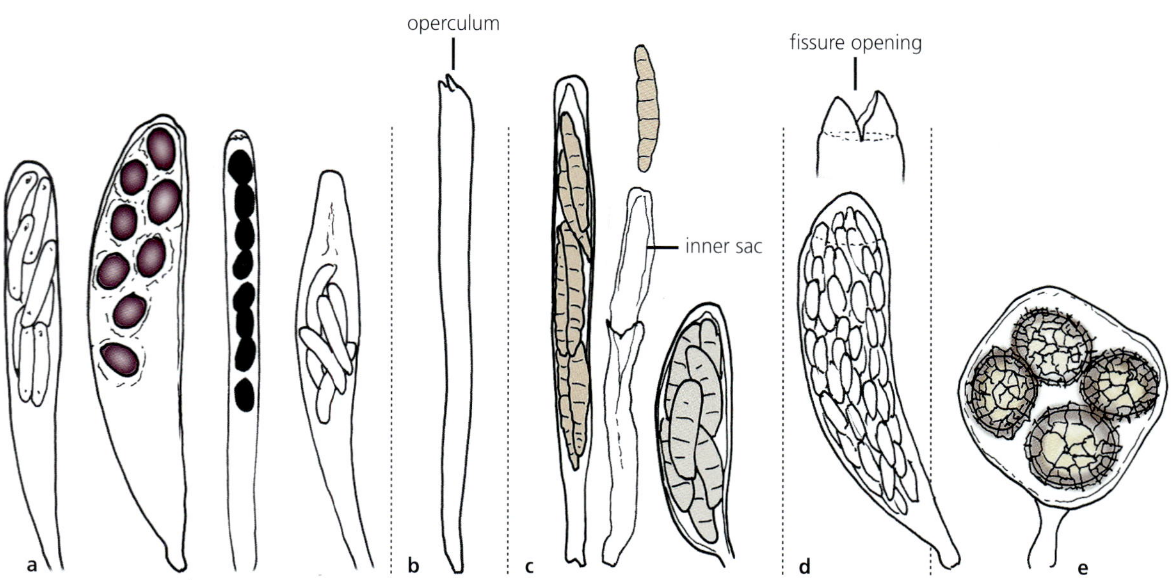

Examples of asci (sing. ascus) with different types of opening mechanism. From the left: (a) four 8-spored asci; (b) an empty operculate ascus; (c) 8-spored bitunicate asci showing the released inner sac and an ejected spore; (d) a multi-spored ascus with fissure opening; (e) a globose ascus with passive dispersal.

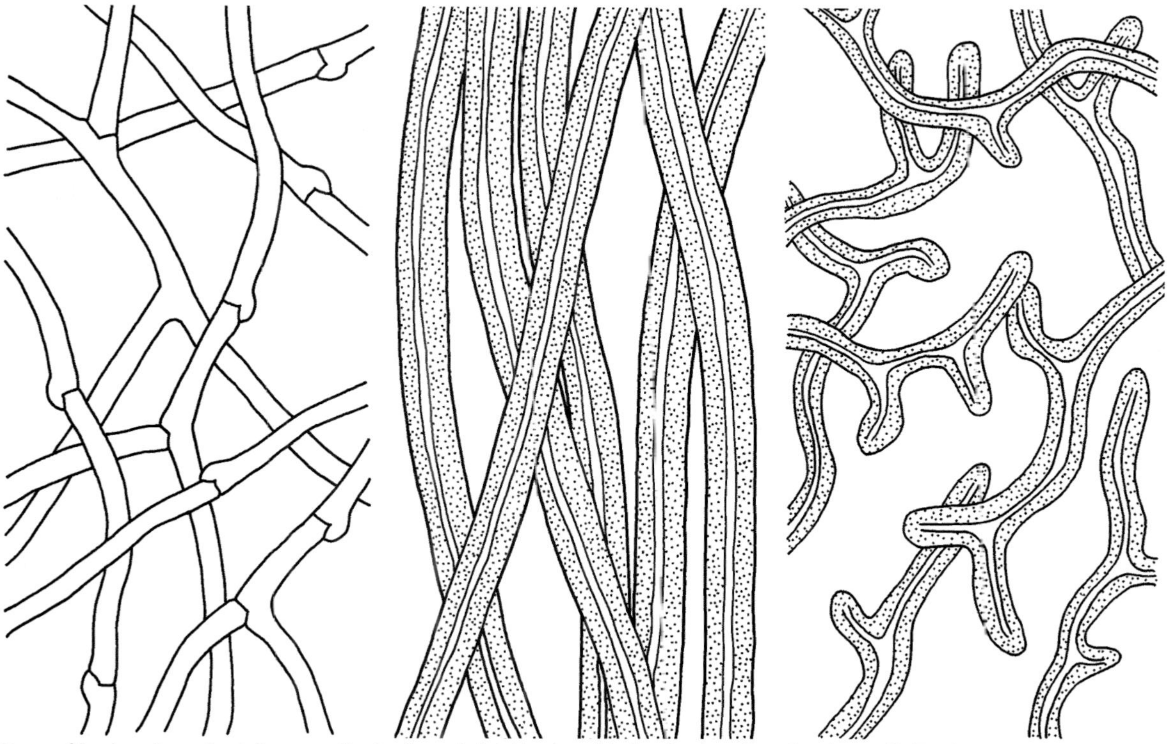

Types of hyphae. From the left: generative hyphae; skeletal hyphae; binding hyphae. Drawing Anette Højlund.

is blown to one side during ejection. Within the inoperculate discomycetes, and many pyrenomycetous fungi, an inner, plug-like structure is blown out and inverted during ejection. In other pyrenomycetous fungi, the outer sac slits open and allows an inner sac to enlarge, after which the spores are ejected (bitunicate ascomycotes). The opening mechanism of asci is key to the understanding of ascomycote classification.

In some species, the ascus wall stains blue in iodine reagents, in others the apical apparatus stains red or blue.

Hyphae

Basidiomycote fruitbodies are generally made up of long, slender or, in part, swollen hyphae, which can be thin- or thick-walled. In some cases the flesh is dominated by globose cells, as in the genera *Russula* and *Lactarius*. The hyphae that give rise to the basidia are called generative hyphae. Species with tough or hard fruitbody flesh can be made up of other types of hyphae: highly branched, ± thick-walled hyphae without septa are called binding hyphae; and unbranched, thick-walled hyphae without septa are called skeletal hyphae. Fruitbodies that have generative hyphae only are termed monomitic; when a further type of hypha is present they are termed dimitic; and if all three are present they are termed trimitic. Puffball fruitbodies contain a branched, ± elastic set of hyphae called a capillitium. Capillitial hyphae straighten after compression, allowing spores to be released in the process.

Ascomycote fruitbodies can also be made up of long hyphae, but more commonly comprise short cells of various shapes. The precise construction of fruitbodies is often used in ascomycote classification and identification.

Part of the life-cycle of basidiomycotes and ascomycotes is dikaryotic – the cells have two separate haploid nuclei from a fusion of two compatible individuals (see p. 11). To maintain this phase, mechanisms of control have evolved. Some basidiomycotes form so-called clamps in connection with some or all septa, whilst some ascomycotes develop a homologous feature at the base of the asci (but never on normal hyphae), termed croziers. Clamps and croziers aid in maintaining the dikaryotic phase. The presence or absence of clamps or croziers is an important diagnostic character in species identification.

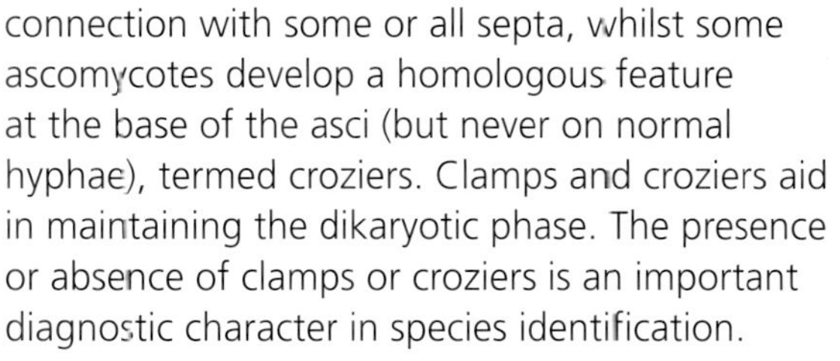

Septa (sing. septum) within *Basidiomycota*: (a) simple; (b) with clamp; (c) with multiple clamps at each septum; (d) with medallion clamp; (e) open clamp at base of basidium; and (f) an ascomycote crozier from the base of an ascus.

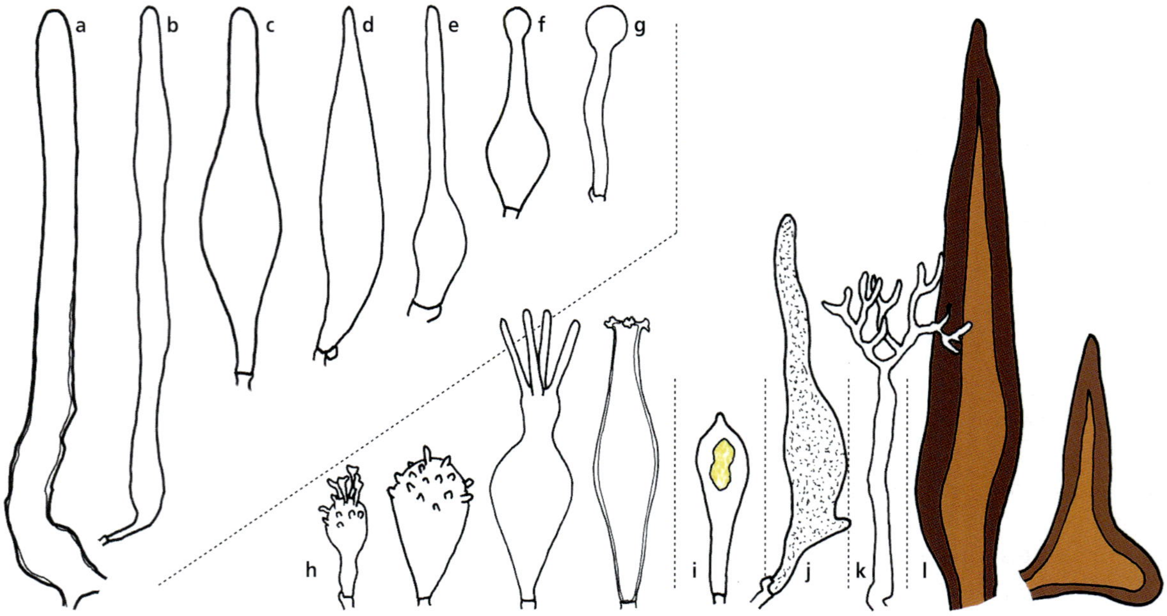

Cystidia. From the left: (a–b) cylindrical; (c) lageniform; (d) fusiform; (e) urticiform (nettle-hair shaped); (f) lecythiform; (g) capitate; (h) cystidia with various appendages; (i) chrysocystidium; (j) oil-filled gloeocystidium; (k) dendrohyphidium; and (l) setae.

Cystidia, pileipellis, hairs, paraphyses

Within fungal fruitbodies, other structures that aid identification can be present. Basidiomycotes may have special surface cells, called cystidia, both on sterile parts or in the hymenium: caulocystidia if on the stem; pileocystidia if on the cap; cheilocystidia if on the gill edge, and pleurocystidia if on the gill sides. The cystidia may have differing contents: if filled with amorphous yellow material they are called chrysocystidia; and if filled with oily substances that stain in sulfo-vanillin they are called gloeocystidia. Some cystidia originate deep down in the flesh and not from surface tissues; these are termed tramal cystidia. Thick-walled, brown, acute, tramal cystidia are known as setae (sing. seta). In some species, particularly among the corticioids, the hymenium can have strongly branched, binding, hyphae-like cells known as dendrohyphidia.

The structure of the cap surface skin (pileipellis/ epicutis) can either have a ± radial orientation of the hyphae, or erect cells in a palisade. Hairs of various types can be present on the cap and/or stem.

Within the hymenia of ascomycotes most species have paraphyses (sterile cells) between the asci. They have various morphologies and contents and can form a protective layer above the asci. This layer can be dominated by excreted waxy substances rather than tips of paraphyses, or can be made up of paraphyses with highly branched tips.

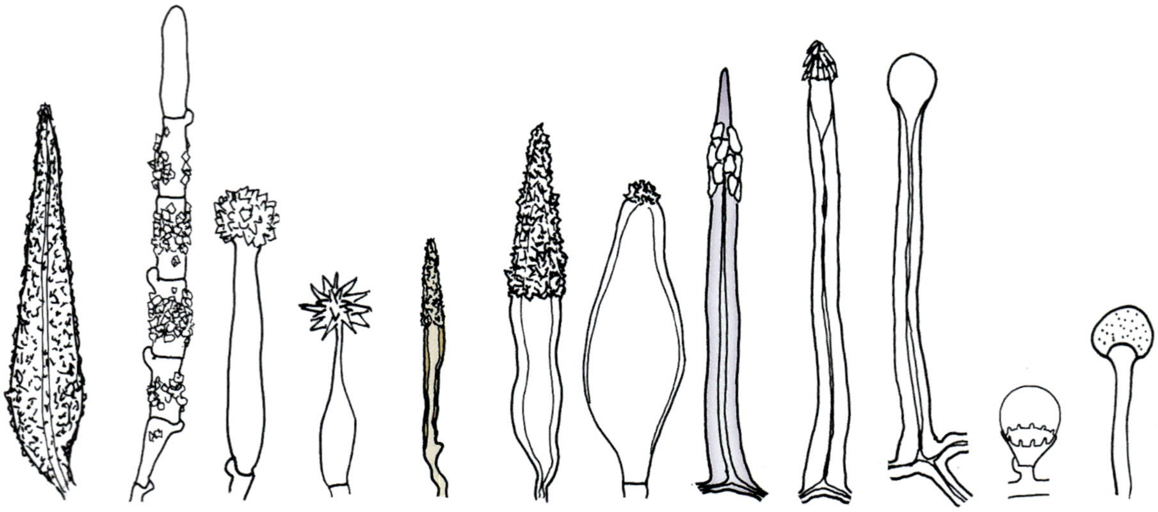

Examples of special cystidia from corticioids: with crystals, thick walls, clamped septa, excretions, *etc.*

Tastes and smells

The smells and tastes of fungi constitute very subjective characters that are hard to describe but, nevertheless, are highly important in the identification process. It is virtually impossible to describe smells in words, and probably the best way forward is to become an apprentice to an experienced mycologist with well-developed and well-trained olfactory senses.

Tastes are easier to describe than smells. They can be mild, or hot like chilli, as in many species of *Russula*. They can also be bitter, sour, peppery or mealy (farinaceous). Some tastes are more distinct, such as like that of the cedar wood found in some pencils.

Bitter tastes can be problematic, since a fair proportion of people cannot sense such tastes, *e.g.* that of *Caloboletus radicans*. In some bitter-tasting fungal species the sample needs to be chewed in the back of the mouth before the taste becomes apparent. Always spit out the sample after testing it.

The field of smells is very complex, and, as with taste, some people are unable to detect anything. The mushroom smell with which most people are familiar is that of the commercial mushroom *Agaricus bisporus.* Another familiar smell is that of certain moulds, *e.g.* some *Aspergillus* and *Penicillium* species. These smells can be found in many other species, often mixed with different odours, creating complex impressions to the sensitive nose. Many polypores and some corticioids have the same sour smell (of polyporic acid) as, *e.g.,* found in *Fomitopsis pinicola*.

A complicating factor is that the smell from unbroken and broken fruitbodies can be very different. Some smells are best sensed from broken tissues, *e.g.* the very important farinaceous smell found in *Calocybe gambosa*. This 'mealy' smell is not like that of wheat flour bought from a supermarket, but more like old, stale, damp bread or that of an overripe watermelon. If in doubt concerning a farinaceous smell, tasting a sample will normally help as all farinaceous-smelling fungi also have a strong farinaceous taste. In some genera, nearly all species have farinaceous smells, *e.g. Clitopilus*, while this type of smell is completely absent in others, *e.g. Russula*.

Another basic type of smell is an earthy or radish-like smell. Many of species of *Hebeloma* have earthy smells with radish components, and the group of species around *Mycena pura* all have strong radish-like smells. On the other hand, numerous species of *Inocybe* have spermatic smells that are also likened to the smell of young camembert cheese.

Within the huge genus *Cortinarius*, a set of very special smells occur, *e.g.* banana-like, maize-like, honey-like, malty or of parsley, and even that of an old steam engine (hot metal). Fruity smells include the rather weak smells of apple, pear and gooseberry found in *Russula*, the strong marzipan (almond) smell in, *e.g., Agaricus arvensis*, the aniseed-like smell of, *e.g., Clitocybe odora* and the strange sweetish-nauseous smell of *Hebeloma sacchariolens* – a smell which recalls that of *Prunus padus* flowers. A group of *Inocybe* species also has strongly perfumed smells, somewhat recalling that of apple cider.

Chemical-like smells are also commonplace, *e.g.* the carbolic smell of *Agaricus xanthodermus*, the iodoform (hospital-like) smell of *Hemileccinum impolitum* (from the stem base) and the chlorine or nitric acid type of smells found in e.g. *Mycena stipata*. Bitter almond or cyanide smells are common in old fruitbodies of *Clitocybe* and *Phaeolepiota aurea*.

We have no thorough knowledge of why these smells are emitted, and many smells are perhaps just a by-product of various processes within the fungus. Some fungi, on the other hand, do clearly emit smells for a reason. The stinkhorns and allies depend upon flies for spore dispersal and mimic the smells of faeces or decaying flesh. These smells can be incredibly strong and difficult to tolerate in a confined space. In *Phallus impudicus*, the flesh itself has a radish-like smell, but the mature spore mass develops a smell that, close up, can resemble hyacinths, but from just a little farther away is experienced as extremely foetid; a smell that can be detected over a considerable distance.

The underground truffles depend very much upon smells in order to be detected, dug up and dispersed. Some produce odours identical to various pheromones, and some of the edible truffles are said to produce testosterone-like smells that attract animals such as pigs and deer. These smells can be extremely pungent and if edible truffles are kept with other foods, these can take up the smell (and taste).

Working with fruitbodies

Collecting and identifying fungi demands a certain degree of skill and some equipment. It is also a great advantage to have good eyesight and well-developed olfactory senses.

Collecting

In principle, the gathering of fruitbodies from a mycelium should not harm the mycelium itself, and should not affect future fruitbody production from the same mycelium. Collecting fruitbodies will, however, diminish spore production, not least if perennial fruitbodies are taken. Only sufficient material for identification of unknown species should be collected.

In many fungi, the variation displayed by fruitbodies from young to old is important to note. Determination of species of *Cortinarius* requires knowledge not only of mature spores but also the colours of the young gills, veils, *etc*. Be careful when picking a fruitbody: do not handle the stem, and try to touch surfaces as little as possible so that accurate notes can be taken of undamaged features. Purposeful bruising, however, may later be required in order to note colour changes.

Small fruitbodies are best collected in small tins or boxes, or in bigger boxes with many tiny compartments. Bigger fruitbodies can be put into larger boxes, small bags or wrapped in waxed paper. Mixing fruitbodies in a basket without wrapping them first should be avoided. Always keep the collected material as cool as possible and out of direct sunlight. Fruitbodies decay very quickly if heated and may not produce spore-deposits.

A sharp knife is an essential tool, and for many fungi fruiting on woody substrates, secateurs and a folding or sliding saw are also invaluable. A good hand lens is obligatory and should magnify in the range 10–20 times. A high-quality ×20 hand lens with an inbuilt led-light can be very useful when studying tiny discomycetes or other small fungi. At home, a dissecting microscope is extremely beneficial, not least for preparing tissue samples for microscopic slides to be viewed through a compound microscope.

Serious fungal identification requires the use of a compound microscope. Good quality models come at a high price; old school microscopes soon fail to do the job required and may discourage the use of microscopes altogether! A new world is revealed when seeing beautiful spore ornaments through quality lenses and decent equipment can often be bought second-hand at more affordable prices. When buying new or second-hand equipment make sure that an oil immersion lens (×100 magnification) is included. Also check that the built-in lighting is good and that one of the oculars has a scale for measuring spores *etc*.

A box with medium-sized compartments for medium-sized fruitbodies.

Chemicals – examples of common types and what they are used for

Tap water – used as a standard medium before adding stains. Spores should be measured in water but if this is not possible a note should be made of the medium in which they were measured.

KOH – 3% in water for studying dried material and 10–20% for chemical reactions.

Ammonia – 5–10% in water is useful for microscopic preparations of dried material.

Melzer's reagent – stains blue (amyloid reaction) or vinaceous to brown (dextrinoid reaction). 1 g iodine and 3 g potassium iodide is dissolved in 40 ml water and 80 g of chloral hydrate is added (poisonous!).

Lugol's reagent (iodine-potassium iodide) – blue stain (amyloid) or red (hemiamyloid). 1 g iodine and 3 g of potassium iodide are dissolved in 100 ml water. (Unlike Melzer's reagent, relatively harmless.)

Congo red – stains hyphal walls red. 1% in water or better in 10% ammonia.

Cotton blue – a blue reaction (cyanophilic). Dissolved in lactic acid.

$FeSO_4$ – for macro reactions of, *e.g.*, russulas and ramarias. As crystals or dissolved in water.

Preparation of microscope slides

When preparing fungal tissues for inspection through a microscope it is fairly easy to make flat, thin slides, since the hyphae are only loosely aggregated and not joined three-dimensionally as are the cells in plants tissues. Nevertheless, only very small tissue samples should be used in order to ensure suitably flat preparations. Place the sample in a drop of water on a slide – it should be so small that it is only just visible with the naked eye. Place a coverslip on top and tap it gently with a rubber or similar to force excess water out and flatten the preparation. Use blotting paper or similar to remove excess water. Place the slide on the microscope stage and check if it is suitable; if not, tap it again. With practice, good results can be obtained. Often several preparations will have to be made for structures to be viewed clearly and in order to get the full picture. Make drawings or take photographs of the objects of interest.

Preparations are best made using a dissecting microscope. Here the tissue that is required can be seen clearly and small cuts can be made. Use a sharp, pointed blade or very fine forceps. If, for example, you want to study the cheilocystidia (those on the gill edge) cut off a tiny piece of a gill edge and, before too much flattening, check the situation through a compound microscope. If necessary, further flattening can take place and regular checks made until the desired features can be found (avoiding them being lost in the process).

Chemicals for living or dead cells

Chemicals (see box above) are used to provoke reactions for macroscopic or microscopic identification. $FeSO_4$ is useful as a macro-chemical reagent for the identification of *Russula* species. Melzer's and Lugol's reagents are used to stain the starch-like chemicals in various tissues or spores blue or reddish to brownish. A hemiamyloid reaction (red) can be obtained with Lugol's reagent but Melzer's reagent will give a negative reaction.

The initial study of fungal material is typically carried out in water mounts. Here the cells will be ± alive and can display a number of characters that will disappear when the spores are killed in various stains. Contents of spores, *e.g.* oil drops, should, where possible, always be studied in fresh material in water mounts.

If the features are difficult to discern, or a specific reaction is needed to determine a character, stains can be added, often directly to the water mount. Excess fluid is removed with

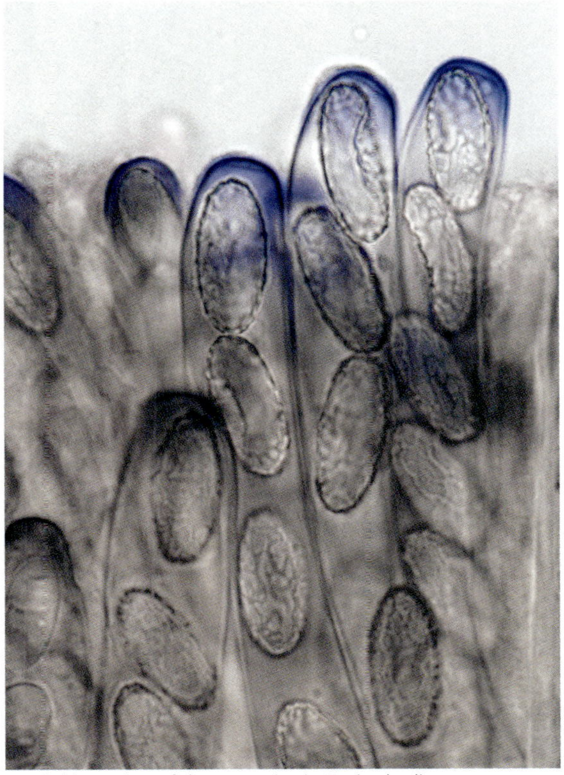

Amyloid reaction of the ascus tips in *Peziza badia*.

blotting paper. Spore ornaments are often best studied in cotton blue.

When dealing with exsiccata (dried material kept as vouchers), mounts can be made in 3% KOH or in 5–10% ammonia. This will restore the hyphae and spores to a ± living shape and size. When using ammonia, other reagents or stains can be added directly, but this is not possible with KOH mounts. It should be noted that internal structures in hyphae, paraphyses, spores *etc.* will be very different from those in living cells in water.

Documentation

Photo-documentation of notable finds is of paramount importance and even a basic camera can provide acceptable evidence. The best compromise between price and usefulness is a mid-range camera with fixed optics and a good macro capability. It is important that aperture and shutter speed can be controlled manually. Use a small aperture (large figure) and a long exposure as standard. This will require the use of a small tripod in order to obtain sharp images free of camera shake. The same camera can, with some practice, be hand-held against one of the oculars of a compound microscope to obtain satisfactory images of microscopic characters.

The best weather for fungal photography is overcast conditions with rather thin cloud cover or hazy sunshine. A piece of white paper can be used as a reflector to lighten the underside of the fruitbodies that are being photographed. In harsh sunlight it may be necessary to shade the subject to avoid hard shadows.

Exciting finds should also have written notes – nowadays often written directly into a database. If the find is an agaric, the size, shape, colour and surface features of the cap, gills and stem should be noted. Remember to note smell and taste, and possible stains when bruised or from added chemicals. Also remember to study the flesh in section. Note down any microscopic characters that have been seen and important details of habitat and substrate, including potential partners or hosts. You will find inspiration for useful terminology in previous sections of this book. If at all possible, try to get a spore-deposit and note, if relevant, the colour. It can be important to keep the material as a voucher for future attempts at identification or documentation. Keep the spore-deposit with the dried sample and use a unique number to identify it. Use the same number on all the photographs.

The collected material should be carefully geo-referenced so that others can return to ± the exact spot. In some countries this can all be done by using online public databases, where just a password is required to enter and access records. These databases, and the mycologists behind them,

Photography is best in overcast weather using a tripod. A sheet of white paper can be used as a reflector to lighten the undersides.

Phylogeny and morphological groups

A **phylogenetic** (natural) group is a group with common ancestry.

A **morphological** group or form group is a (artificial) group of possibly unrelated species sharing common characteristics/traits.

When a group contains species from different evolutionary lines it is called **polyphyletic**.

A group where all members have a common ancestry, and all close relatives are included, is called **monophyletic**.

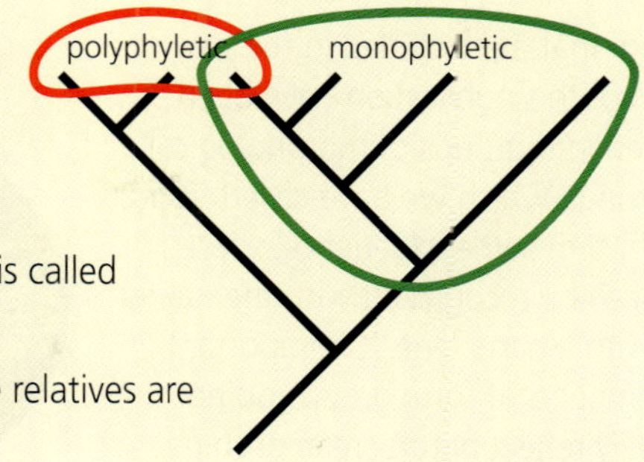

often provide help with identification, which, once confirmed, provides an opportunity to contribute to the scientific record.

Fungaria

The term fungarium is equivalent to herbarium – a collection of conserved fungi rather than conserved plants. Standard conservation involves drying the specimens. They should be dried under moderate temperatures (around 40°C) with free-flowing air. The top of a radiator can be used, but a drying apparatus designed for fruit or fungi is more useful. Small fruitbodies can be put directly on the drier but larger specimens need to be cut lengthwise to speed up the drying process and to make collections less bulky. Tiny fruitbodies may have to be put in small, non-airtight bags before they are dried (otherwise they may be lost). Too high a temperature will destroy the DNA contents and make the specimens less useful for future researchers. Completely dry specimens are packed in standard packets/capsules/ envelopes and/or in sealable plastic bags. Very fragile specimens can be stored in small plastic boxes.

The main hazard for the fungarium apart from humidity is insects, particularly beetles; if not kept under proper conditions, entire fungaria can be eaten within a very short time. If an infestation occurs, deep-freezing is the standard procedure to kill the pests. The best solution is to keep everything in individual, fairly heavy-duty, sealable plastic bags, since this will reduce the risk of any damage.

Use of DNA

The classification of fungi is, as with other organisms, based on natural relationships. The goal is to define monophyletic groups (see box above). Nowadays the most widely used tool in this respect is the analysis of DNA sequences.

It is becoming increasingly problematic to identify many fungal species solely based on morphological characters, and often DNA sequencing of 'bar-code' genes will have to be carried out before an accurate identification is obtained. Many groups of fungi are much more species-rich than previously thought, and many of the 'new' species are ± morphologically identical to other species in the genus. Currently, precise identification of many species of *Cortinarius*, *Inocybe*, *Tomentella* and *Ramaria* is virtually impossible. With a DNA sequence, you might find a matching name, but in many cases only a matching sequence. You will therefore have to use a database number rather than a typical scientific name.

In this publication we have, to a large extent, used gross morphology as our starting point. We have based the layout on form groups rather than natural groups, and have endeavoured to use the latest available names for the species accounts. This approach has the side-effect that the same genus (or family *etc.*) may appear in more than one section; the genus *Resupinatus*, for example, can be found under both the pleurotoid agarics (page 60) and under the cyphelloids (page 1075).

A packet folded from an A4-sheet (use high-quality acid-free paper) and put in a sealable plastic bag. Keep in a dry place.

Form groups and fungal wheels

The fungi that are included in the two volumes of this publication have been organized into groups of morphologically similar fungi, which we have called 'form groups'. With some experience, these groups can be recognized with the naked eye or with a hand lens. It is important to remember that form groups do not necessarily reflect natural relationships, since morphology only reflects such relationships to a certain degree. For example, fungi with polypore morphology have evolved multiple times and, as a result, many of the taxa concerned are not closely related – even though they may look similar. The section on polypores in this book therefore covers all the species with polypore morphology, regardless of their current classification.

There is one major exception to this general rule. We have divided the fungi, irrespective of their macroscopic appearance, into the two major phylogenetic groups: the *Basidiomycota* and the *Ascomycota*. At the end of Volume 2, we have also included sections on asexual fungi, and organisms that belong in other kingdoms but resemble true fungi, *e.g.* the slime moulds.

The fungal wheels are the key feature of this publication. We have adopted a highly pragmatic approach when preparing these wheels, so that, for example, even if you find yourself looking at the wheel that covers cyphelloids (*Basidiomycota*) (p. 1076), you will find cross-references to similar discoid fungi in the *Ascomycota* (p. 1372). For this reason, prior knowledge of fungal systematics is not a prerequisite.

The wheels on this and the following page spread are organized based on how the sexual spores are produced. In order

Fruitbodies with external spore production on an exposed hymenium, see also page 31.

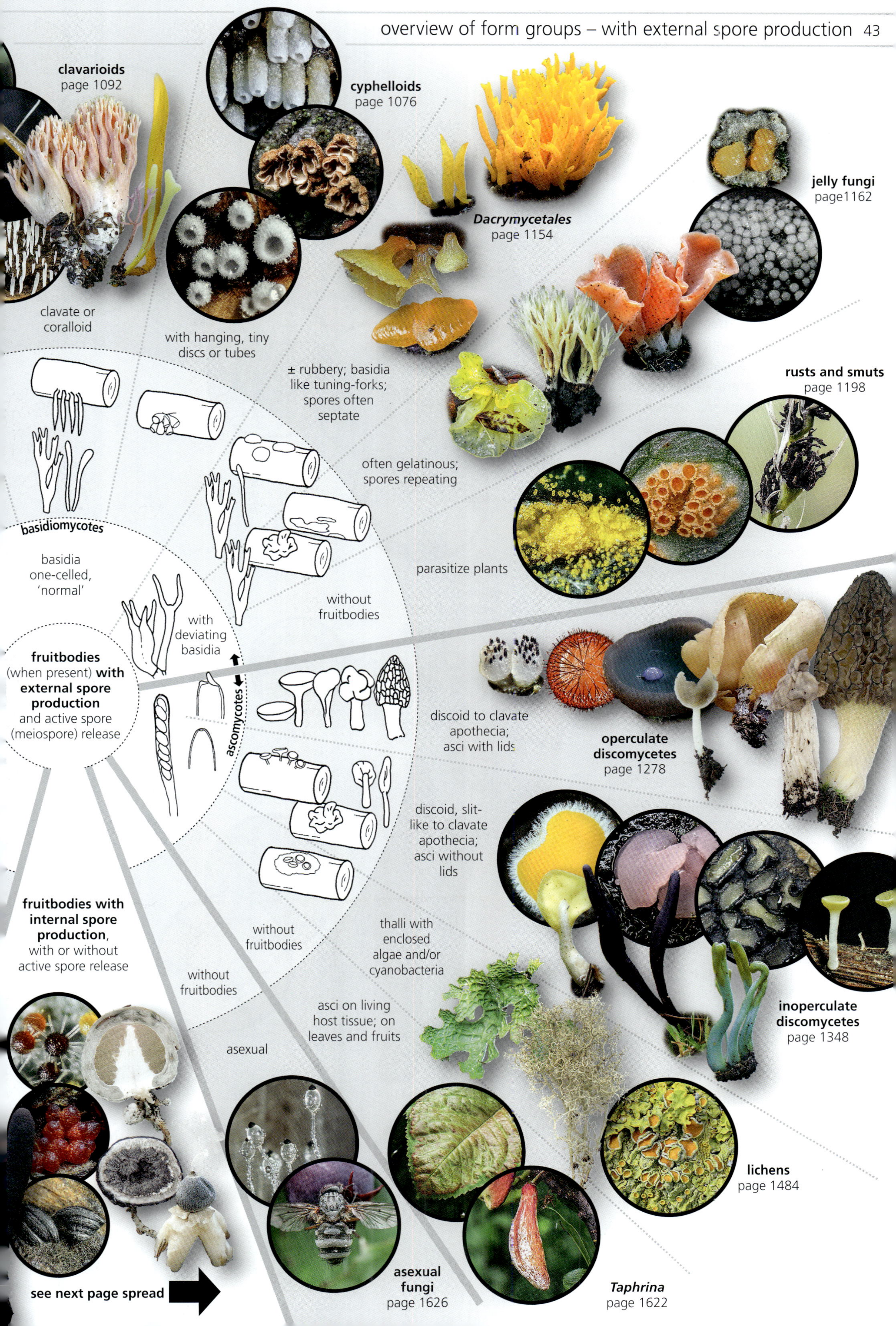
clavarioids
page 1092
cyphelloids
page 1076
Dacrymycetales
page 1154
jelly fungi
page1162
clavate or
coralloid
with hanging, tiny
discs or tubes
± rubbery; basidia
like tuning-forks;
spores often
septate
often gelatinous;
spores repeating
rusts and smuts
page 1198
parasitize plants
basidiomycotes
basidia
one-celled,
'normal'
with
deviating
basidia
without
fruitbodies
fruitbodies
(when present) with
external spore
production
and active spore
(meiospore) release
ascomycotes
discoid to clavate
apothecia;
asci with lids
operculate
discomycetes
page 1278
discoid, slit-
like to clavate
apothecia;
asci without
lids
fruitbodies with
internal spore
production,
with or without
active spore release
without
fruitbodies
thalli with
enclosed
algae and/or
cyanobacteria
without
fruitbodies
inoperculate
discomycetes
page 1348
asci on living
host tissue; on
leaves and fruits
asexual
lichens
page 1484
see next page spread
asexual
fungi
page 1626
Taphrina
page 1622

to decide which wheel to use, you will first have to determine whether your chosen fungus produces spores from an external hymenium (see p. 31 and the illustration p. 42 bottom left). The hymenium consists of a palisade of basidia or asci, and typically forms a smooth surface on the fruitbody. Fruitbodies with internal spore production do not display such a surface. Instead, you may find small openings (ostioles), through which the spores can be ejected. Form groups with external production and active dispersal of the sexual spores (meiospores) from hymenia are found on the previous page spread, and groups with internal spore production are covered on this page spread.

On this page spread, the groups with spores that are actively released are shown at the upper/left side; the groups with fruitbodies that have passive spore dispersal, or have a powdery or slimy inner spore mass, are represented in the lower/right side of the wheel. For passive spore dispersal to take place an external force is often required, *e.g.* rain drops or animal activity.

The wheels here and throughout this publication can be used in various ways. By checking the images in the outer 'rim', a quick overview can be obtained of the variation in the fungal groups covered. Images in the pale blue sections (see, *e.g.*, the wheel on page 46) represent fungi that in some way recall species from the form group covered by the wheel, and which are treated in other wheels in this publication; the relevant page number is given for ease of cross-referencing. The inner part of the wheel shows other features that could be considered, particularly microscopic characters such as spore morphology (see, *e.g.*, the wheel on page 60 and 176).

We recognize that the wheel approach to fungal identification will never reach the same level of accuracy as a traditional analytical key. The distinguishing features of some fungi are not always clear-cut, and several wheels and species accounts may need to be consulted in order to reach a satisfactory conclusion. However, we hope that the approach taken in this publication will be easier to use than trying to follow the often highly technical keys found in other identification guides.

Laboulbeniales
page 1620
externally on living insects and other arthropods
powdery mildews (*Erysiphales*)
page 1500
on leaves and stems; parasitic
Eurotiomycetes
page 1500
on plant litter, feathers, hair *etc.*
asexual fungi
page 1626
conidia dry or in slimy masses
calicioids
page 1494
very small with dusty heads; mainly on wood and bark
cleistothecia-like (chasmothecia)
cleistothecia
without fruitbodies
with active spore release
fruitbodies with internal spore production with or without active spore release
without active spore (meiospore) release
ascomycotes
basidiomycotes
mix
puffballs and the like
page1228
larger, with powdery contents
initially ± onion-shaped, then splitting star-like
earthstars and the like
page 1228
spores in small egg-like structures (peridioles)
bird's nest fungi
page 1216
foetid spore mass; insect-dispersed
stinkhorns and the like
page 1222
± subterranean; animal-dispersed
truffles
page 1256

Chanterelles and the like

The form group have rather fleshy fruitbodies with a ± central stem and a deeply decurrent hymenophore that may be completely smooth or consist of ± branched veins to wrinkles. The fruitbodies can be massive or hollow.

The core group is *Craterellus* and *Cantharellus*. These genera belong in *Cantharellales* and both form ectomycorrhiza.

Gomphus clavatus, which belongs in *Gomphales*, also forms ectomycorrhiza, while *Faerberia carbonaria*, which is situated in *Polyporales*, is a decomposer.

Similar groups:
– clitocyboids have similar shapes to the chanterelles, but have typical gills that may be forked (page 98).
– some omphalinoids have a veined hymenophore, but are smaller with reduced stems (page 128).
– funnel-shaped hygrocyboids and omphalinoids have decurrent gills (pages 128 & 142).
– some mycenas (*e.g. Hemimycena*) may look similar, but are small, fragile and white (page 176).

Further reading: 50, 92, 231, 242.

Chanterelles within the genus ***Craterellus*** have a smooth or veined outer side where the hymenium is situated. They also have a ± hollow stem.

●● ***Craterellus cornucopioides*** is a thin-fleshed, grey-black to brown, rarely yellow-brown, trumpet-shaped fungus with an undulating margin and with an almost smooth, grey hymenium, which is strongly decurrent. The stem is hollow; the cap surface is felty and the smell is very pleasant. The smooth, hyaline spores measure (9–)10–13.5(–16) × (5.5–)6.5–8.5(–9.2) µm. Forms ectomycorrhiza with mostly *Fagus* and *Picea* on better, typically ± clay-rich, calcareous soils.

Craterellus cinereus ▷▷ has obvious wrinkles. The two species occasionally occur close together. A yellowish form has been separated as *C. konradii* (lower image).

Widespread and locally common, becoming scarcer towards the north; mostly September–October.

Jan Vesterholt

Jens Maarbjerg

•• ***Craterellus cinereus*** is a grey-black, rather thin-fleshed, hollow-stemmed chanterelle with an obviously wrinkle-veined hymenophore. The spores measure (7–)8–10.5(–13) × (4–)4.5–6.5(–7) µm. Mostly found on rich soil and usually with *Fagus* – often together with *C. cornucopioides*.

Craterellus cornucopioides ◁◁ has a smooth hymenium. *Faerberia carbonaria* ▷▷ is not mycorrhizal and is found on burnt ground. It has thick-walled cystidia.

Occasional over most of the warmer parts of N Europe, but very rare in some regions; especially August–October.

• ***Craterellus undulatus*** is a very slender, grey-brown chanterelle, mostly with an undulating margin; the rather pale hymenophore varies from almost smooth to somewhat wrinkled-folded. The spores measure (9–)9.5–11.2(–15) × 5–7.5(–8.5) µm. Occurs in large troops on ± clay-rich soils; mycorrhizal with e.g. *Quercus*, *Fagus* and *Corylus*, often with *Craterellus cornucopioides*.

Craterellus cinereus △ is more pure grey-black and has more obvious wrinkles on the hymenophore. *Craterellus cornucopioides* ◁◁ is larger, more trumpet-shaped and darker.

Widespread, occasional, absent from the boreal zone; mostly September–October.

• ***Craterellus melanoxeros*** has a greyish, wrinkled hymenophore and a rather pale, somewhat squamulose-tomentose cap surface. It darkens all over. The spores measure (8.2–)8.5–10(–10.4) × (5–)5.6–6.4(–7.2) µm. Typically occuring on clay-rich soils, mostly with *Fagus*.

The species has been confused with *C. ianthinoxanthus* ⋉, which darkens very slowly, and is a more southern species. May be confused with *C. tubaeformis* ▷ if the discolouration is overlooked.

A rather rare, warmth-loving species, which is possibly somewhat overlooked; mostly found August–October.

•• ***Craterellus lutescens*** is a beautifully golden, hollow-stemmed chanterelle with an almost smooth, rather pale hymenium and a felty-squamulose, brownish cap with a golden glow. The spores measure (9–)10–11.5(–12.5) × (4–)6.5–8(–9) µm, Qav. 1.36–1.59. Mostly with conifers, in clusters in very large troops on moist, calcareous soil with upwelling water.

Craterellus tubaeformis ▽ is less golden and normally has an obviously wrinkled hymenophore; it also has spores with a lower Q value.

Rather scattered, but locally very common; mostly September–November.

Thomas Læssøe

•• ***Craterellus tubaeformis*** is a thin-fleshed, hollow-stemmed chanterelle with brass-yellow stem and, typically, greyish wrinkles below the cap, which is normally brownish and somewhat tomentose-squamulose. The spores measure (9–)9.5–12(–13) × (7–)7.5–9.2(–10.5) µm, Qav. 1.18–1.33. It mostly occurs in clusters and often also in large troops. It is ectomycorrhizal with both deciduous trees and conifers on poor, but also calcareous soils; may occur in large amounts with planted *Abies*.

Craterellus lutescens △ has more golden colours and occurs on more calcareous damp soil. It has a higher spore Q value. *Craterellus melanoxeros* ◁ darkens grey to black in the flesh as the name implies, but, nevertheless, is very similar in appearance.

A widespread and common species, that is less frequent in heavily farmed areas with elevated nitrogen levels; from July, but peaking very late in the season, even after frosts.

Thomas Læssøe

The genus ***Cantharellus*** is characterized by solid, pale flesh and a hymenophore with decurrent wrinkles.

•• ***Cantharellus cibarius*** is a massive, fleshy, yellow or yellow-orange, rarely almost white or pinkish orange, chanterelle with obvious decurrent ribs below the cap that is non-felty when young; it may turn ± brown when bruised. The species belongs to the group with thick-walled hyphae in the cap cuticle. Spores (7.5–)8–9.5(–10) × 4–5.5(–6) µm. Occurs on acid soils with many different mycorrhizal partners (*e.g. Picea*, *Pinus*, *Quercus* and *Fagus*).

Both *Cantharellus pallens* ▷▷ and *C. amethysteus* ▷▷ have the same thick-walled hyphae in the cap cuticle, but are more fleshy, more brown-staining and paler. When young, the latter has an obviously violet cap coating. *Cantharellus friesii* ▷ is more fragile, has thin-walled hyphae in the cap cuticle, and is mostly more reddish; *C. ferruginascens* ▷▷ on rich soils is ± lemon-yellow and stains very strongly brown on the stem; like *C. cibarius* it lacks the pale cap felt when young.

Widespread and common, except in heavily farmed areas; mostly June–August.

Groups of pale or nearly white fruitbodies of *Cantharellus cibarius* are occasionally encountered, but are just mutations of the normal form. A pale form of *C. cibarius* ◁ is depicted to the right. The same species may occur in a pinkish-orange form. Other species may also occur in deviating colour forms.

Thomas Læssøe

•• ***Cantharellus friesii*** is a slender, mostly vivid reddish yellow chanterelle with obvious ribs below the cap. The flesh does not stain brown. The hyphae of the cap cuticle is thin-walled and the flesh reacts ± pinkish with $FeSO_4$. The spores measure (7.5–)8.5–11.5(–13) × 4.5–5.5(–6) µm. Mycorrhizal with deciduous trees (*e.g. Fagus*), mostly on rather acid soils.

Most other chanterelles in temperate Europe have thick-walled hyphae in the cap cuticle, flesh reacting greyish with $FeSO_4$ and are not so vividly coloured, and mostly have more fleshy fruitbodies.

Rather rare and local; summer-autumn.

Joaquín Martín/Claes Ingvert

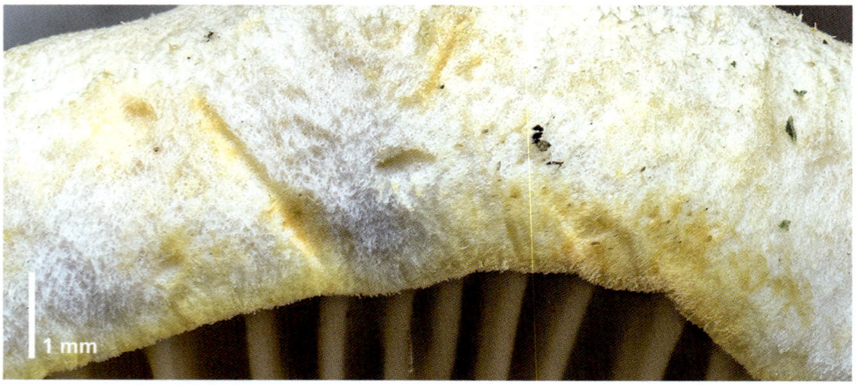

•• ***Cantharellus pallens*** is a large, pale, fleshy chanterelle, mostly with an eye-catching pale stem; the flesh stains strongly brown after bruising; centrally the cap may become grey with age, but also completely yellow after the white surface felt is worn off. The hyphae of the cap cuticle are thick-walled. The spores measure (7–)7.5–9.5(–10) × 3.5–5.5(–6) µm. It is mostly found on clay soils, mostly mycorrhizal with *Fagus*.

Cantharellus cibarius ◁◁ is mostly more yellow, somewhat more fragile, without whitish hyphal felt on the cap and browns less strongly. *Cantharellus pallens* is mostly reported under the name *C. subpruinosus* in southern Europe. *Cantharellus alborufescens* ⋉ is a very similar species associated with evergreeen oaks, which may occur in warmer southern parts of temperate Europe.

Rather common in warmer areas and absent from the boreal zone; mostly June–October.

Thomas Læssøe

●● ***Cantharellus amethysteus*** is a large, sturdy chanterelle with rather pale yellowish colours and, at first, an obvious violet, ± scaly cap coating that may later disappear completely. The flesh stains brown after bruising. The hyphae of the cap cuticle are thick-walled. The spores are voluminous and measure 8.5–13 × 4.5–7.5 µm. It is typically found on clay soils, mostly mycorrhizal with *Fagus* or *Quercus*.

Cantharellus pallens ◁ is a very similar species, but has a white cap coating and has smaller spores, (7–)7.5–9.5(–10) × 3.5–5.5(–6) µm.

A rather uncommon, but possibly overlooked, chanterelle found mainly in warmer parts of N Europe; mostly June–September.

●● ***Cantharellus ferruginascens*** is at first a ± lemon-yellow, rather slender chanterelle with pale hymenophore and a strong tendency to stain brown on the surface and in the flesh. Young fruitbodies may have a thin, pinkish coating on the surface. The spores measure (7–)8–10.5(–11.5) × 4.5–6(–6.5) µm. Occurs with deciduous trees on both acid and calcareous soils.

The more orange, also brown staining, *C. alborufescens* ⋉ occurs with evergreen oaks in southern Europe. *Cantharellus cibarius* ◁◁ and *C. pallens* ◁ have an orange-yellow hymenophore.

Scattered, but possibly overlooked; summer-autumn.

Felice Di Palma

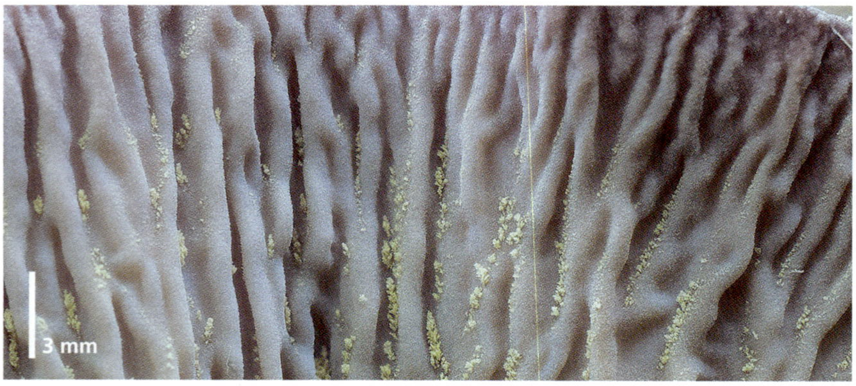

• ***Gomphus clavatus*** is a large and massive, very fleshy, ± violet, chanterelle-like fungus with yellow-brown spores. It may grow alone or in clusters, and sometimes in fairy rings. The verrucose, cyanophilous spores measure 10–15.5 × 4–7 µm. It is ectomycorrhizal with both deciduous trees and conifers (*e.g. Fagus* and *Picea*), mostly on calcareous soils.

There are no obvious identification issues. *Gomphus clavatus* is closely related to *Ramaria* (page 1138), which may display similar colours and similar spores, but the fruitbodies are ± branched.

A rather scattered and local species, that is virtually absent from some of the more heavily farmed regions; mostly August–October.

Jean-Marc Moingeon

Faerberia carbonaria is a rather large omphalinoid to cantharelloid fungus with a dark cap and pale gills. It has thick-walled cystidia with crystals (metuloids) and large cylindrical to allantoid, smooth, inamyloid spores measuring 7–11 × 4–5 µm. Grows on rather old burnt ground, where moss cover has developed.

Myxomphalia maura (page ´03) has smaller, amyloid spores and thin-walled cystidia. Species of *Craterellus* ◁◁ have no cystidia, are mycorrhizal and normally occur on unburnt ground. *Faerberia* is related to the polypores.

A rare species, that has probably declined due to changes in forestry practices; mostly August–October.

Jens Maarbjerg

Agarics

The gilled mushrooms comprise a very large group of macrofungi, with more than 3,000 species in temperate Europe. They are characterized by having gills under a cap, always held in the vertical postion. Most species also have a stem. In the pleurotoid agarics the stem is reduced or absent.

Spore colour

The colour of spore-deposits is a key character for dividing the gilled mushrooms into manageable groups: white, cream to yellow (this page spread) or brownish, rose-brown to black (next page spread). In addition, a few species have green or lilac spore-deposits (also next page spread).

Veil

In some agarics the gills develop freely – a naked development. Others have a protecting tissue over the young gills – a partial veil. Some have the entire fruitbody covered in an universal veil. *E.g. Amanita phalloides* has a membranous version of both veil types: the partial veil is seen as a ring on the stem at maturity, while the universal veil is seen as a volva at the stem base. In *Cortinarius* the two veil types can be thread-like and in others, *e.g. Gomphidius,* they are present as slimy layers.

Gill attachment

The way the gills attach to the stem is an important character for separating species and genera. The gills may be free (not touching the stem), adnexed, adnate, emarginate or decurrent.

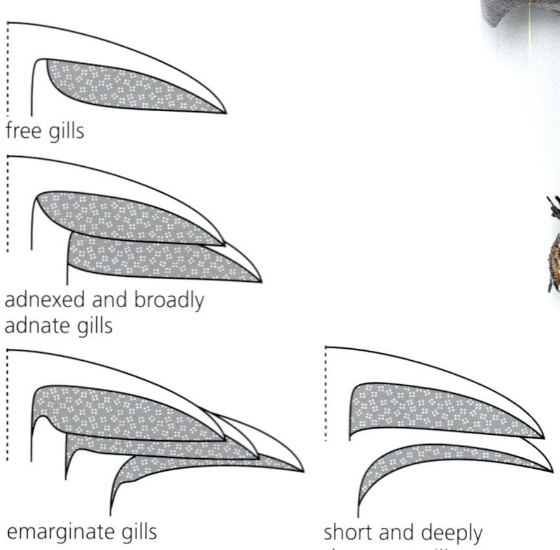

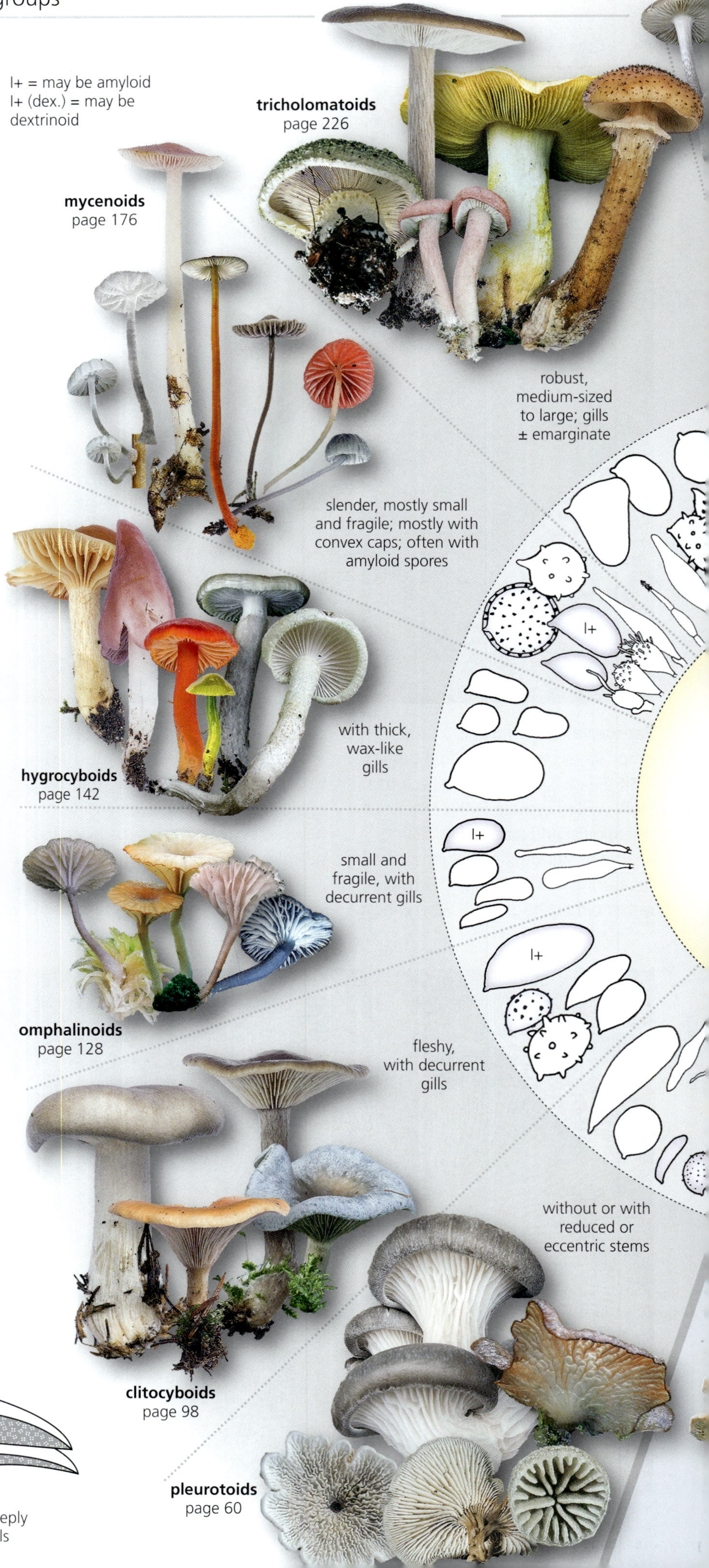

marasmioids
page 302
collybioids
page 274
Cystoderma
and *Cystodermella*
page 314
lepiotoids
page 320
small to medium, rather tough; gills adnexed to adnate
gills mostly adnexed; ± tough, may revive after drying
with grainy surfaces (universal veil)
I+
with free gills; often with partial veil
I+
Limacella and *Chamaemyces*
page 348
I+ (dex.)
I+ (dex.)
with free gills; with slimy universal veil
amanitas
page 352
spore-deposit white to cream
I+
with free or, rarely, adnexed gills and membranous universal veil
agarics with ± soft-fleshed fruitbodies with a cap and with gills below
spore-deposit white to yellowish
with grainy surfaces; universal veil
ornament I+
Phaeolepiota
page 314
other spore colours
other similar fungi
fragile; flesh with spherical cells; without latex
fragile; flesh with spherical cells; with latex
ith decurrent wrinkles,
hanterelles, page 46
Lactarius and *Lactifluus*
page 414
Russula
page 372
leathery, broadly attached fruitbodies
polypores, page 824
next page spread

Microscopical characters

Many species of agarics have cystidia on the stem (caulocystidia), cap surface (pileocystidia) or on the gills. The gill cystidia are divided into those in the hymenium (pleurocystidia) and those on the edges (cheilocystidia). The cystidia are very important in species identification. It it advisable to use a dissection microscope when making preparations, to ensure that the appropriate part of the gill is studied.

Agarics have a huge range of spore types: smooth, angular, to spiny, thin- to thick- or double-walled, with or without germ pore. Some spores stain ± blue with Melzer's reagent (termed amyloid or I+). When studying microscopical characters always check for amyloidity and surface ornaments (use oil immersion).

Relationships

The agarics constitute a 'form group' – an artifical assemblage. The biggest monophyletic groups of agarics are the *Agaricales*, the *Russulales* and the *Hymenochaetales* each of which include many species that are not agarics, *e.g.* corticioids, clavarioids, puffballs and polypores.

OTHER SIMILAR FUNGI:

– chantarelles do not have proper gills but, instead, sinuose, branching veins or wrinkles (page 46).

– certain polypores have gill-like hymenophores, but these structures are normally very tough (page 824).

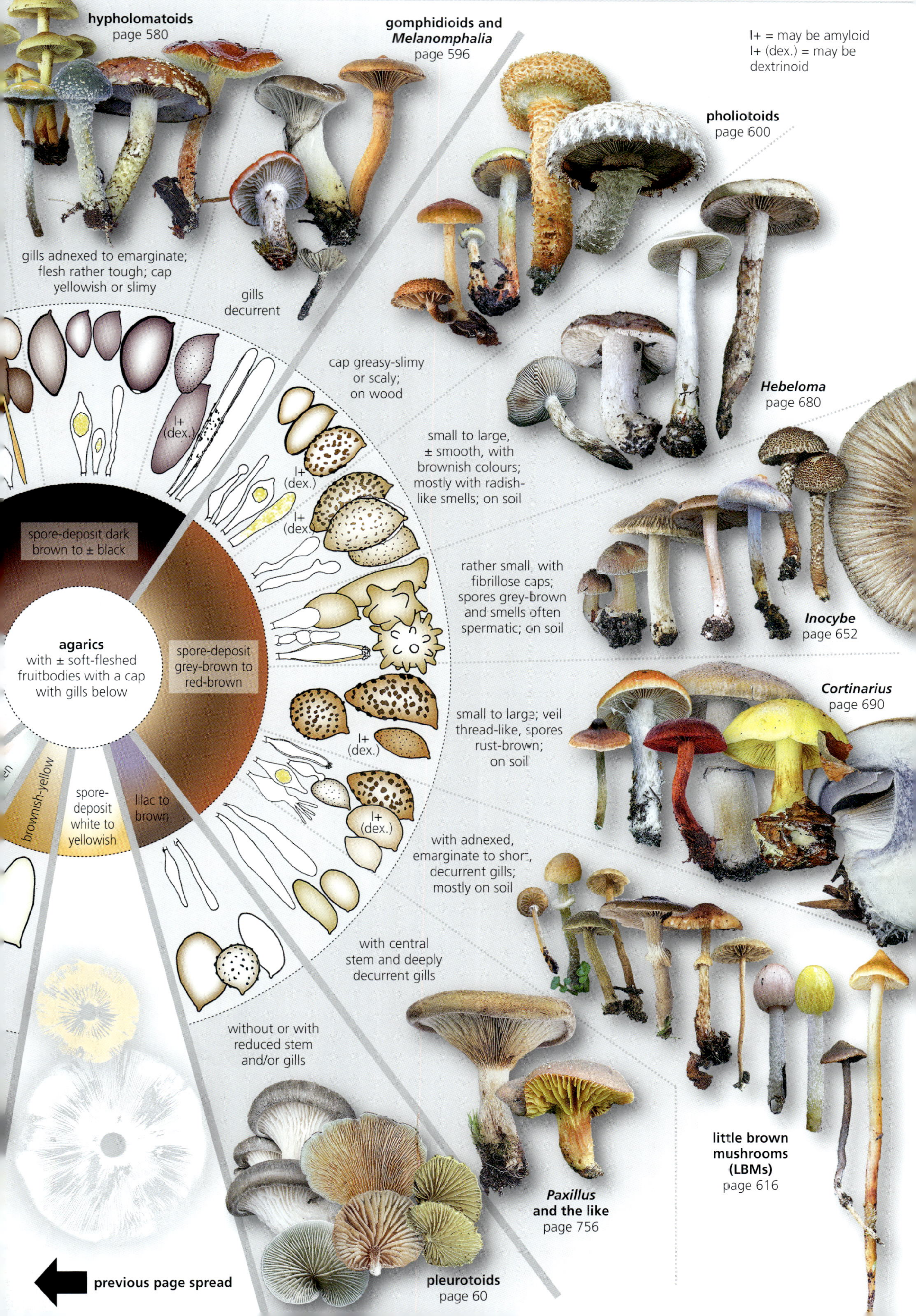

previous page spread

Pleurotoids

The pleurotoid agarics are either fan- or tongue-shaped and lack or have an excentric to lateral stem. Some forms are attached on the backside, with the gills forming a radiating pattern. Some have the gills reduced to veins or wrinkles, or even to a ± smooth hymenophore. Species with a smooth hymenophore and no stem are treated under the cyphelloid fungi (page 1076).

Most species in this group are decomposers and the majority grow on wood, even though some are also nematophagous. A few occur on herbaceous stems: *Gloiocephala*, *Campanella*, and some of the *Hohenbuehelia* species, one species of *Lentinellus*, one *Clitopilus*, several *Entoloma*, some *Deconica* and a few *Crepidotus* species.

Pleurotoid agarics have different spore-deposit colours. The majority are white- to cream-spored, but a smaller group is rose- or brown-spored and *Pleurotus ostreatus* has a greyish-violet spore-deposit. The different spore colours reflects the fact that the species concerned have completely different phylogenies. The spore colour is used in the division of the groups in the wheel.

Some species of pleurotoids are remarkably tough, *e.g. Panus* and *Neofavolus* and also older fruitbodies of *Lentinellus* and *Pleurotus*. Others, *e.g. Schizophyllum*, *Hohenbuehelia* and *Sarcomyxa*, have gelatinous or rubbery flesh.

Other similar fungi:
– the cyphelloid fungi have small fruitbodies without gills and no or hardly any stem. They are typically attached on the upperside of the fruitbody (page 1076).
– the clitocyboids have central stems, but may recall pleurotoids (page 98).

Further reading: 11, 65, 66, 156, 181, 203, 223, 224, 245, 295.

Sarcomyxa
1 species, page 74
Hohenbuehelia
12 species
page 71
Panus
2 species
page 75
Neofavolus
1 species
page 79
Neolentinus
2 species
page 81
Lentinus
2 species
page 80
Lentinellus
6 species
page 76
Pleurocybella
1 species
page 83
Pleurotus
6+ species
page 84
Phyllotopsis
1 species
page 82
Clitopilus
1–4 pleurotoid
species, page 88
Entoloma
≈ 5 pleurotoid species
page 88
Tapinella
2 species
page 96
Crepidotus
≈ 20 species
page 90
Deconica
2 pleurotoids
page 95
Pleuroflammula
1 species, page 96
small to large;
rather tough;
mostly felty
rather large;
± tough; partially
yellowish
or greenish
rather large;
± tough;
base lilac, felty;
with skeletal
hyphae
tough; gills with
hyphal pegs;
smells of aniseed;
with skeletal
hyphae
tough; gills
serrated;
with skeletal
hyphae
tough; gills
serrated;
with skeletal
hyphae
large; gills
smooth-edged, simple,
not anastomosing
large, ± tough; gills
smooth-edged,
anastomosing;
with skeletal hyphae
gill edge entire,
yellow
spores longitudinallly
ridged; with a
farinaceous smell
spores angular
in outline
large; without
obvious
remnants of veil;
thick-fleshed
without obvious
remnants of veil;
small to medium,
± stemless and rather
thin-fleshed
small; with
thread-like veil;
thin-fleshed
with
triangular
remnants of
veil; small
SMALL TO LARGE – GELATINOUS
LARGE – NOT GELATINOUS
LARGE – GELATINOUS
I+
I+
+/-
spore colour
pleurotoids
have no stem or a
± excentric to lateral
stem. Decomposers of
wood or herbaceous
stems
other
similar
fungi

Schizophyllum commune is a very tough and elastic-rubbery, drought-resistant, pale, tomentose-felty, typically lignicolous, pleurotoid fungus with remarkable split 'gills' that curl outwards in dry conditions ('split') but straighten in moist conditions, making them gill-like. The hyaline, smooth spores measure, 4–6 × 1.5–2.5 µm. On woody subtrates, preferably sun-exposed, but also on silage wraps, old whale bones *etc*.

Distinctive and unlikely to be misidentified. *Panellus stipticus* ▷▷ has a small but evident stem and normal gills. See also the cyphelloid *Schizophyllum amplum* (page 1081).

A widespread and common species, except in the northern boreal zone; it can be found year-round.

Schizophyllum commune is consumed in E and SE Asia, and in Ethiopia it is used as chewing-gum. It is also used in genetic experiments and known to be able to grow almost anywhere, unfortunately including in the lungs of immunocompromised patients. The fruitbodies can revive after year-long storage in fungaria.

Plicatura crispa is a rather small, crowded, mussle-shaped, soft but somewhat tough pleurotoid fungus with a veined, white to turquoise or pale blue underside and a finely felty, orange-brown upperside. The margin is inrolled and wavy. The smooth, amyloid spores are allantoid and measure 3–4.5 × 1 µm. It is restricted to bark from various deciduous trees, especially ± standing dead trunks and hanging branches.

This species is also known as *Plicaturopsis crispa*. *Byssomerulius corium* is not mussle-shaped and the hymenophore is less wrinkled. *Panellus stipticus* ▷▷ has normal, regular gills. *Plicatura nivea* (page 966) does not have, or hardly forms, a cap, and is chalk-white and not nearly as wrinkled/veined below. It mostly occurs on *Alnus incana* bark in the northern part of temperate Europe.

Widespread and generally common, but nearly absent from some northern regions; occurs year-round, but at its best during late autumn and in the winter months.

Thomas Læssøe

Thomas Læssøe

Campanella caesia is a rather small, greyish-blue, stemless, white-spored pleurotoid agaric with few, ± obvious gills. The gills are connected by transverse ribs, thus resembling a polypore. Has a gelatinous tissue in the cap flesh, but otherwise much as various marasmioids. The spores measure (6.5–)7.5–9.5(–10) × 4–5.5 (–6) µm and the cheilocystidia are verrucose below. On grasses on hot, often coastal sites.

Gloiocephala menieri ▷▷ may look similar, but has cystidia at the cap margin, longer (15–22 µm) spores and occurs on *Typha*.

Widespread, but rare in the nemoral zone and absent from the boreal zone; mainly August–October.

Arrhenia species are small or rather small and can be omphalinoid or pleurotoid. The pleurotoid species on these pages have veined or wrinkled hymenophores and white spore-deposits. Species with a central stem are covered on page 135.

Arrhenia lobata is a rather large, circular to semicircular (as seen from above), grey-brown, ± translucent pleurotoid agaric with a veined/wrinkled hymenophore and no to only a rudimentary stem. The spores measure 5–10(–15) × 4–8.5 µm. It grows on bryophytes in very wet habitats with springs *etc*.

Well-developed specimens should be easy to separate from other *Arrhenia* species, which are either smaller, have well-formed gills or grow in much drier habitats.

A fairly common species in northern and upland areas, but rare or very rare in the lowlands (easily overlooked in its very wet habitat and due to its early appearance); mostly April–July.

Jan Vesterholt

Arrhenia spathulata is a grey, medium-sized pleurotoid agaric with a short, lateral stem, an almost funnel-shaped, zonate cap and a veined/wrinkled hymenophore. The spores measure (5.5–)6–10(–10.5) × 4–6.5 µm. It probably always grows on the moss *Syntrichia ruraliformis* in dunes or in dune-like habitats.

Arrhenia lobata △ occurs in very wet sites on other bryophytes. *Arrhenia auriscalpia* ▷ is very long-stemmed.

A mostly coastal species with a wide distribution, but also found farther inland in sandy habitats; mainly August–January.

Arrhenia auriscalpia is a ± medium-sized, long-stemmed, spatulate, grey-brown pleurotoid agaric with a decurrent, veined/wrinkled hymenophore. The spores measure (5–)7.5–10(–11) × 4–6.5(–7.5) µm. It is associated with bryophytes in boreal-alpine vegetation.

Arrhenia spathulata ◁ is very short-stemmed, somewhat paler and more southern in its distribution.

Absent from the lowlands of tempereate Europe, but occurs in alpine and boreal areas; summer–autumn.

Arrhenia retiruga is a small, stemless, pale grey pleurotoid agaric with a smooth to somewhat wrinkled hymenophore and a rather felty cap surface. Clamps absent. The ± ellipsoid spores measure (5–)6–9.5(–11.5) × 3–5(–7) µm. Associated with bryophytes in moist places, *e.g.* on shaded lawns.

Species of *Rimbachia* (page 1078) are similar but whiter, typically more cupulate and with clamped hyphae. Greyish forms of *Calyptella* (page 1080) also have clamps and gnarled hyphae in the cap cuticle.

Widespread and common, but ± overlooked; all year, but mainly during winter.

Arrhenia acerosa is a greyish, excentrically stipitate, medium-sized pleurotoid agaric with proper gills. The spores measure (5–)6–9.5(–11) × (3–)4–6.5(–7) µm (4-spored variety). It grows mostly in moist, humus-rich places, within cavities, *etc*.

Pleurotoid, greyish species of *Entoloma* ▷▷ are similar, but have red spore-deposits and angular spores. A 2-spored variety, *A. acerosa* var. *tenella* ⋉, which has larger spores, also occurs.

A widespread, but rather rarely recorded species, with main occurrence June–November.

Thomas Læssøe

Gloiocephala species are small, rather tough agarics with a smooth or wrinkled hymenophore, rather large spores (14–23 µm long) and mostly with cystidia on cap, stem and hymenophore. They typically grow on *Carex* or *Typha*.

Gloiocephala cornelii is a very small, minutely hairy, long-stemmed species, completely without gills and has 12–18 × 3.5–6.5 µm large, almost clavate spores. It only occurs on *Cladium mariscus* stems. Use gloves when looking for this species – *Cladium* cuts deep!

There are no similar fungi on *Cladium*, but *Resinomycena saccharifera* (page 186) with well-developed gills may occur in large quantities on this substrate.

Rarely found, but apparently widespread within the distribution of the host; August–November.

Gloiocephala caricis is a small, white species with an evident, but excentric stem and almost 'proper' gills. The cap cuticle consists of thick-walled ± clavate cells and elongated thick-walled cystidia. Thin-walled, ± clavate cheilo- and pleurocystidia also occur. The spores measure 12–24.5 × 4.5–5 µm and typically form on 2-spored basidia. It is associated with robust *Carex* species, such as *C. acutiformis*.

A different, probably undescribed species, found in Denmark, also on *Carex*, is microscopically more similar to *G. menieri* ▷.

A rarely recorded, possibly rather northern species; August–November.

Gloiocephala menieri is a rather small, pale brown, ± spatulate species with a grey-toned, veined hymenophore. Hymenial cystidia are very large, 32–100 × 5–12(–20) µm. The fusiform, smooth spores measure 15–22 × 5.5–7.5 µm. It is found at water level and only on standing culms of *Typha*.

There are no other similar fungi on *Typha*, but an apparently undescribed, microscopically similar species has been found on *Carex*.

Rarely recorded, but probably overlooked and more widespread than currently realized; mainly June–October.

Cheimonophyllum candidissimum is a small, chalk-white, white-spored, pleurotoid with a small, lateral stem, minutely floccose gill edges and hyphoid cheilocystidia. The smooth, almost spherical spores measure 5–6 × 4.5–5.5 µm. On deciduous wood, often suspended above ground.

Panellus mitis ▷▷ occurs on conifer wood and has a gelatinous cap cuticle. *Pleurocybella porrigens* ▷▷ has similar spores, but is larger and also prefers conifer wood. *Clitopilus hobsonii* ▷▷ has a brownish-rose spore-deposit, no cystidia and longitudinally ridged spores.

Locally common in the boreal-hemiboreal zones to very rare in part of the nemoral zone; August–December.

Chaetocalathus craterellus is a rather small, white, white-spored and completely stemless pleurotoid agaric, covered in long, rigid, thick-walled hairs. The smooth spores measure 7–9 × 5–6 µm.

Unlikely to be confused with any other species in temperate Europe. It is related to *Crinipellis* (page 307).

A southern species found only in the warmer parts of temperate Europe, but it also has records from the UK; mainly autumn–winter.

Marino Zugna

Panellus species are rather small pleurotoid, white-spored, lignicolous agarics that mostly have a sharply delimited, lateral stem, but the stem may also be lacking. Lacks any form of veil on the cap. The spores are amyloid.

Panellus stipticus is a cracking-floccose, whitish to pale brown, bitter-tasting pleurotoid agaric with a very evident and wide, felty, lateral stem, ± brown gills and a very tough texture. The spores are smooth, amyloid and narrowly ellipsoid and measure 4–5 × 2–2.5 µm. Mostly found on hardwoods, particularly *Quercus*, but occasionally also on conifer wood.

Panellus mitis ▽ is paler, mostly smooth and has allantoid spores. *Crepidotus* ▷▷ lacks an obvious stem and has brown spores. *Tectella patellaris* ▷ is stemless and the gills are initially covered by an almost membranous veil.

Widespread and common within the range of *Quercus* and farther north in more Atlantic areas; mostly September–April, but may be found year-round.

Panellus mitis is at first pure white, but with age becomes a somewhat brownish, white-spored pleurotoid agaric with a small, thick, downy, lateral stem. Flesh somewhat gelatinous. Cystidia absent. Spores narrowly allantoid, amyloid, 3.5–6 × 1–1.5 µm. On conifer wood, including thin branches.

Species of *Crepidotus* ▷▷ have brown spore-deposits. *Cheimonophyllum* ◁◁ occurs on hardwood, is not gelatinous and has spherical spores. Species of *Hohenbuehelia* ▷▷ are more felty and have thick-walled cystidia (metuloids).

Common and widespread; mostly October–January.

Panellus ringens is a grey-rose, almost stemless pleurotoid agaric. The small, amyloid spores measure 5–7 × 1–2 µm. Found mostly on deciduous wood, but may also occur on conifer wood.

Panellus violaceofulvus ▽ is a similar colour, but has larger spores, 6.5–9.5 × 2.5–3.5 µm.

Mostly found in northern and subalpine areas and absent from most of the lowlands; mostly autumn and winter.

Panellus violaceofulvus is ± grey-rose pleurotoid agaric with a ± white, felty inner part, with or without an evident stem. Spores rather large, amyloid, 6.5–9.5 × 2.5–3.5 µm. On both deciduous wood and conifer wood.

Panellus ringens △ has smaller spores measuring 5–7 × 1–2 µm.

A rather rare southern species; autumn and winter.

Jacob Heilmann-Clausen

Tectella patellaris is a pale brown, stemless, white-spored pleurotoid agaric with a white veil. The weakly amyloid spores measure 3–5.5 × 1–1.5 µm. It occurs almost exclusively on wood of *Alnus*, mostly in dense, moist scrub or damp Alder forests, mostly together with *Plicatura nivea* (page 966).

Distinctive and unlikely to be misidentified. The species is often included in *Panellus* ◁.

Mostly in boreal or subalpine, moist *Alnus*-dominated habitats; appearing from late summer onwards.

Thomas Læssøe/Jens H. Petersen

Resupinatus species are small, greyish pleurotoids or cyphelloids (page 1084), typically stemless and attached to woody substrates by the upper surface of the cap. The smooth, inamyloid spores are spherical to broadly allantoid and thick-walled cystidia are absent. The flesh is somewhat gelatinous.

Resupinatus applicatus is a greyish, somewhat striate, circular or hoof-shaped, stemless pleurotoid agaric with pale felt at the base of the fruitbody. The ± spherical spores measure (3.5–)4–6 × (3–)3.5–5 µm. Typically found on attached deciduous branches, but may also occur on fallen branches, and rarely on conifer wood.

*Resupinatus trichoti*s ▽ is very similar and young specimens may be impossible to separate. With age it develops a dark felt towards the attachment point. *R. alboniger* ⋉ is darker with dark gills and ± allantoid spores. Some mycologists recognize small striate versions as *R. kavinii* ⋉ and *R. striatulus* ⋉. Small, greyish species of *Hohenbuehelia* ▷ have thick-walled cystidia with ± evident crystals.

Widespread and probably very common, but somewhat overlooked; all year, but mostly late autumn and in mild winters.

Resupinatus trichotis is a dark grey pleurotoid agaric that with age develops a brownish-black, coarse felt at the base of the hoof-shaped or circular, stemless fruitbody. Spores spherical, 4.5–6 × 4–5.5 µm. Usually on bark of deciduous trees and mostly on attached branches in hedges, *etc*.

R. applicatus △ is very similar, and young material may be hard to identify with certainty. Species of *Hohenhuehelia* ▷ have thick-walled cystidia with ± evident crystal cover (metuloids).

Widespread and probably common, but is easily overlooked; all year, but mostly late autumn.

Hohenbuehelia species are small to rather large pleurotoid, white-spored agarics that may be attached without a stem or, in rarer cases, with an excentric to almost central stem. The spores are cylindrical to allantoid, the flesh is gelatinous, and thick-walled cystidia (metuloids) are found throughout. Decomposers on wood, rarely on herbs. Special sticky cells, *e.g.* present on the gill edges, are used to catch and digest nematodes (nematophagous).

Hohenbuehelia cyphelliformis is a small, ± mussle-shaped pleurotoid agaric, with very pale, rather widely spaced gills, a greyish, somewhat felty cap, and 4-spored basidia. The allantoid spores measure 7–10(–11.5) × 3–4.5(–6) µm. Grows typically on coarse, ± lignified herbs, but also on bark of *Sambucus*, *etc*.

Hohenbuehelia fluxilis ▷▷ is typically more hoof-shaped, somewhat more felty, grows mainly on *Salix* and is 2-spored. *Hohenbuehelia unguicularis* ▽ becomes rather dark-gilled with age, but otherwise looks similar.

Widespread, but not common; all year, but most records from August–December.

Hohenbuehelia unguicularis is a small, very dark and hoof-shaped pleurotoid agaric, with whitish, downy hairs on the cap, widely spaced gills that quickly become dark, and thick-walled cystidia with crystals. The spores measure (6.5–)7–9.5(–11.5) × 3.5–5(–6) µm. On deciduous branches in scrub and forest.

Hohenbuehelia cyphelliformis △ is similar, but has white gills. Species of *Resupinatus* ◁ are mostly smaller, not completely hoof-shaped and typically have spherical spores and no thick-walled cystidia.

Rare, but rather widespread; all year but possibly mostly during winter.

Thomas Læssøe

Hohenbuehelia fluxilis is a pale and ± mussle-shaped to hoof-shaped pleurotoid agaric with a rather felty cap, white to cream gills, and 2–spored basidia. The allantoid spores measure 7–10(–11.5) × 3–4.5(–6) µm. Typically on attached branches in *Salix* carrs, rarer on other hardwoods.

Hohenbuehelia cyphelliformis ◁◁ is typically smoother, smaller and 4–spored, and grows mostly on coarse herbs and *Sambucus*. *H. unguicularis* ◁◁ is smaller and the gills tend to become darker with age.

Widespread and common, but under-recorded; mostly June–November.

Thomas Læssøe

Hohenbuehelia mastrucata is a rather large, strongly felty, pale pleurotoid agaric with ± cream to yellowish gills, no stem and a very thick gelatinous layer below the thick felt. The ± ellipsoid spores measure (5.5–)6–8.5(–9.5) × (3–)3.5–4.5 µm. Mostly on trunks of *Fagus*, but also on other types of hardwood.

Similar species are darker and have a thinner gelatinous layer below the felty surface. *Hohenbuehelia fluxalis* △ has 2–spored basidia.

Rare, but rather widespread; mostly August–November.

Hohenbuehelia petaloides is an almost funnel-shaped, large, brown, virtually smooth pleurotoid agaric with thick-walled, crystal-covered, tapering cystidia (metuloids). The ± ellipsoid to ovoid spores measure 5–7.5(–8.5) × 3.5–4.5(–5) µm. Grows on buried wood or wood chips on mull.

Hohenbuehelia tremula ⋉ prefers conifer chips and has somewhat larger spores measuring 7.5–9.5 × 4.5–5 µm.

Rare, but widespread; mostly August–October.

Hohenbuehelia culmicola is a very dark, almost black pleurotoid agaric with a lateral, rooting stem and a fan-shaped to spatulate, felty cap. The spores measure (8–)9–12(–14) × (4.5–)5–7(–7.5) µm. On *Ammophila* and other coarse dune grasses in white or grey dunes.

In the same environment, also at the base of *Ammophila* another *Hohenbuehelia* species can be found, *H. bonii* ⋉, which is currently poorly known; it is somewhat larger and brown rather than blackish.

Rare, but may be somewhat overlooked; September–January.

Steen A. Elborne

Hohenbuehelia auriscalpium is a broadly spatulate, rather smooth, pale brown, largish pleurotoid agaric with thin-walled cystidia in the cap cuticle. The spores measure (4.5–)5–6.5(–7) × 3.5–4.5(–5) µm. It is found on deciduous wood, mostly larger fallen trunks of *Fagus*, in relatively undisturbed forests.

Hohenbuehelia abietina ⋉ is very similar, but grows on conifer wood. The cap cuticle has no thin-walled cystidia, but may have thick-walled cystidia with crystals.

Rather rare, but widespread; August–November.

Thomas Læssøe

Hohenbuehelia longipes is a medium-sized, greyish, clitocyboid pleurotoid agaric with a ± excentric to central stem. It has thick-walled, crystal-covered, tapering cystidia (metuloids). The spores measure (7.5–)8.5–12.5(–14) × (3–)3.5–5.5(–6) µm. On soil at the margins of mires in subalpine sites.

May be confused with species of clitocyboid agarics (page 98) but these all lack thick-walled cystidia. *Faerberia* (page 55) may look similar but grows on burnt soil, has skeletal hyphae, and the hymenophore is more cantharelloid with veins and ribs.

Rare (absent from the lowlands); late summer–autumn.

Thomas Læssøe

Sarcomyxa serotina is a fleshy, somewhat gelatinous, white-spored pleurotoid agaric with a short, thick, dotted, sharply delimited, lateral stem. The cap colour is initially mostly ± orange (the bottom picture shows young primordial fruitbodies), then olive-coloured and, with age, pale grey. The gills are narrow and crowded, pale yellow, and have clavate cheilocystidia. The amyloid, narrowly allantoid spores measure 4–6.5 × 1–2 µm. Mostly on deciduous wood, *e.g. Fagus*, *Alnus*, *Quercus*, *Betula* and *Salix*. It produces a white rot.

Young, fresh specimens are distinctive, but older material may recall small *Pleurotus* ▷▷ specimens. *Pleurotus*, however, lack gel in the tissues, and have a less clearly delimited transition between the gills and the stem, and completely different spores.

Widespread and rather common; mostly October–February.

Panus is a small genus of rather tough, lignicolous, white-spored pleurotoid agarics with lilac and violet colours. Hyphal system dimitic with skeletal hyphae. All species produce a white rot.

Panus conchatus is a rather large, tough-fleshed, downy, white-spored, white-rotting, pleurotoid agaric with violet colours when young, later more brown. The gills have violet to brownish tints and the stem felt stays violet for a long time. The ellipsoid to ± cylindrical spores measure 5.5–7 × 2.5–3.5 µm. On deciduous wood, mostly on sun-baked stumps.

Panus rudis ▽ has similar colours, but is covered by long, rigid hairs on the cap and stem. Species of *Pleurotus* ▷▷ are less tough and lack violet colours. Species of *Lentinus* and *Neolentinus* ▷▷ are also devoid of violet colours and are typically scaly or, in the latter, brown-rotting. Species of *Lentinellus* ▷▷ likewise lack violet colours, and have highly serrated gills, and smaller, echinulate, amyloid spores. *Tapinella panunoides* ▷▷ may have violet felt, but has brown, soft gills and a brown spore-deposit.

Widespread and fairly common; mostly June–October.

Panus rudis is a ± grey-rose to lilac pleurotoid agaric that is covered by long, rigid hairs. It has pale, often somewhat rose-coloured, crowded, deeply decurrent gills and a funnel-shaped cap. The pleurocystidia are thick-walled. The smooth, inamyloid spores measure 4.5–6.5 × 2.5–4 µm. Produces a white rot in deciduous wood.

This species is also known as *Panus neostrigosus*. *Panus conchatus* △ is downy and not covered in long, rigid hairs.

Mostly found in the eastern part of temperate Europe; summer–autumn.

Ladislav Špeta

Lentinellus is characterized by strongly serrated gills, finely echinulate, amyloid spores, skeletal hyphae in the flesh and the production of a white rot.

Species of *Neofavolus* and *Lentinus* have even tougher flesh, and *Neofavolus* also has hyphalk pegs on the gill faces. *Neolentinus* differs in producing a brown rot.

Lentinellus cochleatus is an almost funnel-shaped, clustered, smooth, yellow-brown to red-brown pleurotoid agaric with a strong aniseed smell or, rarely, no odour. It tastes bitter. Typically with large brown chlamydospores in the flesh. The echinulate, amyloid spores measure 4–5 × 3–4 µm. Mainly on deciduous wood, often *Fagus* stumps.

Other *Lentinellus* species are not funnel-shaped and do not smell of aniseed. *L. flabelliformis* ▷ may look similar but has larger spores, 5–6 × 3–4.5 µm. The rather tough texture, very serrated gills, and amyloid spores make identification to genus straightforward. *Neofavolus suavissimus* ▷▷ has a similar smell.

Widespread and rather common; mainly August–November.

Lentinellus flabelliformis is a very variable, rather slender, smooth, deeply umbillicate, ± hygrophanous *Lentinellus* with a long, ± excentric stem. It has finely echinulate, amyloid spores measuring 5–6 × 3–4.5 µm. Mostly recorded on conifer wood.

May include a number of biological species that cannot with certainty be recognized morphologically. The spores are larger than those of *Lentinellus cochleatus* ◁, which may occur in inodorous forms. *Pseudoomphalina kalchbrenneri* (page 103) is also rather similar, but the gill edges are not strongly serrated and the spores are bigger, echinulate and amyloid.

Widespread but rather rare, mostly found in the hemiboreal–boreal/subalpine zones; mainly August–November.

Lentinellus herbarum is a very small, less than 15 mm wide, ± stemless, odourless *Lentinellus*. The hyphae lack clamps. The spores measure 4.5–5.5 × 3.5–4.5 µm. It grows on *Epilobium angustifolium* stems and perhaps on other coarse herbs.

The substrate should make identification straightforward.

Very rarely recorded, but could be overlooked; mainly August–March.

Pierre Roux

Lentinellus ursinus is a pleurotoid, stemless, ± brown-felty *Lentinellus* with strongly amyloid flesh. It may have a dark zone between the cap felt and the flesh, and has ± gnarled, amyloid skeletal hyphae. The spores measure 3.5–4.5 × 2.5–3.5 µm. On deciduous wood and conifer wood, mostly in undisturbed forests.

Young fruitbodies are very pale and may be distinctly wrinkled/veined on the surface and exude 'milk' when cut. Young material has a weak amyloid reaction, since the reacting skeletal hyphae have yet to form. *Lentinellus castoreus* ▷ should have darker fruitbodies, almost without gnarled skeletal hyphae.

Lentinellus vulpinus ▽ has binding hyphae and only somewhat amyloid or inamyloid cap flesh; it lacks the dark zone between the flesh and the cap felt.

Widespread and locally common, possibly expanding its range; mainly August–December.

Thomas Læssøe

Lentinellus vulpinus is a large, rather pale, fleshy, stemless, ± brown-staining *Lentinellus* with a cap surface that may be wrinkled or not. The flesh is inamyloid and lacks a dark line between flesh and cap felt. Lacks skeletal hyphae and at most somewhat dextrinoid or amyloid. Spores 4–5 × 3–4 µm. On larger, fallen trunks of deciduous trees.

Lentinellus ursinus △ and *L. castoreus* ▷ have a strongly blackish-blue amyloid reaction in the cap flesh and a separating line between flesh and felt. They both have skeletal hyphae and at maturity are ± dark brown and felty, at least at the base.

Mostly found in the boreal zone; mainly September–November.

Lentinellus castoreus is a relatively small, brown-haired *Lentinellus*. It may have a dark zone between the cap felt and the flesh. The flesh is ± mild-tasting, with amyloid skeletal hyphae. The spores measure 3.8–4.2(–4.8) × 2.9–3.5 µm. Probably exclusively on deciduous wood.

Lentinellus ursinus ◁ is here defined as a species with distinctly amyloid and gnarled skeletal hyphae. In some published literature the names *L. castoreus* and *L. ursinus* are reversed in interpretation. Also, *L. vulpinus* ◁ has various interpretations – here defined as a species without skeletal hyphae.

Mostly recorded in the southern parts of temperate Europe; autumn.

Thomas Læssøe

Neofavolus suavissimus is a pinkish-buff, rather small, strongly aniseed-scented, tough, white-spored, pleurotoid agaric. The gill sides bear scattered, small, hyphal pegs that are a composite of numerous hyphae, and lack cheilocystidia. The ellipsoid to cylindrical, inamyloid, smooth spores measure 7–8 × 3.5–4 µm. Grows on *Salix* in old, wet *Salix* carrs and is mostly situated quite high in the trees, in a similar manner to *Hypocreopsis lichenoides* (page 1526).

Distinctive and unlikely to be misidentified. It is tougher than *Pleurotus* species, and the odour is so pronounced that it is possible to detect it based on smell alone. *Lentinellus cochleatus* ◁◁ has a similar smell but has serrated gills without hyphal pegs at the sides.

Rare, but also overlooked – apparently absent from the more Atlantic and western parts of temperate Europe; mainly July–October.

Hanne Petra Katballe

Thomas Læssøe

Jesper Hyllemose

Lentinus tigrinus is a tough-fleshed, somewhat clitocyboid, white-spored, odourless, pleurotoid agaric with crowded gills and finely serrated gill edges. The cap and the stem are markedly scaly. The cheilocystidia have an irregular shape. The almost cylindrical spores measure 6–10 × 2–3 µm. Occurs on wet deciduous wood in boggy places, and at lakesides, mostly on completely water-soaked wood in water; it produces a white rot in the decaying wood.

Neolentinus lepideus ▷ is less funnel-shaped, even tougher and produces a brown rot in dry conifer wood. The white-rotting *Pleurotus* species ▷▷ may be felty, but lack the distinct, coloured cap scales found in *Lentinus tigrinus*.

Mostly found in the southern part of temperate Europe, but appears to be spreading northwards; mainly May–October.

Neolentinus is a genus of tough, brown-rotting agarics with serrated gills and excentric stems. The flesh is dimitic with skeletal hyphae.

Neolentinus lepideus is a large, white, very tough and thick-fleshed, pleurotoid agaric with pale grey scales and narrow, crowded, serrated, pale yellowish gills. It forms a brown rot on coniferous wood. May grow in dry and hot places, *e.g.* on old railroad sleepers or on sun-baked stumps.

The white-rotting *Lentinus tigrinus* ◁ occurs on very wet deciduous wood, is more slender and has minute scales. *Pleurotus dryinus* ▷▷ occurs mainly on deciduous wood, where it produces a white rot. It has a veil at the cap margin.

Widespread, occasional; mostly June–October.

Michael Ivan Hansen

Neolentinus schaefferi is a very tough and robust, somewhat clitocyboid, pleurotoid agaric with small scales and pale brown to ± red-brown coloration. It has very narrow, crowded gills and lacks a distinct smell or taste. The ellipsoid, smooth, inamyloid spores measure 10-14 × 3.5–5 µm. Produces a brown rot in deciduous wood, mostly *Populus*.

Panus conchatus ◁◁ has, when young, violet colours, and is more felty and less funnel-shaped.

Very rare, and mainly in the southern parts of temperate Europe; mainly May–July.

Tome Jovanovski

Phyllotopsis nidulans is a medium sized, orange-yellow to orange, somewhat felty, cream- to rose-spored, pleurotoid agaric with an inrolled margin and bright orange-yellow, rather crowded gills that lack conspicuous cheilocystidia. The small, allantoid, smooth, inamyloid spores measure 4–6 × 2–3 µm. May grow on both deciduous and conifer wood, where it produces a white rot. In the lowlands, mostly found on *Fagus*.

Tapinella panuoides ▷▷ has brown spores, less vivid colours and mostly violet-tinted felt. *Crepidotus crocophyllus* ▷▷ has spherical, yellow-brown spores.

Widespread: rather common in some areas, while nearly absent from others; mainly October–February.

† **Pleurocybella porrigens** is a large, chalk-white to pale cream, tongue-shaped to almost funnel-shaped, rather thin-fleshed, white-spored pleurotoid agaric without cystidia. The almost spherical, smooth, hyaline, inamyloid spores measure 5–7.5 × 4–6 µm. Occurs on very decomposed conifer stumps or on wood chips.

Species of *Hohenbuehelia* ◁◁ may look similar, but have a gelatinous cap cuticle and thick-walled cystidia, while species of *Lentinellus* ◁◁ have strongly serrated, tough gills. Species of *Pleurotus* ▷▷ are mostly more greyish and have thicker flesh.

Common in hemiboreal–boreal areas, but much rarer in warmer, lowland regions; mostly August–November.

Pleurocybella porrigens has in recent times been connected to a series of poisonings with fatal outcomes in eastern Asia. Under no circumstances should it be consumed.

Thomas Læssøe

Pleurotus is a genus of ± excentrally stipitate or ± stemless, rather fleshy, somewhat tough, wood-degrading agarics with white, cream to pale lilac spore-deposits. Produces a white rot.

●● ***Pleurotus ostreatus* var. *ostreatus*** is a large and fleshy agaric that normally starts out bluish-grey with an inrolled cap margin, but later turns browner and the margin straightens. The stem is excentric to lateral and of very variable length. The spore-deposit is ± pale lilac and the cylindrical spores measure (7–)8–12.5 × (2–)3–4.5(–5.5) µm. Mostly found on deciduous wood, often trunks of *Fagus*, mostly in undisturbed forests, including old deer parks, *etc*.

Pleurotus pulmonarius ▷ is normally paler and occurs during summer. *Sarcomyxa serotina* ◁◁ can look similar, but has a sharply delimited stem and mostly yellowish coloration. Species of *Crepidotus* ▷▷ are smaller with brown spores. A pale, smaller version of *P. ostreatus* from *Salix* carrs, *etc*. is var. *praecox* ▷.

Widespread and rather common; mostly October–March.

•• ***Pleurotus ostreatus* var. *praecox*** is a smaller, mostly solitary, somewhat brownish-rose variety of *Pleurotus ostreatus* ◁, which is mostly associated with *Salix* carrs.

The main variety (var. *ostreatus*) is larger, darker and normally grows imbricate. *Pleurotus pulmonarius* ▽ is also larger and imbricate.

Rather rare, but poorly known, mostly found in northern parts of temperate Europe; mainly March–September.

•• ***Pleurotus pulmonarius*** is a large, fleshy, almost white to grey-brown, mostly imbricate *Pleurotus*. The spore-deposit is ± cream. The spores measure (7.5–)8–11(–12.5) × 3–4.5 µm. Occurs mostly on large trunks of *Fagus* in open forests.

It probably forms a hybrid population with *Pleurotus ostreatus* ◁ as a result of contamination of wild stock with various cultivated strains. *Pleurotus ostreatus* is, in its typical young form, dark bluish-grey and occurs mostly during winter.

Widespread and rather common; May–September.

Mike Valentine

●● ***Pleurotus cornucopiae*** is a greyish, rather small *Pleurotus* with an almost central stem, a ± depressed cap centre, and deeply decurrent and vein-like anastomosing gills. The spores measure 7–11(–12) × 3.5–5 µm. It mostly occurs on larger hardwood trees, often *Ulmus*.

Pleurotus pulmonarius ⊲⊲ has ± the same colours, but is more similar to *P. ostreatus* ⊲⊲ in shape. A closely related, lemon-yellow species, *P. citrinopileatus* ⋉ is widely cultivated and has been found fruiting outdoors.

Only locally common and mostly very rare; most of the year.

●● ***Pleurotus eryngii*** is a short-stemmed, thick-fleshed, grey-brown *Pleurotus* with a ± central, thick, rooting stem. The spores measure 10-13 × 4–5.5 µm. It is associated with open habitats, where it grows on the roots of *Eryngium*.

If the host is detected there should be no identification problems. In southern Europe a number of similar varieties/species occur on various umbels (*Apiaceae*).

Can be found from the Netherlands southwards; autumn–winter.

Emilio López

• ***Pleurotus dryinus*** is a very thick-fleshed and rather tough *Pleurotus* with ± obvious remnants of a veil on the margin and stem, and a ± felty-scaly, greyish cap surface. The gills are white and crowded. The spore-deposit is ± cream. The almost cylindrical, smooth spores measure 8.5–15.5(–17) × (2.5–)3–5(–5.5) µm. On various deciduous woods, rarely conifer wood. Often seen on fruit trees in gardens.

Hypsizygus ulmarius (page 239) is similar, but lacks remnants of a veil. Other species of *Pleurotus* ± lack a veil and have smooth cap surfaces. *Neolentinus schaefferi* ⊲⊲ is browner and has narrow, crowded, finely serrated gills. Scaly *Neolentinus lepideus* ⊲⊲ has strongly serrated gill edges, a very tough flesh, and more obvious scales on the cap.

Widespread, but generally uncommon; mostly August–November.

Thomas Læssøe

Thomas Læssøe

Clitopilus is a genus of agarics with farinaceous smells and brownish-rose spores, which are ridged lengthwise – best seen in end view. Most species in the genus have centrally stipitate fruitbodies (page 453).

Clitopilus hobsonii is a crepidotoid/pleurotoid *Clitopilus* with a small, thin, excentric stem, a hairy-felty, white cap, rather distant white to pinkish-tinted gills and a farinaceous smell. It occurs rarely in a cyphelloid form without gills. The spores have longitudinal ridges and measure (6.5–)7.5–8.5(–10) × 5–5.5(–6) µm. It may grow on almost anything from rotten straw in dunes to large trunks of deciduous trees.

Clitopilus damsii ⋉ should be identifiable by its larger spores, but is perhaps a somewhat dubious species, as is *C. rhodophyllus* ⋉, which is possibly just a large, luxurious form.

Entoloma jahnii ▷ has spores with an angular outline and is typically more reddish on the gills. Pale species of *Crepidotus* ▷▷ have brown, mostly verrucose spores; they are also normally stemless, with less projecting hairs.

Common and widespread, but under-recorded; all year, but mostly July–November.

Thomas Læssøe

Entoloma species have brownish-rose spores that are angular in outline. Most species are agarics with a central stem (page 448).

Entoloma byssisedum is a greyish, finely and radially felty, crepidotoid/pleurotoid *Entoloma* that lacks cystidia and has hyphae with clamps. The angular spores measure 9.5–12 × 6.5–8 µm. On peat-like soils or strongly decomposed litter or wood.

Entoloma pseudoparasiticum ⋉ and *E. depluens* ⋉ lack clamps. The paler *Clitopilus hobsonii* △ has spores with longitudinal ridges. Species of *Arrhenia* ◁◁ have white spore-deposits and smooth spores.

Widespread, and probably rather common, but often overlooked; August–November.

Entoloma jahnii is a small, white, crepidotoid/pleurotoid *Entoloma* with capitate hairs on the cap surface and on the ± rudimentary, ± excentric stem. The angular spores measure 9.5–13(–15) × 7.5–10(–11.5) µm. Grows directly on soil or on rotten wood, but may also creep out of the soil and form fruitbodies on other fungi (such as on the chanterelle in the picture below).

Other small white entolomatoids with a tendency to display lateral stems lack capitate cystidia. *Clitopilus hobsonii* ◁ has longitudinallly ridged (striate) spores.

Widespread and probably not rare; mainly July–November.

Rudolf Markones

Entoloma albotomentosum is a very small, chalk-white *Entoloma* with a ± excentric, short stem and a silky-fibrillose cap. The gills are distant and adnate, lacking cystidia and clamps. The sharply angled spores measure 9–12.5 × 6–8 µm. On grasses, sedges and rushes in swampy places.

Entoloma jahnii △ has capitate hairs on the cap and stem, and grows mostly on wood. *Clitopilus hobsonii* ◁ has spores with longitudinal ridges and normally a shorter, less evident stem.

Widespread, but distribution poorly known; autumn.

Erik Arnfred Thomsen

Crepidotus is a genus of pleurotoid, mostly stemless agarics with grey-brown spore-deposits, smooth, finely or coarsely verrucose spores, and thin-walled cystidia on the gill edge.

Crepidotus mollis is a large, up to 8 cm wide, hygrophanous, ± smooth *Crepidotus* with an elastic-gelatinous cap cuticle and hyphae without clamps. Cheilocystidia narrow and lageniform. The spores are ovoid to amygdaliform, smooth, thick-walled, rather dark brown and measure 7–10 × 5–6.5 µm. Grows on wood and bark from deciduous trees.

Crepidotus calolepis ▽ is very similar, both macro- and microscopically, but the surface of the cap has fine brown scales. *Crepidotus applanatus* ▷▷ and *C. autochthonus* ▷▷ are also very similar, but are not gelatinous. Species of *Hohenbuehelia* ◁◁ have white spores, a felty cap surface and thick-walled cystidia (metuloids).

A widespread and rather common species; mainly August–January.

Crepidotus calolepis is a sturdy, almost fleshy, brownish, squamulose, clampless *Crepidotus* with a gelatinous layer in the cap cuticle. The smooth spores measure 7–10 × 5–6.5 µm. Grows on the bark of deciduous trees, perhaps mainly on *Populus tremula*.

Crepidotus mollis △ has very similar macro- and microscopical characters, but does not have small, brown scales on the cap. *Crepidotus crocophyllus* ▷▷ has a distinct yellow coloration. Species of *Hohenbuehelia* ◁◁ have hyaline spores and thick-walled cystidia.

Widespread, and locally rather common; June–February.

Thomas Læssøe

Crepidotus luteolus is a medium-sized *Crepidotus*, which genereally has a yellow tint on the inner part of the white, smooth cap (primordia dark cream), and has clamps and narrow, sinuous cheilocystidia. The narrowly amygdaliform to fusiform, very finely ornamented spores measure 7.5–10 × 4–4.5 µm, Q 1.6–2.2. Mostly found on herbs, *e.g.* stems of *Urtica*, but may also grow on bark.

Crepidotus subverrucispermus ⋉ and *C. caspari* ▷▷ have relatively wider spores (lower Q), never show yellowish on the cap, and are not commonly found on herb stems.

A widespread and common species; mostly September–December.

Crepidotus cesatii is a rather small, thin-fleshed, white, downy *Crepidotus* with pale brown, mostly pinkish-tinted, gills, and clamps and branched cheilocystidia up to 50 µm long. Has rather finely, but distinctly ornamented spores, which measure 6–9 × 5.5–7.5 µm, Q 1–1.3. Grows mostly on bark of deciduous trees, usually on attached branches.

Macroscopically much as *C. variabilis* ▷▷ but this species has a Q of 1.6–2.3. *Crepidotus kubickae* ⋉ has a Q value of 1.2–1.45 and typically grows on thin conifer twigs. Also similar to *Clitopilus hobsonii* ◁◁ but that species has pink spores with longitudinal ridges.

A widespread and very common species; mainly August–January.

Jan Vesterholt

Crepidotus variabilis is a thin-fleshed, ± roundish to hoof-shaped, whitish, not hygrophanous or gelatinous *Crepidotus* with ± cinnamon-brown gills. Has clamps and branching cheilocystidia up to 25 µm long. The narrowly ellipsoid to cylindrical spores measure 5.5–7.5 × 2.5–4 µm, Q above 1.4. Mostly on deciduous wood.

Crepidotus cesatii ◁◁ has almost spherical to broadly ellipsoid spores (Q below 1.4) and *C. kubickae* ⋉ likewise has a different spore shape.

A widespread, common species; mostly August–November.

Thomas Læssøe

Crepidotus caspari is a rather robust, somewhat umbonate, whitish, not hygrophanous or elastic *Crepidotus*. Has clamps and unbranched, rather variable cheilocystidia and characteristic hooked hyphae in the cap cuticle. The ovoid to almost amygdaliform, very finely ornamented spores measure 6.5–9.5 × 4.5–6 µm. Mostly on deciduous wood, generally branches and twigs.

Crepidotus subverrucisporus ⋉ has distinctly amygdaliform and significantly more ornamented spores, and narrower cystidia.

A widespread, and rather common species; mainly June–January.

Thomas Læssøe

Crepidotus applanatus is typically a rather large, whitish to pale yellow-brown, rather hygrophanous, somewhat spatulate *Crepidotus* with clavate cystidia and clamps. The rather coarsely verrucose, almost spherical spores measure 4.5–7 × 4.5–6.5 µm. It grows on fallen trunks of deciduous wood, mostly in ± undisturbed forests, but may also occur on conifer wood.

Crepidotus stenocystis ⋉is very similar, but the cystidia are widest at the base.

Widespread and rather common in some areas, but nearly absent from others; August–January.

Crepidotus versutus is a medium-sized (up to 25 mm wide), white, downy, not hygrophanous or gelatinous *Crepidotus* that lacks clamps. Has unbranched, sinuous cheilocystidia. The ellipsoid, almost smooth, rather pale spores measure 8.5–12 × 5–6.5 µm, Q 1.6–2.1.. Mostly on bark of deciduous trees.

The similarly pale-spored and clampless *C. epibryus* ▽ has smaller and more fusiform spores.

Rather common in some, mostly hemiboreal areas, but rare in, or absent from, others; autumn.

Crepidotus epibryus is a small, thin-fleshed, very pale, minutely downy *Crepidotus* that lacks clamps. The cheilocystidia are narrowly lageniform. The pale, narrow, almost smooth spores measure 6.5–9.5 × 2.5–3.5 µm, Q 2.1–3.2. On leaf-litter and herbaceous litter.

Crepidotus versutus △ also lacks clamps but has larger fruitbodies and spores with a lower Q (1.6–2.1). *Clitopilus hobsonii* ◁◁ has spores with longitudinal ridges.

A widespread and at least locally quite common species; mainly August–February.

Crepidotus autochthonus is a large, whitish *Crepidotus* with clamps and clavate cheilocystidia. The smooth, ± amygdaliform spores measure 7–9 × 5–6 µm. It typically fruits directly on clay-rich soils, co-occuring with species of *Cortinarius* subg. *Phlegmacium etc.*, but may also grow on wood.

When growing on soil, there are no significant identification issues, but lignicolous occurrences demand careful microscopy.

A rather local and southern species, that is absent from the boreal zone and very rare in the hemiboreal; mainly July–October.

Jacob Heilmann-Clausen

Crepidotus crocophyllus is a large, ± hairy-scaly, cinnamon-buff *Crepidotus* with yellow gills that stain red with KOH. The hyphae are clamped and have incrusted pigment. The spherical, finely ornamented spores measure 5–6.5(–7) µm in diameter. On deciduous wood.

Crepidotus calolepis ◁◁ has a gelatinous layer in the cap flesh, does not have yellow gills and lacks clamps.

Rather rare and southern, absent from most of the northern and western parts of temperate Europe (one of the most common *Crepidotus* species in North America); summer–autumn.

Crepidotus carpaticus is a small *Crepidotus* that typically produces numerous crowded fruitbodies with ± yellowish cap colours and a very pale gill edge. The hyphae have clamps; with short, 20–30 µm long, coral-like cheilocystidia with incrusted sticky material. The ± spherical, finely echinulate spores measure 5–6.5 × 4.5–6.5 µm. Grows on fine woody debris from deciduous trees.

Crepidotus carpaticus and *C. cristatus* are now regarded as synonyms. The short cystidia are a good character for separating this from similar species.

Widespread, but rather rare; mainly June–November.

Crepidotus cinnabarinus is an orange-red, hairy and spectacular *Crepidotus* even with orange-red gill edges, clamps and simple cheilocystidia. The almost smooth, ovoid spores measure 6–8 × 5–6.5 µm. It grows on the wood of deciduous trees, *e.g.* on the bark of *Populus tremula*, and often with *C. mollis* ◁◁.

A highly distinctive species. The much rarer *Crepidotus roseoornatus* ⋉ has more pinkish colours and concolorous gill edges.

Very rare and apparently absent from many areas; June–November.

Deconica is a genus of small, dark-spored pleurotoid agarics that have spores with germ pores. Most species have central stems (page 548).

Deconica phillipsii is a small, thin-fleshed, pinkish-tinted, translucently striate, somewhat gelatinous, pleurotoid *Deconica* with a short, excentric stem and rather pale brown to almost pinkish gills. The somewhat thick-walled spores have a germ pore and measure 5–7 × 3–4 µm. On grass culms and other herbs, mostly deep in long, wet grass.

Arrhenia acerosa ◁◁ is greyer and white-spored. Various entolomatoids (*Entoloma, Clitopilus*) are paler with red spore-deposits, and *Deconica horizontalis* ▽ is more robust, non-striate and darker, and grows on wood. Species of *Crepidotus* ◁ are, at maturity, mostly completely stemless.

A widespread and common species, but easily overlooked – it can only be found by crawling along the ground; mainly May/ June–October.

Deconica phillipsii has been assigned to eight different genera and is often accepted in *Melanotus* or *Psilocybe*.

Deconica horizontalis is a small, but fleshy, brown-spored, non-gelatinous pleurotoid *Deconica* with a ± excentric, rudimentary stem and a brown, non-striate, ± hygrophanous cap. The smooth, thick-walled spores have a germ pore and measure 6–8 × 4–5 µm. On wood from both coniferous and deciduous trees, and also on textiles.

Species of *Crepidotus* ◁ normally have no stem and have spores without germ pores. *Deconica phillipsii* △ occurs in grassland, is paler, and translucently striate.

A rather common species, which apparently has expanded it range in recent decades; mainly July–March.

Thomas Læssøe

Pleuroflammula ragazziana is a small, yellow-brown, mussle-shaped, lignicolous, pleurotoid agaric with a red-brown spore-deposit and prominent clavate cheilocystidia. The smooth, thick-walled spores have no germ pore and measure 8–10 × 6–7 µm. On rotten hardwood.

Pleurotoid species of *Deconica* ◁◁ have duller spore-deposit colours, and the spores have germ pores.

Rare, with a few records in the Atlantic part of Europe; autumn.

Jan Vesterholt

Tapinella is a genus of rather large, brown-spored, decomposing agarics with excentric or lateral stems. Species in this genus have been included in *Paxillus* (page 756), but that genus is ectomycorrhizal.

Tapinella panuoides is a pale orange-brown, ± fan-shaped, brown-spored, conifer-associated pleurotoid agaric, which may be violet at the base. The cap surface is finely felty to almost smooth, and the margin is pale and ± inrolled. Lacks obviously differentiated cheilocystidia. The smooth, ellipsoid spores measure 4–6 × 3–4 µm. On brown-rotted conifer wood. It may also attack moist conifer timber in houses, where a violet felt on the surface of the affected timber may be evident. Similar signs may, however, result from infestation by *Serpula himantioides* (page 984).

Phyllotopsis nidulans ◁◁ has a more vivid coloration on both upper- and underside, and a white spore-deposit.

Fairly common in temperate Europe; mostly August–November.

Tapinella atrotomentosa is a large to very large, fleshy, brown-spored, bitter-tasting, pleurotoid agaric with an excentric, thick and dark brown felty stem. The gills are decurrent, wide and soft, and may be loosened from the cap flesh. The ellipsoid, smooth spores measure 5–6 × 4 µm. Occurs on conifer stumps and on buried conifer wood in forests and plantations.

Distinctive and unlikely to be misidentified. Species of *Pleurotus* ⊲⊲ have white to greyish-lilac spore-deposits and lack thick, brown felt on the stem. Small, young specimens of *Tapinella atrotomentosa* may perhaps be confused with *T. panuoides* ⊲ but that species lacks dark brown basal felt.

Common and widespread; July–November, mostly early in the season.

Clitocyboids

Clitocyboid agarics have short to deeply decurrent gills. The cap is umbonate, applanate or funnel-shaped and the flesh is mostly obviously fibrillose but not very tough. Some species form small fruitbodies, but most are medium-sized to very large. Most clitocyboids have white to cream spore-deposits, but in some species belonging to the genera *Clitocybe* and *Paralepista* they are clay-pink.

Almost all clitocyboids are decomposers of litter or soil. Exceptions include *Ossicaulis* and *Omphalotus* that are lignicolous. *Hygrophoropsis* and *Aphroditeola* may also grow on remnants of wood. *Catathelasma* forms ectomycorrhiza with conifers.

The generic division of the clitocyboid agarics has gone through large changes, with many species, previously accepted in *Leucopaxillus* and *Clitocybe,* now being assigned to a number of new genera.

Other similar fungi:
– omphalinoids may look rather similar, but are mostly more fragile and smaller. Some have a biotrophic association with mosses, *etc*., others are decomposers (page 128).
– pleurotoids likewise have deeply decurrent gills but have excentric stems. Many of these (*e.g. Neolentinus* and *Lentinellus*) have distinctly tough flesh (page 60).
– entolomatoids may be shaped like clitocyboids, but have brownish-rose, verrucose, striate or angular spores (page 448).
– *Paxillus* and others are shaped like clitocyboids but have brownish spore-deposits (page 756).

Litteratur: 8, 9, 10, 27, 110, 156, 174, 181, 183, 340, 344.

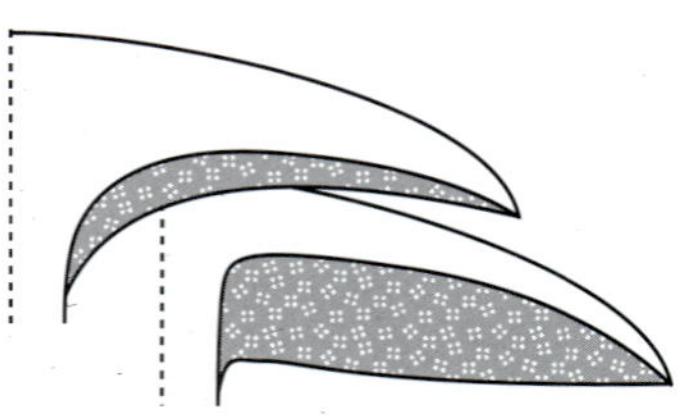

deep and short decurrent gills

Micro-drawings: spores first, followed on the inside by the cystidia

Approximate species number applies to temperate Europe

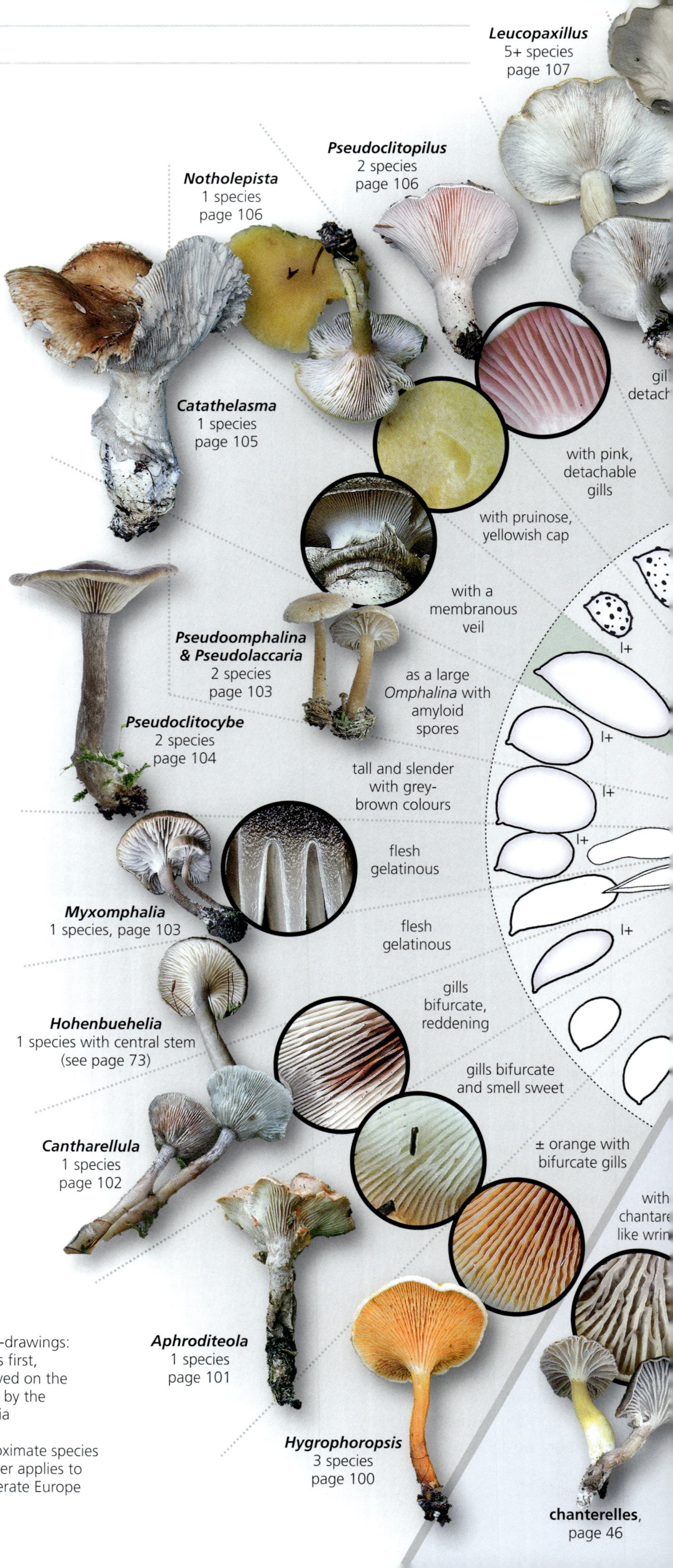

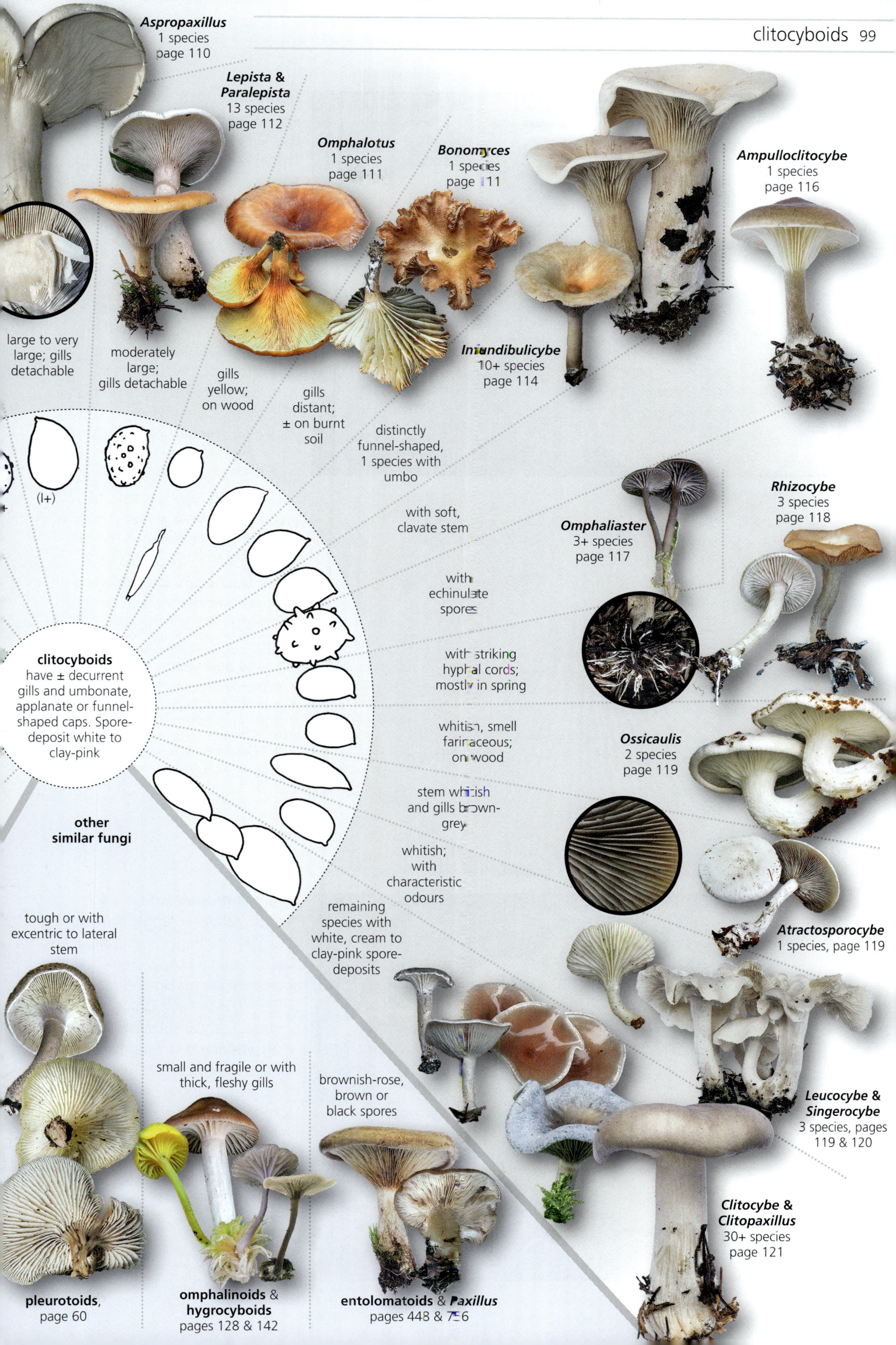
Aspropaxillus
1 species
page 110
Lepista & Paralepista
13 species
page 112
Omphalotus
1 species
page 111
Bonomyces
1 species
page 111
Ampulloclitocybe
1 species
page 116
Infundibulicybe
10+ species
page 114
large to very large; gills detachable
moderately large; gills detachable
gills yellow; on wood
gills distant; ± on burnt soil
distinctly funnel-shaped, 1 species with umbo
with soft, clavate stem
with echinulate spores
with striking hyphal cords; mostly in spring
whitish, smell farinaceous; on wood
stem whitish and gills brown-grey
whitish; with characteristic odours
remaining species with white, cream to clay-pink spore-deposits
(I+)
Omphaliaster
3+ species
page 117
Rhizocybe
3 species
page 118
Ossicaulis
2 species
page 119
Atractosporocybe
1 species, page 119
Leucocybe & Singerocybe
3 species, pages 119 & 120
Clitocybe & Clitopaxillus
30+ species
page 121
clitocyboids
have ± decurrent gills and umbonate, applanate or funnel-shaped caps. Spore-deposit white to clay-pink
other similar fungi
tough or with excentric to lateral stem
small and fragile or with thick, fleshy gills
brownish-rose, brown or black spores
pleurotoids,
page 60
omphalinoids & hygrocyboids
pages 128 & 142
entolomatoids & Paxillus
pages 448 &

Hygrophoropsis is a genus of agarics with gills that bifurcate regularly towards the cap margin. The smooth spores are dextrinoid. All are decomposers. *Cantharellula umbonata* ▷▷ has similar gills. The species were earlier considered edible, but should be avoided, as they may cause an upset stomach.

Hygrophoropsis aurantiaca is a rather thin-fleshed, soft and dry, orange clitocyboid with narrow, bifurcate, soft, decurrent gills and a ± hollow stem. There are no cystidia. The smooth, somewhat thick-walled, dextrinoid spores measure 5–7 × 3.5–4.5 µm. It is found on acid litter, *e.g.* needle debris, but also on decomposed woody substrates, *e.g.* on old piles of wood chips, both in forests and in more open habitats.

Chanterelles (page 46) have firmer fruitbodies and branching thick veins rather than proper gills, and the spores are non-dextrinoid. *Hygrophoropsis pallida* ▷ is more felty and buff-yellow. *H. rufa* ▽ has a distinctly brown, thick cap felt. The genus belongs to the *Boletales*, like *e.g. Paxillus*.

Widespread and very common; mainly July–January.

Thomas Kehlet

Hygrophoropsis rufa is a soft-fleshed, reddish-orange, brown-felty clitocyboid with decurrent, bifurcate, soft gills and a hollow stem. The spores are dextrinoid and measure 5–6.5(–7) × 3–4 µm. On old wood chips and similar substrates.

Hygrophoropsis aurantiaca △ is not brown-felty, and *H. pallida* ▷ is much paler. Various other clitocyboids lack the dense, brown cap felt and have non-dextrinoid spores.

A rather rare, but widespread species; mainly August–October.

Hygrophoropsis pallida is a soft-fleshed, beige clitocyboid with decurrent, pale, bifurcate gills. The cap may become greyish and rather felty with age. The somewhat thick-walled, dextrinoid spores measure 6–9(–10) × 4–5 µm. Typically found in open places and mostly in rather moist sites. It is possible that more than one species may be involved.

Other species of *Hygrophoropsis* are more vividly coloured. Chanterelles (page 46) have firm flesh and the spores are non-dextrinoid.

Found occasionally across most of temperate Europe; mainly September–November.

Jan Vesterholt

Aphroditeola olida is a rather small, pale, soft-fleshed, pale-spored clitocyboid with decurrent, bifurcate gills and with pink shades, especially on the cap. Has a strong, nauseous-sweet smell. The gills have no cystidia. The somewhat dextrinoid, broadly ellipsoid spores measure 3.5–4.5 × 2.5–3 µm. It is mostly found in old-growth calcareous *Picea* forests, but may also occur in nutrient-poor *Pinus* stands.

Also known as *Hygrophoropsis morganii* and *H. olida*. Species of *Hygrophoropsis* ◁ lack pink colours and are odourless. *Bonomyces sinopicus* ▷▷ has a strong farinaceous smell.

A rare species, mostly found in the boreal zone of temperate Europe and very rare in the nemoral zone; late summer–autumn.

Jan Vesterholt

Cantharellula umbonata is a grey and pruinose-chalky clitocyboid agaric with obviously bifurcate gills, which redden with age. Cystidia are absent. The smooth, fusiform, amyloid spores measure 8–11 × 3–4 µm. It is associated with acidic soil and mosses, perhaps exclusively *Polytrichum*, which it may parasitize.

Species of *Hygrophoropsis* ◁◁ have similar bifurcate gills, but are ± orange. *Pseudoclitocybe* ▷▷ also has amyloid spores, but the gills are not bifurcate and do not stain red with age.

A rather common fungus in the hemiboreal–boreal zone, but apparently rare in the more polluted lowlands, where it has shown a decline; mainly August–December.

Pseudoomphalina kalchbrenneri is a rather compact, omphalinoid-clitocyboid agaric with white spores, a farinaceous taste and, typically, a furrowed and compressed, smooth stem. Lacks cystidia. Has smooth, amyloid spores measuring 6.5–9.5 × 3.5–5 µm. On calcareous soils in open forests and scrub.

Pseudolaccaria pachyphylla ▽ is bitter-tasting and has very indistinctly amyloid spores. It tends to prefer more acid conditions.

Rarely reported and exact distribution is not known – fairly frequent in parts of Scandinavia; mainly September–November.

Jan Vesterholt

Pseudolaccaria pachyphylla is a somewhat bitter-tasting, ± farinaceous, rather compact omphalinoid-clitocyboid agaric with a somewhat felty cap surface and a compressed, furrowed stem. Lacks cystidia. Spores smooth, somewhat amyloid, 6.5–9 × 4.5–6 µm. In deciduous and coniferous forest on sandy soils; also on burnt soils and in grassland.

Pseudoomphalina kalchbrenneri △ is very smilar, but it is not bitter-tasting and has distinctly amyloid spores and, apparently, a tendency to grow in more calcareous habitats.

Rarely recorded, but widespread; mainly July–November.

Lucie Zibarova

Myxomphalia maura is a dark brown, omphalinoid-clitocyboid, spermatic- or farinaceous-smelling omphalinoid-clitocyboid agaric with pale gills and gelatinous cap flesh (inserted picture). Has large ± clavate cheilocystidia and pleurocystidia. The smooth, amyloid spores measure 5.5–7 × 4–5.5(–6) µm. On burnt soils in conifer plantations.

The ± similar *Faerberia carbonaria* (page 55) has a cantharelloid hymenophore and thick-walled cystidia. Burnt soil-associated species of *Lyophyllum* (page 287) are less omphalinoid, do not have amyloid spores and lack cystidia.

Widespread, but has declined due to changes in forestry practices; mostly October–November.

Pseudoclitocybe is a small clitocyboid genus characterized by grey-brown, rather tall fruitbodies and amyloid spores. Lacks cystidia. Saprotrophic.

Pseudoclitocybe cyathiformis is a rather large and relatively robust, dark brown, pale-spored, clitocyboid agaric that quickly dries to a more greyish colour. The cap is deeply depressed at the centre and the margin remains inrolled for a long period of time. The gills are decurrent, rather crowded and pale greyish. Lacks cystidia. The smooth and distinctly amyloid spores measure 8–12 × 5–6.5 µm. It occurs on litter and on decomposed wood, often partly moss-covered, mostly in deciduous forests, but also in conifer plantations and among tall herbs.

The very closely related species *P. expallens* ▽ is smaller and occurs in grasslands, but it is not always possible to separate the two species. Species of *Clitocybe* ▷▷, *Infundibulicybe* ▷▷ and *Lepista* ▷▷ do not have amyloid spores.

A widespread and common species; September–March.

Pseudoclitocybe expallens is a brown, but quickly drying to grey, white-spored, clitocyboid agaric with amyloid, smooth spores, 7–10 × 5–7 µm. It is found in dry, unfertilized grasslands.

Pseudoclitocybe cyathiformis △ is typically somewhat larger, less pale in the dried out condition, and is usually found in forests. Nevertheless, the separation remains difficult. *Pseudoclitocybe expallens* may have smaller spores.

Widespread and probably not rare; mostly October–December.

Catathelasma imperiale is a large, very fleshy, white-spored, brown, tricholomatoid-clitocyboid agaric, with a slimy and membranous universal veil and a dry, membranous partial veil that is left as a ring on the stem. The ring may be ± double (elements from both veils). The universal veil may also leave scaly remnants on the cap. In comparison to *Tricholoma* species, the gills are deeply decurrent. The stem is tapering and somewhat rooting. The odour is farinaceous, and the taste is bitter and farinaceous. Has simple cheilocystidia and no pleurocystidia. The smooth, amyloid spores measure 11–14 × 5–6.5 µm. It is mycorrhizal with *Pinus* and *Picea*, mostly in somewhat older forests on ± calcareous soils.

Large, annulate species of *Tricholoma* (page 258) may look similar, but they have ± emarginate gills and inamyloid spores.

An uncommon, hemiboreal–boreal/subalpine species; it is on the red list in several countries; late summer–autumn.

Pseudoclitopilus rhodoleucus is a silky-white, medium-sized clitocyboid agaric with rose-coloured gills and occasionally with a similar rosy mycelium at the base of the stem. The colour may change to dark cream on very mature specimens. The amyloid and verrucose spores measure 6.5–7.5 × 4.5–5.5 µm. Occurs on light, calcareous soils, mostly in scrub and typically in conspicuous fairy rings. It almost always occurs together with a number of other interesting species, *e.g.* species of *Geastrum* (page 1248), and various species of *Lepiota*, *Cystolepiota* and *Echinoderma* (page 320).

Other ± white species have less distinctly pink gills, although *P. salmonifolius* ⋉ is rather similar but should have smaller spores with a more coarse ornament. *Clitocella fallax* (page 452) is less pink, less fleshy and smells farinaceous. *Leucopaxillus alboalutaceus* ▷▷ is bitter-tasting and less pink.

A rare species that is absent from the boreal zone; mainly August–January.

Karl Soop/Pertti Salo

Notholepista subzonalis is a small, yellowish, hygrophanous and pruinose clitocyboid agaric. It is without a distinct smell. The amyloid, verrucose spores measure 4–6 × 3.5–4.5 µm. On litter in conifer forests on rich soil.

In the field, this species may recall a species of *Hygrophoropsis* ◁◁, but the gills are not forked and the spores are amyloid and verrucose. It may also resemble *Lepista gilva* ▷▷ but that species has inamyloid, somewhat ornamented spores.

Very rare and mostly with records from the hemiboreal–boreal zone; autumn.

Thomas Læssøe

Leucopaxillus form large, fleshy fruitbodies with decurrent gills that mostly detach fairly easily from the cap flesh. The spores are amyloid and verrucose. All are saprotrophs.

Leucopaxillus tricolor is a very large and fleshy, somewhat nauseous- smelling, farinaceous clitocyboid agaric with a yellow-brown to brownish-red tinted cap, yellowish gills (± purple when dry) and a thick, yellowish stem. The verrucose, amyloid spores measure 6–8(–9) × 4–5(–5.5) µm. Occurs on calcareous soil in deciduous forests with a long continuity.

Also known under the wrongly applied name *L. compactus*. *Leucopaxillus gentianus* ▷▷ is somewhat smaller, more red-brown, and the taste is very bitter. The large *Tricholoma acerbum* (page 261) has more emarginate gills, a more wrinkled cap margin, and the spores are not amyloid.

A rare and very local fungus, absent from the boreal zone and apparently also from more Atlantic areas; mainly August–October.

Thomas Læssøe

Jan Vesterholt

Leucopaxillus gentianeus is a very bitter-tasting, farinaceous, large clitocyboid agaric with a red-brown, somewhat felty cap and pale, narrowly adnate to emarginate gills. The verrucose, amyloid spores measure 4–6 × 3.5–5 µm. Occurs on calcareous soil in deciduous forests and scrub.

This species should be easy to recognize, but may superficially resemble a *Tricholoma* (page 258). However, species in that genus have inamyloid, smooth spores and emarginate gills.

A rare species, possibly with a preference for continental, climatic conditions; it is absent from the boreal zone; mainly September–October.

Jan Vesterholt

Leucopaxillus alboalutaceus is a bitter-tasting, almost white, but brown-staining clitocyboid agaric with cream to yellowish gills. The amyloid spores have scattered warts and measure 3.5–6.5 × 3–4.5 µm. In calcareous conifer forest and plantations.

Aspropaxillus giganteus (incl. *A. candidus*) ▷▷ is typically larger, with a mild taste and smooth spores. *Pseudoclitopilus rhodoleucus* ◁◁ has distinctly pink gills and larger spores, while *Leucopaxillus cerealis* ▷ is mild-tasting.

A rarely recorded, but apparently widespread species; mainly August–November.

Leucopaxillus nauseosodulcis is a large, rather robust, very unpleasant-smelling, mild-tasting, yellow-brown clitocyboid agaric with cream gills and mostly with an excentric stem. The amyloid, verrucose spores measure 7–8 × 5–5.5 µm. Can be found in conifer forests and plantations, perhaps mostly on calcareous soils.

Leucopaxillus cerealis ▽ normally has a central stem and lacks the unpleasant nauseous smell. *Leucopaxillus alboalutaceus* ◁ tastes bitter.

A very rare, mostly northern species; autumn.

***Leucopaxillus cerealis* s.l.** is a mild-tasting and pleasant-smelling, whitish to pale yellow-brown clitocyboid agaric with white gills. The verrucose, amyloid spores measure 5.5–8.5 × 4–5.5 µm. On rich soil in older conifer forests.

Leucopaxillus alboalutaceus ◁ tastes bitter, while *L. nauseosodulcis* △ has an unpleasant smell. Several closely related, ill-defined species occur. One of them should have a cracking cap cuticle.

A rather rare fungus, mostly recorded in the hemiboreal zone; late summer–autumn.

Thomas Læssøe

• ***Aspropaxillus giganteus*** is a large to very large, almost white, cream to pale grey-brown clitocyboid agaric with crowded, decurrent, somewhat off-white gills, which can be loosened from the cap flesh, and a rather short, thick stem. The cap margin remains inrolled for a long time and can be crenulate. The smooth, amyloid spores measure (5.5-)6–9(–11) × (3.5–)4–5.5 µm. Occurs typically in fairy rings on various grassland, more rarely in scrubs and forests. The characteristic rings with wide, green and dead zones are easily seen on aerial photographs.

The smooth, amyloid spores clearly separate this species from *Clitocybe* and others ▷▷, and from species in the genera *Leucopaxillus* ◁◁ and *Pseudoclitopilus* ◁◁. *Aspropaxillus candidus*, is recognized as a species by some, but there do no seem to be reliable separating characters.

A widespread and rather common species; July–December.

† ***Omphalotus illudens*** is a large, clustered, fleshy, orange-yellow to yellow-brown, mostly smooth clitocyboid agaric with a tendency to have an excentric stem and with golden, deeply decurrent gills. The cap is funnel-shaped with an inrolled margin that persists for a long time. The odour is distinctive and may be somewhat rancid. Has a buff-yellow spore-deposit. The surface becomes dark green with ammonia. Has smooth, inamyloid spores that measure (4.5–)5–7 × 4–6(–6.5) µm. On deciduous wood in thermophilic forests. The fruitbodies are ± luminescent.

Omphalotus olearius ⨯ is a darker, more southern species, which mostly can be found on olive trees. The ammonia reaction is paler green. Funnel-shaped species of *Lepista/Paralepista* (e.g. *Paralepista flaccida* ▷▷) are very similar, but they occur on leaf-litter. *Omphalotus illudens* has also been confused with chanterelles (page 46), but these have branching thick veins rather than proper gills and form ectomycorrhiza.

Known from the Netherlands and farther south; late summer–autumn.

Jan Vesterholt

Etienne Charles

Bonomyces sinopicus is a vivid, brownish-orange clitocyboid agaric with very pale gills and a strong farinaceous smell. The cap centre may be finely felty-scaly. Spores 7–10.5 × 5–6 µm. It is typically associated with burnt soils or otherwise disturbed mineral soil.

The combination of a farinaceous smell, brownish-orange colours and large spores should make a correct identification straightforward. Various species of *Infundibulicybe* ▷▷ have a different smell and narrower spores. *Omphalina pyxidata* (page 139) is smaller and, likewise, does not have a farinaceous smell.

Rather rare, but widespread; mainly April–November.

Frede Scheye

Paralepista* & *Lepista are medium-sized clitocyboid decomposers with decurrent or emarginate gills. They are mostly recognized by their greasy cap surface and cream to grey-rose spore-deposits and rough spore walls (see also page 249).

Paralepista flaccida is a clitocyboid agaric with ± red-brown colours, somewhat paler, deeply decurrent, crowded gills, no drop-like blotches and a rather neutral smell. The cap surface is greasy. The spore-deposit is cream to pale pinkish-buff. Lacks cystidia on the gills. The small, broadly ellipsoid, echinulate spores measure 4–5 × 3–4.5 µm. Grows in rings on leaf and needle debris.

Paraleista gilva ▽ has distinct drop-like blotches on the somewhat yellower cap. *Infundibulicybe gibba* ▷▷ has a more flesh-pink, non-greasy cap and smooth spores. The poisonous *Omphalotus illudens* ◁◁ occurs on hardwood in southern parts of temperate Europe.

A widespread and common species; mainly August–January.

Jan Vesterholt

Paralepista gilva is an apricot-orange, clitocyboid agaric with distinct drop-like blotches on the cap, deeply decurrent gills, a neutral smell and a pale, almost white spore-deposit. The broadly ellipsoid, echinulate spores measure 4–5 × 3–4.5 µm. On litter on rich soils, mostly in forests.

Clitocyboids have smooth spores, and *P. flaccida* △ is browner and has no drop-like blotches. *Notholepista subzonalis* ◁◁ has amyloid spores.

A widespread and rather common species; mostly fruiting in the autumn.

Lepista densifolia is a rather large, brownish, not hygrophanous clitocyboid agaric with very crowded, pale brown to pinkish-tinted, decurrent gills, a neutral smell and a somewhat reddening flesh. Smell rather neutral. Spore-deposit grey-rose. The spores are broadly ellipsoid, verrucose and measure 4–4.5 × 3–3.5 µm. On litter in in coniferous or mixed forests.

Lepista ovispora ▽ is more red-brown on the cap.

Mostly recorded in northern parts of temperate Europe; autumn.

Lepista ovispora is a rather long-stemmed, distinctly clustered, red-brown clitocyboid agaric with decurrent gills, a neutral smell and a grey-rose spore-deposit. The ellipsoid, almost smooth spores measure 5.5–8.5 × 3.5–4.5 µm. On rich soil and litter in open habitats.

Species of *Lyophyllum* (page 254) have white spore-deposits. *Lepista densifolia* △ is paler brown and has very crowded gills and mostly grows on acidic litter below *Picea*.

A rather rarely recorded species with a northern tendency in its distribution; mainly August–November.

Jan Vesterholt

Infundibulicybe has rather large and fleshy, funnel-shaped fruitbodies with whitish, smooth, inamyloid spores.

• ***Infundibulicybe geotropa*** is a large, up to 15 cm wide, tall, fleshy, pale cream, mostly fairy ring forming woodland species with a small umbo in the otherwise ± funnel-shaped cap with a furrowed margin. It has a sweetish-aromatic smell and a white spore-deposit. The inamyloid, smooth spores measure 7.5–9.5 × 6–7 µm. It typically forms fairy rings on thick leaf-litter, mostly in rich deciduous forests, but may also grow with conifers.

Aspropaxillus giganteus ◁◁ lacks the small umbo and has amyloid spores. *Clitopaxillus alexandri* ▷▷ and *Clitocybe nebularis* ▷▷ are both more colourful, greyish to grey-brown species.

Rather common and widespread; mainly July–January.

Infundibulicybe squamulosa is a rather fleshy, medium-sized, indistinctly, dark squamulose, cinnamon-brown, not umbonate or marginally furrowed clitocyboid agaric. The cap cuticle consists of hairs (a trichoderm) with incrusted pigment. On dry material the cap cuticle stains dark brown with KOH. The drop-shaped spores measure 7–9.5 × 4–5 µm. It mostly occurs along roads in conifer plantations on poor soils.

Infundibulicybe costata ▷ is similar, but has somewhat smaller spores and is felty rather than squamulose.

Widespread and fairly common in some regions, mainly in northern parts; mainly May–November.

Infundibulicybe gibba is a rather large, but thin-fleshed, pinkish-buff to flesh-pink, smooth to finely felty, ± marginally furrowed, not farinaceous clitocyboid agaric with a white spore-deposit and a pale stem, typically paler than the cap. The cap surface on dry material does not stain dark brown with KOH and the adpressed hyphae in the cap cuticle has finely incrusted pigment. Spores measure 5.5–8 × 3.5–5 µm and are drop-shaped. On rich soil and litter, mostly in deciduous forests.

Infundibulicybe splendoides ⋉ is very similar, but does not have incrusted pigment on the hyphae. *Infundibulicybe costata* ▽ is somewhat more colourful, mostly on the stem, and dried specimens stain dark brown with KOH.

Common and widespread; mainly June–November.

Infundibulicybe costata is a sturdy, fleshy, pale cinnamon-brown, marginally furrowed, finely felty clitocyboid agaric with a ± unicoloured cap and stem. Dry material stains dark brown with KOH. Spores measure 6–8 × 3.5–4.5 µm and are drop-shaped. Both in deciduous and coniferous forests, mostly at road sides.

Infundibulicybe gibba △ is somewhat paler, especially at the stem, and dry material does not stain dark brown with KOH. *Infundibulicybe squamulosa* ◁ has small cap squamules and is not furrowed; *Infundibulicybe glareosa* ⋉ has a darker cap.

Widespread and rather common towards the north; June–November.

Jens Maarbjerg

† ***Ampulloclitocybe clavipes*** has a peculiar soft flesh, white spores, a dry, grey-brown cap, mostly with drop-like blotches, and an umbo. The gills are deeply decurrent and cream, and the clavate stem has a spongy texture. Has hyphae with clamps but there are no cystidia. The smooth, inamyloid, ellipsoid spores measure 6.5–9 × 4–5 µm. Found in forests and in more open habitats, both on disturbed, rather rich soils and more acid sites.

May recall a small version of the firmer, fairy ring forming *Clitocybe nebularis* ▷▷, which, in addition, has less decurrent gills. *Lepista panaeolus* (page 249) has similar coloration and drop-like blotches, but may be recognized by its ± emarginate gills and ornamented spores.

Common and widespread; mainly July–December.

Possibly causing an antabuse-like reaction when eaten.

Jan Vesterholt

Omphaliaster asterosporus is a rather small, grey-brown, ± striate, hygrophanous, farinaceous clitocyboid agaric. It lacks cystidia and clamps. The inamyloid spores are almost spherical, measure 5.5–7.5 × 5–6.5 µm and have long spines. It is mostly associated with mossy conifer forests, but may also occur in somewhat open, mossy habitats.

Recalls a small, marginally striate and farinaceous *Clitocybe* ▷▷, but is easily recognized by the echinulate spores. *Omphaliaster borealis* ⋉ is darker, more robust and very northerly in its distribution.

Widespread and fairly common, but easy to overlook; almost all year, except in hot summer months.

Jean-Marc Moingeon

Rhizocybe is recognizable by the very conspicuous basal hyphal cords and its main occurrence being from winter to late spring.

Rhizocybe pruinosa is a rather small pruinose-chalky, not hygrophanous or farinaceous clitocyboid agaric with conspicuous, white hyphal cords. The smooth spores measure 5–6.5 × 2–4 µm and do not occur in tetrads. In conifer forests.

Rhizocybe vermicularis ▽ occurs in the same habitats and season, but is not chalky pruinose and the spores are somewhat smaller.

Has a hemiboreal–boreal/subalpine distribution in temperate Europe; winter–early spring.

Rhizocybe vermicularis is a rather small, but thick-fleshed, almost odourless, hygrophanous, but non-striate, smooth clitocyboid agaric with conspicuous white hyphal cords. The smooth, inamyloid spores measure 3.5–5.5 × 2–3 µm. Occurs on litter in conifer forests and plantations, mostly on calcareous soils, more rarely in deciduous forests.

Rhizocybe pruinosa △ has a pale, somewhat splitting coating and somewhat larger spores (5–6.5 × 2–4 µm). *Bonomyces sinopicus* ◁◁ smells strongly farinaceous.

Mainly a hemiboreal–boreal-subalpine species; January–June.

Thomas Læssøe

Thomas Kehlet/David Boertmann

Ossicaulis lignatilis is a rather large, white-spored, very farinaceous, lignicolous clitocyboid agaric with very crowded gills and a mostly excentric stem. The spores measure 4.4–6 × 2.4–3.6 µm. On older dying or dead deciduous trees, mostly in cavities and mostly in places with a lot of decaying, larger trunks.

A very likely confusion species is the recently recognized *Ossicaulis lachnopus* ⋉, which can be distinguished by its smaller spores (2–4 × 1.5–2.4 µm) and possibly more greyish coloration. Lignicolous species of *Clitocybe* with a pale, somewhat splitting coating on the cap (e.g *C. truncicola* ⋉) may also recall species of *Ossicaulis*, but the smells are not rancid farinaceous.

Local to rare; mainly August–December.

Svend Aage Olsen/Jan Vesterholt

Atractosporocybe inornata is a somewhat atypical, pale grey-brown, somewhat felty-scaly, marginally furrowed, rather large clitocyboid agaric with a whitish, felty stem and an unpleasant smell recalling half-rotten fish. The ± fusiform spores measure 8–10.5 × 3–4 µm. It occurs in both deciduous forests and with conifers, mostly on ± calcareous soils.

This species should cause no identification problems when using a microscope, but may macroscopically recall some species of *Lyophyllum* (page 254).

Rather local, nowhere common; mainly August–December.

Leucocybe connata is a ± clustered, large, up to 10 cm wide, with a pale, somewhat chalcy coating, not hygrophanous, almost white, ± clitocyboid to tricholomatoid agaric with crowded, almost white, cream to pale grey gills. Smell characteristic: perfumed sweetish-nauseous, or like bitter almond. The surface and the gills stain violet with ferrous sulfate. The spore-deposit is white. The ellipsoid spores do not form tetrads and measure 5.5–6.5 x 3.5–4 µm. Occurs on rather rich, mostly disturbed soil in moist places along forest roads *etc.*

Clitocybe phyllophila >> is less clustered, has a coloured spore-deposit and smells spicy, but not perfumed. '*Clitocybe*' *dilatata* ⋉, if a genuine species, should be very similar, but should be inert to ferrous sulfate.

Widespread and common; mainly July–January.

Leucocybe houghtonii is a small, slender, somewhat rose-tinted, hygrophanous clitocyboid agaric with a smell of crushed tomato leaves. The spore-deposit is white. The smooth, inamyloid spores measure 7–9 × 4–5 µm. On litter on nutrient rich forest soil.

There should be no obvious identification issues provided the smell and pink colouration are evident and the spore measurements are within the correct range.

A rare species with most records from the British Isles; mainly September–November.

Thomas Læssøe

Jens Maarbjerg

Leucocybe candicans is a rather small, thin-fleshed, whitish clitocyboid agaric with a pale, somewhat splitting coating at the cap surface. The odour is variable, from almost absent to somewhat sour, grass-like or like crushed tomato leaves. Has a white spore-deposit. The ellipsoid spores, not forming tetrads, measure 5–6.5 × 3–3.5 µm. On litter in forests.

Clitocybe rivulosa ▷ is normally found in grasslands. It has a somewhat thicker stem, somewhat smaller spores, and dried specimens have brownish gills.

Widespread and rather common; mainly July–December.

Singerocybe phaeophthalma is a rather characteristic, hygrophanous, smooth clitocyboid agaric with an unpleasant smell like a chicken run or a wet chicken (has also been described as smelling like honey); with rather pale colours and highly characteristic, narrow, balloon-shaped cells with refractive contents in the cap flesh. Spores measure 4.5–6.5 × 3–4.5 µm, not sticking together. On litter in nutritious deciduous forests.

Even without noting the smell, the species can easily be recognized microscopically via the special cells in the cap flesh.

Widespread and fairly common; mainly August–November.

Clitocybe has ± applanate to funnel-shaped caps with ± decurrent gills and white to vinaceous-brown, spore-deposits. Some species are poisonous as they contain muscarine.

Clitocybe phyllophila is a rather large, pale, somewhat hygrophanous *Clitocybe* with a with pale somewhat splitting chalcy coating, a vinaceous-brown spore-deposit and a somewhat spicy smell. The mostly agglutinated spores measure 4–5.5 x 3–4 µm. It occurs in places with plenty of leaf-litter, mostly on rich soil, in deciduous and coniferous forest.

Leucocybe connata ◁ has a white spore-deposit and separate spores, and is typically more clustered. It also has a characteristic, perfumed smell. *Leucocybe candicans* ◁ is smaller and similarly has a white spore-deposit. *Clitocybe phyllophila* belongs to a difficult species complex, and more species are probably involved.

Widespread and common; mainly August–January.

††***Clitocybe rivulosa*** is a convex-applanate, non-striate, medium-sized *Clitocybe* with a pale, somewhat splitting coating and off-white, very short, decurrent gills, a white spore-deposit and a neutral to somewhat sweetish smell. The spores measure 4.5–6 × 3–4 µm, mostly not agglutinated. In open, rich and grassy habitats.

Leucocybe candicans ◁ is more confined to woodland and has a thinner stem (2–4 mm versus 3–8 mm in *Clitocybe rivulosa*). *Clitocybe phyllophila* △ has a brownish-rose spore-deposit and occurs in woodland. *Clitocybe augeana* ⋉ and similar species smell farinaceous.

Widespread and common; mainly; mainly July–January.

Clitocybe ditopus is a medium-sized, ± grey, strongly farinaceous *Clitocybe* with a somewhat zonate, hygrophanous cap surface, sometimes with a pale coating, and with a white spore-deposit. The very small, almost spherical spores measure 3–4 x 2–3 µm. It mostly grows on needle debris in conifer plantations, but may also occur with deciduous trees.

Clitocybe vibecina ▽ is also farinaceous, but marginally striate and is without a pale coating. It has larger spores. *Clitocybe metachroa* ▷ does not smell farinaceous and is without a pale coating. Species of *Rhodocybe* s.l. (page 448) have brownish-rose spore-deposits and small rounded protuberances on the spores.

Widespread and rather common; mainly August–January.

Clitocybe vibecina is a rancid-smelling, brown-grey, ± marginally striate *Clitocybe* with somewhat to obviously decurrent gills and a stem with whitish floccules, mostly near the top. The spore-deposit is white and the cap cuticle hyphae are ± incrusted with pigment. Spores measure 4.5–7 x 3.5–4.5 µm, ± agglutinated. Mostly on peaty soils, on needle debris *etc.*, both in open and in forested habitats.

Clitocybe ditopus △ has a pale, somewhat splitting coating at the cap surface. Similar species with adnate gills include *C. subspadicea* ▷▷ but that species has a faint nitrous smell.

Widespread and common; April–January.

Clitocybe metachroa is a ± brown-grey, hygrophanous, not farinaceous *Clitocybe*, which typically has a stem that becomes dark below, and a dark spot in the centre of the cap. Has a white spore-deposit. Cap cuticle without, or hardly with, incrusted pigment. The spores measure 5.5–8.5 × 3.5–5 µm and are mostly agglutinated. On deep leaf-litter in all kinds of woodland.

Clitocybe ditopus ◄ smells farinaceous. *Clitocybe metachroides* ⋉ has distinctly incrusted pigment in the cap cuticle and is generally darker brown. There is also a yellow-brown form, *Clitocybe metachroa* var. *aquosoumbrina* ►. Correct identification demands careful microscopy.

Widespread and common; mainly August–January.

Clitocybe agrestis is a rather slender, pale pinkish-buff, hygrophanous, smooth *Clitocybe* with a somewhat chalky margin. Has adnate to somewhat decurrent, ± cream gills and a smell of coumarin. Spore-deposit ± cream. Spores mostly agglutinated, measure 4–6 × 2.5–4 µm. Occurs mostly in dry grassy places.

The smell is characteristic. *Clitocybe marginella* ⋉ has a darker cap centre, more saturated colours and a more colourful spore-deposit. Whitish species of *Cuphophyllus* (page 163) have thicker gills, different smells and larger spores.

Widespread and probably common; mainly August–November.

Jan Vesterholt

Clitocybe subspadicea is a dark grey-brown, hygrophanous, smooth, non-striate, umbillicate, somewhat nitrous-smelling *Clitocybe* with a uniformly coloured stem, except for a pale zone at the very apex. The spore-deposit is white. The cap cuticle is without incrusted pigment. The spores measure 6–9 × 3.5–4.5 µm. Occurs amongst litter in both deciduous and coniferous forests, mostly in places with rich soils.

The faint but, nevertheless, distinct nitrous smell and the pale apical zone on the stem are good distinguishing characters. Other similar grey-brown species have, in addition, mostly incrusted pigment in the cap cuticle.

Widespread, but often overlooked; mainly August–October.

Thomas Læssøe

Clitocybe barbularum is a small, dark, hygrophanous, striate, farinaceous, somewhat omphalinoid *Clitocybe* with rather dark gills and a somewhat gelatinous cap cuticle that has incrusted pigment. The spores measure 5–7 × 3–4 µm. It is mostly found on or with moss, in calcareous dunes or on other sandy soils.

Clitocybe vibecina ◁ is mostly larger and has only somewhat incrusted pigment and at most a somewhat greasy cap surface; *Clitocybe legaliae* ⋉ is a very similar, but non-striate species.

Rather rare and scattered; mainly October–December.

††***Clitocybe diatreta*** is a relatively slender, ± flesh-pink, smooth, somewhat greasy, hygrophanous, not clustered *Clitocybe* with pale gills, a pale, somewhat chalky margin and a ± vinaceous-brown spore-deposit. Smell somewhat aromatic. The spores measure 4–6 × 2.5–4 µm and are mostly agglutinated. Found both in deciduous and coniferous forests and in more open environments, *e.g.* dunes.

Clitocybe leucodiatreta ⋉ has a white spore-deposit and separate spores. *Clitocybe marginella* ⋉ has a dark cap centre.

Widespread and common; mainly August–January.

Clitocybe fragrans is an aniseed-scented, rather slender and pale, mostly somewhat 'dark-eyed', hygrophanous, smooth *Clitocybe* with short, decurrent, pale gills. The spore-deposit is not completely white. Has large, mostly agglutinated spores measuring 6.5–9 × 3.5–5 µm. On litter in both coniferous and deciduous forests, often on rich soils.

Clitocybe acicola ⋉ has a similar smell, but generally has a completely white spore-deposit and is less hygrophanous. The two species are, however, often treated as one. *Clitocybe odora* ▷▷ is more sturdy and has mostly greenish colours.

Widespread and common; all year, but mainly late in the autumn.

• ***Clitocybe odora*** is a fleshy, ± convex to slightly depressed, ± pale turquoise, aniseed-scented *Clitocybe* with only somewhat decurrent gills. The cap is, with age, ± yellowish around the centre. The spore-deposit is not completely white. The spores measure 6–8.5 × 4–5.5 µm. On litter in deciduous and coniferous forests, more rarely in open habitats.

Typical turquoise, fragrant specimens are readily identified, but washed-out, less fragrant material, may cause difficulties. *Clitocybe fragrans* ⊲⊲ produces a much more slender fruitbody.

Widespread and common; mainly August–November.

Clitocybe nebularis is a very large, fleshy, convex to slightly depressed *Clitocybe* with a somewhat inrolled margin, only somewhat decurrent, crowded gills and a ± perfumed smell. Spore-deposit sordid cream. Spores ellipsoid, measuring 6.5–8 × 3.5–4 µm. In large fairy rings on leaf-litter.

Clitopaxillus alexandri ⊳ has a felty-squamulose cap. Species of *Lepista* ⊲⊲ have ornamented spores. Deformed, white-spotted fruitbodies may be parasitized by *Volvariella surrecta* (page 490).

Widespread and common; mostly September–January.

Clitocybe trulliformis is a ± convex to slightly depressed, felty-squamulose, brown-grey, farinaceous *Clitocybe* with rather pale, deeply decurrent, occasionally bifurcate gills. Spore-deposit white. Spores 5–7.5 × 3–4 µm, mostly agglutinated. On rich mull in deciduous forests, mostly at road sides; rarely on very rotten wood.

This species belongs to a poorly understood complex of small, grey-brown clitocyboids with felty caps. *Clitocybe fuscosquamula* ⋉ has narrow, fusiform spores, 6.5–9 × 2–3 µm, and *C. collina* ⋉ almost ellipsoid spores, 4.5–6 × 2.5–3 µm.

Rather rare and absent from cooler parts of temperate Europe; July–November.

Clitopaxillus fibulatus is a big, very fleshy, convex-umbonate, non-hygrophanous, dirty yellowish grey to pale brown, felty to squamulose *Clitocybe*-like, white-spored agaric with a inrolled cap margin, a felty stem base and a ± aromatic smell. The hyphae in the cap cuticle have lots of clamps. Spores smooth, 5.7-6.4 × 4.1-4.6 µm, av. 6.1 x 4.4, Q 1.30-1.47, Qav. 1.39 On needle beds in conifer forests and plantations on fairly rich, often calcareous soils.

Clitopaxillus alexandri ⋉ has slightly smaller spores and lacks clamps in the cap cuticle. *Clitocybe nebularis* ◁ is non-squamulose and has a sordid cream spore deposit and more grey tinges. Similar species of *Lyophyllum* (page 254) stain black or blue or have globose or otherwise deviating spores.

Mainly found in northern parts of the nemoral zone and further north, rare to occasional; mainly August-October.

Jan Vesterholt/Jens H. Petersen

Omphalinoids

The omphalinoid fungi are a loosely defined form group of white-spored agarics with relatively small, mostly dry, somewhat fragile and elegant fruitbodies and rather deeply decurrent gills. Although the name suggests that all are umbilicate, this is far from the case. They can also be convex, umbonate or applanate without an umbilicus.

Some omphalinoid fungi are biotrophic, either as lichenized with algae or as parasites on mosses (left side of the wheel). Others are decomposers (saprotrophs) of litter, herbs or wood (right side of the wheel).

OTHER SIMILAR FUNGI:
– some hygrocyboids (*Hygrocybe* and others) with decurrent gills look distinctly omphalinoid. However, the hygrocyboid fungi have, in general, very slender basidia (6–9 times as long as wide) and live mostly as biotrophs amongst phanerogams (page 142).
– some mycenas with decurrent gills are also omphalinoid. Many mycenas can be separataed by their amyloid spores, but *e.g. Phloeomana speirea* and related species and the genus *Hemimycena* are inamyloid (page 176).
– the clitocyboids have predominantly decurrent gills, but are mostly larger and tougher than the omphalinoids (page 98).
– some species of entolomatoids also look similar, but these have brownish rose spore-deposits, and the spores are ridged or angular (page 448).

LITTERATUR: 156, 181, 262.

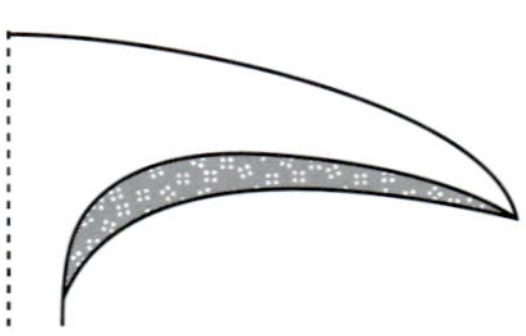
deeply decurrent gills

Micro-drawings: spores and gill cystidia.

Approximate species number applies to temperate Europe.

Lichenomphalia species live in mutualistic symbiosis with green algae. They form a grainy or leaf-shaped, green thallus from where the fruitbodies arise (see page 19).

Lichenomphalia umbellifera is an omphalinoid fungus with an umbillicate cap, deeply decurrent gills and a ± greyish-violet stem apex. At the base it has a grainy, green thallus. The almost spherical to broadly ellipsoid, smooth spores measure 7–10.5(–11) × (4.5–)6–8 µm. On moist peaty soil or on wet, partly moss covered, fallen trunks.

Lichenomphalia velutina ▽ is more grey-brown on the cap and stem and occurs typically in drier and more open habitats. *Lichenomphalia hudsoniana* ▷ has a leaf-shaped thallus and is more yellow on the fruitbody.

Widespread and common; may be found with fruitbodies almost year-round.

Lichenomphalia velutina is a rather anonymous, grey-brown omphalinoid with a translucently striate cap, a densely downy stem surface, and at the base of the stem a grainy, green thallus. The spores measure 7–10(–11) × (3–)4–6 µm. In shaded grasslands on disturbed, exposed soil, along streams *etc.*

Lichenomphalia umbellifera △ is more yellow-brown with a smooth stem and dark stem apex. Species of *Arrhenia* ▷▷ lack the green, grainy thallus at the stem base.

Mostly found in uplands and rather rare in lowlands; mainly June–December.

Lichenomphalia alpina is a bright yellow omphalinoid agaric with a downy stem and a grainy thallus. The spores measure (6.5–)7–13.5(–16) × (3.5–)4–5.5(–6.5) µm. Occurs in ± alpine regions on peat-like soil.

A distinct species with no obvious identification issues. Yellow species of *Chromosera* (page 148) are similar, but ± viscid or slimy.

Although this species occurs outside the geographic scope of this book, it is included here for comparison with other similar species. All *Lichenomphalia* species in Europe are included in this way. It is widespread in mountains above the treeline; summer–autumn.

Steen A. Elborne

Lichenomphalia hudsoniana is a yellow-capped omphalinoid fungus, with a downy, paler stem that at the base has a leaf-shaped, blue-greeen thallus with pale margins. The smooth, ellipsoid spores measure 6.5–9.5(–10) × 4.5–5.5 µm. On moist, peat-like soil and on nutrient-poor, sandy slopes.

Other *Lichenomphalia* species have a grainy thallus.
Rarely with fruitbodies in the lowlands, but rather frequent in the uplands; may be found almost year-round.

Blasiphalia pseudogrisella is a small, grey-olive, translucently striate omphalinoid fungus with conspicuous, thin-walled cystidia on cap, gill edges and stem. The spores measure (6–)6.5–8.5(–9.5) × 3–5 (–6.5) µm. It parasitizes the liverwort *Blasia* via so-called appressoria on the 'roots' of the host. It is found along rivers and streams in the boreal-subarctic zone, but also rarely in the nemoral zone.

Other small, grey omphalinoids are not associated with *Blasia*, and most species are darker or more colourful. The numerous cystidia point to a relationship with *Rickenella* ▷▷ nested in the *Hymenochaetales*.

Fairly common in the uplands, but very rare in, or absent from, the lowlands of the nemoral zone; mostly autumn.

Arne Aronsen

Gyroflexus brevibasidiatum is a small and slender, pale cream to clay-buff omphalinoid with deeply decurrent, white gills and a white stem. Completely lacks cystidia. The inamyloid spores measure 5.5–7 (–7.5) × 3–4.5 µm. It is associated with *Sphagnum* in mires in the boreal zone. It is probably biotrophic and is related to other omphalinoid moss- and liverwort-associated agarics in the *Hymenochaetales*, and not the *Agaricales*, where most similar fungi are placed phylogenetically.

Distinctive and unlikely to be misidentified.

The species is ± boreal in its distribution and is probably completely absent from the nemoral zone; summer–autumn.

Loreleia marchantiae is a small, orange omphalinoid with a ± crenulate-furrowed cap margin. It has no conspicuous cystidia and no incrusted pigment in the cap cuticle but does have oil-containing hyphae in the flesh. The spores measure (7.5–)8.5–10.5 × (4.5–)5–7 µm. It is associated with the liverwort *Marchantia*, which grows mostly in shady places on disturbed soil, along walls, in plant nurseries, in connections with railroads, *etc*.

Loreleia postii ⋉ is a very similar species that also occurs in similar places. It has a ± smooth cap margin, no oleiferous hyphae in the flesh, and has incrusted pigment in the cap cuticle. *Rickenella fibula* ▷▷ occurs on bryophytes, not liverworts, and the fruitbody is covered in long cystidia (hand lens required!).

Widespread and probably rather common, except in upland areas, but is easily overlooked; mainly May–November.

Thomas Læssøe

Cantharellopsis prescotii is a small, rather tough, not hygrophanous, somewhat cantarelloid or omphalinoid agaric with buff cap colours and a fairly silk-like surface. The gills are thick, bifurcate, and white like the stem. Lacks cystidia and has smooth, inamyloid spores that measure 4.5–6(–6.5) × 2.5–4 µm. Associated with mosses on calcareous rocks.

The thick, bifurcate gills combined with its small size and habitat should make this species easy to recognize. Chanterelles (page 46) have even more ribbed to veined gills, which are never pure white. Mycenoids with veins (page 188) are smaller and more thin-fleshed.

A rare fungus, mostly recorded in the hemiboreal zone; summer–autumn.

Thomas Læssøe

Arne Aronsen

Rickenella is a small genus of omphalinoids with decurrent gills and with all parts covered by projecting cystidia (hand lens required!). They are parasites on bryophytes.

Rickenella fibula is a translucently striate, ± orange to orange-brown (washing out to whitish) omphalinoid with small hairs (cystidia) on its stem and cap, as well as on the deeply decurrent, distant and pale gills. The inamyloid, smooth spores measure (4–)5.5–7(–7.5) × 2–3(-4) µm. Always parasitic on mosses, mostly in moist grassland, but also on moss-covered trunks and in other damp places. It often occurs with *R. swartzii* ▽.

Mycena acicula (page 193) does not have decurrent gills and is two-coloured with more red in the cap and more yellow in the stem. The rather poorly known *R. mellea* ⋉ has less vivid colouration, possibly less obvious hairs on the fruitbody and probably somewhat larger spores.

Widespread and very common; mostly June–November.

Rickenella swartzii is a dark grey-brown to red-brown omphalinoid agaric with deeply decurrent, pale gills, a brown to greyish-violet stem apex, and a yellow-brown stem. The entire fruitbody is covered in small projecting hairs (cystidia). The spores measure (4.5–)5–7 × 2–3 µm. A parasite on bryophytes, mostly in moist lawns and similar places.

Distinctive and unlikely to be misidentified. *Lichenomphalia umbellifera* ◁◁ may have somewhat similar colours, but lacks the fine hairs on the cap and gills, and has small green thallus grains at the base.

Widespread and common; mainly May–December.

Arrhenia is a genus of partly omphalinoid agarics with grey-brown, greenish or reddish colours. Some species are parasites on mosses and lichens, while others are saprotrophs (see also page 64).

Arrhenia gerardiana is a rather large, grey, striate (to the centre), funnel-shaped *Arrhenia* with small and discrete dark cap scales, and decurrent gills with mostly dark edges. The rather narrowly cylindrical spores measure (8–)9–13(–14) × (3–)3.5–5 µm. Associated with *Sphagnum*.

The similarly *Sphagnum*-associated *A. philonotis* ⋉ has concolorous cap squamules and a lower spore Q. *Arrhenia onisca* ⋉ lacks scales, is typically somewhat larger and has a lower spore Q. *Sphagnurus paluster* (page 286) is more mycenoid and smooth-capped.

Widespread and fairly common, except in intensively farmed areas; mainly May–September.

Arrhenia peltigerina is a slender, funnel-shaped *Arrhenia* with grey-brown gills (not least after drying) and a ± downy stem. Basidia (2–)4-spored, with clamps. The spores measure 7–11 × 4–5.5 µm. Associated with thalli of *Peltigera* in sandy areas.

If the connection with the lichen *Peltigera* is noted, and the jizz is as shown here, it is unlikely to be misidentified. The species was described from North America. Based on the wording in the original description the European species would appear to be something else.

Fairly scattered records, many new; both spring and autumn and probably also in wet summers.

Jens H. Petersen/Guillaume Eyssartier

Arrhenia rickenii is a rather slender, grey-brown, funnel-shaped *Arrhenia* with ribbed/veined gills and a smooth stem. The hyphae are clampless and the basidia 4-spored. The drop-shaped to ellipsoid spores measure 6.5–10 × 3.5–5.5 µm. It is found in sandy, mossy habitats amongst mosses.

Arrhenia griseopallida ⋉ has hyphae with clamps. Other similar species typically have hairy stems and gills that are less irregular.

Rare or overlooked, with rather few records; both spring and autumn.

Thomas Læssøe

Arrhenia obscurata is a grey-brown to dark brown, translucently striate, funnel-shaped *Arrhenia* with a smooth, or at least not prominently hairy, stem. Has 2–4-spored basidia and clamped hyphae. The spores have a rather high Q value and measure 7.5–11(–13) × (4–)4.5–6.5 (–7.5) µm. It is associated with open ± disturbed, nutrient-poor habitats with scattered mosses.

This species belongs to a difficult group of small grey-brown omphalinoid fungi. However, its ± smooth stem, should point to a correct identification.

Widespread and probably rather common in some regions if not all; both spring and autumn.

Arrhenia velutipes is a small, dark grey-brown, striate, downy-stemmed, *Pelargonium*-smelling, ± funnel-shaped *Arrhenia*. Has clamps and long and somewhat thick-walled hairs on the stem. The ± ellipsoid spores measure 7–9 × 4–6 µm. In open vegetation, amongst mosses, *etc*., mostly on clay-rich soils.

Lichenomphalia velutipes ◁◁ looks similar, but is paler and always has small, green, grainy thalli at the stem base. Other small, grey *Arrhenia* species mostly have ± smooth stems.

Widespread and, at least in some places, rather common; possible year-round.

Thomas Læssøe

Arrhenia epichysium is a small to large, brown-grey, marginally striate, smooth, funnel-shaped, rather long-stemmed *Arrhenia*. It has 4-spored basidia and clamped hyphae. The spores measure 6–9(–11) × 3.5–5 (–6) µm. It occurs on rotten, mostly moss-covered fallen trunks in ± unmanaged forests.

A rather unremarkable-looking omphalinoid fungus, which in principle could be macroscopically confused with omphalinoid but pink-spored species of *Entoloma* (page 448). *Phloeomana speirea* (page 192) also occurs on rotten wood but is not umbillicate, is typically smaller, and has distinct cheilocystidia.

A rare fungus generally, but quite common in areas with old growth forests; mainly August–November, rare in the spring.

Thomas Læssøe/Jens H. Petersen

Arrhenia chlorocyanea is a rather small ± turquoise to steel-blue, even greenish, ± funnel-shaped *Arrhenia* with deeply decurrent gills. The spores measure (6.5–)7–9.5(–11) × (3–)3.5–5.5(–6) µm. It occurs on open gravelly/sandy, disturbed soils with scattered mosses, *Juncus etc.*, *e.g.* at the margins of new artificial ponds created in connection with road building or nature conservation projects.

A few species of *Entoloma* (page 448) have blue colours and decurrent gills, but the spore-deposit is pinkish in comparison to the white deposit of *Arrhenia chlorocyanea*, and the spores are angular.

Appears to be fairly frequent in the northwestern lowlands of temperate Europe – with few old, but many new records; mostly March–April, but may be found year-round.

Thomas Læssøe

Arrhenia discorosea is a large, funnel-shaped, grey-rose to brownish-red, finely scaly-felty *Arrhenia*. It is the only *Arrhenia* with a distinctly pinkish spore-deposit. The egg-shaped to ellipsoid spores measure 7–8 × 3.5–4.5 µm. On large dead decidous trees, *e.g.* within cavities of standing dead trunks, *e.g. Populus*.

Distinctive and unlikely to be misidentified, since there are no other lignicolous omphalinoids, clitocyboids or entolomatoids with these colours.

Rare and mostly found in eastern parts of temperate Europe; summer–autumn.

Contumyces rosellus is a small, flesh-pink, funnel-shaped omphalinoid agaric with deeply decurrent gills. Has cheilocystidia and clamps. The inamyloid, smooth spores measure (7–)8.5–10.5 × 4–6 µm. On soil amongst moss and various herbs.

Omphalina lilacinicolor ▽, mainly recorded from dunes, is darker brownish pink. *Laccaria tortilis* (page 278) has less decurrent gills and subglobose, warty spores.

Recorded in nemoral parts of temperate Europe, very rare or overlooked; late summer–autumn.

Leif Örstadius

Omphalina pyxidata is a red-brown, 4-spored omphalinoid with pale gills tinted the same colour as the cap that is striate to the centre. Stem ± smooth to minutely hairy. The hairs on the stem are less than 50 µm long; with scattered or no cheilocystidia. The spores measure 7–8 × 5–6 µm. On open, disturbed, mostly gravelly soils with low mosses.

This species belongs to a difficult complex. *Omphalina subhepatica* ⋉ is only striate at the very margin and has somewhat smaller spores, no cystidia and possibly somewhat longer hairs on the stem.

Widespread and common; mostly late autumn.

Omphalina lilacinicolor is a flesh-pink to brownish-rose omphalinoid with thick, occationally bifurcate, pale pinkish gills. Has scattered cystidia of variable shape. Basidia 2–4-spored. With or without cheilocystidia. The spores measure 6–9 × 5–7 µm. In dunes on open, disturbed sandy soils with low moss cover.

Other species of *Omphalina* and *Arrhenia* ◁ in dunes do not have lilac-tinted fruitbodies. Also known as *O. galericolor* var. *lilacinicolor* and *Arrhenia lilacinicolor*.

A rather rarely recorded species, but frequent in the Dutch dunes; October–March.

Menno Boomsluiter

Omphalina mutila is a white to pale cream, rather large *Omphalina* with a minutely downy stem. Has 4-spored basidia and no cystidia. The spores measure 6.5–10 × 4–6 µm. Found in damp heathland with various ericaceous plants.

Species of *Clitopilus* (page 453) smell farinaceous, have pinkish spore-deposits and longitudinally ridged spores. Material suspected to be this species should be examined very carefully in order to rule out aberrant pale forms of species belonging to other genera that are normally more pigmented.

Generally very rare, but locally quite common, *e.g.* in the Netherlands; mainly August–September.

Haasiella venustissima is a yellow-orange to orange, rather small omphalinoid agaric with a strigose-hairy cap margin, normally occuring in small clusters. The yellow pigment consists of carotenoids. The spore-deposit is somewhat yellowish. The gill edge is fertile, without cystidia. Basidia 1–4-spored, with or without clamps. The rather thick-walled spores measure 5.5–8 × 4–5(–6) µm. Occurs in scrub and at forest edges with ± calcareous soil on wood (sometimes buried) of deciduous trees, *e.g. Sambucus* and *Syringa*.

Distinctive and unlikely to be misidentified. Other yellow to orange omphalinoid agarics grow on conifer wood, or are moss- or liverwort-associated. Species of *Hygrophoropsis* (page 100) have forked gills. The 4-spored form of *Haasiella venustissima* was once regarded as a separate species, with the name *H. splendidissima*.

A very rare species probably mostly found in central and eastern parts of temperate Europe; mainly September–October and probably also in mild winter periods.

Marek Brtník

Michal Mikšík

Chrysomphalina is a small genus associated with conifer wood and with yellow to yellow-brown colours.

Chrysomphalina chrysophylla is a large, golden, omphalinoid agaric with a somewhat sordid yellow-brown cap and bright orange-yellow, deeply decurrent gills. Has 4-spored basidia, no clamps and no cystidia. The smooth, ellipsoid to amygdaliform spores measure 9–12 × 5–6 µm. On conifer wood in moist, moss-rich *Picea*-forests.

In colour, may recall *Craterellus tubaeformis* (page 49), but that species has a veined/wrinkled hymenophore rather than narrow gills. *Chrysomphalina (Gerronema) strombodes* ⋉ occurs in central Europe and is recognized by its greyish cap and less vividly coloured gills.

Mostly a hemiboreal–boreal species but also found in more southern uplands; late summer–autumn.

Jens Maarbjerg

Jens Maarbjerg

Chrysomphalina grossula is a greenish-yellow, pale lemon-yellow to yellowish white, smooth or somewhat felty omphalinoid agaric with greenish-yellow to almost white, deeply decurrent gills. No cystidia and no clamps. The smooth, inamyloid spores measure 7–10(–11) × (4.5–)5–6 µm. On wood of various conifers.

Completely washed-out specimens may cause identification problems, but there is mostly a yellowish shade left on the gills.

A widespread species, which seems to have become more common; mostly in late autumn and during mild winter periods.

Hygrocyboids

The hygrocyboid agarics (waxcaps and others) are recognized by their thick, wax-like and mostly rather distant gills, and many species have very vivid colours. Microscopically, most hygrocyboids have unusually long and slender basidia, typically 6–9 times as long as wide. The spore-deposits are whitish and the spores smooth and inamyloid; as a general rule cystidia are lacking.

Most hygrocyboids are thought to be biotrophic, with a poorly understood symbiosis with herbs. Using DNA-analysis techniques, living hyphae have been detected inside tissues and seeds of *Plantago*. Species of *Hygrophorus* are also proven to be biotrotrophic, but they form ectomycorrhiza with a number of woody partners. One species (*Hygrophorus exiguus* ⋉) appears to be associated with *Tricholoma* mycorrhiza.

Most hygrocyboid agarics have very demanding habitat requirements. They are particularly sensitive to fertilizers and many favour habitats with a long continuity. The hygrocyboids share the same habitats and have similar preferences to a number of clavarioids, entolomatoids and earthtongues; they are all good indicators of sites of special nature conservation value.

In most of Europe the typical habitat for hygrocyboids, except *Hygrophorus* and *Hodophilus*, is old, unfertilized grassland – a habitat that has declined dramatically over the past 50 years. In other parts of the world, the hygrocyboid agarics are predominantly to be found in forests with long continuity. Species of *Hodophilus* tend to prefer thorny thickets on clay soils.

Other similar fungi:
– omphalinoids may also have thick, fleshy gills; many are parasites on mosses or are lichenized, but some are saprotrophs (page 128).
– mycenoids generally have 'typical' gills and most have cystidia and amyloid spores. They are all believed to be saprotrophs (page 176).
– entolomatoids may be colourful, but have angular, pinkish spores (page 448).

Litteratur: 1, 2, 3, 43, 57, 156, 172, 178, 183.

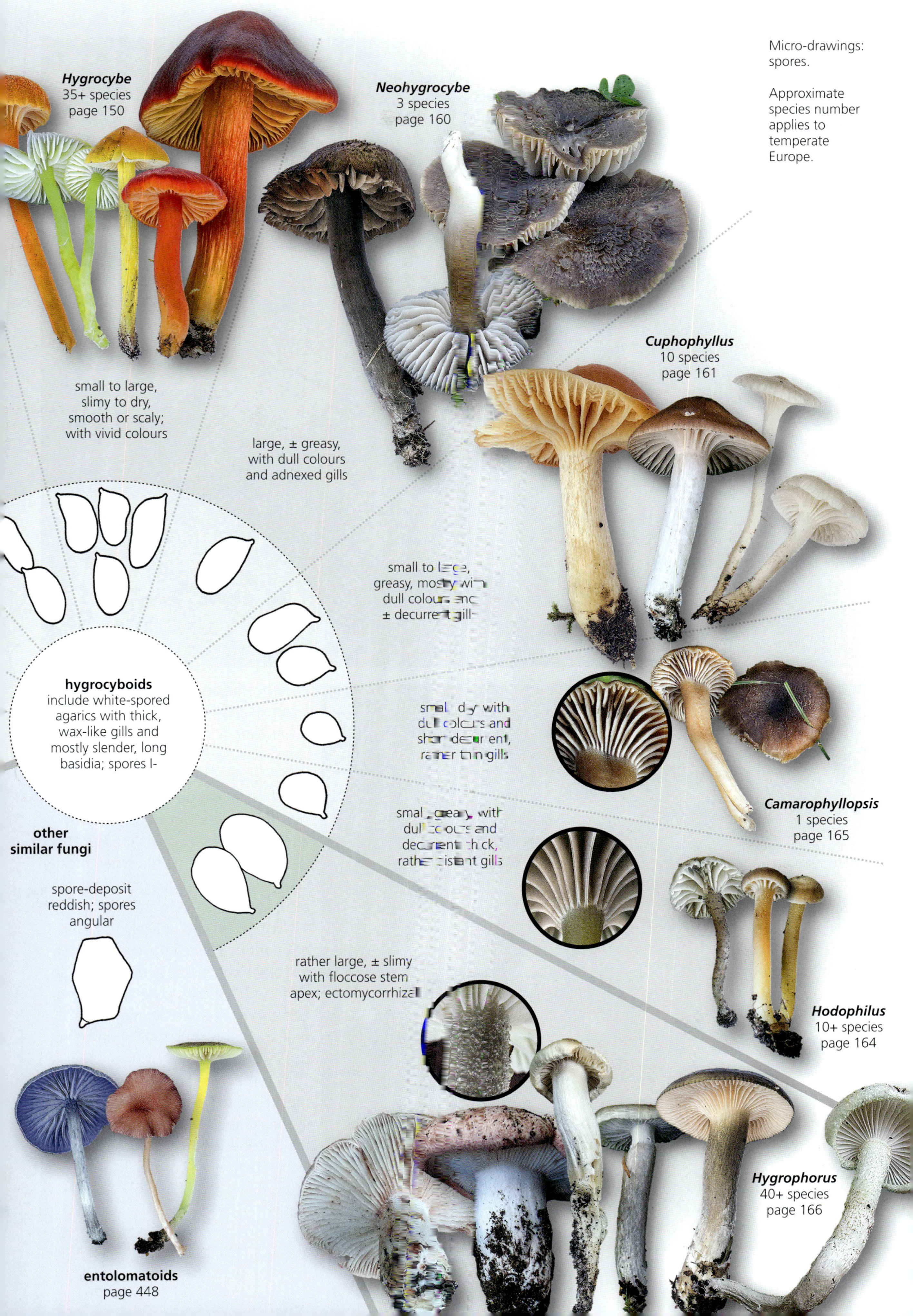
Micro-drawings: spores.
Approximate species number applies to temperate Europe.
Hygrocybe
35+ species
page 150
small to large, slimy to dry, smooth or scaly; with vivid colours
Neohygrocybe
3 species
page 160
large, ± greasy, with dull colours and adnexed gills
Cuphophyllus
10 species
page 161
small to large, greasy, mostly with dull colours and ± decurrent gills
small, dry, with dull colours and short decurrent, rather thin gills
Camarophyllopsis
1 species
page 165
small, greasy, with dull colours and decurrent, thick, rather distant gills
Hodophilus
10+ species
page 164
hygrocyboids
include white-spored agarics with thick, wax-like gills and mostly slender, long basidia; spores I-
other similar fungi
spore-deposit reddish; spores angular
rather large, ± slimy with floccose stem apex; ectomycorrhizal
Hygrophorus
40+ species
page 166
entolomatoids
page 448

Jacob Heilmann-Clausen

Porpolomopsis calyptriformis is a large, acutely conical, rose-coloured, ± dry waxcap with a whitish or somewhat rose stem and a radially split margin. The fruitbodies are very fragile and do not blacken, but the colours may wash out with age. The smooth spores measure 6–9 × 4–7 µm. Occurs in older, mostly clay-rich or calcareous grasslands, including old churchyards.

Distinctive and unlikely to be misidentified, since the combination of colour and shape is unique. It can be found in a rare, pure white form. *Gliophorus reginae* ▷▷ may have similar colours, but is slimy on both cap and stem.

A rather rare and local species that tends to be more common in the Atlantic areas, especially in the British Isles; mainly August–November.

The hygrocyboid fungi in the genus ***Gliophorus*** are slimy on both cap and stem and some are very brightly coloured. One is even slimy on the gill edge. They may have strange smells.

Gliophorus laetus is a medium-sized waxcap with somewhat dull colours, decurrent, pale greyish gills with a slimy edge, and a strange smell like burnt rubber. The stem apex is mostly greyish-violet or olive and below is brass-yellow. The smooth spores measure (5–)6–7 (–8.5) × (3.5–)4–4.5(–5) µm. It is also found in an almost pure yellow form. Occurs in unfertilized grasslands, mostly on somewhat acidic soils, or on heathlands.

This species is also known as *Hygrocybe laeta*. *Gloioxanthomyces vitellinus* ▷▷ also has a slime-thread on the gill edge, but is ± odourless, smaller and bright yellow. *Gliophorus psittacinus* ▷▷ does not have decurrent gills, and likewise no slime-thread on the gill edge; it is mostly greener.

A rather common and widespread species; mainly August–November.

Gliophorus irrigatus is a medium-sized, very slimy, grey to grey-brown, rarely almost white, waxcap with pale, somewhat decurrent gills. It does not have a distinctive smell, and the gill edge is not slimy. The spores measure (5–)6.5–8(–9) x (3.5–)4.5–5(–6) µm. Occurs in unfertilized grasslands, mostly with other hygrocyboids, clavarioids and earthtongues, and occasionally also in forests.

Distinctive and unlikely to be misidentified.

A widespread, rather common species; mainly July–November.

Gliophorus psittacinus is a spectacular, very slimy, variegated green-yellow waxcap with ± emarginate gills without a slimy edge, and without a distinct smell. The colours are very variable, often in response to exposure – specimens shielded from the sun are mostly a deep bottle-green and those exposed are more yellowish. Yellow versions can change back to green when kept in the shade. The smooth spores measure 7–9.5 × 5–6 µm. Occurs in unfertilized grasslands and in moist forests on clay-rich soils with scattered mosses.

Gliophorus perplexa ▷is similar, but has brownish-red colours, and *G. reginae* ▷ has almost brownish-violet colours. *Gliophorus laetus* ◁◁ has a slimy gill edge, decurrent gills and a strong, rubbery smell.

A widespread and rather common species; mainly June–January.

Thomas Læssøe

Jan Vesterholt

Gliophorus perplexus is a brownish-red version of *G. psittacinus* completely without green colours, but with similar gill attachment and thick slime layers on cap and stem. The spores measure 7–9.5 × 5–6 µm. In unfertilized grasslands with long continuity, *e.g.* in old deer parks and other old grasslands.

Gliophorus reginae ▽ has more reddish-violet colours. *Gliophorus laetus* ◁◁ has distinctly decurrent gills with a slimy edge.

Widespread but rather rare; mainly May– November.

Gliophorus reginae is a reddish-violet to greyish-purple, medium-sized, slimy waxcap with emarginate gills and a pale stem. The gills are greyish without a slimy edge. The spores measure 7–9.5 × 5–6 µm. Occurs in unfertilized grasslands with other hygrocyboids, clavarioids *etc*.

Gliophorus perplexus △ is more brownish-red on the cap and stem.

Described from the British Isles with a few additional records from continental Europe; mainly July–November.

David Boertmann

Chromosera species are small, brightly coloured, ± omphalinoid hygrocyboids with decurrent gills.

Chromosera lilacina is, at first, a slimy, small waxcap with a violet cap and stem, and yellow, deeply decurrent gills. It later becomes almost dry. The spores measure 8–10(–12) × 5–7 µm. Occurs on peaty soil in open, moist habitats, mostly in alpine-arctic habitats.

Chromosera viola ▽ is dry, has more distinct blue colours on the cap and stem, and occurs in deciduous forests.

Very rare in the lowlands of temperate Europe; summer–autumn.

Lucie Zíbarová

Chromosera cyanophylla is an omphalinoid, lignicolous, dry waxcap with lilac-tinted to lavender, deeply decurrent gills, but other parts have yellowish colours. The spores are smooth, inamyloid, while the flesh in the gills is dextrinoid. Grows on mossy, very decomposed trunks of conifers in forests with high conservation value.

Baeospora myriadophylla (page 283) also occurs on the trunks of conifers, but has crowded, adnexed, greyish-lilac gills. However, *C. cyanophylla* is distinctive and unlikely to be confused with other species on trunks of conifers.

Rare and eastern; mainly autumn.

Gianluigi Boerio

Chromosera viola is a small, very striking, violet, dry, ± striate, ± umbonate or umbillicate waxcap with decurrent, somewhat paler gills. Occurs on clay-rich, ± moss covered soils in deciduous forests.

Chromosera lilacina ⋉ is, at least when young, distinctly slimy and occurs in open, wet habitats.

A very rare, rarely reported, mainly nemoreal species, not known from the boreal zone; summer–autumn.

Chromosera citrinopallida is a small, yellow to yellowish-white, somewhat slimy waxcap with deeply decurrent, chrome-yellow gills without slimy edges. The stem is typically almost pure white. Occurs on peat in moist habitats, mostly in alpine-subarctic environments.

Gloioxanthomyces vitellinus ▽ is more slimy and has a slimy gill edge.

Very rare in the lowlands of temperate Europe; summer–autumn.

Gloioxanthomyces vitellinus is a rather small, very slimy, chrome-yellow waxcap with decurrent gills with a slimy edge. It is ± odourless. The colours may fade to almost white in older specimens. The smooth spores measure 6.5–8(–9) × 5–7.5 µm. Occurs in damp grassland, juniper thickets, open old *Salix* carrs and at the border of *Alnus* carrs, typically on soil, but may also be found on twigs of *Juniperus*. It is often well concealed in dense vegetation.

Species of *Gliophorus* ◁◁ are just as slimy, but only *G. laetus* ◁◁ has a slimy gill edge; it also smells strongly of burnt rubber and has relatively longer spores. Other slimy hygrocyboids do not have decurrent gills with a slimy edge.

Widespread, with the highest frequency in more Atlantic areas, but nearly absent from the boreal zone; mainly July–November.

The hygrocyboid fungi in the genus ***Hygrocybe*** are mostly brightly coloured. Within a species the colour may vary considerably. Many species have dry caps, and the cap surface can be smooth or minutely scaly (hand lens required!). The gill attachment spans from adnexed to deeply decurrent.

Hygrocybe chlorophana is a medium-sized, yellow to pale yellow-orange waxcap with a convex, sticky cap and a dry to somewhat glutinous, fragile stem, which mostly have small floccules near the top. The gills are adnexed to emarginate and yellowish. The odour is not distinct. The flesh in the gills consists of more than 1 mm long cells. The spores measure 7–9(–10.5) × 4–5.5(–6.5) µm. Occurs amongst herbs and grass in unfertilized grasslands, more rarely in scrub and damp forests.

Hygrocybe acutoconica ▽ has a more conical cap and a fibrillose stem. Other yellow species have adnate to decurrent gills.

Rather common and widespread; mainly June–November.

Hygrocybe acutoconica is a rather large, yellow to orange-yellow, conical, not blackening waxcap with a somewhat fibrillose, dry stem and a smooth, dry to ± viscid cap. The spores are cylindrical to bean-shaped and measure 10–12.5 × 4.5–6.5 µm. Typically occurs in calcareous, dry, unfertilized grasslands but can also be found in less calcareous sites.

Hygrocybe konradii ⋉ (sometimes regarded as a variety of *H. acutoconica*) has a spore Q of 1–1.7 compared with 1.5–2.5 in *H. acutoconica*. *Hygrocybe conica* s.l. ▷▷ has black-staining flesh.

Widespread and rather common; mainly May–November.

Hygrocybe ceracea is a rather small, yellow, sticky waxcap with pale, rather thin, adnate to somewhat decurrent gills. The stem is yellow and dry. Occasionally there may be orange shades on the cap or on lower parts of the stem. The odour is insignificant. The spores measure (5.5–)6-8(–10) × 3–4(–5.5) µm. Occurs in unfertilized grasslands, rarer in forest, but is not particularly discriminating and can quite often be found along road sides.

Hygrocybe chlorophana ◁ has adnexed to emarginate gills and is usually somewhat more robust.

Widespread and common; mainly July–December.

Hygrocybe quieta is a medium-sized, dry or almost dry, ± yellow-orange waxcap with wide, mostly adnexed, mostly transversely veined gills, which typically are more intensely coloured than the cap and stem. The stem is smooth and dry. It smells like *Lactarius quietus* (page 432) or stink bugs. The spores are ± constricted and measure (7–)7.5–9(–11) × 4–5 (–6) µm and the flesh hyphae are short. Amongst moss in clay-rich, damp forests and scrub or in more open, unfertilized grasslands.

Other dry, orange species mostly have a fibrillose stem surface.

Widespread, but not very common; mainly July–December.

Jan Vesterholt

Hygrocybe glutinipes is a small, vividly coloured, yellow-orange to red waxcap with a slimy cap and stem, and typically a translucent 'eye-spot' at the centre of the translucently striate cap. The gills are mostly adnate to somewhat decurrent. The stem is mostly long in relation to the cap width. The spores measure 7–8(–10) × 4–5 µm. Occurs in unfertilized, mossy grasslands or scrub, mostly on clay-rich soils.

Hygrocybe insipida ▷▷ is similar, but has a moist, rather than slimy stem.

Widespread and locally quite common; mainly June–November.

Hygrocybe coccineocrenata is a rather small, red to orange-red, dry, scaly waxcap with deeply decurrent gills and with cap scales that with age become grey to black. The smooth spores measure (8–)9.5–11.5(–13) × (5–)5.5–6.5(–8) µm and are not constricted. Grows directly in *Sphagnum* or on moist peaty soils.

Hygrocybe turunda ⋉ is very similar, but occurs in dry, unfertilized grasslands. *Hygrocybe cantharellus* ▽ does not develop dark cap scales.

Widespread but has declined due to drainage and eutrophication; mainly May–December.

Thomas Læssøe

Hygrocybe cantharellus is a bright red to orange-red, pale scaly (hand lens required!), dry waxcap with decurrent, rather pale, distant gills. Smell insignificant. The spores measure (7.5–)9–10.5(–11.5) × (5–)5.5–7(–9) µm and are not constricted. Occurs in damp woodland including *Alnus* carrs, fens and unfertilized grasslands.

Hygrocybe coccineocrenata △ and *H. turunda* ⋉ both have grey-brown to grey-black staining cap scales. *Hygrocybe substrangulata* ▽ has adnate to somewhat decurrent gills and spores that commonly have a central constriction.

Widespread and locally quite common; mainly June–November.

Hygrocybe substrangulata is a rather small to medium-sized, dry and finely scaly (hand lens required!), orange waxcap with adnate to somewhat decurrent gills. The large spores are typically constricted at the centre and measure (7–)9–12(–14.5) x (4–)5–6.5(–8.5) µm. Occurs mostly in dune slacks and in damp depressions in heathlands, usually with *Salix repens*, but also in meadows with upwelling water.

Hygrocybe miniata ▷ has smaller, violin case-shaped spores and normally occurs in less moist sites.

Mostly coastal and Atlantic, locally quite common; mainly August–November.

Hygrocybe miniata is a rather small, orange-red to orange-yellow, finely scaly (hand lens required!), dry waxcap with adnate to somewhat narrowly decurrent gills. The cap scales are the same colour as the cap or somewhat paler. The spores are mostly constricted like a violin case and measure (6–)7–9(–10) x (4–)4.5–5.5(–6.5) µm; the cells of the gill flesh are short. Occurs in a range of unfertilized grassland mostly with *Hieracium pilosella* – and is one of the waxcaps with less discriminating habitat requirements.

Hygrocybe helobia ▷▷ is typically more bright red, has long cells in the gill flesh, more decurrent gills and smells like garlic. *Hygrocybe calciphila* ▷▷ is found in more calcareous environments and has almost spherical to broadly ellipsoid spores without constrictions.

Widespread and common; mainly July–December.

Hygrocybe intermedia is a rather large, bright orange, dry and heavily felty-scaly, somewhat conical waxcap. The stem is longitudinally fibrillose and golden. The flesh is fragile. The odour is not distinct. Has rather long cells in the gill flesh and spores that measure 8–10 × 5–6.5 µm. Occurs in unfertilized grasslands.

Hygrocybe splendidissima ▷▷ and *H. punicea* ▷▷ have smooth cap surfaces and the former also a smooth stem surface. Both tend to occur later in the season.

Widespread but rare in most areas; June–October, mostly found early in the season.

Jan Vesterholt/Jacob Heilmann-Clausen

Jan Vesterholt

Hygrocybe reidii is a rather small to medium-sized, completely dry and smooth, ± orange to orange-red waxcap that during decay emits a characteristic honey-like smell. The gills are rather pale and adnate to somewhat decurrent. The spores are only rarely constricted and measure (6–)6.5–8.5(–10) × 4–5(–7.5) µm. Occurs in unfertilized grasslands, mostly species-rich sites with long continuity.

Hygrocybe miniata ◁◁ has a finely scaly cap surface, no smell and the spores are constricted, shaped like a violin case.

Widespread and fairly common; mainly June–November.

Hygrocybe mucronella is a small, initially sticky, but quickly dry, orange-red waxcap with a distinctly bitter taste. The cap margin is mostly paler than the centre, and when drying out the surface becomes finely felty. The gills are adnate to somewhat decurrent. The spores may be almost triangular and measure (6–)7–8.5(–10.5) × (4–)4.5–6(–7) µm. Occurs in unfertilized grasslands, mostly moist places at the borders of bogs and fens, but also in scrub.

Hygrocybe insipida ▽ is similar, but lacks the bitter taste and the spores are different.

Widespread but rather local; mainly August–January.

Hygrocybe insipida is a small, orange-yellow to orange-red waxcap with a sticky cap surface, but a moist, non-sticky stem. The gills are yellow to orange and adnate to somewhat decurrent. Smell insignificant. Has small spores, (5–)6–7.5 × 3–4(–5) µm and short cells in the flesh. Occurs in unfertilized grasslands or on moist clay soils in forests.

Hygrocybe mucronella △ is similar, but tastes bitter. *Hygrocybe glutinipes* ◁◁ has a thick slime layer on the cap and stem. *Hygrocybe ceracea* ◁◁ has a dry stem.

Widespread and rather common; mainly July–November.

Hygrocybe coccinea is a rather large, somewhat sticky, bright red, convex or almost bell-shaped/umbonate, finely rugose waxcap. The cap margin can be yellow. The gills are thick and distant, reddish at the base and ± emarginate-decurrent. Does not have a distinct smell. The spores are rather variable in shape, but normally not constricted, (6.5–)7.5–10(–13) × (4–)4.5–5.5(–7) µm; with short cells in the gill flesh. Occurs in unfertilized grasslands, more rarely in forests. This species has been shown to be an endophyte in *Plantago lanceolata*, including the seeds.

Hygrocybe punicea ▷▷ has a distinctly fibrillose stem and *H. splendidissima* ▷▷ has adnexed gills and a honey-like smell during the decay process.

Widespread and rather common; mainly July–December.

Hygrocybe phaeococcinea is a rather small, bright red, rarely yellow, dry waxcap that with age develops a blackish covering on the cap. The gills are adnate to somewhat decurrent and rather variable in colour, but mostly ± salmon. The spores measure (6–)7.5–10(–12.5) × (4–)4.5–5.5(–7.5) µm. On moist, unfertilized sites such as dune slacks or even on lake sediments in dry summers in association with *Littorella* (*Plantago*) *uniflora*.

Hygrocybe constrictospora ⋉ is similar and often occurs in the same type of habitat, but it does not develop a black cap surface.

Widespread but rather local; mainly July–November.

David Boertmann

Hygrocybe splendidissima is a rather large waxcap with a bright red, dry, mostly bell-shaped to conical cap, a shining, longitudinally striate stem, and adnexed and emarginate gills. Has a honey-like smell during the decay process. The spores measure (6.5–) 7.5–9(–10) × (4–)4.5–5.5(–6) µm, and are not constricted; the cells of the gill flesh are short. Occurs only in species-rich, unfertilized grasslands with other rare and discriminating grassland fungi.

Hygrocybe coccinea ◁◁ is more fragile, has a completely smooth stem, a sticky cap and broadly adnate gills. *Hygrocybe punicea* ▷ is large with a coarsely fibrillose stem. *Hygrocybe intermedia* ◁◁ has a felty-scaly cap and a coarsely fibrillose stem. All three lack the smell of honey.

Widespread, but declining due to the loss of its pristine habitat; mainly July–December.

H. punicea *H. intermedia.* Photo Knud Knudsen *H. splendidissima.* Photo Knud Knudsen *H. coccinea*

Hygrocybe punicea is a large, fleshy waxcap with a brownish-orange to bright red, sticky, pallescent cap, a coarsely fibrillose, thick-set stem and emarginate, yellowish to reddish, thick gills. The stem is typically tapering and yellow near the base. Smell insignificant. The spores measure (7–)8.5–10(–11) × (4–)4.5–5.5(–7) µm and are mostly constricted; the cells of the gill flesh are short. Occurs in unfertilized, species-rich grassland and in scrub. A good indicator of high conservation value grasslands.

Hygrocybe spendidissima ◁ is similar, but has a dry, less pallescent cap, a shiny, striate stem and a distinct honey-like smell when kept in a box for a while. *Hygrocybe coccinea* ◁◁ is more fragile and has a completely smooth stem and broadly adnate gills. *Hygrocybe intermedia* ◁◁ has a felty-scaly cap.

Widespread, but in decline; mainly August–December.

Thomas Læssøe

† ***Hygrocybe conica* s.l.** is a very variable, conical, blackening, yellow to red waxcap. The cap and stem surfaces are dry and somewhat fibrillose. When handled, with age or after frost it stains completely black. The spores measure 8.5–10(–12) × 5–6.5(–8) µm and the gill flesh consists of more than 1 mm long cells. Occurs in unfertilized grasslands, but also along gravel roads, in quarries, in lawns, in dune slacks and in thickets.

The species is known to include scores of 'phylospecies', but at present it is not possible to unravel this variation. A tiny form from swampy places is known as *H. conica* var. *conicopalustris* ⋉; a very large and robust form is sometimes called *H. conica* var. *pseudoconica* ⋉; and a yellow form, *H. conica* var. *chloroides* ◁, also occurs. The closely related *H. conicoides* ▷, with reddish gills and slightly different spores, occurs in dunes.

Widespread and common; mainly June–December.

† ***Hygrocybe conicoides*** is a waxcap closely related to *H. conica* s.l. with orange to reddish gills (at least basally) and ± cylindrical spores at 10–12.5 × 5–6 µm (Q above 2). Like other species in this complex it stains black. Occurs in dunes.

Other *H. conica* s.l. forms ◁ from dunes have either other colours or wider spores (Q below 2), but the taxonomy of this group is still unresolved.

Widespread in dune areas; mainly June–November.

Jan Vesterholt

Hygrocybe spadicea is a rather large, very distinctive, somewhat conical waxcap with yellow gills, yellow stem and a grey-brown to dark brown cap. The cap surface is radially fibrillose, and the stem is likewise obviously fibrillose. The flesh is not blackening. The smooth spores measure 9–12 × 5–7 µm. Occurs in unfertilized grasslands.

A form with white gills is known as *H. spadicea* var. *albifolia*. This species is unlikely to be misidentified.

Widespread, but rare and seemingly absent from the boreal zone; mainly August–October.

Thomas Læssøe

Jan Vesterholt

The large brown to nearly white hygrocyboids in the genera **Neohygrocybe** and **Cuphophyllus** mostly have greasy caps and greyish to white gills.

Neohygrocybe ovina is a sturdy, thick-fleshed, dark reddish-grey, dry waxcap, which reddens in the flesh. The cap is at first smooth, but may with age crack into scales. It smells rather unpleasant, ± like *Tricholoma stiparophyllum* (page 270) or faintly nitrous. The gills are somewhat paler, distant, wide, thick and adnexed. The stem becomes ± fibrillose with age. The spores measure 9–12 × 5–7 µm. Occurs in dry grasslands of high conservation value – typically with other hygrocyboids *etc.*

Neohygrocybe ingrata ▽ is a browner, somewhat smaller fungus with an evident nitrous smell, but also with reddening flesh. *Pseudotricholoma metapodium* (page 253) has more crowded gills and smaller, amyloid spores.

Rare and declining due to eutrofication, urbanization *etc.*; mainly July–November.

Rasmus Ejrnæs/David Boertmann

Neohygrocybe ingrata is a rather large, pale yellow-brown, ± reddening, dry, smooth to somewhat scaly waxcap with a distinct nitrous smell. The flesh reddens rather slowly and the gills are almost white but also redden after bruising. The spores measure (6–)7–9(–10) × (4–)5–6.5 (–7) µm. Occurs in very species-rich, unfertilized grasslands.

Neohygrocybe ovina △ is a more robust, darker species with hardly any nitrous smell. *Neohygrocybe nitrata* ▷ is greyer and does not redden.

Rare and threatened; mainly July–November).

This species is on the IUCN red list.

Neohygrocybe nitrata is a rather large, grey-brown, dry waxcap that becomes somewhat scaly with age. It has a distinct nitrous smell. The broad gills are somewhat emarginate and do not redden. The spores measure (7–)8–9(–10) × (4–)4.5–5.5(–6) µm. Occurs in unfertilized grasslands.

Species of *Tricholoma* (page 226) typically have thinner, more crowded gills and do not smell nitrous. *Neohygrocybe ingrata* ◁ is pale yellow-brown and has reddening flesh and gills.

Widespread and fairly common; mainly June–November.

Cuphophyllus flavipes is a greyish to almost blue-grey, sometimes marginally striate, greasy waxcap with distant, pale grey, decurrent gills. The stem is typically somewhat yellowish near the base. The spores measure 6–7.5 × 5–6 µm and have an average Q value of 1.1–1.3. Occurs in unfertilized grasslands, typically those of high conservation value.

Cuphophyllus lacmus ⋉ has a smooth stem but lacks yellow.

Widespread but rather rare; mainly July–November.

Cuphophyllus fornicatus is a rather tricholomatoid, ± conical, whitish to greyish, slightly reddening waxcap. The cap is finely innately fibrillose to fibrillose-scaly toward the centre. Has ± emarginate, pale gills. The odour is ± unpleasant, similar to that of *Tricholoma stiparophyllum* (page 270). The spores measure (6–)7–8.5(–9.5) × (4–)4.5–5.5(–6) µm. Occurs in unfertilized, ± calcareous grasslands or in scrub.

Grey species of *Tricholoma* (page 26) and *Pseudoporpoloma pes-caprae* (page 253) mostly smell farinaceous or have veils.

Widespread but uncommon; mainly August–November.

Thomas Kehlet

• **Cuphophyllus pratensis** is a rather compact and fleshy, pale brownish-orange, almost dry and smooth waxcap with orange-tinted, decurrent, distant and thick gills. The stem is dry, firm and almost white. Does not have a distinctive smell or taste. The smooth spores measure (5–)5.5–6.5(–7.5) × (3.5–)4–5(–5.5) µm. Occurs in unfertilized grasslands and in scrub. It does not have especially excating habitat requirtements and is thus not a very good indicator species.

Cuphophyllus hygrocyboides ⋉ is similar, but is smaller and hygrophanous. The pale cream *C. berkeleyi* ⋉ is also similar, but ± cream coloured. The almost dry mycorrhizal *Hygrophorus nemoreus* ▷▷ has a floccose stem apex.

Widespread and common; mainly August–January.

Jan Vesterholt

Cuphophyllus colemannianus is a rich red-brown to orange-brown, somewhat greasy, more or less marginally striate, convex-umbonate waxcap with distant, somewhat brown-tinted decurrent gills and a dry, firm and white stem. The spores measure (6–)7.5–9(–10) × (4–)5–6(–7) µm. Occurs on calcareous, unfertilized grasslands and in calcareous thickets.

Cuphophyllus fuscescens ⋉ has paler colours with a browner cap centre.

Rather rare and local; mostly September–November.

Cuphophyllus virgineus is a rather variable (size-wise), hygrophanous, whitish to pale greyish-cream, ± greasy waxcap with white, distant, decurrent gills. It does not have a distinctive smell, but infected specimens may smell like coconut cookies. The stem is firm and white, but an infection often causes a reddish stain in the base. The spores measure (6–)7–8.5(–9) × (4–)4.5–5(–6) µm. Occurs in unfertilized grasslands, including old lawns – but not particularly discriminating and therefore not a good indicator of grasslands of high conservation value.

There is a complex of difficult species around *C. virgineus*. *Cuphophyllus fuscescens* ⋉ is recognized by a brown cap centre; *C. ochraceopallida* ⋉ by the more yellowish colours. *Cuphophyllus russocoriaceus* ▽ is also more yellowish and smells of cedar wood (a smell that some people unfortunately cannot sense).

Widespread and common; mostly August–January.

Cuphophyllus russocoriaceus is a medium–large, sordid yellowish-white, greasy waxcap with rather pale, decurrent gills and a distinct smell of cedar wood. The spores measure (6–)7.5–9(–10) × 4.5–5.5 (–6) µm. Occurs in unfertilized grassland, but is not very discriminating.

This species is hard to identify (and particularly for the minority of people who cannot sense the smell). *Cuphophyllus ochraceopallidus* ⋉ from calcareous sites is very similar, but lacks the smell.

Widespread and fairly common; mainly September–December, with a late peak.

Hodophilus species are small, omphalinoid hygrocyboids with thick gills and a cap cuticle that ± consists of clavate cells. They are mostly found on clay soils in thickets, but may also occur in grasslands.

Hodophilus foetens is a foetid (smelling of napthalene), grey-brown *Hodophilus*, which mostly has scattered, small, dark scales on the stem. Has deeply decurrent, at first pale but later darker gills (with 18–30 entire gills). With age the fruitbodies discolour to almost black. The cap cuticle consists mostly of erect pear-shaped cells, which may cause the surface to crack in an irregular pattern. The spores are smooth and measure (4.5–)5.1–5.9(–6.6) × (3.8–)4.2–4.7(–5.4) µm (av. 5.5 × 4.5 µm). Occurs mostly on clay-rich soils in thickets, but may also be found in open, unfertilized grasslands.

There are four other European species with the same foetid smell as *H. foetens* s.str. *Hodophilus tenuicystidiatus* ⋉ has no dark squamules on the stem, darker gills and more, narrower caulocystidia; *H. albofloccipes* has yellow tints in the fruitbodies; *H. pallidus* and *H. subfoetens* have more distant gills (10–20 entire gills).

Widespread, but rather rare, to some extent overlooked; July–November.

Hodophilus atropunctus is a typical, but rather pale *Hodophilus* with deeply decurrent gills, and black squamules on the stem. The cap becomes almost white when dry. The odour is not distinct. The nearly globose spores measure 4.5–6 × 3.5–4.5 µm. Occurs mostly on clay-rich soils in thickets, on riverbanks or on steep, forested slopes.

Hodophilus variabilipes ⋉ is very similar, but has darker colours and may have a smooth stipe. Some of the species with foetid odours may also have dark squamules on the stem.

Rare and local; mainly August–November.

Hodophilus micaceus is a typical *Hodophilus* with a yellow, smooth stem and a pale yellow-brown cap. The gills are rather pale and deeply decurrent. The smooth, broadly ellipsoid spores measure (3.5–) 4.6–5.4(–6) × (3.1–)3.7–4.3(–4.7). It grows alone or in clusters, and mainly occurs on clay-rich soils in thorny, often coastal thickets.

At least three other odourless species have ± yellow colours, incl. *H. phaeoxanthus* ⋉. All have less uniformly coloured stems. They have the same ecology and a range of *Hodophilus* species may co-occur.

Rather rare, but quite widespread; August–November.

Camarophyllopsis schulzeri is a somewhat aberrant, rather small to medium–large waxcap. It can be recognized by its orange-brown colours, sometimes with a somewhat olive tint, and the short decurrent, rather thin (for a waxcap), pale gills. The cap cuticle may have partly swollen elements, but mainly comprises regular horizontal hyphae. The spores are almost spherical and measure 4.5–5 × 3.5–5 µm. Occurs both in unfertilized grasslands and in scrub on clay-rich soils.

Camarophyllopsis atrovelutina ⋉ is very similar, but is somewhat darker and has even smaller spores 3.1–4.2 × 2.8–3.8 µm. *Lamelloclavaria petersenii* ⋉ has more oblong spores, (5–)5.2–5.9(–6.2) × (2.5–)2.7–2.9 (–3) µm and a simple epicutis. *Camarophyllopsis schulzeri* may also recall species in the genus *Neohygrocybe* ◁◁, but is typically smaller, more or less odourless and has smaller spores. *Dermoloma* species (page 245) smell strongly farinaceous and have a finely cracking cap cuticle.

Widespread, but uncommon; mainly July–November.

Hygrophorus species are, when immature, surrounded by a slimy universal veil. They mostly have thick, decurrent gills and a floccose stem apex. All are probably ectomycorrhizal.

Hygrophorus eburneus is a white to yellowish, very slimy, somewhat smelly *Hygrophorus* with gills that do not stain upon drying or after KOH treatment. The gills are, as in typical *Hygroporus* species, distinctly decurrent and the stem apex finely floccose. The spores measure (6.5-) 7–9(–9.5) × (4–)4.5–5 µm. On mull-rich soils with *Fagus*.

Hygrophorus discoxanthus ▷ is similar when young, but discolours brown relatively quickly and the gills stain brown during drying and after a drop of KOH is added. *Hygrophorus cossus* ⋉ is also very similar, but associates with *Quercus*.

Widespread and common, but absent from the boreal zone; mainly August–December.

Thomas Læssøe

Hygrophorus piceae is a white, very slimy *Hygrophorus* with cream gills and a rather weak smell. The stem apex is floccose. The spores measure (6–)7–9 × (4–)4.5–5.5 µm. It only forms mycorrhiza with *Picea* on nutrient-rich soils.

This species is very similar to a number of other white *Hygrophorus* species, and when the mycorrhizal partner is in doubt it can difficult to identify with certainty. The cream gills are a good indicator.

Common in the hemiboreal–boreal zone, but very rare in, or absent from, the nemoreal zone; mainly August–November.

Hygrophorus discoxanthus is a white, yellowish to yellow-brown, aromatic smelling *Hygrophorus*, that aquires brown gills when dry and with KOH treatment. The spores measure 7–9 × 5–6 µm. Occurs with *Fagus* on ± mull-rich soils.

Hygrophorus eburneus ◁ is more or less inert to KOH treatment (stains yellowish). *Hygrophorus cossus* ⋉ is strictly associated with *Quercus* and recalls *H. eburneus* more than it does *H. discoxanthus*.

Widespread and common but absent from the boreal zone; mainly July–December.

Hygrophorus chrysodon is a medium-sized, yellow-white, odourless *Hygrophorus*, which, when young and fresh, has very obvious, bright yellow floccules, and drops at the cap margin and on the stem apex. It turns brown by drying, but not with KOH. The spores measure 7–9(–10) × 3.5–4.5(–5) µm. Occurs mostly with *Fagus* on nutrient-rich soils. Other partners may indicate the presence of closely related, as yet unnamed species.

Hygrophorus discoxanthus △ may also display yellowish colours, but not as fine floccules, smells strongly and stains brown with KOH.

Widespread but local, absent from the boreal zone; August–November.

• ***Hygrophorus penarius*** is a sturdy, short-stemmed, almost white to cream, dry to almost dry *Hygrophorus*. The gills are distant, cream and only somewhat decurrent. The odour is weak, rather sweetish. The spores measure 6–9 × 4–5 µm, Q=1.8. Occurs with *Fagus*.

Hygrophorus penarioides ⋉ is a sister species, which is associated with *Quercus*. It is even more robust than *H. penarius* and has spores with a lower Q value at 1.4–1.7.

Widespread and locally common, absent from the boreal zone; August–November.

Jens Maarbjerg

Hygrophorus erubescens is a large, mostly long-stemmed, pale, bitter-tasting and yellowing *Hygrophorus* with conspicuous red spots and a more reddish cap centre. The stem is floccose-scaly and whitish with a red tint, and the gills are similar in coloration. The spores measure 8–11 × 4–5(–6) µm. It is associated with moist, mossy coniferous forest and form mycorrhiza with *Picea abies*.

The rather similar *Hygrophorus russula* ▽ is mycorrhizal with *Quercus*, while the conifer-associated, but mild-tasting *H. purpurascens* ⋉ and *H. persicolor* ⋉ do not stain yellow. *Hygrophorus purpurascens* has a distinctly white veil. All four species are somewhat preferential to calcareous soils.

Common in hemiboreal–boreal *Picea* forests, but rare in the nemoreal zone; mainly July–October.

Jan Vesterholt

Hygrophorus russula has a sturdy, robust appearance and is almost completely dry and mild-tasting. It has conspicuous wine-red colours and a pale margin; the gills are only slightly decurrent. The spores measure 7–8(–9) × 4–5(–6) µm. It is associated with *Quercus* on calcareous soils and is possibly also mycorrhizal with other deciduous trees farther south in Europe.

The occurence with deciduous trees and the large, almost dry fruitbodies should ensure an accurate identification.

Rare in Scandinavia, but rather common farther south in Europe; mainly August–December.

Hygrophorus hypothejus is a late-occuring, very slimy *Hygrophorus* with ± yellow, distinctly decurrent gills, a grey-brown to olive-brown cap, which is darkest at the centre, and a very slimy, ± yellow to olive-brown tinted stem. The form *H. hypothejus* f. *aureus* ⋉ has golden colours on the cap, rather than the olive-brown colours of f. *hypothejus*. The spores measure 7–9 × 4–5 µm. This species is mycorrhizal with *Pinus* and occurs mostly on poor, sandy soils partially covered in lichens.

The dark form (f. *hypothejus*) is distinctive and unlikely to be misidentified. The orange-yellow form (f. *aureus*) may perhaps recall *H. lucorum* ▽ although that species is mycorrhizal with *Larix*.

Widespread and fairly common; mainly October–January, after the first frosts.

Hygrophorus lucorum is a ± pale yellow, very slimy *Hygrophorus* with decurrent and more sordid yellow gills. The spores measure 7–10 × 4–6 µm. It forms mycorrhiza with *Larix* on rich soil.

Hygrophorus lucorum var. *speciosus* ⋉ is separated as a reddish variety. The golden form of *H. hypothejus*, f. *aureus* ⋉, is more orange and grows with *Pinus*.

Mostly found in the southern parts of temperate Europe, very rare towards the north and absent from parts of Scandinavia; mainly September–January.

Karel Tejkal

Claes Ingvert

Hygrophorus poetarum is a very large and fleshy, pinkish-buff to pale apricot-orange, almost dry, fruit-smelling *Hygrophorus*, which has an inrolled cap margin when young and only shortly decurrent gills. The taste is mild. The spores measure (6–)7–8 × 4.5–5.5 µm. The species is mycorrhizal with *Fagus* on clay-rich, ± calcareous soils. It occurs typically in sites with numerous other rare ectomycorrhizal fungi.

Hygrophorus penarius ◁◁ is smaller and paler and lacks the fruity smell, which some regard as nauseous. *Hygrophorus poetarum* may also recall some species of *Tricholoma* (page 258) but normally has emarginate gills.

A widespread, but southern and rather rare species; mostly September–October.

Jan Vesterholt

Hygrophorus nemoreus is a ± completely dry, radially felty, farinaceous, mild-tasting *Hygrophorus* with cinnamon-buff to yellow-brown colours and decurrent gills tinted the colour of the cap. The spores measure 6–8 × 3.5–5 µm. It mostly occurs on somewhat sandy soils with *Quercus*.

Cuphophyllus pratensis ◁◁ has a ± smooth and greasy cap surface, no farinaceous smell and a different hyphal structure in the gills. *Hygrophorus arbustivus* ⋉ is a more fragile, white-margined, somewhat reddish, *Quercus*-associated species, which probably prefers richer soils.

Local and rather rare; mostly August–October.

Hygrophorus unicolor is a very slimy *Hygrophorus*, which, when young and seen from above almost recalls a freshly fried egg. When dry the contrast between centre and margin is less pronounced and the colours less shiny. Lacks a distinct smell. The spores measure 6.5–9 × 4–5 µm. It forms mycorrhiza with *Fagus*, mostly on calcareous soils.

Two other *Hygrophorus* species are very similar, but have different partners: *H. lindtneri* ▽ is accociated with *Corylus* and *H. carpini* ⋉ with *Carpinus*.

Widespread, but absent from the boreal zone; August–December.

Jan Vesterholt

Hygrophorus lindtneri is a pale orange-brown, very slimy *Hygrophorus* with a darker centre. The stem is ± floccose near the apex and tinted the same colour as the cap. Lacks a distinct smell or taste. The ellipsoid spores measure 7.5–9.5 × 5–6(–6.5). On rich, ± calcareous soils, where it forms ectomycorrhiza with *Corylus*.

Hygrophorus unicolor △ is a very similar but *Fagus*-associated species. *Hygrophorus carpini* ⋉ is also similar but associated with *Carpinus*.

Mostly found in southern and eastern parts of temperate Europe; autumn.

Michal Mikšík

Hygrophorus discoideus is a rather small *Hygrophorus* with an orange-brown cap that has a paler margin. The stem is pale and floccose above and has a very indistinct ring zone. The gills are only slightly decurrent and tinted the same colour as the cap. There is no distinct smell or taste. The spores measure 6–9 × 3–5 µm. Calciphilous and *Picea*-associated.

When the mycorrhizal partner is known, this species is unlikely to be misidentified: it is darker than the above hardwood-associated species and has a brown cap centre.

Mostly found in the hemiboreal–boreal zones; August–December.

Martin Kříž

Hygrophorus agathosmus is a rather large, not very slimy, grey, rarely white *Hygrophorus* with a strong smell of almond soap. The stem and gills are paler than the cap, and the stem is strongly floccose towards the apex. The spores measure 8–11 × (4.5–)5–6 µm. It is associated with *Picea*, mostly on rich soil, and occurs mostly in large troops.

Hygrophorus pustulatus ⋉ is somewhat smaller, has obvious dark dots on the stem and is odourless. *Hygrophorus exiguus* ⋉ is very small and also odourless. It occurs with *Tricholoma inamoenum* (page 271). *Hygrophorus secretanii* ⋉, which is absent from the nemoreal zone, has a similar smell, but is more yellow to reddish on the cap. Grey species of *Tricholoma* (page 264) are completely dry and have a different smell.

Widespread and common; mainly August–December.

Hygrophorus mesotephrus is a medium-sized, slimy, brown-grey, ± odourless *Hygrophorus*, which typically has a two-coloured cap, a pale to greyish stem without conspicuous belts and ± white gills. The spores measure 8.5–12 × 5.5–7.5(–8.5) µm. It is mycorrhizal with *Fagus* on rich soils, mostly in sites with many species of *Cortinarius* subg. *Phlegmacium* and other mycorrhizals. Farther south is it also reported with *Quercus*.

If found with the correct hosts, typical material is unlikely to be misidentified.

Widespread, but uncommon in many regions (with poor soils) to locally common; August–December.

Hygrophorus pustulatus is a rather slender to medium-sized, grey, somewhat slimy *Hygrophorus* with distinct dark dots on the stem and without a distinct smell. The cap is darkest at the centre and the margin appears somewhat crenulate/dentate. The gills are rather pale and decurrent. The spores measure 7–10 × 4.5–6 µm. It forms mycorrhiza with *Picea*, mostly on rich soils.

Hygrophorus agathosmus ◁ has a white-flocculose stem, more slime on the cap and a strong smell of almond soap. *Hygrophorus suaveolens* is very similar, but is mycorrhizal with *Pinus*.

Widespread and common; mainly September–January.

Hygrophorus latitabundus is a large and robust, grey-brown to olive-brown, very slimy *Hygrophorus*. The cap margin remains inrolled for a long time. The stem has greyish veil zones. The odour is ± sweetish and the taste mild. The spores measure (8–)9–12 × 5–7(–8) µm. It is mycorrhizal with *Pinus* on calcareous soils.

Hygrophorus persoonii ▷▷ is slightly smaller and is associated with *Quercus*.

Rare in Scandinavia, but widespread and more common farther south in Europe; autumn.

Jean-Marc Moingeon

Hygrophorus olivaceoalbus is a rather dark, grey-brown to grey-olive, very slimy *Hygrophorus*. The stem is mostly tinted the same colour as the cap and may be distinctly belted in olive-brown colours, or can be also almost unicoloured. The gills are white. The spores measure 10–16 × 7–8.5 µm. It forms mycorrhiza with *Picea*, probably mostly on rich soils.

Hygrophorus korhonenii ⋉ is somewhat more robust and darker and has smaller spores, 10–14 × 5–7.5 µm. It is mainly found in the boreal zone. *Hygrophorus persoonii* ▽ is typically larger and occurs with *Quercus*.

Widespread and fairly common; mainly August–December.

Hygrophorus persoonii is a sturdy, slimy, dark grey-brown to grey-olive *Hygrophorus* with distinct coloured belts on the stipe below a white, floccose apex. The gills are white. KOH stains the cap surface turquoise. The spores measure 9–12 × 5–7 µm. It forms mycorrhiza with *Quercus* on mostly clay-rich soils.

Hygrophorus korhonenii ⋉ looks similar, but is *Picea*-associated and does not stain turquoise with KOH.

A rather local, but widespread species, except in the boreal zone where it is absent; mainly August–November.

Hygrophorus camarophyllus is a dark brown-grey, rather dry, fleshy, mostly somewhat funnel-shaped *Hygrophorus*, with pale, thick and distant gills and a somewhat innately fibrillose cap cuticle. The spores measure 7–9 × 4–5 µm. It forms mycorrhiza with *Picea* on poor, acidic soils, *e.g.* *Vaccinium*-dominated forests.

Ampulloclitocybe clavipes (page 116) has more crowded gills, an obviously swollen stem base and a softer texture.

Quite common in the hemiboreal–boreal zone, but nearly absent from the nemoreal zone; mainly August–November.

Thomas Læssøe

Hygrophorus atramentosus is a sturdy, dark, grey-black to greyish blue, at first somewhat slimy but later dry *Hygrophorus* with radiating, innate fibres in the cap cuticle. The gills are white and the stem grey. It lacks a distinctive odour. The spores measure 8–9 × 4–5 µm. It forms mycorrhiza with *Picea*, typically on calcareous soil.

Hygrophorus camarophyllus △ lacks the greyish shades and occurs on more nutrient-poor soils.

Tricholoma portentosum (page 268) has a yellowish shade on the stem and more crowded gills.

Rather widespread and common in the hemiboreal–boreal zone; autumn.

Jean-Marc Moingeon

Mycenoids

Mycenoids are small, rather fragile or somewhat tough agarics with white spore-deposits. Most species have a bell-shaped or convex cap, but may become applanate or even somewhat funnel-shaped with age. The gill attachment comes in all forms, except completely free. Smell and surface features – dry versus slimy *etc.* – are important characters. In the field it is important to note whether the fruitbodies have a distinctive odour, *e.g.* radish-like, like iodoform or nitrous, and whether parts of the fruitbody are slimy.

The mycenoid fungi often have smooth, amyloid spores. Many species have characteristic cystidia in the hymenium or in other places.

The mycenoid fungi are mainly decomposers and often occur on leaf or needle debris, on dead wood or on the bark of living trees. A few species (*e.g. Mycena galopus*, page 205) may form mycorrhiza-like associations with ericaceous plants.

Mycena is the most speciose mycenoid genus in temperate Europe, and the next is *Hemimycena*. The latter has inamyloid spores and the fruitbodies are small or very small and whitish; some are without gills and form a link to the cyphelloid fungi (page 1076).

OTHER SIMILAR FUNGI:
– omphalinoids and hygrocyboids with decurrent gills may look rather similar (pages 128 & 142).
– collybioids and marasmioids may also look similar, but these are mostly somewhat tougher or can revive after desiccation (pages 274 & 302).
– similar entolomatoids have rose-coloured spore-deposits (page 448).
– similar *Panaeolus* species have blackish spore-deposits (page 554).
– similar little brown mushrooms (LBMs) have grey-brown to red-brown spore-deposits (page 616).

FURTHER READING: 12, 15, 104, 156, 183, 206, 272, 273.

Micro-drawings: spores and cystidia.

Approximate species count applies to temperate Europe.

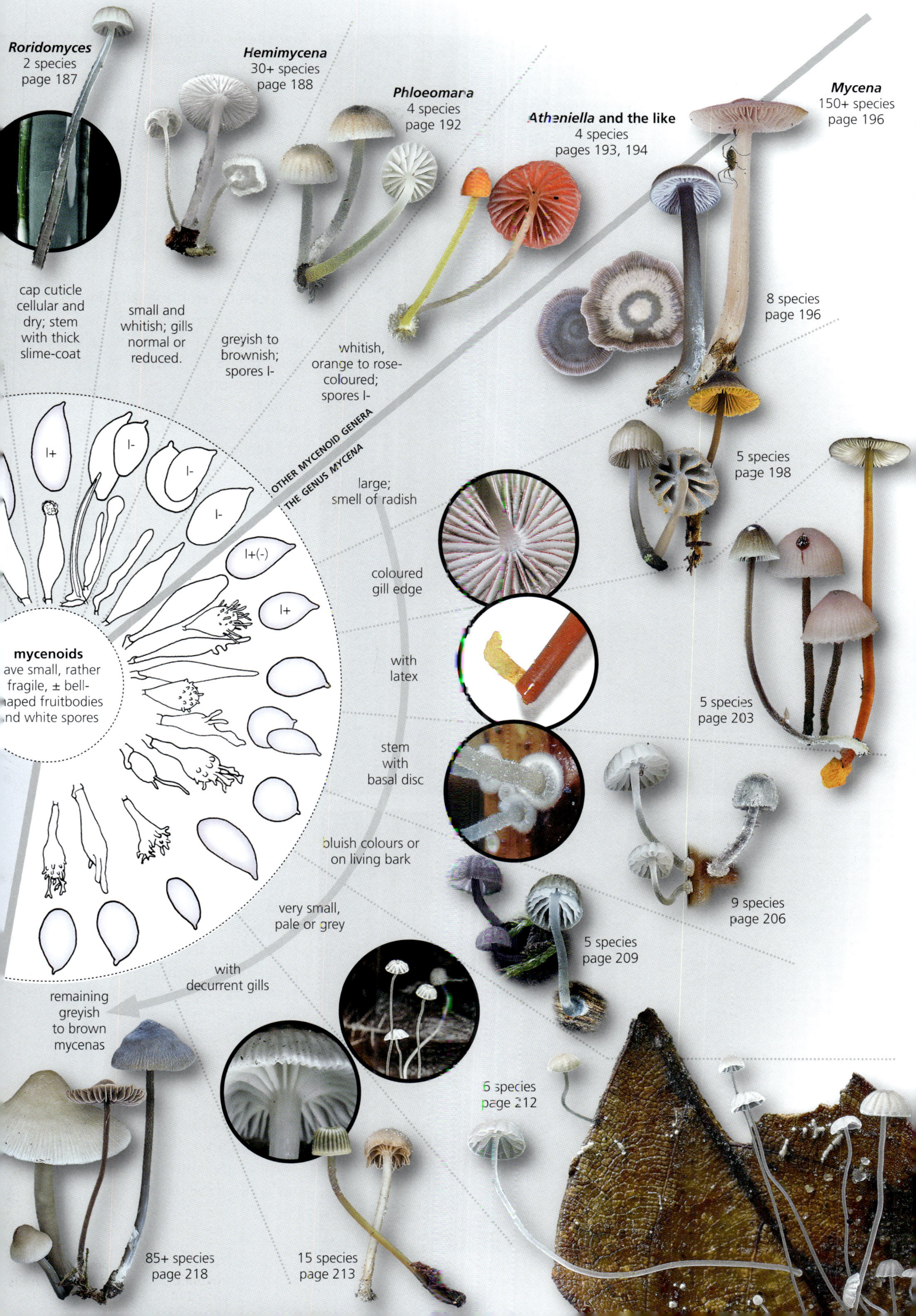

Roridomyces
2 species
page 187
Hemimycena
30+ species
page 188
Phloeomana
4 species
page 192
Atheniella and the like
4 species
pages 193, 194
Mycena
150+ species
page 196
cap cuticle cellular and dry; stem with thick slime-coat
small and whitish; gills normal or reduced.
greyish to brownish; spores I-
whitish, orange to rose-coloured; spores I-
8 species
page 196
OTHER MYCENOID GENERA
THE GENUS MYCENA
large; smell of radish
5 species
page 198
coloured gill edge
with latex
5 species
page 203
stem with basal disc
bluish colours or on living bark
9 species
page 206
very small, pale or grey
5 species
page 209
with decurrent gills
remaining greyish to brown mycenas
6 species
page 212
85+ species
page 218
15 species
page 213
mycenoids
ave small, rather fragile, ± bell-
aped fruitbodies
nd white spores
I+
I-
I-
I-
I+(-)
I+

Thomas Læssøe

Mycenella is a mycenoid genus with ± pruinose cap surface and a highly pruinose-downy stem. All species also have conspicuous cystidia on the gill sides and inamyloid, smooth or nodulose spores. They require microscopy for safe identification.

Mycenella lasiosperma is a rather anonymous, grey, striate mycenoid agaric with a very distinct pruinose-downy stem. It has unique cheilocystidia with a small thin-fingered 'brush' at the top. The inamyloid, spherical, verrucose spores measure 6–8 × 5.5–8 µm. Mostly found in moist, rich places, *e.g.* in *Carex* tussocks, but may appear in a broad range of habitats, including on very decomposed wood.

Here, *M. margaritispora* is considered a synonym of *M. lasiosperma*, but there is no consensus on this issue. Other species of *Mycenella* do not have cystidia with a 'brush' at the top, and species of *Mycena* do not have verrucose spores. *Mycena leptocephala* (s.l.) ▷▷ can be almost as downy on the stem, but has a nitrous smell.

Widespread and probably fairly common in most of temperate Europe, but is clearly under-recorded; mainly July–November.

Jan Vesterholt

Mycenella trachyspora is a grey-brown to brown, almost non-striate agaric with a distinctly pruinose-floccose stem and pale, adnexed gills. The hymenial cystidia are large, widened basally and tapering upwards. The spores are inamyoid and spherical, verrucose-nodulose, and measure 4.5–6 × 4.5–5.5 µm. Dry material has fine red dots on the gills. On calcareous soil, mostly with a scattered moss cover.

Mycenella bryophila ▷ is very similar, but the necks of the cystidia are more cylindrical and it lacks red dots when dry.

Widespread but rather rare; mainly July–November.

Mycenella bryophila is a rather small mycenoid agaric with floccose-pruinose surfaces and dull grey-brown colours but whitish, adnexed gills. The somewhat thick-walled hymenial cystidia are large and have a long, ± cylindrical neck. Dry material lacks red dots on the gills. The inamyloid, almost spherical spores measure 6–9.5 × 5–8 µm and have a rather coarse, rounded ornament. On calcareous, ± moss-covered soil in deciduous forests.

Mycenella trachyspora ◁ is very similar, but has more tapering cystidia, and dry material is typically dotted with red on the gills.

Widespread but rare; mainly August–November.

Thomas Læssøe

Mycenella salicina is a rather robust, dark-capped mycenoid agaric with paler gills and a very pruinose-floccose, greyish stem. The slightly thick-walled, inamyloid, spherical to almost triangular spores measure 5–7 × 5–6.5 µm and lack warts. The cystidia are large, lageniform and lack protuberances at the tips. On ± calcareous soil, in both rather open and dense, thorny thickets, mostly in very dry places such as south-facing coastal slopes.

Other species of *Mycenella* have verrucose spores. Species of *Dermoloma* (page 245) have a strong, farinaceous smell and lack cystidia.

Rather rare, but probably under-recorded; mainly August–November.

Thomas Læssøe

Gamundia striatula is a white-spored, very variable mycenoid agaric with translucently striate, hygrophanous, ± felty, slimy to dry cap and pinkish-buff, grey-brown to red-brown colours. The cream to buff gills are rather distant, arched and wide with varying attachment. The stem is dry, shiny and mostly turns darker towards the base; the apex is minutely floccose. The cheilocystidia are mixed with basidia, somewhat sinuous and 35–100 µm long. The spores are ellipsoid, finely echinulate, inamyloid and measure 6–8.5 × 4–6 µm. Occurs on rather decomposed litter, not least old *Peltigera* thalli, but has a broad habitat range and may be found both in forests, parks and in open heathland.

A very anonymous agaric that easily is overlooked. A safe identification requires microscopy. In the past, two species were recognized, *G. striatula* and *G. leucophylla*, but these are now considered to be synonyms. Similar clitocyboids (page 98) do not have cheilocystidia. Similar small *Lyophyllum* s.l. (page 287) normally also lack cystidia.

Widespread and probably fairly common in most of temperate Europe; mainly September–January, but sometimes in spring.

Jan Vesterholt

Fayodia is a mycenoid genus with very distinctive spherical, uniquely double-walled spores, the inner with an echinulate ornament. The outer wall is amyloid. Identification requires microscopy.

Fayodia bisphaerigera is a greyish to pale red-brown, striate, white-spored mycenoid agaric with slightly decurrent gills. Has rather large, cylindrical to clavate cystidia; with clamps. Basidia 2-spored. The spores are spherical, measure 8–12 µm and have an 'inner ornament' and an outer, amyloid, smooth wall. Amongst moss and litter in damp forests or in more open habitats.

Fayodia anthracobia ▽ is somewhat darker with rather smaller spores and occurs on burnt or peaty soils. It lacks clamps, but is also 2-spored.

Rarely recorded, but probably somewhat overlooked; mainly August–November.

Fayodia anthracobia is a ± dark brown, pale-spored mycenoid agaric with paler gills. With large, ± clavate cystidia, no clamps and 2-spored basidia. Spores spherical, 6–8.5 µm, with an inner, echinulate ornament and an outer, smooth, amyloid wall. Occurs on burnt soils or on peaty soils.

Fayodia bisphaerigera △ is also 2-spored, but has clamps and is a paler, more striate fungus. *Fayodia anthracobia* is devided in two varieties: one on burnt soils and with an applanate cap (var. *anthracobia*) and an umbonate variety on peat soil, *F. anthracobia* var. *bisphaerigerella* ⋉.

Rare or overlooked; mainly October–November.

Erik Arnfred Thomsen

Hydropus is a heterogeneous assemblage of mycenoid, collybioid or omphalinoid, white-spored agarics, mostly with large pleurocystidia and some with amyloid, smooth spores.

Hydropus subalpinus is a mycenoid-collybioid *Hydropus*, with wide, arched, adnexed, white gills that have an irregular, undulating margin. Has a silk-like, radially fibrillose, ± cinnamon-buff cap and a pale, ± rooting stem. With large pleuro- and cheilocystidia. The allantoid, inamyloid spores measure 6.5–10 × 2.5–4.5 µm. Occurs on buried wood in rich deciduous forests and parks, mostly with *Fagus*.

The allantoid spores are a unique feature. *Hymenopellis radicata* (page 299) is larger and has a straight, often dark gill edge and spherical spores. Also recalls an elegant version of *Megacollybia platyphylla* (page 238), but this has thick hyphal cords and ± globose spores.

Widespread in the nemoral zone, except in more Atlantic regions; appears early in the season, April–January.

Hydropus marginella is a medium-sized, greyish *Hydropus* with ± decurrent gills that have a grey-black edge. The cheilocystidia are large, 50–95 µm long. The spores are smooth, amyloid and measure 6–8 (–9) × 3.5–5 µm. It occurs on rather decomposed conifer wood, mostly in forests with large amounts of dead wood and long continuity.

Hydropus moserianus ▷▷ is a somewhat smaller agaric with uncoloured gill edges.

Known from *e.g.* Germany, Poland and Finland, but not from other parts of Scandinavia or farther west; mainly July–October.

Hydropus trichoderma is a dark *Hydropus* with a pubescent cap surface and white, adnexed gills. The cap cuticle has conspicuous cystidia with both vacuolar and incrusted brown pigment. The spores are slightly amyloid and measure (7.5–)8.5–10 × (5–)6–7 µm. On rich soils in ± calcareous deciduous or mixed forests.

Hydropus/Mycopan scabripes ⋉ is less pubescent and lacks incrusted pigment in the cap cuticle. *Hydropus floccipes* ▷▷ is more gracile, grows directly on wood, has dark dots on the stem and spherical spores.

A rare fungus in the north, most records from subalpine areas farther south; mainly August–November.

Thomas Læssøe

Hydropus atramentosus is a ± collybioid, blackening, grey-olive *Hydropus* with pale, crowded, adnate gills. Has ± fusiform cheilo-, pileo- and caulocystidia. The almost spherical, amyloid spores measure 3.5–6.5 × 3.5–6 µm. Occurs on dead conifer trunks.

A unique species, which cannot be confused with any other European fungus.

A rare fungus mostly recorded in the eastern parts of temperate Europe; summer–autumn.

Jean-Marc Moingeon

Thomas Læssøe

Hydropus floccipes is an elegant mycenoid *Hygropus* with an almost velvety surface on the brownish cap and distinct but fine, dark dots on the almost translucent stem. The cap is conical to convex and the fruitbodies are rather fragile. It has no significant smell. The hymenial cystidia are ± lageniform, the basidia are 4-spored, and clamps are present. The spores are ± spherical, inamyloid and measure 5–7 µm in diameter. Occurs in moist places with a lot of dead wood or big mossy trunks of deciduous trees, *e.g.* in old *Alnus* carrs but also in old deer parks *etc.*

Similar species of *Mycena* ▷▷ or *Hydropus* do not have spherical, inamyloid spores. *Mycenella salicina* ◁◁ occurs on soil, and the floccules of the stem are not brown. Sometimes co-occurs with *Mycena erubescens* ▷▷ but that species has amyloid spores and a brown-staining stem base.

Very rare in the Nordic countries, but more common in southern and Atlantic parts of temperate Europe; mainly July–October.

Thomas Læssøe

Hydropus moserianus is a small, somewhat hygrocyboid-omphalinoid *Hydropus* with strongly decurrent, white and distant gills and a dry, finely pruinose, greyish, striate cap. With long-necked pleuro- and cheilocystidia and ± ellipsoid, amyloid spores that measure 7.5–11 × 4–6 µm. Occurs on turf in open coastal meadows or in dune slacks.

Similar species of *Mycena* ▷▷ have a dextrinoid reaction in the gill flesh. Omphalinoids and hygrocyboids (pages 128 & 142) do not have amyloid spores.

Apparently very rare, with scattered records from *e.g.* the Netherlands, Germany and Denmark; mainly September–October.

Xeromphalina is a genus of ± yellow-brown mycenoid-marasmioid agarics with darker stems and yellow-brown felt at the base (such is also seen in *Mycena aurantiomarginata* ▷▷ and *M. picta* ▷▷). The spores are amyloid and all have cheilocystidia. The texture is rather tough.The fungus copes well with dry conditions.

Xeromphalina campanella is a lignicolous *Xeromphalina* with a yellow-brown to brownish-orange cap, arched-decurrent gills, wine-reddish stem, and a small apricot-orange clump of felt at the base. Cheilocystidia are clavate. Spores ± drop-shaped and measure 6–8 × 3–4 µm, amyloid, smooth. In large troops on rotten conifer stumps.

Most other *Xeromphalina* species occur on litter in smaller numbers and are typically taller. They also have ± branching cheilocystidia.

Rather common in the hemiboreal-zone and much rarer in the nemoral; fruits almost year-round with most records in the spring.

Xeromphalina cauticinalis is a ± vivid brownish-orange, long-stemmed, bitter-tasting *Xeromphalina* with cystidia along the full length of the stem and with a yellow-brown felt at stem base. The cheilocystidia are gnarled-branched. The spores measure 5–7 × 3–4 µm. Occurs mainly on litter from conifers, perhaps mostly from *Pinus*.

Xeromphalina fraxinophila ⋉ is mild-tasting and occurs typically in deciduous forests. *Xeromphalina cornui* ⋉ should be thinner-stemmed and smoother and should occur in damp habitats.

Widespread and locally quite common, but completely absent from some areas; mainly July–January.

Delicatula integrella is a small and elegant, but rather robust, ± translucent to porcelain-white, finely floccose, odourless mycenoid agaric with a veined hymenophore and ± bell-shaped to applanate, or even umbilicate, cap. Often with a hyaline central 'eyespot'. With a hand lens a fine veil can be detected as triangular protrusions at the the cap margin (inset in lower picture). Very young fruitbodies are pixy-hat-shaped with a veil (lower picture) with a silky veil covering. The stem base is ± bulbous and hairy. The spores are amyloid, smooth, and measure 6.5–9 × 4–5 µm. Occurs mostly in rather damp or wet habitats, such as *Alnus* carrs, often on very rotten, moss-covered wood, but also on turfy soil, and even in old lawns, or coastal *Salix* thickets *etc*.

May recall species of *Hemimycena* or other white mycenas ▷▷, which typically have narrower spores and/or thick-walled hairs, or amyloid spores and no veil.

Widespread and probably rather common in most of temperate Europe; mostly June–October.

Resinomycena saccharifera is a small, white, very glistening, grainy-powdery mycenoid agaric with distant, wide and broadly adnate gills. With cystidia with sticky, widened tops. Spores are amyloid and measure 8.5–14(–16.5) × 4–6(–6.5) µm. Occurs in fens and other boggy or damp areas, mostly on monocots, but also on *e.g. Rubus idaeus*.

The presence of amyloid spores excludes *Hemimycena* ▷▷. Species of *Mycena* ▷▷ do not have sticky cystidia on all surfaces. Species of *Marasmiellus* (page 305) are more tough-fleshed.

Rather common and widespread; mainly May–November.

Roridomyces roridus is a rather small, brown-grey mycenoid agaric with decurrent, rather distant gills that have downy edges, a thick slimecoat on the stem and a dry cap with a palisade of erect, clavate cells. The cap may seem grainy at a high magnification. The cheilocystidia are somewhat bowling pin-shaped. The spores are smooth and amyloid and measure 8–13 × 3.5–6 µm. Often occurs on acidic litter, *e.g.* in dense *Vaccinium myrtillus* stands and in other heathy habitats.

Mycena vulgaris ⊳⊳ has a sticky cap surface, a detachable slime-thread on the gill edges and branched cystidia. *Mycena clavicularis* ⊳⊳ is larger, and has smooth gill edges and a thinner slimecoat on the stem. *Roridomyces appendiculatus* ⋉, described from southern Germany and later reported from Italy, should be recognizable by a thin extension on the pear-shaped cells in the cap cuticle.

Widespread and fairly common in most parts of temperate Europe; mainly July–November.

The mycenoid genus ***Hemimycena*** is an artificial group (polyphyletic) of small, whitish mycenoids. Some species have deeply decurrent gills; in some the gills are reduced or even absent. All parts are inamyloid/indextrinoid and the spores often narrow. Many have prominent caulo- and pileocystidia, some capitate or thick-walled.

Hemimycena tortuosa is a short- to long-stemmed, rather large, odourless *Hemimycena* with rather well-developed, relatively crowded gills. It may occur in a pleurotoid form. It retains many water droplets, which are caught in the helical, capitate caulo- and pileocystidia, and has pointed cheilocystidia. The spores measure 8.5–11 × 2.5–4.4 µm. Occurs mostly on moist bark of deciduous trees, *e.g.* from *Quercus* and *Fagus*.

Hemimycena cephalotricha ⋉ has similar hairs, but completely different, ± capitate cheilocystidia.

Especially prevalent in the Atlantic areas, where it can be common; mainly August–November.

Hemimycena mauretanica is a small *Hemimycena* with long, tapering, thin- to somewhat thick-walled hairs on its stem and a cap with reduced, ± decurrent gills. It is odourless. Basidia 4-spored with clamps. The spores measure 6.5–10 × 2–3.5(–4) – Q approx. 3.2. The species has a broad habitat spectrum, but mostly occurs on moist plant litter.

A number of varieties have been described, some of which may prove to be full species. These may be 2-spored without clamps, have capitate cystidia, be with or without cheilocystidia, or have longer spores than usual.

Widespread and probably common; mainly June–February.

Hemimycena epichloë is a small, almost trumpet-shaped, completely white, odourless *Hemimycena* that lacks or has reduced to almost normal, decurrent gills. With clamps, but lacks thick-walled or pointed cystidia. Cap cuticle with numerous protuberances but with almost smooth stem cortex hyphae. Caulocystidia rather simple and thin-walled. The spores are almost fusiform and measure 8–12 × (2.5–) 3–4.5 µm, with a Q value of 2.7–3.4. Has a preference for decaying grass, sedges *etc.* in moist habitats.

Hemimycena ignobilis ⋉ occurs directly on soil and has a smooth cap cuticle and more pronounced gills. Other similar species have either cheilocystidia and/or pointed, sometimes thick-walled caulo- or pileocystidia.

Probably rather common and widespread, but rarely reported; mainly June–October.

Hemimycena subtilis is a small, white, almost gill-less *Hemimycena* with long, setose hairs on the stem, and with few to numerous similar pileocystida, which arise from a finely digitate cap cuticle. Basidia 2-spored. The spores are narrowly drop-shaped and measure 10–16 × 3.5–5 µm. On plant litter, mostly in moist places.

A number of other species of *Hemimycena* with thick-walled (setose) stem- and cap hairs also occur, but *H. subtilis* is the only 2-spored species with long, thick-walled caulo- and pileocystidia.

Distribution not clear, but probably not rare; mainly May–November.

Hemimycena mairei is a very atypical, grey-brown mycenoid-omphalinoid *Hemimycena* with a white stem and decurrent, pale gills. Cheilo- or caulocystidia are clavate. The inamyloid citriform spores measure 6–8 × 3–5 µm. On rich, often disturbed soil, with a preference for lawns and road verges.

May superficially recall a small *Cuphophyllus* (page 163). Two very similar species have been described: *H. sordida* ⋉ with conspicuous, ± lecythiform cheilocystidia and a more downy stem, and *H. juncinum* with no clamps and 2-spored basidia.

Widespread and locally quite common, but absent from the boreal zone; mainly June–October.

Hemimycena delectabilis is a rather robust *Hemimycena* with well-developed, distant, decurrent gills and a surprisingly strong nitrous smell. Has well-developed long-necked cheilocystidia and spores that measure 7–9 x 4.5–5.5 µm and have a relatively low Q value of 1.7–1.9. Occurs in stands of coarse herbs and in grassland, but may also occur on needle debris and on wood.

As well as the typical 4-spored form, a 2-spored form without clamps can be found. Distinctive and unlikely to be misidentified.

Widespread and probably common; mainly July–December.

Hemimycena cucullata is a large, white to yellowish *Hemimycena* with crowded gills. Cheilo- and pileocystidia short and pointed. Basidia 2-spored. Typically found at soil level, *e.g.* on buried twigs, mostly in deciduous woodlands, but also in conifer forests.

Hemimycena lactea ▷, on thick litter, has more distant gills and small heads on the cheilocystidia; *H. pseudolactea* ⋉ has large pleurocystidia and more distant gills. *Atheniella flavoalba* ▷▷ has more distant gills and spherical elements on the stem cortex hyphae.

Widespread and fairly common, except in the boreal zone; July–November.

Hemimycena lactea is a chalk-white *Hemimycena* with rather crowded, normal gills and minutely downy surfaces. With capitate pileo-, caulo- and cheilocystidia. The capitate caulocystidia are mixed with cystidia with other shapes. It is predominantly 2-spored and the hyphae have clamps. On conifer debris, mostly in moist places or late in the season.

Hemimycena pseudolactea ⋉ has prominent, large pleurocystidia. *Hemimycena cucullata* ◁ has more crowded gills, and tapering, pointed cystidia, while *H. gracilis* ⋉ lacks the capitate cystidia and has more decurrent gills. *Atheniella flavoalba* has different cystidia and globose cells on the stem cortex. A form of *H. lactea* on leaf-litter from *Fagus* produces very large fruitbodies; in the past this was known as '*Mycena echinipes*' and possibly warrants species or varietal status.

Common and widespread; mainly September–January.

Thomas Læssøe

The mycenoid genus ***Phloeomana*** includes a number of small, lignicolous mycenoid agarics with inamyloid spores and simple cystidia.

Phloeomana alba is a small, whitish to pale greyish, dry mycenoid agaric, with ± decurrent gills and no basal disc. Spores ± spherical, inamyloid, and measure (7–)8–9(–11) × 6.5–8 (–9.2) µm. Occurs on high pH bark of deciduous trees, mostly with other mycenas and bryophytes.

Phloeomana hiemalis ⊳⊳ may look similar, but the mature spores are never as spherical. '*Mycena*' *olida* ⋉ is more conical, often yellow-tinted, ± radially fibrillose and mostly occurs in moss at the base of large tree trunks.

Widespread, but local and almost absent from some areas; late autumn.

Phloeomana speirea is a small to rather small, dry, pale grey-brown, rarely darker, inodorous mycenoid agaric with arched, decurrent, rather distant, whitish gills. The stem is very variable in length depending on the position of the fruitbody, and is typically strigose/hairy at the base. The stem base and apex may be yellow and such examples are sometimes referred to as *P. speirea* var. *camptophylla*. The cheilocystidia are cylindrical to lageniform and the gill flesh do not react with Melzer's reagent. The spores are inamyloid and measure 6.5–10 × 4–6 µm with a Q value of 1.6–1.9. Usually occurs in moist habitats on twigs and bark, mostly from deciduous trees, but may also grow on conifer debris. It is a very common species in *Salix* carrs.

Phloeomana hiemalis ▽ is similar, but has adnexed to adnate gills and a Q value of 1.3–1.4, but likewise has inamyloid spores. *Phloeomana clavata* ▷ has brownish gills and incrusted hyphae in the cap cuticle.

Very common and widespread; may be found almost year-round in areas with a milder climate.

Jens H. Petersen/Thomas Læssøe

Phloeomana hiemalis is a rather small, grey-brown, smooth and dry mycenoid agaric with adnexed to adnate, rather distant, whitish gills. The stem is rather pale and pruinose with a hairy base. The spores are inamyloid and measure 6–10 × 5–7 µm with a Q value of 1.3–1.4. It occurs on moist bark of deciduous trees, often with moss and other mycenoids.

Phloeomana speirea △ also has inamyloid spores but decurrent gills and a higher spore Q.

Rather common and widespread; June–March.

Phloeomana clavata is a rather small, pinkish-buff, dry mycenoid agaric with similar coloured, decurrent gills. With simple cheilocystidia. The spores are inamyloid, measure 6–10 × 5–8 µm, and have a Q of 1.2–1.5. The cap cuticle has incrusted, brown pigment. Occurs mostly on conifer twigs and bark, including on *Juniperus*.

Phloeomana speirea ◁ is very similar but has paler gills, no incrusted pigment and a higher spore Q value.

An apparently local species that seems to be almost absent from the Atlantic parts of temperate Europe, but is probably under-recorded; mainly August–December.

Mycena acicula is a small, elegant, vividly coloured, completely dry, inodorous mycenoid agaric with a yellow, pruinose stem, an orange, bell-shaped cap and adnexed, whitish gills. Spores and gill flesh react with Melzer's reagent. The spores measure 8.5–11 × 3.5–5 µm. Occurs in somewhat moist, rich, often disturbed places, *e.g.* deep in *Urtica* stands, at the border of carrs, *etc.* It decays ± buried twigs and bark fragments, mostly from deciduous trees, and often occurs with *Phloeomana speirea* ◁ and *Mycena mirata* ⋉.

Distinctive and unlikely to be misidentified. *Atheniella leptophylla* ▷▷ has a white stem that stains brown from the base upwards. *Rickenella fibula* (page 134) has deeply decurrent gills and less contrast between cap and stem. Hygrocyboids (page 142) are more fleshy with thick, waxy gills. *Mycena acicula* is not very closely related to the other mycenoids and should probably be recognized in a genus of its own.

Widespread and common in most regions of temperate Europe but less frequent towards the north; mainly May–November.

Thomas Læssøe

The mycenoid genus ***Atheniella*** includes medium-sized mycenoids with yellow to reddish cap colours, inamyloid spores and simple cystidia.

Atheniella adonis is a striking, small to medium-sized, ± dry mycenoid agaric with a shiny, vivid red cap and a white to red stem. The cap washes out to a pinkish white but never attains yellow colours. The gills may be tinted the same colour as the cap, especially at the base, and the gill flesh does not react with iodine. The inamyloid spores measure 7.2–9.5 × 5–5.5 µm. Occurs mostly in moist, mossy places, both on open land and in woodland. It may grow directly on soil or on twigs, both from conifers and deciduous trees.

It is not clear whether the red-stemmed form that mostly occurs on conifer twigs should be afforded varietal or species status. The typical form is larger and mainly occurs in mossy, open habitats. The pink form of *A. flavoalba* ▷ fades quickly to yellow and has spherical cystidia on the stem cortex.

Widespread but not very common; September–November.

Atheniella leptophylla is a medium-sized, vividly coloured, dry mycenoid agaric with an apricot-orange, matt cap and a whitish, pruinose stem that stains brown from below and after bruising. No reactions with iodine. The spores measure 6.8–9.5(–10.5) × 5.5–7.2 um. It is mostly found deeply rooting into deciduous wood on clay-rich soils, *e.g.* on forested riverbanks.

Typical specimens are easily recognized, but more faded versions could be confused with other mycenoids and collybioids. The brown staining feature is a good characteristic.

Rare and eastern; mainly August–October.

Thomas Læssøe

Atheniella flavoalba is a pale yellowish buff or rarely pink, completely dry mycenoid agaric with no iodine reactions in the spores or gill flesh. Has ± spherical caulocystidia and lageniform cheilocystidia. The spores measure 6.5–9 × 3–4.5 µm. Occurs mostly in open grassland, including in ± unfertilized lawns, but may also occur on litter in forests and plantations.

The pink form was previously regarded as a separate species, *Mycena floridula*. Recent phylogenetic studies indicate that there are two species involved but it is not known whether both can be pink. Species of *Hemimycena* ◁◁ can be similar. They are best recognized by microscopical features, but tend to be more pure white. *Hemimycena cucullata* ◁◁ has short, pointed cheilocystidia and *H. lactea* ◁◁ has small heads on the cystidia. '*Mycena*' *olida* ⋉ is also iodine inert and is a woodland species. It has caulocystidia with a wavy outline and no globose cells on the stem cortex.

Common and widespread; mainly September–December.

Arne Aronsen/Jens H. Petersen

Thomas Læssøe

Jan Vesterholt

The genus **Mycena** is a large genus of very small to medium-sized saprotrophic agarics, often with rather fragile fruitbodies, ± bell-shaped to convex caps, ± branching cystidia and amyloid spores.

Mycena pura s.l. is a very variable, radish-smelling, somewhat collybioid *Mycena*. The stem is ± finely to coarsely floccose-fibrillose and mostly tinted the same colour as the cap. The cap appears in numerous grey-rose to pale blue colour shades, but may also be completely white. Has large, clavate, smooth pleurocystidia. The amyloid spores measure 6–9.5 × 4–5.2 µm. Occurs on litter, mostly in forests and plantations, less often in more open habitats.

Mycena pura is a species complex, where only some of the species have been named. *Mycena rosea* ▷ is larger, more umbonate and has uniform pinkish to rose colours and a thick, clavate, pale stem. *Mycena diosma* ▽ has darker greyish-violet gills, a very zonate cap and a cedar wood-like smell before bruising. *Mycena luteovariegata* ⋉ has, with age, a yellow cap and occurs in unfertilized dry grassland. *Mycena pearsoniana* ▷ has decurrent gills, small fruitbodies and ± inamyloid spores.

Very common and widespread; June–December.

Mycena diosma is a cedar wood (cigar box)-smelling, very zonate *Mycena* with greyish-violet gills and a dark stem. When bruised it smells of radish. Has clavate cheilocystidia. Pleurocystidia are absent or only very sparse near the edge. Spores amyloid and measure 7.4–9.8 × 3.6–5.4 µm. Occurs on humus and litter, mostly on rich soils in *Fagus* woodland.

The less zonate *M. pura* s.l. △ also smells of radish before bruising but has pleurocystidia and paler gills. *Mycena pelianthina* ▷▷ has similar colours, but very dark, serrate gill edges.

Widespread and locally common but probably absent from the boreal zone; mainly August–November.

Mycena pearsoniana is a medium-sized, grey-rose to greyish-violet, dry, radish-smelling *Mycena* with decurrent, pale greyish-violet gills. Has clavate cheilocystia but pleurocystidia are absent. The inamyloid, or first after a long time somewhat amyloid spores measure 6–9 × 3.5–4.5 µm. Grows on acidic litter, mostly in moist conifer forests, but may also occur in deciduous woodland.

M. pura s.l. ◁ has ± emarginate gills and amyloid spores; *M. kuehneriana* ⋉ should also have amyloid spores, but in other characters it should resemble *M. pearsoniana*. The taxonomy of this complex is, however, not yet fully understood.

Widespread, uncommon – probably under-recorded; July–December.

Arne Aronsen/Jens H. Petersen

† ***Mycena rosea*** is a very fleshy, somewhat collybioid to almost tricholomatiod, radish-smelling, dry *Mycena* with attractive rose colours on the cap and a paler, long and clavate stem tinted the same colour. It also occurs in pure white forms. The cheilo- and pleurocystidia are large, clavate and smooth. The amyloid spores measure 7–8.5 × 4–5 µm. Occurs predominantly on humus and leaf-litter in *Fagus* woodland.

Mycena pura s.l. ◁ is not pure rose and it is typically smaller with a more cylindrical stem. *Porpolomopsis calyptriformis* (page 144) has similar colours but a conical cap, and occurs in unfertilized grassland.

Widespread and common south of the boreal-hemiboreal zone; July–December.

Mycena pelianthina is a large, ± collybioid, radish-smelling, dry to somewhat greasy, hygrophanous, pale greyish-violet *Mycena* with violet gills and a serrate, dark brownish-violet gill edge. The stem is pale and floccose above. The large clavate cheilocystidia have coloured contents. Spores amyloid and measure 6–7.5 (–8) × 3.1–4.1(–4.5) µm. Occurs on humus and leaf-litter, mostly in *Fagus* woodland, but occasionally also in other types of woodland.

The dark, serrated gill edge should make identification straightforward, although the closely related, northern sister species *M. lammiensis* ⋉ looks similar but has slightly wider spores (4–5 µm). *Mycena diosma* ⊲⊲ has concolorous, smooth gill edges. Species of *Entoloma* (page 459) have pinkish, angular spores.

Rather common and widespread; June–November.

Mycena olivaceomarginata is a medium-sized, very variable, dry *Mycena* with a red-brown to brownish-olive gill edge and variable grey-brown to vinaceous-brown cap and stem colours. The cheilocystidia are ± lageniform but sometimes with a few 'fingers'. The amyloid spores measure 9.5–11.5 × 4.5–6.5 µm. Favours dry, open, grass-dominated, unfertilized habitats.

Various other species with a coloured gill edge normally occur in woodland, *e.g.* the closely related *M. citrinomarginata* ▽.

Common and widespread; mainly August–December.

Mycena citrinomarginata is a tall and slender, ± yellow-brown to greenish-yellow *Mycena* with a ± lemon-yellow gill edge. The stem generally has pale hairs on the lower half. The odour is faint, ± radish-like. Has yellowish, lageniform cheilocystidia. The amyloid spores measure 8–12(–14.5) × 4.5–6 (–6.5) µm. Occurs on litter in damp woodlands.

Other species with a yellowish gill edge have different types of cystidia. *Mycena olivaceomarginata* △ has a darker gill edge and prefes dry and open habitats.

Widespread and fairly common; mainly August–December.

Arne Aronson/Jens H. Petersen

Mycena aurantiomarginata is a medium-sized, rather dark *Mycena* with a very distinct orange-coloured gill edge and basal stem felt with the same colour. Cheilocystidia clavate with small, ± evenly distributed protuberances. Spores are amyloid and measure 7.5–10.5 × 4–5.5 µm. On needle debris, including needles of *Juniperus*.

Distinctive and unlikely to be misidentified, but *Xeromphalina* ◁◁ and *M. picta* ▷▷ also have orange felt at the stem base.

Widespread but rather scattered; August–December.

Mycena pterigena is a pale, small, but very distinct *Mycena* with arcuate decurrent gills with ± pink to orange gill edge, and a similar coloured cap margin and stem base. The cheilocystidia have orange coloured contents and are broadly clavate with rather long, thin protuberances. The spores are amyloid and measure (7.5–)9.5–12.5 × 4–5 µm. Occurs exclusively on dead fern fronds in damp woodland.

There are no obvious identification issues. *Mycena lohwagii* ⨯ is also preferential to ferns, but is a somewhat larger, greyish fungus without coloured gill edge.

Common and widespread; August–December.

Arne Aronsen

Mycena rosella is a rather slender, elegant, trooping, strikingly rose-red, translucently striate *Mycena* with arcuate decurrent gills and dark rose-red gill edge. The stem is tinted the same colour as the cap. The cheilocystidia are clavate with small protuberances. Spores amyloid, 7.5–10 × 4–5 µm. On damp needle debris in mossy, conifer forests.

Other mycenoids with red gill edges on leaf-litter are smaller and have different gill attachments.

Widespread and rather common in the north and east, but rare to absent from the nemoral zone; August–November.

Mycena renati is a clustered, dry, pink-capped *Mycena* with yellow stem and a nitrous smell. The gill edge may be slightly yellowish. The cheilocystidia are lageniform to fusiform. The spores are amyloid and measure 8–10.2 × 4.9–6 µm. It occurs on trunks and thick branches of deciduous trees, mostly *Fagus*.

Distinctive and unlikely to be misidentified. *Mycena inclinata* ▷▷ lacks the pink shades on the cap and has a different smell.

Generally widespread and locally common, but very rare in north-western parts of temperate Europe; mainly May–October, often appearing early in the season.

Thomas Læssøe

Mycena capillaripes is a rather slender, mostly tall, dry *Mycena*, which is recognized by a combination of its nitrous smell and brownish-red gill edge (hand lens required!). The clavate hymenial cystidia have coloured contents. The spores are amyloid and measure 8–11.5 × 4.5–6 µm. Occurs in large troops on needle debris.

Mycena sanguinolenta ▷▷ and *M. rubromarginata* ▷▷ do not have a nitrous smell and the latter typically occurs on conifer twigs.

Rather common and widespread, but easily overlooked; mainly (June–) August–November.

Mycena purpureofusca is a medium-sized to rather large, darkish, dry to somewhat greasy *Mycena* with adnexed gills that have a distinct, greyish-purple edge. The cheilocystidia are irregular-clavate to broadly lageniform. The amyloid spores measure (7.5–)9–12.2 × (5.5–) 6.2–8.2(–9) µm. Mostly without clamps. Occurs typically on conifer bark and wood in damp woodland.

Mycena rubromarginata ▷▷ has a more brownish-red gill edge and numerous clamps and narrower cheilocystidia.

Widespread, seemingly with most records from the hemiboreal zone; may be found almost year-round.

Mycena rubromarginata is a rather small to medium-sized *Mycena* with a grey-brown to red-brown, ± striate cap and rather distant, adnexed gills with a distinct red-brown edge. The cap cuticle has coralloid protuberances. The cheilocystidia are lageniform with narrow necks. The spores are amyloid and measure 9.2–13.4 × 6–9.4 µm. Almost always on conifer wood, both twigs, rotten bark and wood.

Mycena sanguinolenta ▷ is mostly more slender, has more red on the cap and reddish latex, and the cystidia are more tapering. *Mycena purpureofusca* ◁◁ has purple gill edges and more rounded cheilocystidia. *Mycena capillaripes* ▷▷ is generally more slender, has a nitrous smell and occurs on needle debris.

Rather common and widespread; mainly July–November.

Jan Vesterholt/Pietro Curti

Mycena seynii is a medium-sized, reddish-brown, dry *Mycena* with ± brownish-red gill edge. With few simple protuberances on the cap cuticle. The amyloid spores measure 10.3–14 × 6–7 µm. Occurs on coarse pine cones.

Mycena rubromarginata △ is very similar, but has coralloid protuberances on the cap cuticle.

Known from southern England and from Belgium southwards; October–April.

Mycena sanguinolenta is a rather small to almost medium-sized, translucently striate, dry, pale red-brown *Mycena* with brownish-red latex and rather distant, adnexed gills with a brownish-red edge. The stem is mostly smooth and tinted brownish-red. The cystidia are slender with tapering necks and coloured contents. The spores are amyloid and measure 7–10 × 4.5–6 µm. Occurs on litter and wood, mostly mossy, often in damp places, both in deciduous and coniferous woodland.

If the latex is not evident, may be confused with *M. rubromarginata* ◁◁, but that species has less tapering cystidia.

Widespread and common to very common in most areas; mainly July–November.

Jens H. Petersen/Michael Krikorev

Mycena haematopus is a rather large, typically clustered, greyish-pruinose, brownish-red *Mycena* with copious brownish-red latex and a dentate, pale cap margin. The gills mostly have a brownish edge, especially near the cap margin. The stem is dark, pruinose and in fresh condition bleeds profusely after bruising. The cheilocystidia have a long, thin neck. The spores are amyloid and measure 8–9.5 × 5–6.5 µm. Typically occurs on wood from deciduous trees, mostly of larger dimensions.

Distinctive and unlikely to be misidentified. *Mycena inclinata* ▷▷ is a tougher, paler fungus without latex. *Mycena sanguinolenta* △ is much smaller, translucently striate and does not grow in clusters.

Common and widespread; mainly August—December.

Thomas Læssøe/Jens H. Petersen

Mycena crocata is a very characteristic, typically rather large *Mycena*, which, after bruising, bleeds an orange-yellow latex, both from the cap/gills and stem. All parts are dry, and the stem is ± orange-red, rarely almost pure grey. The cystidia are clavate with rather few thin protuberances. The amyloid spores measure 8.2–11.5 × 5–7 µm. Occurs almost exclusively on wood of *Fagus* and mostly in rather rich woodland. Often co-occurs with *Mycetinis alliaceus* (page 303).

Distinctive and unlikely to be misidentified due to its unique latex colour. *Mycena haematopus* ◁◁ has a powdery cap surface and brownish-red latex.

Common in parts of the nemoral zone, less so to the west, and absent from the hemiboeal-boreal zones; mainly August–December.

Mycena erubescens is a bitter-tasting, dark, pruinose, rather small *Mycena* with whitish but watery latex, that quickly discolours red-brown. The stem is minutely downy. Has two types of cheilocystidia: clavate with thin protuberances, and fusiform with grainy contents. The amyloid spores measure 8–12 × 6.5–9 µm. It occurs on the mossy bark of deciduous trees, often *Quercus*, but very occasionally is found rooting into buried wood.

If taste and latex are not recorded the distinctive cheilocystidia may help to confirm identification.

Widespread and fairly common, although seemingly less so in some areas; mainly September–January.

Jens H. Petersen/Thomas Læssøe

Mycena galopus is a very variable, dry, medium-sized *Mycena* with white latex (only fresh, turgid specimens!). Cap and stem colours vary from white (var. *candida*) via greyish shades (var. *galopus*) to almost black (var. *leucogala*). The gills are adnexed or emarginate and rather distant. Stem ± distinctly floccose. The odour is insignificant. Cheilocystidia are very long-necked and the spores are amyloid and measure 10–14 × 5–6 µm. Occurs on litter and humus in a range of habitat types, including directly in *Sphagnum* or on burnt soils. It can also form a mycorrhizal-like association with ericaceous plants.

If latex cannot be detected the strikingly distant gills are a good field mark. Further studies are required to determine whether the variation within this species actually relates to a range of cryptic species that have yet to recognized. The fungi pictured below may, for example, be a separate species that could be referred to as *M. leucogala*.

Very common and widespread; mainly July–January.

Mycena bulbosa is a rather small, greyish, inodourous *Mycena* with a somewhat large, downy basal disc and smooth, striate cap. The gills loosen easily from the stem, but are joined in a kind of collarium. The gill edge is slimy and may be removed as a thread. Cheilocystidia have long, narrow protuberances. The spores are inamyloid and measure 7.5–11 × 4–5(–6) µm. Associated with wet meadows, fens and bogs, where it typically occurs deep down in tussocks of *Juncus*, but also on other monocots.

Mycena stylobates ▷ is larger, and normally does not occur on *Juncus*; it also has erect 'spines' on the cap and a wide, distinctly striate basal disc. *Mycena mucor* ⋉ occurs on leaf-litter.

Widespread and probably common in most of temperate Europe (although formerly regarded as rare, in reality this is one of the commonest mycenas, but is only likely to be found when specifically searched for); June–November.

Thomas Læssøe

Mycena clavularis is a rather small, pale grey, nitrous *Mycena* with a downy, rather thick basal disc and a somewhat downy cap surface. The cheilocystidia have very long, thin protuberances. The spores are amyloid, almost spherical and measure 8–11.5 × 6.5–11 µm. Occurs on the bark of deciduous trees in damp woodland.

Mycena tenerrima ▷▷ has a grainy, not downy, cap surface, and is inodorous; its spores are narrow and the echinulate cystidia have a long beak.

Widespread but not very common; mainly October–January.

Mycena stylobates is a rather small to almost medium-sized, whitish to greyish *Mycena* with scattered 'spines' on the cap and a large, furrowed, marginally fimbriate-hairy basal disc. The 'spines' are made up of clusters of thin-walled, smooth hyphae. The cheilocystidia have swollen protuberances. The spores are amyloid and measure 7.5–10.5 × 3.5–5 µm. Occurs on all kinds of leaf-litter, mostly in woodland, but also in boggy areas and in scrub.

Mycena tenuispinosa ⋉ has even more 'spines' (with echinulate hyphae) on the cap, a somewhat smaller basal disc and, generally, smaller fruitbodies. The cheilo-cystidia have narrow protuberances. Other similar species have smooth caps and *M. aciculata* ▽ has thick-walled hairs (setae).

Widespread and rather common; May–November.

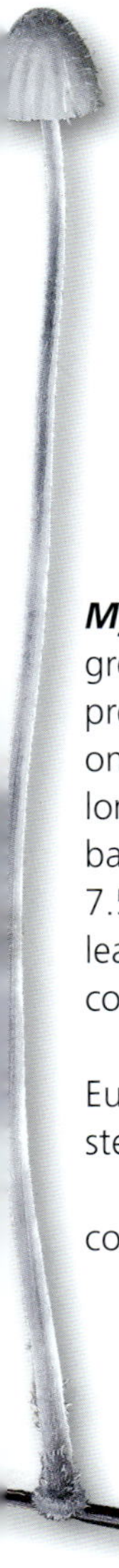

Mycena aciculata is a pale greyish, rather small *Mycena* with projecting, thick-walled hairs (setae) on the cap and stem. The typically long stem springs from a distinct basal disc. The spores are amyloid, 7.5–10.5 × 3.8–4.7 µm. On acidic leaf-litter from both deciduous and coniferous trees.

The only *Mycena* in temperate Europe with long setae on cap and stem.

Widespread and locally quite common; June–November.

Mycena tenerrima is a small, very powdery, white to pale greyish *Mycena*, which typically has rather long, scattered hairs and fine grains on the stem, and a powdery, small and rounded basal disc. The cheilocystidia are finely verrucose, but some have a long, smooth beak. The spores are ± ellipsoid, amyloid, and measure 8–10.5 × 4.5–6 µm. On bark, twigs, dry fruits and other woody fragments in the litter layer from both deciduous and coniferous trees; more rarely on herbaceous stems.

This species was previously also known as *M. adscendens*. *Mycena clavularis* ◁◁ has a minutely downy cap surface, a larger basal disc, no long hairs on the stem, is greyer, and has spherical spores. The very downy-grainy, pure white *M. corynephora* ▽ has no basal disc and also has spherical spores. *Mycena rhenana* ⋉ has a more applanate and plicate cap and lacks cheilocystidia.

Widespread and very common in most parts of temperate Europe; more or less year-round.

Thomas Læssøe

Mycena corynephora is a white, very powdery-downy, small, odourless *Mycena* without a basal disc, but sometimes seemingly tickened basally by a powdery-downy tuft. The relatively thick stem is downy from top to bottom and also finely striate lengthwise. Cheilocystidia have fine, dense protuberances. Caulocystidia are ± allantoid with small spines on the surface. The amyloid, ± spherical spores measure 6–10 × 6–8 µm. On mossy bark of deciduous trees in damp woodland, rarely on conifers.

Mycena tenerrima △ has a thinner stem with scattered, smooth hairs, a small basal disc and completely different cystidia and spores.

Common in much of the British Isles and also occurs in other western and central parts of temperate Europe. To the north-east, there is a single record from Denmark; summer–autumn.

Mycena amicta is a medium-sized *Mycena* with a detachable, gelatinous cap cuticle. The colours are typically rather dark grey. Often has a bluish or blue-green hue to the cap margin and also at the base of the rooting stem, which, in addition, is distinctly downy-hairy. The amyloid spores measure 7.5–10.7 × 4.5–6 µm. Occurs on bark fragments *etc.* in deciduous and coniferous forests. Sometimes a prominent blue mycelium can be seen on the substrate.

Mycena cyanorrhiza ▽ is smaller, paler and has a striking sky blue stem base.

Widespread and fairly common in most of the area; May–December.

Mycena cyanorrhiza is a rather small, somewhat pale *Mycena* with a ± convex cap and a pale grey, downy stem with a striking sky blue, somewhat widened stem base. The gills are pale and ± adnate, but may loosen from the stem. The cheilocystidia have narrow, thin protuberances. The spores are drop-shaped and measure 6.5–9 × 4–5 µm. Occurs on conifer bark, *e.g.* on flakes of pine bark deep in grass or the brushwood of *Picea*, *Pinus* or *Larix*. Possibly mostly found on calcareous soils.

Mycena amicta △ is typically larger and darker, and the blue shades are never distinctly sky blue.

Mostly in southern parts of temperate Europe, but also with scattered occurrences in *e.g.* the Nordic countries; summer–autumn.

Thomas Læssøe

Mycena pseudocorticola is a rather small, mostly somewhat short-stemmed, greyish-blue, dry, finely pruinose *Mycena*. With almost white, adnate, distant gills. Occasionally, is it almost grey or slightly tinged green. The cheilocystidia have branched, dense protuberances. The spores are almost spherical and amyloid, and measure 8–10.5 × 7.5–10 µm (2-spored basidia), to 13 × 12 µm (4-spored basidia). It occurs on mostly mossy bark of deciduous trees, typically living trees, and very occasionally on *Juniperus* twigs and other conifers.

Faded specimens may recall washed-out specimens of *M. supina* ⋉ and *M. meliigena* ▷. Such specimens, and also whitish specimens, require very careful microscopy.

Common and widespread; July–January, mainly late in the season.

The species on this page spread are closely related; they all have almost spherical, amyloid spores.

Mycena meliigena is a smallish, ± vinaceous-red, dry, finely pruinose *Mycena* with rather wide, pale gills. The cheilocystidia have branched protuberances. The amyloid spores are almost spherical and measure 8–11 × 8–9.5 µm. Occurs on ± mossy bark of deciduous trees, mostly together with other mycenas.

Faded specimens of the closely related *M. supina* ⋉ and *M. pseudocorticola* ◁ may be difficult to identify and require detailed microscopy. However, *Mycena supina* has only short and unbranched protuberances on the cheilocystidia and its spores are smaller (7–9 × 6.5–9 µm).

Widespread, but not very common; August–March, mainly late in the season.

Mycena juniperina is a small, apricot-orange to grey-brown, finely powdery *Mycena* with an almost semiglobose cap, adnate, distant gills, and a short, typically curved, finely powdery stem. The cheilocystidia have numerous thin, ± branched protuberances. The amyloid spores are almost spherical and measure 8.8–12.3 × 8.1–10.5 µm. Occurs on twigs of *Juniperus*, typically deep inside the bush; outside temperate Europe also occurs on other *Cupressaceae*.

Mycena pseudocorticola ◁ and *M. meliigena* △ recall *M. juniperina* in shape and in microscopical characters. However, both only very rarely occur on *Juniperus*, and on fresh material the colours are different.

Locally common and quite widespread; August–March.

Jens H. Petersen/Thomas Læssøe

Mycena capillaris is a very small, pale grey to white *Mycena*, with adnexed gills. On very young specimens the stem apex is mostly black. Spores are amyloid and large, 8.5–11.5(–13.5) × 2.8–4.5 µm. The cheilocystidia have fine protuberances. Occurs almost exclusively on *Fagus* leaves.

May recall one of various species of *Hemimycena* ⊲⊲, which have inamyloid spores and, mostly, different spore shapes or distinctive hairs. *Rhizomarasmius setosus* (page 304) has a dark stem base and long hairs on the rigid stem.

Very common and widespread in areas where *Fagus* occurs; August–December, mainly late in the season.

Mycena polyadelpha is a very small, white *Mycena* with few, or sometimes none, arcuate-decurrent gills. The stem is thread-like and white. The cheilocystidia have short, fine protuberances. The amyloid spores measure 9–11.8 × 3.5–5 µm (from 4-spored basidia) or 10–13.5 × 3.5–5 µm (from 2-spored basidia). Occurs almost exclusively on moist, lying *Quercus* leaves.

Mycena capillaris △ has adnexed gills, and species of *Hemimycena* ⊲⊲ have inamyloid spores.

Widespread, but often overlooked; mainly October–January.

Arne Aronnsen

Mycena smithiana is a very small, pinkish-tinted *Mycena* with adnexed gills. The cheilocystidia have small protuberances. The spores are amyloid and measure 10–12.5 × 3.8–5 µm. Occurs probably exclusively on *Quercus* leaves in the litter.

White forms are very hard to distinguish from *M. capillaris* △, but the substrate Involved may provide a helpful clue.

Widespread in the nemoral zone, occasional, but probably overlooked; September–January, mainly late in the season.

Steen A. Elborne

Mycena tubarioides is a small, ± pinkish-tinted, finely pruinose *Mycena* with a gelatinous cap cuticle, a crenulate-striate cap and a gelatinous gill edge. The gills are distant and arcuate-decurrent. The cheilocystidia are clavate with thin, simple to somewhat branched protuberances. The narrowly ellipsoid spores are amyloid and measure (9–)9.8–14 × 4–5(–5.5) µm. Occurs on various monocots at lake-sides and in fens and bogs.

Other small, pinkish-tinted mycenas on monocots do not have gelatinous gill edges.

A rarely reported, possibly eastern species, that is probably somewhat overlooked; mainly September–November.

Mycena riparia is a small, grey-rose to greyish, dry *Mycena* with deeply decurrent, rather distant, ± grey-rose tinted gills. The cheilocystidia have rather long, narrow, ± branched protuberances. The amyloid spores measure 7.8–10.5 × 3.8–4.5 µm. The hyphae lack clamps. Occurs on coarse *Carex* culms in fens and *Alnus* carrs.

The similar *M. juncicola* ⋉ has clamps, and *M. tubarioides* △ has gelatinous gill edges.

Widespread, apparently very rare, but probably overlooked; mainly October–November.

Thomas Læssøe

Jan Vesterholt

Mycena tintinnabulum is a medium-sized, clustered, robust, rather dark *Mycena* with moderately decurrent gills. It smells somewhat spermatic, and the taste is astringent. The cheilocystidia are almost coral-like. The spores are small, measuring 3–6 × 2–3 µm, and amyloid. Occurs mostly on stumps and fallen trunks of *Fagus*, but is also known from other deciduous trees.

It is the only large *Mycena* that is able to tolerate cold winter temperatures for prolonged periods. *Mycena maculata* ▷▷ typically occurs on *Quercus* and has larger spores (7–9 × 4.5–6 µm).

Fairly common in parts of Scandinavia and central Europe, but rare in western parts of temperate Europe; September–May, mainly in mild winter perioids.

Mycena cinerella is a rather small to medium-sized, dry *Mycena* with greyish to grey-brown colours, narrow, ± decurrent gills and a strong farinaceous smell. The cheilocystidia are strongly branched and coral-like. The amyloid spores measure 7–12.5 × 3.5–7 µm. Occurs in acidic habitats, on leaf-litter, both in deciduous, mixed and conifer woodlands.

The superficially similar *M. vulgaris* ▷▷ and *M. clavicularis* ▷▷ may occur in the same habitats but have ± sticky fruitbodies. *Hemimycena mairei* ◁◁ occurs in grassy turf and lacks any smell.

Rather common and widespread; mainly October–December.

Mycena picta is a small, strange-looking *Mycena* with a helmet-shaped, striate, grey-brown cap and a small umbilicus in the centre. The stem base has rusty strigose hairs. Gills adnate to somewhat decurrent and so high that they almost look like elongated pores. The cheilocystidia are formed in chains and have finger-like protuberances. The spores are amyloid, ± drop-shaped and measure 6.5–10(–11.5) × 4–5 µm. Occurs on litter in both deciduous and coniferous forests, but also on large rotten trunks of deciduous trees in damp woodland.

Distinctive and unlikely to be confused with any other European fungus. In the past, this species was accepted in *Xeromphalina* mainly based on the characteristic hairs at the stem base. Within *Mycena* it occupies an isolated position but with some tropical relatives.

Widespread and rare, but also easily overlooked; mainly July–October.

Thomas Læssøe

Mycena pseudopicta is a medium-sized, dry *Mycena* with arcuate, decurrent gills and grey-brown colours. The cheilocystidia are very finely branched. With 2-spored basidia. The spores are amyloid and measure10.6–13(–14) × 5–6.5(–7.2) µm. Occurs in dry, unfertilized grasslands, *e.g.* south-facing coastal slopes – perhaps somewhat calciphilous.

Mycena agrestis ⋉ may occur in the same habitats, but has a gill edge that may be detached as a slimy thread.

Widespread and rare, but probably also somewhat overlooked; mainly October–November.

Thomas Læssøe

Mycena epipterygia is a higly variable, ± yellow-stemmed, very slimy *Mycena* with somewhat emarginate-decurrent gills that have a detachable, slimy thread on the edge. With age it develops rusty spots and the stem becomes mostly brown at the base. The cheilocystidia are very branched. The spores are amyloid and measure 8–11 × 4.5–8 µm. Grows on acid litter, mostly from conifers, including rotten wood.

A somewhat greenish form, var. *epipterygioides* ⋉, has 2-spored basidia and typically grows on *Pinus* litter; it is most likely a species in its own right. *Mycena pelliculosa* ⋉ occurs in open grasslands but lacks, or has only slight, yellow tones on the stem apex; it is often considered a variety of *M. epipterygia*. It is likely that further molecular studies will reveal additional species in the *M. epipterygia* complex.

Very common and widespread; August–January, mainly late in the season.

Mycena vulgaris is a medium-sized, sticky, grey-brown *Mycena* with arcuate decurrent gills that have a detachable, slimy thread on the edge. The spores are amyloid and measure 7–9(–11) × 3.5–4.5(–5) µm. Occurs in large troops on thick needle debris.

Mycena epipterygia △ has forms that almost lack yellow colours, but the gills are emarginate-decurrent. *Mycena clavicularis* ▷ does not have a slimy gill edge. *Mycena agrestis* ⋉ is very similar to *M. vulgaris* but occurs in grasslands.

Widespread and rather common; August–March, mainly late in the season.

Mycena belliae is a very unusual, medium-sized, grey-brown, striate and umbilicate *Mycena* with deeply decurrent gills and a tough cap cuticle. It mostly occurs in small clusters. The stem is finely white flocculose and typically swollen at the base. The cheilocystidia are large and somewhat digitate. The large, amyloid spores measure 10.5–15.5 × 6–8.5 µm. Occurs exclusively on *Phragmites*, typically on standing, dead culms in rather deep water. It may occasionally occur completely submersed. It often produces mass fruitings.

Distinctive and unlikely to be misidentified, since both the morphology and habitat are particular to this species.

Widespread and probably rather common in much of temperate Europe, but so far not found in *e.g.* Norway; September–February, especially late October onwards. In England it can also be found during the summer.

Thomas Læssøe

Mycena clavicularis is a rather compact, greyish, slimy-stemmed *Mycena* with ± decurrent gills that do not have a slimy edge. The cheilocystidia are clavate with thin, unbranched protuberances. The spores are amyloid and measure 7–9 × 3.5–5 µm. Typically occurs on needle litter from *Pinus* in sandy places, but may also grow on *Picea* litter.

Mycena vulgaris ◁ is very similar, but is more slender and has a slimy gill edge.

Widespread and fairly common; July–January.

Thomas Læssøe/Jens H. Petersen

Thomas Læssøe

Mycena stipata is a rather dark, medium-sized, greasy, ± clustered, ± conical *Mycena* with a strong nitrous smell (like a swimming pool). Cheilocystidia with long necks or with irregular protuberances. Basidia 4-spored. The spores are amyloid and measure 9–11.5 × 5.3–6.5 µm. Occurs on rotten trunks of conifers.

Mycena viridimarginata ⋉ has typically ± olive-coloured gill edges. The very similar 2-spored *M. silvae-nigrae* ⋉ occurs early in the season. *Mycena leptocephala* ▽ is not greasy on the cap and does not occur in clusters.

Widespread and locally rather common, especially in the boreal zone; mainly July–November.

Arne Aronsen

Mycena abramsii is an anonymous, fragile, grey *Mycena*, which has a faint nitrous smell when picked. It exudes a colourless liquid from the stem base. The cheilocystidia are clavate with a beak. Lacks caulocystidia. The spores are oblong, amyloid, and measure 7.5–13 × 4–6(–6.5) µm. Occurs both on hardwood and conifer wood.

There are numerous similar grey mycenas but all have ± distinct caulocystidia. *Mycena niveipes* ⋉ smells similar but is larger, more silvery-shiny and is early fruiting. *Mycena leptocephala* ▽ has a stronger smell.

Rather common and widespread; April–January.

Thomas Læssøe

Mycena leptocephala is a variable, grey to grey-brown *Mycena* that can be recognized by its nitrous smell combined with a dry cap, lageniform cheilocystidia and well developed, somewhat swollen caulocystidia. The spores are amyloid, and measure 8–11.2 × 4–5.8 µm. Occurs in grassland, but also in woodland, on wood chips, twigs *etc.*

Mycena abramsii △ lacks swollen caulocystidia and has a beak on the cheilocystidia. *Mycena aetites* ▷ is greyer on the gills and smells radishy. *Mycena scirpicola* ⋉ and *M. cretata* ⋉ have long, tapering caulocystidia.

Widespread and very common; April–January.

Mycena plumipes is a rather dark, grey-brown to brown, greasy, robust and nitrous-smelling *Mycena* with broad, emarginate gills. The cheilocystidia are ± fusiform. The amyloid spores measure 7–9.6 × 3.6–5.8 µm. Occurs on *Picea abies* cones amongst mosses in damp conifer forests.

Mycena silvae-nigrae ⋉ also occurs in the spring, but is more brown-grey and has 2-spored basidia and protuberances on the cystidia; it normally occurs on conifer wood.

Mainly an upland, hemiboreal–boreal species, absent from the nemoral lowlands; March–May.

Arne Aronsen

Mycena aetites is a medium-sized, grey, rather fragile *Mycena* with greyish gills and a smell of radish. The spores are amyloid, and measure 7–11 × 5–6.5 µm. The cheilocystidia are large, simple-clavate to lageniform or with a few 'fingers'. Occurs on soil, mostly in mossy lawns and other grasslands, but also in amongst tall herbs and in open forests.

Mycena leptocephala ◁smells strongly nitrous. The rare *M. latifolia* ▷▷ has more distant gills, no smell and large cystidia that have warts at the base.

Widespread and common; July–December.

Mycena flavescens is a medium-sized *Mycena* with a pale cap and a smooth, greyish stem. The odour after breaking is distinctly earthy. Gill edge concolorous or may be somewhat yellow. The large, balloon-shaped cheilocystidia have small protuberances. The amyloid spores measure 7.5–10 × 4–5 µm. On litter in different types of woodland, rarely in more open habitats. May occur in dense clusters on *Fagus* leaf-litter.

The smell, combined with the contrast between cap and stem colour are the main characteristics, but it is often necessary to resort to microscopy.

Widespread and rather common; mainly August–November.

Michael Krikorev

Mycena septentrionalis is a dark, ± conical-umbonate, pruinose, *Pelargonium*-smelling *Mycena* with adnexed to emarginate gills. Cheilocystidia are 'hedgehog-like', but with a constriction in the middle. Spores amyloid and measure (7.5–)8–9(–10.5) × (3.5–)4–5.1 µm. On needle debris, mostly from *Picea*, but also found with *Juniperus* and *Abies*.

The distinctive smell and constricted cystidia are characteristic. Dark forms of *M. filopes* ▷ may be recognized by the different smell and by having unconstricted, clavate cystidia.

Rare or overlooked, mostly recorded in the Nordic countries; September–January.

Thomas Læssøe

Mycena arcangeliana is a very variable *Mycena*, which, when young, typically has a dark grey stem and a ± olive cap. The stem base is covered in long, rigid, white hairs. It later becomes more anonymous, with greyish colours or with yellow tints in the cap. Smells strongly of iodoform (keep in a box first). The cheilocystidia are 'hedgehog-like'. The amyloid spores measure (7–)8–9 × 4.5–5 µm. Occurs on wood of all sizes, *e.g.* in small bushes.

Other mycenas with a smell of iodoform and 'hedgehog-like' cystidia have different colours and habitats.

Rather common and widespread, but rare in the north; mainly September–January.

Mycena chlorantha smells of iodoform when drying and often occurs in clusters. The cap and stem are yellowish-grey and the gill edge is ± yellowish to greenish. The cheilocystidia are 'hedgehog-like'. The spores are amyloid and measure (8–)9–11 × 5.2–6.5(–7) µm. Occurs exclusively in dunes or similar sandy places, mostly directly in tussocks of *Ammophila* in white dunes.

The combination of specific habitat type, smell and cystidial form should exclude other possibilities.

Rather common in suitable sites along exposed coasts, but apparently absent from the north; August–January, mainly late in the season.

Mycena filopes is a tall, fragile, iodoform-smelling *Mycena* with a pale, splitting coating, rather small 'hedgehog-like' cystidia and amyloid spores that measure 9–11.5 × 5.5–6.5 µm. Grows on humus and litter in both deciduous and conifer forest, mostly on disturbed, rich soil at road sides.

This species belongs to a complex that is not fully understood. A number of 'forms' occur in different habitats and vary in colour, but grade towards being pinkish-tinted, but otherwise very similar to *M. metata* ▽.

Widespread and very common; May–January, mainly late in the season.

Thomas Læssøe

Mycena metata is a ± greyish-red, fragile, iodoform-smelling, ± conical, dry *Mycena* with up to 70 µm-long, 'hedgehog-like' cystidia on the gill edge. Spores amyloid, measure 8–12.5(–14.5) × 5–6.3 µm (from 2-spored basidia) or 8.5–11.8 × 4.5–6.5 µm (from 4-spored basidia). Mostly found on conifer debris, but in damp woodland it can occur on fallen trunks. May also occur on other types of acid litter.

Mycena filopes △ is typically greyer, may have a pale coating, and has smaller cheilocystidia.

Very common and widespread; mainly September–January.

Mycena inclinata is a large, clustered, somewhat greasy, pale grey to grey-brown *Mycena* with emarginate, pale gills, a strong smell of wet paint and a floccose stem, which, with age, becomes brownish-orange towards the base. Cheilocystidia have irregular, thin protuberances. Spores amyloid, 7–9 (–11) × 5–6(–8) µm. Almost exclusively on *Quercus* trunks and stumps.

Mycena maculata ▷▷ has a smooth, more wine-red tinted stem and a less distinct smell. *Mycena silvae-pristinae* ▷▷ does not develop orange colours on the stem, has no smell, and has large cystidia with very small, projecting spines at the very top.

Common and widespread; August–February, mainly late in the season.

Thomas Læssøe

Mycena maculata is a clustered, dry, grey-brown to red-brown, odourless, medium-sized to rather large *Mycena*, which, with age, becomes ± reddish-brown speckled. The amyloid spores measure 7.5–10 × (4.5–)5–6 µm. Occurs typically on old *Quercus* trunks and stumps, but the same, or a closely related species, is also known from conifer wood in the boreal zone.

Its habit and habitat make this species easy to confuse with *M. inclinata* ◁◁, but that species has a distinct smell of wet paint, a floccose stem, usually a dentate cap margin and, with age, more orange colour down the stem. *Mycena galericulata* ▷▷ has somewhat larger spores (9–12 × 6–7 µm), and greyer colours.

Widespread, but not common, although also somewhat overlooked; mainly September–December.

Mycena zephirus is a trooping, rather large, pale, later reddish-brown-speckled *Mycena* that does not have a distinct smell. The stem is distinctly white flocculose and fragile. Cheilocystidia rather large and variable in shape, simple or with a varying number of protuberances. Spores amyloid, measure 8.5–12.2 × 4–5 (–6) µm. Occurs on needle debris, mostly from *Picea* in ± rich habitats. When seen with deciduous trees, needles are often present in the litter.

The fragile, floccose stem makes this species easy to recognize.

Widespread and fairly common, but rare towards the west; mainly August–November.

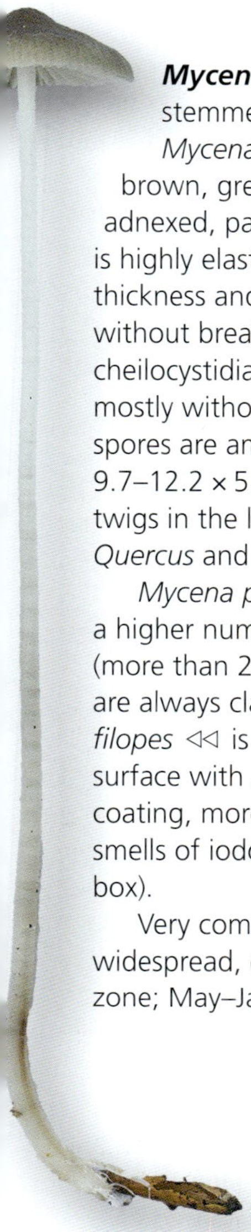

Mycena vitilis is a smooth-stemmed, robust, odourless *Mycena* with a grey-brown, greasy cap surface and adnexed, pale gills. The stem is highly elastic considering its thickness and can be swung without breaking. The ± 'fingered' cheilocystidia are clavate; mostly without clamps. The spores are amyloid and measure 9.7–12.2 × 5.5–7.5 µm. Occurs on twigs in the litter, especially from *Quercus* and *Fagus*.

Mycena polygramma ▷▷ has a higher number of entire gills (more than 23) and the hyphae are always clamped. *Mycena filopes* ◁◁ is fragile, has a cap surface with a pale, splitting coating, more crowded gills and smells of iodoform (if kept in a box).

Very common and widespread, except in the boreal zone; May–January.

Jens Maarbjerg

Mycena latifolia is a grey-brown *Mycena* with rather distant, broad, somewhat decurrent, rather thick, ± transversely ribbed gills. Has large cheilocystidia that are smooth above, and which have small, short, wart-like protuberances from the middle and downwards. The amyloid spores measure 7.2–10 × 4–5.5 µm. Occurs in dry grassland, in mires or on litter in conifer plantations.

The distinctive cheilocystidia make this species easy to recognize. However, *Mycena aetites* ◁◁ is macroscopically somewhat similar.

Very rare, but possibly overlooked; August to November.

Arna Aronsen

Mycena polygramma is typically a rather large and long-stemmed, greyish *Mycena* with a silver-shiny, longitudinally striate stem. It can, however, also occur in smaller, almost non-striate forms, as well as a pure white form. The cheilocystidia are lageniform to somewhat branched with long 'fingers'. The spores are amyloid and measure 8–10 × 5.5–7 µm. It typically occurs on ± buried wood from deciduous trees, but may also fruit at the base of standing trunks, and in such situations often occurs in dwarf forms.

Large specimens of *M. vitilis* ◁◁ may easily be confused with indistinctly striate specimens of *M. polygramma*. The latter has, typically, more than 23 entire gills and always has clamps; *M. vitilis* mostly lacks clamps and has fewer gills.

Common and widespread; August–January, mainly late in the season.

Thomas Læssøe

Mycena silvae-pristinae is a rather large, greyish, ± floccose-stemmed *Mycena* that lacks a distinct smell. The cheilo- and pleurocystidia are ± broad but with a rather long neck, which, above, is covered by very thin and short protuberances (difficult to see on dried material). The spores are amyloid, 7–10 × 6–7.5 µm. On old, fallen *Quercus* trunks and stumps.

Young specimens of *M. inclinata* ◁◁ can be distinguished either by their wet paint smell or branching cystidia.

Very rarely recorded, but recently found to be quite common in the New Forest in England; September–December.

Mycena galericulata is a very tough-fleshed, large, greyish to grey-brown *Mycena* with a somewhat greasy and often wrinkled cap surface. The gills are rather distant and adnexed-emarginate and, with age, mostly have transverse ribs and often a pink tinge. The stem is completely smooth and, typically, somewhat rooting. The cheilocystidia are highly branched. The spores are large and amyloid, and measure 10–12 × 7–9 µm (from 2-spored basidia) or 9–12 × 6–7 µm (from 4-spored basidia). It occurs on wood from deciduous trees, including large trunks, stumps and buried wood, but also occasionally on conifer wood.

Mycena megaspora ⋉ is a very similar species but is darker and browner, and occurs in very acidic habitats such as mires. Other similar species have very fragile flesh.

Very common and widespread; mainly May–January.

Tricholomatoids

The tricholomatoid fungi are rather fleshy agarics with ± solid stems, ± emarginate gills and whitish or, rarely, brownish-rose spore-deposits. The genera *Tricholoma* and *Catathelasma* form ectomycorrhiza, while species of *Squamanita* are parasites on other agarics. The remaining genera are decomposers (saprotrophs), although *Armillaria* may kill trees and bushes before degrading the wood (necrotrophs).

The tricholomatoid fungi have rather few reliable characters for separating the genera. Besides the spore morphology and the presense of veils, identification very much depends upon experience and gut feeling. It is, for example, relatively straightforward for an experienced field mycologist to recognize the elegant melanoleucas as such, but it is very difficult to put into words the macroscopical differences between a *Melanoleuca* and a *Lyophyllum*.

OTHER SIMILAR FUNGI:

– the large, brown hygrocyboids in the genus *Neohygrocybe* are rather similar but have thick, wax-like gills (page 160).

– collybioids may look very similar, but are mostly more fragile with hollow stems, and typically have narrowly adnate gills (page 274).

– species of *Entoloma* with robust fruitbodies are very similar, but have pinkish, angular spores (page 478).

– *Hebeloma* and *Cortinarius* may have the same stature, but brown spores (pages 680 & 690).

FURTHER READING: 28, 64, 88, 111, 156, 157, 181, 183, 285, 342, 343, 345.

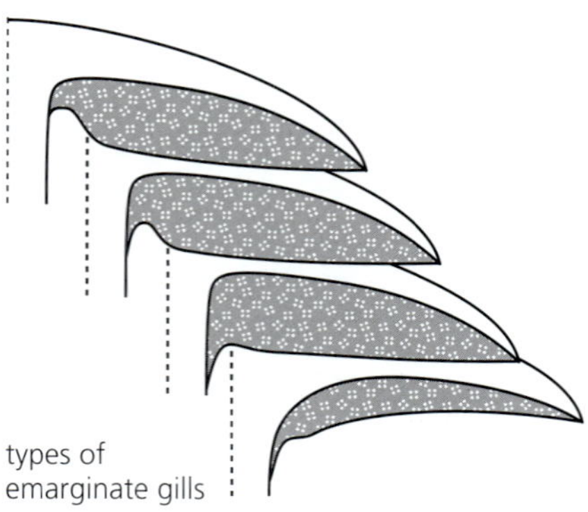

types of emarginate gills

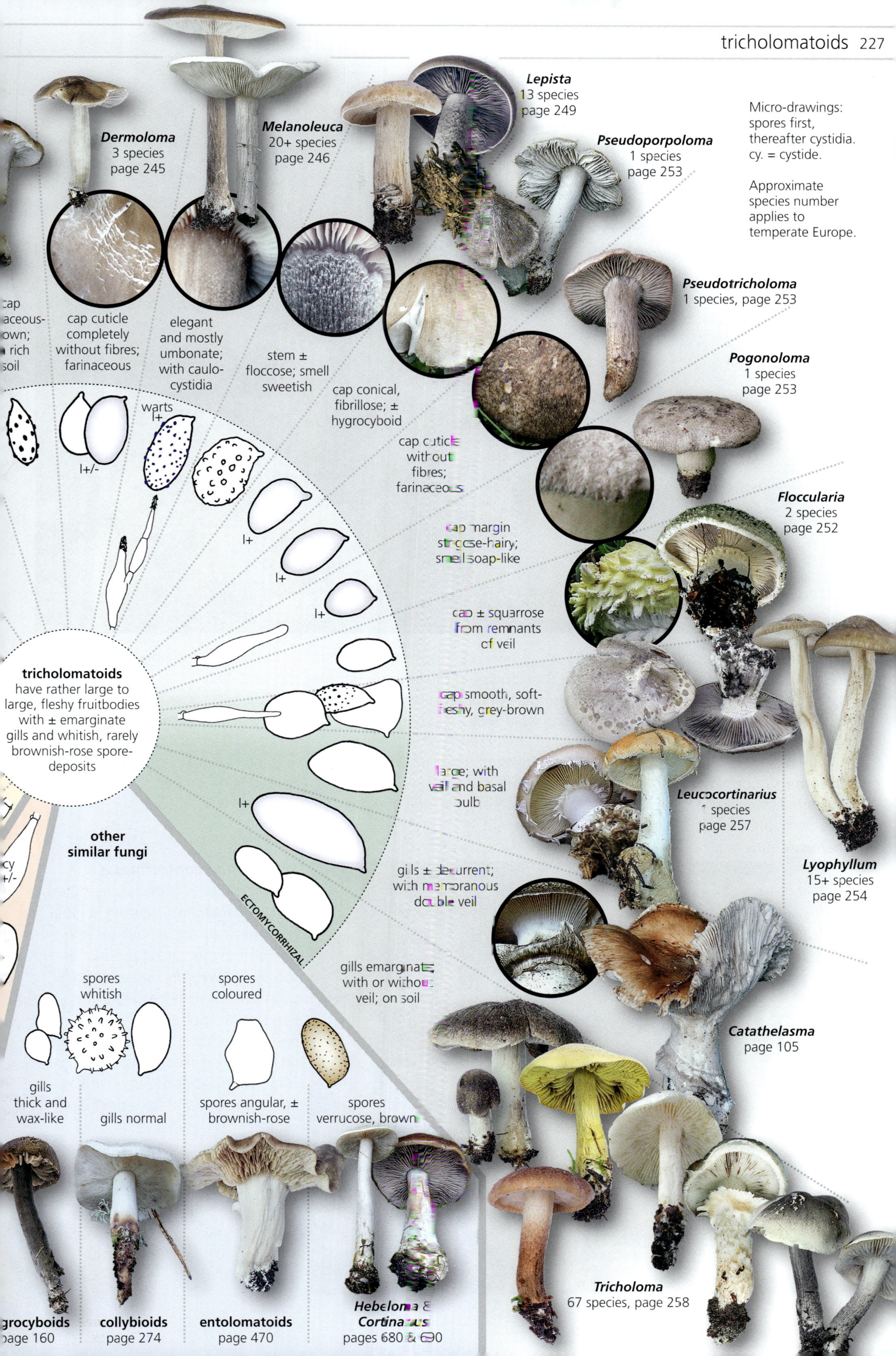
Dermoloma
3 species
page 245
Melanoleuca
20+ species
page 246
Lepista
13 species
page 249
Pseudoporpoloma
1 species
page 253
Micro-drawings: spores first, thereafter cystidia. cy. = cystide.
Approximate species number applies to temperate Europe.
Pseudotricholoma
1 species, page 253
Pogonoloma
1 species
page 253
cap
aceous-
own;
rich
soil
cap cuticle completely without fibres; farinaceous
elegant and mostly umbonate; with caulo-cystidia
stem ± floccose; smell sweetish
cap conical, fibrillose; ± hygrocyboid
cap cuticle without fibres; farinaceous
cap margin strigose-hairy; smell soap-like
cap ± squarrose from remnants of veil
cap smooth, soft-fleshy, grey-brown
large; with veil and basal bulb
gills ± decurrent; with membranous double veil
gills emarginate, with or without veil; on soil
warts
I+
I+/-
I+
I+
I+
I+
ECTOMYCORRHIZAL
Floccularia
2 species
page 252
Leucocortinarius
1 species
page 257
Lyophyllum
15+ species
page 254
Catathelasma
page 105
tricholomatoids
have rather large to large, fleshy fruitbodies with ± emarginate gills and whitish, rarely brownish-rose spore-deposits
other similar fungi
cy
+/-
spores whitish
spores coloured
gills thick and wax-like
gills normal
spores angular, ± brownish-rose
spores verrucose, brown
grocyboids
page 160
collybioids
page 274
entolomatoids
page 470
Hebeloma &
Cortinarius
pages 680 & 690
Tricholoma
67 species, page 258

Squamanita includes agarics that transform other agaric fruitbodies, so that the stipe or 'bulbs' partly consist of the host fungus, while the upper part is pure *Squamanita*.

Squamanita odorata has a strongly perfumed smell and a greyish-violet, scaly cap and stem, which arise from yellow 'bulbs'. These have the radish-like smell of the host when cut. The inamyloid spores measure 6.5–9.5 x 4–6 µm. Parasitic on *Hebeloma mesophaeum* (page 682) and may occur both in deciduous and conifer forests and in parks and gardens.

Distinctive and unlikely to be misidentified.

Rare and rather meteoric in its occurrence, with most records from northern Germany, the Netherlands and the Nordic countries; mainly September–October.

Squamanita contortipes is a slender, violet or greyish-blue-tinted, scaly *Squamanita*. The stem is finely fibrillose-scaly above, but smooth and buff in the parasitized part and somewhat swollen. The amyloid spores measure 5–7 x 5–6.5 µm. Parasitic on species of *Galerina*, perhaps exclusively *G. pumila* (page 641) in unfertilized grasslands.

Similar species of *Squamanita* are more scaly and have dextrinoid spores.

Very rare and scattered, perhaps more common in the hemiboreal–boreal zones; mainly September–October.

John Bingham/Jens H. Petersen

Squamanita paradoxa is a greyish-violet, ± appressed-scaly *Squamanita*, which appears directly from the stem of *Cystoderma* fruitbodies. The spores are dextrinoid and measure (7–)8–10(–11) x (4–)4.5–5(–6) µm. Found mostly in unfertilized grasslands, typically together with the host *Cystoderma amianthinum* (page 316).

Squamanita pearsonii × has erect scales on cap and stem, and the host relationship with *Cystoderma* is less obvious. *Squamanita contortipes* ◁ parasitizes species of *Galerina* and has amyloid spores.

Very rare, but possibly somewhat overlcoked; autumn.

Squamanita schreieri is a large, fleshy, golden-brown *Squamanita* with a ± golden-brown, scaly stem. The white base (the parasitized tissue) is swollen and somewhat rooting. Gills pale and adnexed. The inamyloid, smooth spores measure (4–)5–7 x (3.5–)4–5 µm. It is considered to be a parasite of *Amanita solitaria* (page 355) and *A. strobiliformis* (page 356).

Distinctive and unlikely to be misidentified

Widespread but rare in central Europe, absent elsewhere in temperate Europe; autumn.

Paolo Caciagli

Thomas Læssøe/Morten Christensen

Leucopholiota decorosa is a large, yellowish-brown, squarrose-scaly, pale-spored agaric with emarginate, crowded gills. The stem is smooth above the ring-zone and densely covered in pointed, projecting brown scales below. Has ± fusiform cheilocystidia. The smooth, ellipsoid, amyloid spores measure 5.5–6(–7) x 3.5–4 µm. Occurs on large trunks of deciduous wood, towards the north of temperate Europe mostly on *Betula*.

Echinoderma asperum (page 334) has free gills, dextrinoid spores and occurs on rich soil. Species of *Pholiota* (page 600) have brown spores.

A very rare fungus in central Europe, also known from Finland and farther east; late summer–autumn.

•• ***Lentinula edodes*** (or shiitake) is a brown, ± white-scaly, dry, tough-fleshed, white-spored agaric with a ± floccose-scaly short stem. The crowded gills are cream and adnexed. The hyphal system is monomitic. Lacks proper cystidia. The smooth, inamyloid spores measure 5–6.5 x 2.5–3 µm. It is lignicolous and is cultivated commercially on *Quercus* logs, but nowadays also on other compressed cellulose sources.

Species of *Pholiota* and *Hemipholiota* (page 600) have dark spores. *Neolentinus* (page 81) is even tougher.

An Asian species that has escaped from cultivation in parts of temperate Europe; mostly autumn.

Margaretha Liebmann

Rhodotus palmatus is a fleshy, very distinctive agaric that appears somewhat intermediate between a *Pleurotus* and a *Pluteus*. The cap is greasy, coral-red to ± pinkish and reticulate-veined. The gills are pale reddish and adnexed to emarginate. The odour is perfumed-fruity and the taste bitter. The cheilocystidia are lageniform. The reddish, spherical, verrucose, inamyloid spores measure 5.5–7 x 5–7 µm. Occurs on large, dead trunks of deciduous trees, *e.g. Ulmus* and *Fagus*.

Could possibly be confused with a *Pluteus* (page 486), although these have free gills and smooth, pinkish spores. Microscopy should be used if there is any doubt.

Rare and local, and has almost disappeared from some areas due to the die-back of elms following the latest major outbreak of Dutch elm disease; mainly August–November.

Dan Molter/Wikipedia

Dan Molter/Wikipedia

Armillaria is a genus of rather large, wooly-scaly, white-spored agarics with emarginate to almost decurrent gills. They degrade wood, where they form thick, flattened, black hyphal cords. They have no cystidia and the spores are inamyloid and smooth. They may cause damage in forestry and parks and gardens. Edible species must be cooked thoroughly before being eaten, but may still cause stomach problems, albeit rarely.

• ***Armillaria mellea*** is a very long-stemmed and densely clustered, ± honey-coloured to green tinted *Armillaria* with a large, pale ring and a cylindrical stem without dark scales. The hyphae lack clamps. The spores measure 7–9 x 5–6 µm. Occurs on deciduous wood, mostly on and around stumps or at the base of dying trees.

Armillaria lutea ▷, which also typically occurs on deciduous wood, has a shorter stem that is yellowish and widened at the base, a more scaly cap, and a more floccose ring.

Widespread and rather common, except in the hemiboreal–boreal zones; mainly September–November.

• **Armillaria lutea** is an orange-brown *Armillaria* with a floccose, ± yellowish ring with yellowish to dark scales, and a clavate, mostly distinctly yellow, stem base. The young cap has yellow-brown to yellow scales all over the surface, and the margin is inrolled. Hyphae have clamps. The spores measure 7–9 x 4.5–5.5 µm. Often grows in clusters on wood, including buried wood, but may also occur individually. Mostly found in deciduous woodlands.

The yellow-green-tinted *A. mellea* ◁ has a longer, more cylindrical stem and a large, projecting ring. *Armillaria ostoyae* ▷▷ has a more robust, dark-scaled ring and mostly occurs on conifer wood. Several other similar species occur but are difficult to identify using just morphological characters.

A very common and widespread species in temperate Europe, rarer towards the north; August–December.

Armillaria ostoyae has a very distinct, well-defined ring with dark scales, and a cap with similar contrasting dark scales. The hyphae have clamps. The spores measure 8–10 x 5–6 µm. In large or small clusters, rarely solitary. It is mostly associated with conifers and may occur in vast quantites in plantations, where it can have an adverse effect on timber production.

Armillaria lutea ⊲⊲ has more yellowish colours and a quickly disappearing ring. *Armillaria borealis* ▽ is smaller, with distinctly pale scales on the ring and stem, and occurs mostly from July to August.

Widespread and very common; mainly September–December.

The species was earlier considered edible after cooking. It may, however, cause stomach upset and vomiting and eating is not recommended.

Håkon Holien

Armillaria borealis is an early fruiting species, which typically grows in dense clusters. The scales are relatively pale and ± the same colour as the cap. The ring is relatively well defined, and the veil white. The stem is cylindrical to clavate. Has clamps. The spores measure 7–8.5 x 4.5–5.5 µm. On deciduous wood.

Armillaria mellea ⊲⊲ lacks clamps and typically has a more greenish-yellow tone and an even smoother, longer stem. *Armillaria lutea* ⊲⊲ has clamps but usually shows distinct yellow veil remnants.

Mostly found in the hemiboreal–boreal zones, with few records from farther south; mainly June–August.

Jens Maarbjerg

Desarmillaria includes *Armillaria*-like species with no rings on the stem and no black hyphal cords. They are probably decomposers.

Desarmillaria ectypa has no veil and is of rather slender stature. It is relatively pale yellow-brown to red-brown, but has somewhat darker, fine scales towards the cap centre. The gills have a pinkish tint. The stem is ringless, somewhat swollen towards the base and darkens from below. The hyphae have clamps. The ellipsoid spores measure 7–9 x 5.5–6.5 µm. Occurs alone or in small clusters in rich fens and mires.

Similar *Armillaria* species ◁ all have distinct veils. The specific habitat and rather large, fleshy fruitbodies with yellow-brown colours should also help to ensure an accurate identification. The more southern *D. tabescens* ▽ occurs in clusters on deciduous wood, mostly *Quercus*.

Widespread but rare to very rare; mainly June–September.

The species is included on the IUCN international red list, primarily due to the fast decline in suitable habitats.

Desarmillaria tabescens is a yellow-brown, brown-scaly, clustered *Armillaria*-like species. The stem is ringless and somewhat tapering towards the base; it lacks a veil. The gills quickly become pale red-brown. The hyphae have clamps. The spores measure 7–9 x 5–6.5 um. It is associated with deciduous wood, mostly *Quercus*.

Old, weathered specimens of *Armillaria* ◁ can almost lack remnants of a veil and appear to have a ringless stem. *Desarmillaria ectypa* △ only occurs in fens and mires.

Mostly found in central Europe and the British Isles, absent from the hemiboreal–boreal zones; autumn.

Jan Vesterholt

Tricholomopsis includes wood-degrading white-spored agarics with yellow gills, yellow flesh and caps that are almost entirely covered in dark scales.

Tricholomopsis rutilans is a purple to wine-red, scaly, tricholomatoid agaric with vivid yellow, ± emarginate gills and a stem that is ± the same colour as the cap. The colour may wash out, so the fruitbodies can appear pale yellow with only very faint reddish tinges. The cheilocystidia are large and club-shaped to fusiform. The flesh is yellow and lacks a distinctive smell or taste. The spores are ellipsoid, smooth, inamyloid, and measure 5–7.5 x 4.5–6 µm. Occurs on rotten conifer wood.

Tricholomopsis flammula ▷ is very similar to a small *T. rutilans*, but the stem is yellow. *Tricholomopsis decora* ▷ and *T. sulphureoides* ▷ completely lack the purplish colours and have fine, dark scales on a yellow background.

Widespread and common; July–December.

Tricholomopsis decora is a fleshy, yellow tricholomatoid agaric with dark, hair-like scales, especially near the cap centre. The stem has small, scattered scales. The cheilocystidia are various shapes. Spores measue 6–8 x 4–5.5 µm. On conifer wood, mostly in damp forests with plentiful dead wood.

Tricholomopsis rutilans ◁ always has ± distinct purplish colours on cap and stem; *T. flammula* ▽ is smaller and also mostly ± purple on the cap. *Tricholomopsis sulphureoides* ▽ is an almost smooth, yellow species that is restricted to eastern Europe.

Widespread and common in the hemiboreal–boreal zones, less so in the nemoral zone; mainly July–October.

Tricholomopsis sulphureoides is a ± fleshy, orange-brown to yellow-brown tricholomatoid agaric with ± smooth, slightly felty or scaly cap, pale yellow gills and pale yellow stem that becomes apricot-orange towards the base. The spores measure 6–9.5 x 4.5–6 µm. Occurs on conifer wood.

Tricholomopsis decora △ has distinct, dark, hair-like scales on the cap and some on the stem.

Known from conifer forest in eastern Europe, *e.g.* Estonia and Slovakia (and from North America); late summer–autumn.

Jukke Vauras

Tricholomopsis flammula is a rather slender tricholomatoid agaric with ± distinct purple to wine-red hair-like scales on the cap and a pure yellow stem. The spores measure 5.2–8 x 3.2–4.8 µm. Occurs both on conifers and on deciduous wood in various forest types and plantations.

Apart from the distinctive stem colour, this species may be separated from *T. rutilans* ◁ on the spore shape (Q value 1.51–1.63 in *T. flammula* and 1.3–1.4 in *T. rutilans*).

A rarely recorded species with scattered records, mainly from upland areas; probably absent from the lowland nemoral zone; late summer to autumn.

Megacollybia platyphylla is a sturdy, dry, grey-brown, radially fibrillose collybioid-tricholomatoid agaric with broad, white, emarginate gills and a greyish, fibrillose, hollow, clavate stem. At the base there are thick, tough, white hyphal cords, which may extend several metres through the litter layer from one fallen branch to the next. The cheilocystidia are large and clavate. The spores are inamyloid, smooth, ± spherical, and measure 7–10 x 6–8.5 µm. Occurs on ± buried wood on ± rich soil, mostly in deciduous forests.

There should be no obvious identification issues. However, the spore colour and gill attachment need careful examination in order to rule out *Pluteus cervinus* (page 491). Small specimens could recall *Hydropus subalpinus* (page 182), but that species is yellower, has allantoid spores, and no prominent hyphal cords. Species of *Clitocybula* ▽ and *Tricholoma* ▷▷ have ± solid stem flesh and no thick hyphal cords.

A very common and widespread species, becoming less frequent towards the north; May–November.

Jan Vesterholt

Clitocybula lacerata is a collybioid-tricholomatoid, ± clustered, dry agaric with no prominent hyphal cords. The cap is finely floccose, radially fibrillose (the margin is almost fimbriate), and very thin-fleshed, but only indistinctly striate. The smell recalls that of *Hygrophorus discoxanthus* (page 167). The amyloid spores are spherical and measure 4–6 x 4–5 µm. Mostly on conifer wood.

Megacollybia platyphylla △ is larger and has hyphal cords and inamyloid spores. *Clitocybula familia* var. *compressa* ⋉ is characterized by its almost smooth cap and somewhat decurrent gills.

Mostly found in central Europe but also known from England; autumn.

• ***Hypsizygus ulmarius*** is a large, whitish to pale grey, white-spored, mild-tasting, non-farinaceous agaric with a somewhat excentric stem. The gills are adnexed to emarginate. The hyphae have clamps, and the hymenium has no cystidia. The almost spherical, smooth, inamyloid spores measure (5)5.5–6(7) x 5–5.5(6) µm. Almost exclusively associated with dying *Ulmus* trees, where it produces a white rot.

The very closely related *H. tessulatus* ▽ has distinct drop-like blotches on the cap, is more clustered and occurs on other deciduous hosts. It also has somewhat smaller spores. The two species are usually treated as one and often in the genus *Lyophyllum* ▷▷. The cultivated species *H. marmoreus* ⋉ also has conspicuous drop-like blotches. *Ossicaulis lignatilis* (page 118) and *O. lachnopus* ⋉ smell and taste very strongly farinaceous.

Widespread, but rare as a consequence of most large elm tress being lost due to Dutch elm disease; June–December.

• ***Hypsizygus tessulatus*** is a rather large and fleshy, greyish agaric with an excentric stem and distinct, dark, drop-like blotches on the cap. The gills are adnexed to emarginate. The spores are almost spherical and measure 4–5(6) x 4–4.5 µm. On standing trunks of deciduous trees, mostly in grassland habitats with scattered trees, such as deer parks.

Hypsizygus ulmarius △ is very similar, but lacks drop-like blotches and occurs mostly on *Ulmus*. The spores of *H. tessulatus* should also be smaller, but the two species are very closely related or perhaps even identical. The cultivated *H. marmoreus* ⋉ is regarded by some as a synonym of *H. tessulatus*.

Mostly found in the hemiboreal–boreal zones, with few nemoral records; mainly September–October.

Seppo Kytöharju

Thomas Læssøe

Calocybe are small to large, white-spored, saprotrophic, terrestrial agarics with convex caps that lack a radial structure.

• ***Calocybe gambosa*** is a very fleshy, strongly farinaceous, white-spored *Calocybe* that is typically greyish-white, but may also occur in a pale orange-brown form (f. *graveolens*, middle picture). The gills are crowded and somewhat emarginate. The fruitbodies often stain ± brown with age. Lacks cystidia, but has clamps. The spores are smooth, inamyloid and measure 5–7 x 2–4 µm. Occurs mostly as fairy rings in both deciduous and conifer forests, and in scrub and in lawns, often in rich, ± calcareous habitats.

Vernal, ± farinaceous, whitish species of *Entoloma* (page 485) have pinkish spore-deposits and angular spores. *Calocybe gambosa* f. *graveolens* recalls *Gerhardtia incarnatobrunnea* ▷▷, but that is an autumn-fruiting species with finely ornamented spores.

Common and widespread, except in the boreal zone; April–June, and rarely during autumn.

Calocybe carnea is a rather small, fleshy, rose-coloured to grey-rose *Calocybe* with crowded, white gills. It may grow alone or in small clusters, and the stem base may be ± hairy-felty. The odour is somewhat perfumed, not farinaceous. The smooth and narrowly ellipsoid-fusiform spores measure 4.5–7(–8) x 2–3 µm. A rather tolerant not very demanding grassland agaric, which may also occur newly established lawns.

This species was formerly included in the genus *Rugosomyces*, which is currently 'lumped' with *Calocybe*. A clustered form with a hairy stem base is treated by some as *C. persicolor*. *Entoloma* species (page 464) with similar colours are more thin-fleshed and produce reddish spore-deposits.

Rather common and widespread; June–November.

Calocybe obscurissima is a dull-coloured *Calocybe* with pale, crowded gills. The cap is ± radially felty and may be tinted violet at the cap margin and on the stem apex. The smell and taste are farinaceous. The spores are smooth, ellipsoid to fusiform, and measure 5–6 x 3–3.5(–4.5) µm. Occurs on rich, often disturbed soil in deciduous and coniferous forests.

Species of *Melanoleuca* ▷▷ have amyloid, verrucose spores and cystidia. *Calocybe favrei* ▷▷ is larger with yellow gills and a reddening flesh. See also *C. ionides* ▷▷.

Rare to locally fairly common, absent from the boreal zone; mainly August–November.

Calocybe ionides is a very characteristic, rather fleshy *Calocybe* with a violet-blue stem and cap, and ± cream, emarginate, rather crowded gills. It smells and tastes farinaceous. Lacks cystidia. The smooth, ellipsoid, inamyloid spores measure 5–5.5 x 2.5–3 µm. It is mostly associated with rich soils in deciduous forests, mostly *Fagus* forests. It often co-occurs with phlegmacia (page 698–721).

Until recently treated in the genus *Rugosomyces*, which is now considered a synonym of *Calocybe*. *Calocybe obscurissima* ⊲⊲ is very similar in stature and ecology, but lacks the intense violet-blue colours: it has been regarded as a variety of *C. ionides*. *Calocybe favrei* ▷ is a larger fungus with distinctly yellow gills and reddening flesh.

Widespread but uncommon, and very rare in more northern areas; mainly August–October.

Christian Lange

Calocybe cerina is a rather small, almost unicoloured curry-yellow *Calocybe* with crowded, yellow gills and a farinaceous smell. Cap cuticle with a radial structure. The spores measure 3.5–4(–4.5) x 2–2.5(–3) µm. It occurs primarily in calcareous habitats and mostly in conifer plantations, but has also been reported from dunes with *Salix etc.*

It is debatable how many yellow *Calocybe* species should be recognized and which names should be applied. In principle, *C. cerina* (also known as *C. fallax*) differs from *C. chrysenteron* ⋉ by possessing a cellular cap cuticle, while the latter has a radial structure.

Widespread but rare; mainly September–November.

Calocybe hebelomoides is a yellow-brown, rather anonymous, medium-sized, ± inodourous *Calocybe* with adnexed to emarginate gills. The smooth spores measure 4–6 x 2–3 µm. It occurs mostly in groups and rings on light, calcareous soil in thickets with *Quercus* and *Crataegus*.

Lepista martiorum ⋉, apparently shares some characters with *C. hebelomoides* but has a pinkish-brown spore deposit. Both may look similar to a *Hebeloma* ▷▷, but *L. martiorum* apparently have a more noticeable odour.

Very rare and only known from a few nemoral localities; July–September.

Thomas Læssøe

Calocybe favrei is a large, fleshy *Calocybe* with a greyish-blue cap and rather vivid yellow, ± emarginate, crowded gills. The yellowish, firm flesh becomes wine-red when cut and subsequently blackens. The smell and taste are strongly farinaceous. The gill edge has hyphae-like cystidia. The spores are smooth and measure 3.5–5 x 2.5–3.5 µm. Occurs on calcareous soil in both deciduous and coniferous forests.

Calocybe ionides ◁ and *C. obscurissima* ◁◁ are smaller than *C. favrei* and do not have reddening flesh.

A rare to very rare fungus that is absent from the Nordic countries and the Netherlands. It is known from just a single county in England. It is, however, more frequent in the Prealps; late summer–autumn.

Jan Vesterholt

Seppo Kytöharju

Tricholomella constricta is a fleshy, rather small, somewhat rooting, chalk-white, white-spored tricholomatoid agaric with a strong farinaceous smell. The cap surface is finely felty, but may also be somewhat sticky. The stem has a ± evident, membranous but narrow ring. There are no cystidia on the gills. The inamyloid spores are ornamented with warts and measure (6–)7–9(–10.5) x 5–6.5(–7) µm. Occurs on enriched areas of ground, *e.g.* where animals have urinated, both along forest edges and in more open habitats.

Ringless specimens may perhaps recall small specimens of *Calocybe gambosa* ◁◁, which normally only occurs in spring, or even a small, white *Tricholoma* ▷▷. However, both these genera have smooth spores.

Widespread but rather rare, not present in the boreal zone; mainly June–December.

Gerhardtia incarnatobrunnea is a fleshy, pale clay-buff to vinaceous-brown, cream-spored tricholomatoid agaric with a faint farinaceous smell. Lacks clamps. The very finely ornamented, inamyloid spores measure 4.5–6.5 x 2.5–3.5 µm. Mostly in forests, but also in open habitats.

The orange-brown form of *Calocybe gambosa* ◁◁ has completely smooth spores and a very strong farinaceous smell. *Rhodocybe gemina* (page 451) has a pinkish-brown spore-deposit, bumpy spores and a perfumed smell.

Rare, but perhaps somewhat overlooked, not recorded in the boreal zone; August–November.

Dermoloma are small, grey-brown, fleshy, tricholomatoid agarics with distinctly cellular, mostly cracking cap cuticles. They are found in scrub and unfertilized grasslands.

Dermoloma cuneifolium is a robust, but rather small tricholomatoid, white-spored agaric with greyish, broad gills and a very strong farinaceous smell. The grey to grey-brown cap cracks irregularly when drying out. The spores are inamyloid, although occasionally dextrinoid, and measure 5–6.5 x 4–5 µm. Occurs in unfertilized grasslands and thorny thickets, often co-occuring with species of *Hodophilus* (page 164), mostly on clay-rich or calcareous soils.

Other dermolomas have amyloid spores. Species of *Hodophilus* have a different smell and smaller spores.

Widespread and fairly common although rare in the boreal zone; mainly August–November.

Dermoloma pseudocuneifolium is a rather slender, though rather fleshy, dark grey-brown *Dermoloma* with a ± cracking cap cuticle. The amyloid, ellipsoid spores measure 6–8 x 3.5–5 µm, Q value 1.5–1.7. Occurs in unfertilized grasslands, tall grass and scrub, mostly on clay-rich and/or calcareous soils, often in the company of clavarioids, earthtongues etc.

Dermoloma josserandii ⨯ has a lower spore Q (1.2–1.4) and pale grey colours. *Dermoloma cuneifolium* △ has inamyloid or dextrinoid spores.

Widespread but rather rare; mainly June–November.

Jens Maarbjerg

Melanoleuca are ± applanate-umbonate white-spored agarics with cylindrical ± downy stems. Typically, they have lageniform, fusiform or urticiform hymenial cystidia and warty, amyloid spores. A difficult genus with many poorly defined species.

Melanoleuca cognata is a medium-sized to large *Melanoleuca*, mostly characterized by the cognac-tinted, very crowded gills. The cystidia are one-celled and ± lageniform with crystals at the top. The amyloid, verrucose spores measure 7–9.5 x 4–6 µm. Grows on rich soil in a variety of habitats, but mostly in disturbed places, *e.g.* roadsides.

Typical specimens are distinctive and unlikely to be misidentified, but it is possible that more than one species is involved as there is considerable variation in fruitbody morphology. Species of *Pluteus* (page 491) have completely free gills with a pinkish tint.

Rather common and widespread; March–January.

Jan Vesterholt

Melanoleuca polioleuca is a medium-sized, rather dark *Melanoleuca* with stem flesh that is blackish at the base. The cheilocystidia are ± conical, lack 'shoulders' and are single-celled. The spores measure 6.5–8.5 x 4–5 µm. Occurs on ± disturbed soil, newly sown lawns, ± fertilized grasslands, roadsides, *etc.*

Melanoleuca polioleuca f. *pusilla* ⋉ has a <40 mm diameter cap and *M. polioleuca* f. *langei* ⋉ has a stem that is shorter than the cap diameter. *Melanoleuca friesii* ⋉ has cystidia with distinct 'shoulders', but is otherwise very similar.

Widespread and common; mainly September–November.

Melanoleuca exscissa is a medium-sized, greyish to grey-brown *Melanoleuca* with rather pale, but mostly tinted gills. The stem is floccose, but not distinctly lengthwise-striate. The cheilocystidia are of the 2-celled type with a cylindrical upper part. The spores are amyloid and warty, and measure 7.5–9.5 x 5–6.5 µm. Occurs along roadsides and in similar places, both in the open and in woodlands.

The variety *M. exscissa* var. *iris* ⋉ has a smell that recalls that of *Lepista irina* ▷▷.

Widespread and fairly common; April–December.

Melanoleuca microcephala is a small and slender *Melanoleuca* with pale gills. It has small spores that measure 6–8 x 4–5 µm, and 2-celled cystidia with a cylindrical upper part. It occurs in ± open, grassy habitats, but the precise ecology is not fully understood.

Melanoleuca rasilis ⋉ also has small fruitbodies and small spores but the gills are darker.

Rare or overlooked; July–November.

Jan Vesterholt

Melanoleuca grammopodia is, typically, a large, mostly rather short-stemmed, grey-brown, umbonate *Melanoleuca* with a distinct floccose-striate stem. The cheilocystidia are of the 2-celled type with a ± cylindrical upper part. The spores measure 7.5–10 x 4–6 µm. Occurs mostly as fairy rings in forests and parks.

This species belongs to a poorly understood complex and includes part of what used to be called *M. brevipes*. *Melanoleuca cognata* has more colourful gills and different cystidia.

Widespread and rather common; April–December.

Jan Vesterholt

Thomas Kehlet

Melanoleuca verrucipes is one of the few easily recognizable species of *Melanoleuca*, characterized by whitish colours and a cylindrical stem that has conspicuous brownish-black scales on a white background. It smells somewhat like aniseed. The cystidia are 2-celled, with crystals at the top. The verrucose, amyloid spores measure 8–10.5 x 5.5–6 µm. Often occurs on piles of wood chips in forests and parks, but can also grow in more natural settings, such as in conifer forests.

Distinctive and unlikely to be misidentified.

Widespread and rather rare to locally common, although clearly increasing in the nemoral zone of temperate Europe; May–November.

Jens Maarbjerg

Melanoleuca strictipes is a very pale and relatively long-stemmed, medium-sized *Melanoleuca* with a floccose, somewhat lengthwise-striate, grey-brown-tinted stem. The gills are white and may have a flesh-pink reflex. The cheilocystidia are single-celled and ± tapering. The amyloid, verrucose spores measure 7–10 x 4.5–6 µm. Occurs in open grassland or in open forests.

Melanoleuca subalpina ⋉ has a shorter stem and less tapering cheilocystidia. Species of *Tricholoma* ▷▷ have less crowded gills, less floccose stems and inamyloid spores.

Widespread but rare to locally common; April–October.

Lepista are medium-sized to large, tricholomatoid decomposers with ± emarginate to decurrent gills. Most species are characterized by a greasy cap surface, ± sweetish odours, cream to grey-rose spore-deposits and ± verrucose spores. See also page 112.

Lepista irina is a rather variable, beige, pale grey-brown to more reddish *Lepista* with ± emarginate gills and a strongly perfumed smell like that of *Iris* roots. The cap is ± hygrophanous. The spore-deposit is brownish-rose. The ellipsoid, indistinctly verrucose spores measure 6.5–9 x 4–5 µm. Occurs in fairy rings on rich soils in forests and scrub, rarely in open grassland.

Clitocybe nebularis (page126) has a different smell, decurrent gills, smooth spores and a whitish-cream spore-deposit. Various species of *Lyophyllum* ▷▷ have paler, non-verrucose spores. Species of *Tricholoma* ▷▷ also have paler, smooth spores and emarginate gills.

Widespread and fairly common, but almost absent from the boreal zone; autumn, mostly late in the season.

Lepista panaeolus has grey-brown colours and, typically, concentric drop-like blotches on the slightly felty cap surface. The odour is ± perfumed. The spore-deposit is brownish-rose. The spores are rough and measure 5–6.5 x 3.5–5 µm. Occurs in fairy rings and small clusters in rather rich, unfertilized grasslands.

Lepista irina △ mostly occurs in woodland, smells stronger, and normally lacks drop-like blotches on the cap. *Lepista densifolia* (page 113), and *Ampulloclitocybe clavipes* (page 116) have decurrent gills.

Widespread and fairly commom; mainly September–November.

Thomas Læssøe

• ***Lepista saeva*** is a sturdy and fleshy *Lepista* with violet to lilac colours on the lower part of the fibrous stem, and with a pale greyish white to grey-brown (never violet), somewhat greasy and smooth cap. The spore-deposit is brownish-rose. The finely ornamented spores measure 6–9 x 4–6 µm. Occurs in fertilized or unfertilized grassland, mostly on clay-rich and/or calcareous soil, including grass verges in towns; more rarely in scrub or on bare soil.

Lepista nuda ▷ has paler violet colours that are evenly distributed on the stem and a darker, often ± violet cap. *Lepista sordida* ▽ is smaller and distinctly hygrophanous, and usually has violet colours in the cap.

Widespread and rather common, but rare in cooler regions; occurs mostly late in the season and on into winter if the weather is mild.

• ***Lepista sordida*** is a slender, very variable, typically ± violet, hygrophanous *Lepista*. The gills are ± emarginate and ± violet-tinted. The spore-deposit is brownish-rose. Spores are ellipsoid, finely verrucose, and measure 5.5–7 x 3.5–4 µm. Can be found in a range of habitats, including grasslands, roadsides in forest *etc.*

The more fleshy *L. nuda* ▷ is mostly not hygrophanous. Grey forms of *L. sordida* are identified by their spore colour and spore ornament. Species of *Calocybe* ◁◁ have white to yellowish gills and smooth spores.

Widespread and rather common; mainly July–January.

• ***Lepista nuda*** is a large, fleshy, ± violet to violet-brown, only slightly hygrophanous *Lepista* with a perfumed smell and a brownish-rose spore-deposit. The rough spores measure 6–8.5 x 4–5 µm. It is mostly found in fairy rings on thick leaf and needle debris or on old compost, old wood chip piles, *etc.,* but may also occur in grassland.

Lepista glaucocana ⋉ is much paler in all parts. *Lepista sordida* ◁ is distinctly hygrophanous and smaller, but is very variable. *Lepista saeva* ◁ has very pale, never violet, cap and a violet to lilac tint on the lower part of the stem. Similar bluish-violet *Cortinarius* species (page 707) have a thread-like veil and red-brown spore-deposits.

Common and widespread; occurs almost year-round but mostly late in the autumn.

Jean-Luc Fasciotto

Floccularia is a genus of tricholomatoid/amanitoid agarics with a veil left as scales on the cap, and a ring zone on the stem. The gills are adnexed to emarginate and the spores amyloid. It is debatable whether they are decomposers or ectomycorrhizals.

Floccularia luteovirens is a spectacular yellow-scaled, white-spored tricholomatoid/amanitoid agaric with whitish to yellow, narrowly adnate to emarginate gills. The cap is white but the colour is obscured by the yellow, upright, conical scales. Lacks cystidia. The smooth, amyloid spores measure 6–9 x 4.5–5 µm. It is found in dry, thermophilous grasslands on ± calcareous soils.

Floccularia rickenii ▽ has smaller spores (4.5–6 x 3–3.5 µm).

A very rare, mostly eastern fungus in temperate Europe; autumn.

In the Himalayas, this or a similar species is regarded as a choice edible, but in Europe *F. luteovirens* is usually considered to be of no culinary value.

Christian Lange

Floccularia rickenii is a sturdy, pale yellow-scaled, tricholomatoid/amanitoid agaric with narrowly adnate to emarginate gills. There are no cystidia on the gills. The smooth, amyloid spores measure 4.5–6 x 3–3.5 µm. It occurs in steppe-like, open habitats.

Floccularia luteovirens △ is very similar, but the spores are a different shape and it is apparently somewhat more yellow.

A very rarely occuring, eastern species in temperate Europe; autumn.

Pogonoloma spinulosum is a large, yellow-staining tricholomatoid agaric with white, emarginate, medium-crowded gills. Cap pale greyish with small hair-like scales; margin furrowed, inrolled and strigose-hairy (inserted image). Smell strong, perfumed, soap-like. Cheilocystidia irregularly clavate to fusiform. Spores broadly ellipsoid, amyloid, smooth, 4.5–6 x 3–4 µm. On ± clay-rich soil in older deciduous forests. Probably a decomposer.

Also known as *Porpoloma spinulosum*. Species of *Tricholoma* ▷▷ are inamyloid and lack combination of a strigose margin and perfumed smell.

Rare and ± absent from the north; mainly August–October.

Jan Vesterholt/Thomas Læssøe

Pseudotricholoma metapodium is a fleshy, white-spored, reddening, yellow-brown to grey-brown tricholomatoid agaric. Gills rather distant. Cap surface finely felty and cracks irregularly on drying and with age. With or without cheilocystidia. The smooth, amyloid spores measure 6–8 x 3–4 µm. Occurs in old unfertilized grasslands (mostly with red listed fungi). Probably a decomposer.

Also known as *Porpoloma metapodium*. Species of *Neohygrocybe* (page 160) have inamyloid spores and thicker gills. *Lepista* ◁◁ do not redden and have a brownish-rose spore-deposit.

A rare or locally more common species; July–November.

Jan Vesterholt/Thomas Læssøe

Pseudoporpoloma pes-caprae is a ± tricholomatoid or hygrocyboid, radially fibrillose, conical-umbonate agaric with ± grey-brown colours on the cap centre and a paler margin. The stem is almost white with a ± yellow tinge. Smell and taste farinaceous. Has clamps but no cystidia. The smooth, amyloid spores measure 6.5–8.5 x 4.5–5.5 µm. In open, calcareous, unfertilized grasslands. Probably a decomposer.

Also known as *Porpoloma pes-caprae*. Species of *Tricholoma* ▷▷ and *Cuphophyllus* (page 161) do not have amyloid spores.

Rare and local; autumn.

Nicolò Oppicelli

Lyophyllum is a genus of rather large, soft-fleshed, greyish, white-spored tricholomatoid agarics with emarginate gills.

Some species are considered to be decomposers, while others apparently form ectomycorrhiza.

• ***Lyophyllum fumosum* s.l.** is a clustered, firm-fleshed *Lyophyllum* that is not distinctly rooting and does not darken. The spores are smooth, spherical, and measure 5–6.5 µm. Occurs on rich soil in deciduous forests and parks.

It is currently not possible to differentiate the species in the *L. fumosum* complex. *Lyophyllum fumosum* sensu stricto should have an acidic-mealy, spermaticis smell and is apparently separable from *L. decastes* ⋉ by the latter's lacking smell, but molecular investigations have shown that more species, possibly five, are involved but not yet named. In the Nordic countries, the similar *L. shimeji* ▷ has been recognized as a result of molecular studies. It is said to form ectomycorrhiza, unlike the closely related saprotrophic species.

Widespread and common (when treated as a species complex); mostly August–November.

●● ***Lyophyllum shimeji*** (or hon-shimeji) is a fleshy, ± clustered, ± rooting, not darkening, grey-brown to dark grey-brown *Lyophyllum*. The smooth, spherical spores measure 5–6.5 µm. Typically occurs on poor, *Pinus*- and *Cladonia*-dominated heath forests in the boreal zone; apparently a facultative ectomycorrhiza-former.

The habitat and tendency to form clavate stems separates this species from others in the *Lyophyllum fumosum* complex ◁. Other large, grey tricholomatoids in *Pinus* forests, *e.g. Tricholoma sudum* ▷▷, have strong farinaceous or soapy odours.

Almost confined to the hemiboreal–boreal zones, very rare in the nemoral zone (Denmark); autumn.

It is a well-known edible in far east Asia.

Lyophyllum paelochroum is a rather small, fleshy, ± grey-brown, pale-stemmed, bluing to blackening *Lyophyllum*. The ± spherical, smooth spores measure 4–7 x 4–6.5 µm. Occurs on clay-rich soil below various deciduous trees, *e.g.* in old parks, churchyards, *etc.*; usually found with *Tilia*, perhaps forming ectomycorrhiza.

Lyophyllum eustygium ▷▷, which perhaps is more associated with typical phlegmacia (pages 698–721) hotspots, is ± a large version of *L. paelochroum*.

Rather rare and local; June–October.

David Boertmann

Lyophyllum semitale is a rather fleshy, tall, ± solitary, hygrophanous, grey-brown, blackening *Lyophyllum*. The ± fusiform spores measure 6–8(–9) x 4–5 µm. On rich soil in conifer forests and perhaps forming ectomycorrhiza.

Lyophyllum amygdalosporum ⋉ has somewhat larger and even more broadly fusiform spores. *Lyophyllum deliberatum* ⋉ has larger spores, which are almost rhombic seen in front view.

Rather rare, but locally quite common; August–November.

Thomas Læssøe

Thomas Kehlet

Lyophyllum eustygium is a large, fleshy and robust, pale grey to brown-grey *Lyophyllum* with greyish, adnexed gills and ± evident blackening or bluing surfaces. The stem is often tapering and sometimes somewhat rooting. The taste is ± bitter or farinaceous. The almost spherical spores measure 5–7 x 4.5–6 µm. Occurs on rich soil, *e.g.* in phlegmacia (pages 698–721) hotspots in *Fagus* forests, and is perhaps mycorrhizal.

Lyophyllum paelochroum ◁◁ is very similar but smaller. *Lyophyllum caerulescens* ⋉ apparently have an even more tapering stem.

Rare and local; mainly August–October.

Jan Vesterholt

Leucocortinarius bulbiger is a fleshy tricholomatoid agaric with a marginate, white bulb and a pale orange-brown cap with a white, floccose universal veil. Also has a ± well-developed partial veil in the shape of a floccose ring or ring zone on the white stem. The odour is insignificant. The spore-deposit is cream or slightly pinkish. Cystidia absent. The smooth, inamyloid spores measure 7–9 x 4–5 µm. Occurs mostly on rich soil with conifers, probably forming mycorrhiza.

This species is surprisingly similar to the not closely related bulbous species of *Cortinarius*, the so-called phlegmacia (page 698-721), but those species are distinguished by their rusty brown spore-deposit.

Rather rare and local, and almost absent from some areas; July–October.

Thomas Læssøe

Thomas Læssøe

Tricholoma are rather large, white-spored agarics with ± emarginate gills. They can be naked or have veils that may be slimy, thread-like or membranous. The spores are smooth and inamyloid. All species are ectomycorrhizal.

Tricholoma aurantium is a medium-sized to large, easily recognizable *Tricholoma* with yellow-brown to orange colours on the cap and a transversely cracking covering on the stem. The colour in young fruitbodies may be very saturated and in this stage they are ± sticky. The spores measure 4.5–5.5 x 3.5–4 µm. Occurs on calcareous soils, both in deciduous and conifer forests.

The orange colours on the cap and stem make this species distinctive and unlikely to be misidentified. *Tricholoma batschii* ⋉ and *T. ustaloides* ⋉ may look similar, but both have duller, more brownish colours and lack transverse cracks on the stem. *Tricholoma focale* ▽ has a ring.

Rare to uncommon, locally abundant; July–November.

Tricholoma focale is a large *Tricholoma* with a brownish-orange to red-brown, greasy cap and a woolly to membranous ring. The stem has rather indistinct red-brown cracks below the ring. The spores measure 4.5–5.5(–6) x 3–4 µm. The odour is farinaceous. Occurs in *Pinus* and *Picea* forests and plantations on nutrient-poor soils.

This species is easily told by its orange-brown cap, prominent ring and farinaceous smell. It is macroscopically rather similar to the *T. matsutake* complex ▷, but those species are dry, have scaly caps, and a much stronger, perfumed-sweetish smell.

Fairly common in the north, rarer towards the nemoral zone; autumn.

Jukka Vauras

●● ***Tricholoma matsutake*** is a very fleshy, dark red-brown *Tricholoma* with dry scales on the cap and stem and a ± well-defined ring. The odour is perfumed, as is the taste. The spores measure 6.5–8.5 x 5.5–6.5 µm. Occurs on nutrient-poor sand in *Pinus* forests and in dune plantations with *Pinus* and *Picea*.

Tricholoma focale ◁ has more orange tinges on the cap and a strong farinaceous smell. *Tricholoma caligatum* ⋉ and *T. dulciolens* ⋉ belong to a complex alongside *T. matsutake*. They are, however, somewhat smaller than *T. matsutake* and have shorter spores.

Locally common in the boreal zone, elsewhere absent or very rare; mainly August–October.

Tricholoma matsutake and closely related species are extremely popular and expensive edibles in Japan, where they are imported from around the world. In the Himalaya, young fruitbodies are picked in remote forests and transported to Japan within 24 hours in order to fetch the highest price.

Tricholoma colossus is a very large, fleshy *Tricholoma* with a pale pinkish-buff to brick-red cap and a solid, mostly bulbous stem with an ephemeral ring zone. The odour is weak and the taste mild. The spores measure (7–)7.5–9(–10) x 5–6.5 µm. Mycorrhizal with *Pinus*, mostly on calcareous sandy soil.

The very large and sturdy fruitbodies are characteristic of this species. It could be confused with *T. focale* ◁ and *T. batschii* ⋉, but the stem of those species is rarely wider than 30 mm and the spores are smaller.

Rare and local, almost absent in the nemoral zone; autumn.

Tricholoma vaccinum is a red-brown, dry, fibrillose-scaly *Tricholoma*, which, when young, has a bearded-woolly cap margin. The cap surface recalls certain species of *Inocybe*, page 652. The pale gills become brown-spotted with age. The spores measure 5.5–6.5(–7) x 4–5.5 µm. Smell and taste ± farinaceous. Mostly associated with *Picea*, but may occasionally occur with other conifers.

Tricholoma imbricatum ▽ is less fibrillose and is not bearded at the margin; it is mycorrhizal with *Pinus*.

Common in the hemiboreal–boreal zones, rarer southwards; mainly September–December.

Tricholoma imbricatum is a dry, grey-brown to red-brown, coarsely felty *Tricholoma* with white gills and lacks a felty-bearded margin. The stem is ± brownish, without a veil. The spores measure (5–)5.5–7(–7.5) x 4–5.5(–6) µm. Smell insignificant; taste farinaceous. Probably exclusively mycorrhizal with *Pinus*, mostly on poor soils.

Tricholoma vaccinum △ is more felty-scaly, is almost bearded at the cap margin, and occurs with *Picea*. *Tricholoma albobrunneum* ▷▷ has a sticky cap surface.

Common in the hemiboreal–boreal zones, rarer southwards; mainly September–November.

Tricholoma psammopus is a rather small to medium-sized, dry, yellow-brown *Tricholoma*. The stem is grainy-scaly, with a red-brown base and a rather distinct pale top. The ± emarginate gills are mostly tinted the same colour as the cap. The spores measure (5–)5.5–6.5(–7) x 4–5(–5.5) µm. The odour is insignificant. Mostly mycorrhizal with *Larix* on clay-rich or ± calcareous soils, but may also occur with other conifers.

A rather aberrant *Tricholoma* that is unlikely to be misidentified.

Widespread but rather rare; August–November.

Jan Vesterholt

Tricholoma acerbum is a large and robust *Tricholoma* with crowded, ± yellowish gills and a smooth to felty, beige to pale yellow cap, which has an almost permanently inrolled, smooth to furrrowed margin. The solid stem is pale and ± finely grainy. The spores measure 4–6 x 3.5–4(–4.5) µm. The odour is indistinct, but the taste is bitter. Occurs on ± calcareous clay in thermophilic sites with *Quercus* and *Fagus*.

Typical specimens should be easy to recognize. *Leucopaxillus tricolor* (page 107), has distinctly yellow-toned, somewhat decurrent gills and amyloid spores. A similar species, *T. roseoacerbum* ⋉, is found with *Pinus* in the boreal zone and also with decidous trees in the south. It has a more sticky cap surface.

Rare and local; mainly September–November.

The species is included on the IUCN international red list.

Tricholoma apium is a rather pale, yellow-brown, finely felty to grainy, dry, robust and fleshy *Tricholoma* with an inrolled cap margin and a strong, lovage-like smell. The spores are very small and measure 3.5–5 x 3–4 µm. It is mainly mycorrhizal with *Pinus*, but also with *Picea*, on sandy soils, *e.g.* in coastal plantations.

The distinct smell, inrolled cap margin and rather pale colours make this species distinctive and it is unlikely to be misidentified.

Rare and local, more common in northern parts of temperate Europe; mainly August–October.

The species is included on the IUCN international red list.

Jan Vesterholt

Tricholoma ustale is a medium-sized, sticky to slimy *Tricholoma* with an orange-brown to yellow-brown cap, no ring zone on the stem, and a mild, somewhat farinaceous taste. In dry weather the cap surface is dry and shiny. The odour is faint to ± farinaceous. The cream, emarginate gills develop red-brown spots with age. The spores measure (5.5–)6–7.5 x (4.5–)5–6 µm. Mycorrhizal with *Fagus* and *Quercus*, mostly on fairly rich soils.

The very similar *T. ustaloides* ⋉ is also *Fagus*-associated but has a bitter taste in the cap cuticle, generally more red-brown cap colours and a more clearly delimited, white stem apex. There are also some similar but conifer-associated species, *e.g. T. pessundatum* ▷ and *T. albobrunneum* ▷.

Widespread and common, decreasing towards the north; August–November.

Tricholoma fulvum is a medium-sized, rather slender, somewhat sticky to dry, ± radially felty, red-brown *Tricholoma* with yellowish gills that become brown-spotted with age. The cap margin is ± wrinkled/furrowed; the stem is ± red-brown and the flesh in the stem is ± yellow. The taste is mealy to somewhat bitter. The spores measure 5–6.5(–7) x 4–5 µm. Mostly mycorrhizal with *Betula*, but also with *Picea* and *Abies*.

Similar red-brown species do not have yellowish stem flesh.

Common and widespread; July–November.

Tricholoma pessundatum is a rather compact, somewhat sticky, red-brown *Tricholoma* with dark drop-like blotches on the cap surface and a smooth margin. Stem and gills are whitish, but become darker with age. The flesh is white. Smell and taste are farinaceous. The spores measure 3.5–5 x 2.5–3.5 µm. Mostly mycorrhizal with *Pinus*, but also with *Picea* and *Abies*.

The conifer-associated *T. stans* ⋉ has a furrowed, paler margin and normally no drop-like blotches on the cap. *Tricholoma albobrunneum* ▽ has a sharply delimited pale stem apex.

Fairly common to rare in the nemoral zone; August–November.

Tricholoma albobrunneum is a medium-sized, ± sticky *Tricholoma* with a grey-brown to red-brown, radially fibrillose cap with no drop-like blotches, and a brownish stem that is whitish at the sharply delimited apex. Smell farinaceous. The spores measure 5–7 x 3–4 µm. Mycorrhizal with *Pinus*, typically on sandy soils.

Best told from other conifer-associated species by the distinctly radially fibrous cap without drop-like blotches, and its sharply delimited stem apex. *Tricholoma fulvum* ◁ is similar, but has yellowish flesh and mostly yellow-toned gills.

Fairly common to rare in the nemoral zone; August–November.

Jan Vesterholt

Tricholoma populinum is a medium-sized, ± sticky, pinkish-buff to cinnamon-brown, finely fibrillose *Tricholoma*. It does not have drop-like blotches on the cap or delimited colours at the stem apex. The cap margin is smooth and paler than the centre. Smell and taste distinctly farinaceous. The spores measure (4–)4.5–6(–6.5) x (3–)3.5–4 µm. Exclusively mycorrhizal with *Populus*, mainly in open habitats.

Tricholoma ustale ◁ is typically somewhat darker, is associated with *Fagus*, and does not smell strongly farinaceous.

Occasional to locally common; August–November.

Jean-Marc Moingeon

† *Tricholoma pardinum* is a fleshy, grey-scaled *Tricholoma* with the scales organized in a radiating pattern on a very pale background. The odour is somewhat sweetish, but mealy after bruising and the surface stains yellow-brown. The hyphae have clamps. The spores measure (6.5–)7–9.5(–10) x 5–7 µm. Occurs in thermophilic forests, mostly with *Fagus* and *Abies*.

Recalls other large, grey-scaled tricholomas, particularly *T. filamentosum* ⋉, which also has clamps but occurs farther north and is less contrasty. Other similar species are clampless.

A rather southern species that occurs south of the Nordic countries and is absent from the northwestern, parts of the nemoral oceanic zone; autumn.

Morten Christensen

Tricholoma squarrulosum is a medium-sized, grey-black, scaly, dry *Tricholoma* with a grey-scaled stem and a distinct pepper-like smell. The cap margin is woolly in young specimens. It does not stain blue, but may redden somewhat at the stem base. Basal mycelium white. The spores measure (5–)6–8(–10.5) x (3–)4–5.5(–6.5) µm. Mycorrhizal with deciduous trees, mostly *Fagus,* on ± calcareous, mostly clay-rich soils.

Tricholoma orirubens ▷ lacks the conspicuous smell, has yellow basal mycelium, and may stain blue.

Widespread, occasional, absent from the boreal zone; August–November.

Tricholoma orirubens is a sturdy, dark, grey-scaled, dry *Tricholoma* with lemon-yellow hyphal cords and basal mycelium, and a tendency to stain blue in the stem base and red on the gills. The odour is honey-like and, when bruised, farinaceous. The spores measure 4–6(–7) x 3–4.5(–5) µm. Mycorrhizal with deciduous trees, usually *Fagus*, mostly on clay-rich, ± calcareous soils, often in places with many phlegmacia (pages 698–721).

Tricholoma squarrulosum ◁ also occurs with deciduous trees but has white basal mycelium and a spicy, pepper-like smell; *T. atrosquamosum* ⋉ occurs with conifers, has white basal mycelium and more recurved cap scales.

Widespread and locally common, absent from the boreal zone; August–November.

Tricholoma terreum is a rather small to medium-sized, somewhat dark, blackish-grey, woolly-felty *Tricholoma*, which is mostly odourless. The flesh does not stain with age, and the stem has no dark scales or fibres. The spores measure (4.5–)5–7.5(–8) x 3–5(–5.5) µm, Q value below 1.7. Mostly mycorrhizal with *Pinus*, but may also occur with other partners.

The similar species in the *T. scalpturatum* ▷▷ complex, have rancid-farinaceous smells. The conifer-associated *T. triste* ⋉ and *T. bonii* ⋉ have greyish fibres or scales on the stem.

Common and widespread; May–January.

Tricholoma cingulatum is a rather small, grey to grey-brown, dry, strongly felty-scaly *Tricholoma* with a distinct woolly-membranous, whitish ring, and a felty covering below. Stains somewhat yellow with age and smells rancid-farinaceous. The spores measure (4.5–)5–6 x (2–)2.5–3.5 µm. Mycorrhizal with *Salix*, mostly on disturbed, waterlogged soils.

Tricholoma argyraceum ▽ never has a well-developed ring.

Widespread, but occasional; mainly September–November.

Jens Maarbjerg/Jacob Heilmann-Clausen

Tricholoma argyraceum is a small to medium-sized *Tricholoma* with a conical to umbonate cap, which is whitish to brown-grey and felty to minutely scaly. The odour is farinaceous and the threadlike veil strongly developed. It may stain yellow with age. The spores measure (3.5–)4–6(–7) x 2–3.5 µm, and Qav. is higher than 1.6. Mycorrhizal with deciduous trees, mostly with *Betula*, more rarely with conifers.

Tricholoma inocybeoides ▽ is very similar but has a less-developed veil, as has *T. scalpturatum* ▷, which also has a more convex cap and wider spores.

Widespread and rather common; May–November.

Tricholoma inocybeoides is a small, rather pale, conical to acutely conical, dry *Tricholoma* with a tendency to stain yellow with age. The veil is poorly developed. Odour rancid-farinaceous. The spores measure (4–)4.5–6.5(–7) x (2–)2.5–4 µm and Qav. is higher than 1.6. Mycorrhizal with various deciduous trees, *e.g.* *Betula* and *Populus* in parks and gardens.

Tricholoma argyraceum △ is darker and has a well-defined veil. *Tricholoma scalpturatum* ▷ is more applanate and has a spore Qav. below 1.55.

Widespread and rather common; May–December.

Tricholoma scalpturatum is a rather small, somewhat pale, grey-scaled *Tricholoma* with a tendency to stain yellow with age. It has a very limited veil and the cap soon becomes applanate. The odour is rancid-farinaceous. Spores 4–6 (–6.5) x (2.5–)3–4(–4.5) µm, with a Qav. below 1.55. Mycorrhizal with various deciduous trees, mostly on rather disturbed soils.

The closely related *T. inocybeoides* ◁ and *T. argyraceum* ◁ are typically more conical and umbonate, and have a Qav. above 1.6. The latter has more distinct veil remnants.

Common and widespread; May–December.

Tricholoma sciodes is a rather compact, somewhat shiny, radially fibrillose, grey, dry *Tricholoma* that has greyish gills with eroded, ± grey-black edges. The odour is somewhat earthy and the taste unpleasantly bitter to acrid. The spores measure (6–)6.5–7.5(–8) x 5–6.5 µm. Mycorrhizal with *Fagus*, mostly on rich soils.

Other grey tricholomas in *Fagus* forests are either more scaly or have a different smell and taste.

Common in the nemoral zone, decreasing towards the north; mainly August–November.

Jan Vesterholt/Jens Maarbjerg

Tricholoma virgatum is a medium-sized, silvery, radially fibrillose, conical to acutely umbonate, dry *Tricholoma* with a bitter to acrid taste. Spores measure 6.5–7.5(–8) x 5–6 µm. Mycorrhizal, mostly with *Picea* on rich soils, but also with *Pinus* and *Betula*.

Tricholoma sciodes △ is darker, with eroded, ± dark gill edges and an association with *Fagus*.

Common in the hemiboreal–boreal zone, but rare in the nemoral zone; mainly August–November.

Jan Vesterholt/Jens Maarbjerg

Tricholoma portentosum is a medium-sized to large, sticky, innately fibrillose, convex-conical, brown-grey *Tricholoma* with pale grey, ± yellowish gills and a somewhat yellow toned, pale stem. Taste mild and farinaceous. Spores measure (5.5–)6–7.5 x (3–)3.5–4.5(–5) µm. Mycorrhizal with *Pinus*, *Picea* and *Fagus*, although it mostly asscociates with conifers; it prefers sandy or gravelly soils.

Readily identified if the yellow colours are present on the stem. However, young specimens that lack yellow may recall the stinkbug-smelling *T. josserandii* ⋉.

Fairly common to rather rare in the nemoral zone; mainly September–November.

Tricholoma sudum is a sturdy, somewhat greasy, grey to grey-brown, slightly felty *Tricholoma* with reddening flesh. The gills are pale, but redden more or less with age. The stem becomes greyish and fibrous towards the base and attains ± flesh-pink spots with age. The odour is somewhat soap-like, but becomes more mealy after bruising. The spores measure 5.5–7(–7.5) x 3.5–4.5 µm. Mycorrhizal with *Pinus* on poor, sandy soils, including in large conifer plantations.

Tricholoma saponaceum ▷ does not smell farinaceous when cut and has greenish components in the cap colour. *Lyophyllum shimeji* ⊲⊲ similarly has no farinaceous smell.

Fairly common in the hemiboreal–boreal zone but rare in the nemoral zone; August–November.

Tricholoma saponaceum is a somewhat compact, rather dark, mostly olive-toned, greasy, smooth *Tricholoma* with an evident soapy smell. The stem is smooth to ± dark- scaled, and the extent of green pigmentation is very variable. The flesh discolours brownish after bruising. The spores measure (4–)4.5–6(–6.5) x 3–4(–4.5) µm. Occurs mostly with deciduous trees, and it is possible that records from conifers relate to other closely related species.

Tricholoma saponaceum forms a difficult species complex, where some of the forms are treated as separate species: *T. boudieri* ⋉ is recognized by its very strong, nauseous smell and a very reddening flesh; *T. rapipes* ⋉ has a rooting stem and yellowish cap colours, and is mycorrhizal with conifers; and *T. sudum* ◁ has a mealy smell after bruising and no green tones in the cap.

Widespread and fairly common; August–November.

Tricholoma stiparophyllum is a rather compact, mostly rather long-stemmed, whitish to yellowish-white, dry, almost smooth, nauseous-smelling *Tricholoma*. The cap margin is somewhat furrowed and remains ± inrolled for a long time. The gills are cream. The spores measure 5–7 x 3.5–4.5 µm. Mycorrhizal with *Betula*, mostly on somewhat waterlogged soils.

Tricholoma album ⋉ is associated with *Quercus,* is smaller and smells honey-like. *Leucocybe connata* (page 119) is even whiter, grows in clusters and smells more perfumed.

Widespread and common; August–December.

Tricholoma lascivum is a rather small to medium-sized, pale greyish-white to yellowish-grey, dry, smooth and matt *Tricholoma* with a nauseous smell. The spores measure 6–8(–9.5) x 3.5–5 µm. Mycorrhizal with *Fagus*, typically on mull soils.

Tricholoma album ⋉ is whiter, has a honey-like smell and is associated with *Quercus*. The somewhat larger *T. stiparophyllum* △ has a ± furrowed cap margin and is mycorrhizal with *Betula*.

Common and widespread, but decreases towards the north; August–November.

Tricholoma columbetta is a medium-sized, white, silky-fibrillose *Tricholoma*, which mostly develops reddish and bluish spots with age. The gills are ± cream with a ± distinct pinkish tint. The odour is ± distinctly farinaceous, as is the taste. The spores measure (5–)5.5–7(–7.5) x (3.5–)4–5(–5.5) µm. It is mycorrhizal with deciduous trees and considered to be calciphobic.

Other whitish tricholomas do not have a radial, silk-like structure on the cap and typically smell more unpleasant.

Occasional and widespread but decreases towards the north; August–November.

Tricholoma inamoenum is a ± buff to pale grey-brown *Tricholoma* with distant, pale gills, an almost white, long stem and a strong gaseous smell. The spores measure 9.5–12 (–13) x 6–7.5(–8) µm. Occurs in damp, mostly calcareous *Picea* forests or plantations, mostly in thick moss. May be found together with the tiny *Hygrophorus exiguus* ⋉.

The prominent, distant gills and strong smell make this species easily recognizable. *Tricholoma sulphureum* ▽ has ± the same smell but is a completely different colour.

Quite common in the hemiboreal–boreal zone but rare in the nemoral zone; July–November.

Tricholoma sulphureum is a rather small, mostly almost unicoloured, warm yellow to greenish-yellow, dry *Tricholoma* with an extremely unpleasant, penetrating gaseous smell. The spores measure (8–)8.5–11(–12) x 5–6.5(–7) µm. Forms with more red-brown colours on the cap also occur. Usually mycorrhizal with deciduous trees, mostly on rich soils, but may also be found with various conifers.

Tricholoma sulphureum forms a complex of species that requires a thorough taxonomic review. Forms with more reddish cap centres are sometimes separated as *T. bufonium* ⋉, but this treatment is not fully supported by molecular data.

Common and widespread; mainly August–November.

Tricholoma arvernense is a medium-sized to large, dry, innately fibrillose *Tricholoma*, with an olive to sordid yellow cap, white gills and a rancid-farinaceous taste. The hyphae have clamps, and the spores are small, measuring 4.5–6 x 3.5–5 µm. Associated with sandy conifer forests and coastal plantations, where it is mycorrhizal with *Pinus* and, rarely, *Abies*.

No other tricholomas with yellowish to olive cap colours and white gills occur in the same habitat.

Quite common in the hemiboreal–boreal zones, but rare or absent from the nemoral zone; July–November.

Tricholoma sejunctum is a rather large, dry, innately radially fibrillose *Tricholoma* with a greenish-yellow to olive-grey cap and white gills. The taste is farinaceous, as is the smell after bruising. The spores measure (5.5–)6–8(–8.5) x 4–6.5(–7) µm. Occurs in deciduous forests on clay-rich soils, mostly with *Fagus*, often on sites with numerous other rare species.

Tricholoma umbonatum ⋉ has an acute umbo and a broad, whitish cap margin. *Tricholoma viridifucatum* ⋉ is less fibrillose and paler.

Rather rare to occasional, and local; August–November.

Tricholoma aestuans is a medium-sized, at first sticky, smooth to minutely scaly *Tricholoma* with a conical, yellowish cap, pale yellow gills and bitter-tasting flesh. The spores measure 6.5–7.5 x 4–5 (–5.5) µm. It is mycorrhizal with *Picea* and *Pinus*, mostly on sandy soils.

Tricholoma equestre ▷ and *T. frondosae* ▷ are recognized by the more distinct yellow colours on the gills and by the mild-tasting flesh.

Occasional and rare in the nemoral zone; July–November.

†† ***Tricholoma equestre*** is a sturdy, fleshy, farinaceous, mild-tasting, ± yellow, sticky and shiny *Tricholoma*. The gills are crowded and vividly yellow. The cap may also be vivid yellow, but becomes more reddish-brown and minutely scaly with age and exposure. The fruitbody is mostly ± immersed in sand. The spores measure (6–)6.5–7.5(–8) x 3.5–5 µm. Mycorrhizal with *Pinus*, generally on sandy soils.

Tricholoma frondosae ▽ is a more slender and associated with *Populus*. This and *T. equestre* belong to a poorly understood group of yellow, farinaceous, mild-tasting tricholomas. *Tricholoma aestuans* ◁ is more conical and tastes very bitter.

Common in areas with sandy soils and *Pinus*; mainly August–November.

Tricholoma equestre was formerly considered a choice edible but following a series of poisonings, some lethal, it is no longer accepted as such. The symptoms include the breakdown of the muscles – rhabdomyolysis – but the precise mechanism is not fully understood.

††? ***Tricholoma frondosae*** is a rather large, slender, farinaceous, ± yellow *tricholoma* with vividly yellow gills and an umbonate and, with age, ± greyish to brownish, minutely scaly cap. The spores measure 5–6.5 x 3.5–4.5 µm. Mycorrhizal with species of *Populus*, including *P. tremula*.

Tricholoma frondosae is part of a species complex with *T. equestre* △, but that species is more robust, hardly as dark scaly, and is mycorrhizal with *Pinus*. *Tricholoma aestuans* ◁ is bitter-tasting and associated with conifers.

Widespread, but local; September–November.

Collybioids

The collybioids are characterized by convex or umbonate to applanate, silky-fibrillose, greasy, hairy or slimy caps, adnexed to broadly adnate gills and white spore-deposits. *Rhodocollybia* has a pale grey-rose deposit and *Macrocystidia* may have a ± brown deposit. Most species are moderately tough. The spores are, with a few exceptions, inert to iodine.

The collybioids function predominantly as decomposers. A number of species have long, rooting stems, which arise from buried substrates such as tree roots or buried cones. *Laccaria* differs by being important ectomycorrhizals.

In older literature many collybioid species were included in the genera *Lyophyllum*, *Collybia* and *Xerula*. However, molecular phylogenetic studies have shown that these genera are all strongly polyphyletic, and this has led to much 'splitting', as can be seen from the many unfamilar names.

Other similar fungi:
– mycenoids with narrowly adnate gills may look very similar. However, they mostly have campanulate-umbonate caps, tend to be more fragile, and most species have amyloid spores (page 176).
– marasmioids are rather similar to the collybioids in shape and habitat, but most can tolerate desiccation. The stem is typically darker with a pale apex. They are also somewhat tougher (page 302).
– tricholomatioids are more fleshy and have ± emarginate gills (page 226).
– entolomatoids have brownish-rose spore-deposits and angular spores (page 448).
– LBMs (little brown mushrooms) may have collybioid shapes, but are brown-spored (page 616).

Further reading: 4, 13, 24, 25, 30, 156, 181, 183, 246, 348.

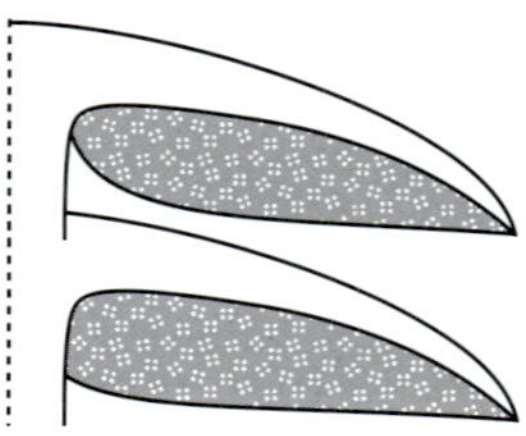

adnexed to broadly adnate gills

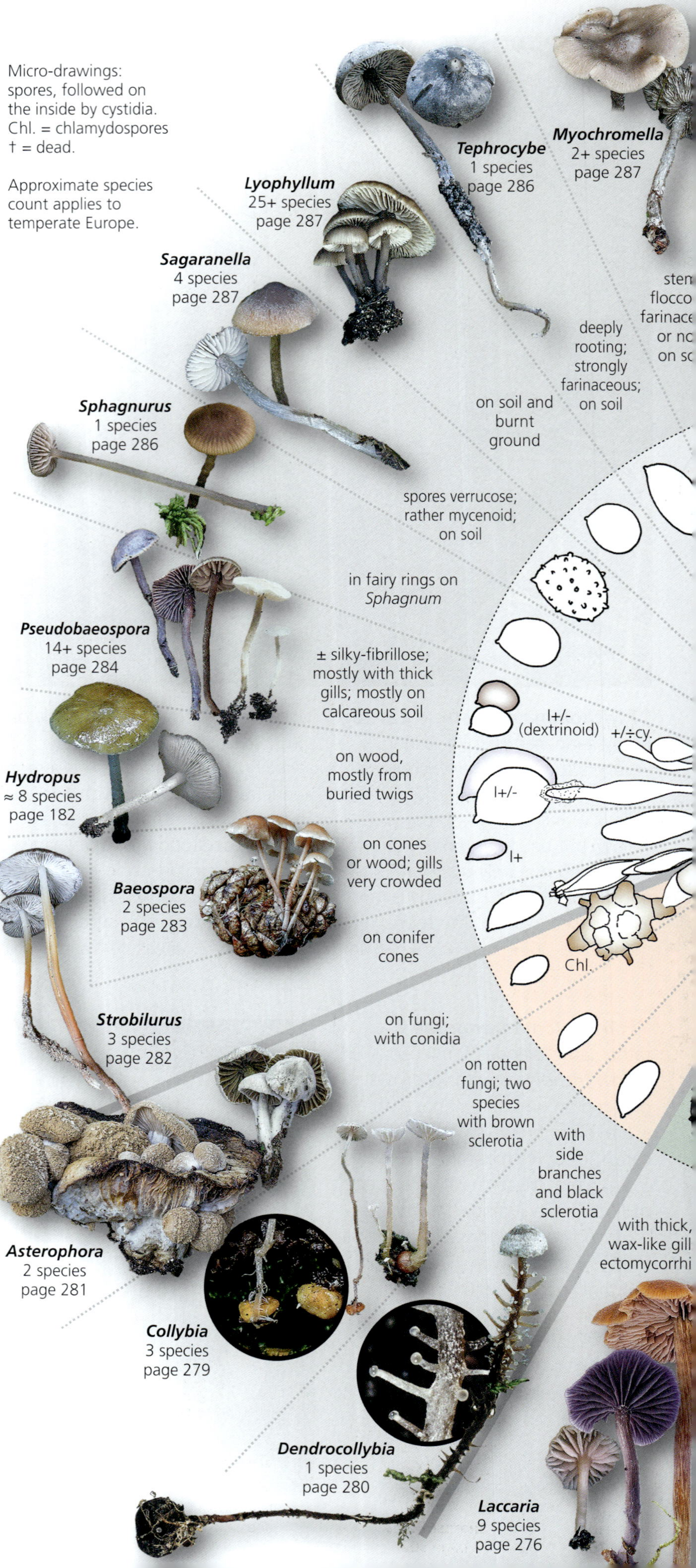

allistosporium
2+ species
page 289
sordid yellow-brown; dead spores have orange 'drop'; on soil or wood
Macrocystidia
1 species
page 288
stem and gills with projecting cystidia; smell of linseed; on soil or wood chips
Clitocybula
and the likes
4 species
(see page 238)
cap surface radially cracking; on wood
Collybiopsis, Connopus & Gymnopus
30+ species, page 290
gills distant to crowded; cap greasy; on litter
Rhodocollybia
5+ species, page 296
gills ± crowded; cap greasy; spores ± thick-walled and deposit pale grey-rose; on litter
I+/-
(dextrinoid)
Xerula
2 species
page 298
gills distant; cap and stem hairy; from roots
Paraxerula
1 species
page 298
gills distant; cap pruinose; from roots
gills distant, mostly dark-edged; cap greasy; spores large; from roots
Hymenopellis
1 species
page 299
cap greasy-slimy; stem dark, tomentose; on wood
Flammulina
6 species
page 300
cap slimy; stem with ring; on wood
Mucidula
1 species
page 301
collybioids
are rather tough, have convex to applanate caps, narrowly to broadly adnate gills and, mostly, whitish spore-deposits
other similar fungi
cap mostly campanulate-umbonate; gill attachment variable
mycenoids
page 176
tough and revive after drying; gills narrowly to broadly adnate or with collarium
marasmioids
page 302
with coloured spores
entolomatoids and **LBMs**
pages 448 & 616

Laccaria is recognized by the thick, wax-like gills and large, echinulate spores. It is ectomycorrhizal.

Laccaria amethystina is a ± violet *Laccaria* with thick, rather distant, emarginate or adnate, violet gills and a rather tough, somewhat fibrillose stem. Older specimens discolour to a strange greyish-violet colour. The spores are echinulate, almost spherical and measure 7.5–10.5 × 7.5–10 µm. Mycorrhizal with both deciduous trees and conifers on a range of soil types.

The thick and distant gills should prevent confusion with *e.g. Mycena pura* (page 196), or the brown-spored *Inocybe lilacina* (page 671). *Mycena pura* is also distinctly hygrophanous in comparison to *L. amethystina* and smells strongly of radish.

A common to very common and widespread species; mainly July–November.

Laccaria bicolor is a brownish-red, somewhat scaly *Laccaria* with violet-tinted gills and stem base, but otherwise is similar to *L. proxima* ▷. It may have ± differentiated cheilocystidia. The spores measure 7–9.5 × 6–8.5 µm, Qav. 1.1–1.25, with spines up to 1.5 µm long. It is mostly found in sandy conifer forests and plantations and forms ectomycorrhiza with *e.g. Pinus* and *Picea*.

Distinctive, provided the colour of the felt at the stem base is evident.

Widespread and common to fairly common; mainly June–November.

Laccaria proxima is a sturdy, often tall, minutely scaly, ± brownish-red *Laccaria* with a rather thick, coarsely fibrillose stem that lacks violet felt at the base. The echinulate spores are not completely spherical and measure 7.5–11 × 6–9 µm, Qav. 1.2–1.35. Typically found on somewhat peaty soils, where it is mycorrhizal with both deciduous trees and conifers.

Laccaria laccata ▽ is less scaly and usually has spherical spores. *Laccaria bicolor* ◁ has violet-tinted gills and violet-tinted felt at the base of the stem.

Widespread and common; mainly July–November.

Laccaria laccata is a rather slender, ± brownish-red to coral-red *Laccaria* with a finely fibrillose stem and an almost smooth to minutely scaly cap, which normally is somewhat striate. Basidia are 4-spored. The ± spherical spores measure 7.5–11.5 × 7–9.5 µm, with Qav. 1–1.1(–1.15) in the most common form. Occurs in numerous habitats, but usually in damp places and mostly with deciduous trees.

Laccaria fraterna ⋉ is 2-spored; *L. proxima* △ is more scaly, more coarsely fibrillose and the spores have a higher Q. *Tubaria furfuracea* (page 645) is brown-spored with thinner gills.

Widespread and very common; mainly June–November.

Laccaria maritima is a rather compact, ± brownish-red *Laccaria* with a deeply rooting stem that holds a lot of sand. Basidia are 4-spored. The short, echinulate, hyaline, ellipsoid spores measure 12–17 × 7–11 µm. Mycorrhizal with *Salix repens* in dunes, typically on almost bare, damp sand; possibly also on sandy river banks.

May be confused with other species of *Laccaria* or possibly with hygrocyboids (page 142), but the extreme habitat combined with the large, minutely ornamented, ellipsoid spores should ensure an accurate identification.

Rather rare (or overlooked) and apparently almost exclusively coastal; mainly September–November.

Laccaria tortilis is a small, slender, ± flesh-pink *Laccaria*, often with the cap almost at soil level. The cap is distinctly striate to the centre and smooth. The basidia are 2-spored. The spherical spores have 2–3 µm long, echinulate ornaments and measure 10.5–15 µm. Occurs in damp sites, mostly with *Salix*.

All other laccarias have spore ornaments less than 2 µm long. Small specimens of *L. laccata* ⊲⊲ can look similar but have 4-spored basidia. *Omphalina demissa* ⋉ and *Contumyces rosellus* (page 139), have smooth spores.

Rather common and widespread, but easily overlooked; mainly June–November.

Species of ***Collybia*** & ***Dendro-collybia*** are small, white-spored agarics growing on the remnants of rotten fungal fruitbodies (fungicolous). Three species produce fruitbodies from sclerotia, the fourth from darkened host tissue.

Collybia tuberosa is a small, whitish collybioid, which produces fruitbodies from rather large, pointed, at first yellow-brown, later red-brown, sclerotia that were formed during the previous season in the fruitbodies of other fungi. Lacks conspicuous cystidia. The smooth, ellipsoid, inamyloid spores measure 4.5–6.5 × 2.5–3.5 µm. Develops in fungal fruitbodies, which may be totally decomposed when the *Collybia* fruitbodies emerge from the sclerotia. Predominantly found in wet habitats, including mires with *Sphagnum*. Common hosts include species of *Lactarius* and *Russula*.

Collybia cookei ▽ is very similar in habit but has yellower and more rounded sclerotia; it also prefers drier habitats. *C. cirrhata* ▷▷ also prefers drier habitats and produces fruitbodies from remnants of the host – not from sclerotia.

Widespread and fairly common; mainly August–October.

Thomas Læssøe

Collybia cookei is a small, white collybioid with a silk-like appearance to the cap surface. Fruitbodies are formed from rounded, bulb-shaped, pale yellow-brown sclerotia. The smooth, inamyloid spores measure 4.5–6(–7) × (2.5–)3–3.5 µm. Occurs both in deciduous and coniferous forests; *Meripilus* is a common host.

In order to ensure an accurate identification, the shape and colour of the sclerotium must be determined. The sclerotium is red-brown in *C. tuberosa* △ and recalls a *Fagus* bud. In *C. cirrhata* ▷▷ no sclerotia are formed.

Widespread and common; mostly August–November.

Collybia cirrhata is a small, whitish collybioid with a somewhat chalk-like cap surface, pale, adnexed gills and a slightly buff-tinted stem. The smooth, inamyloid spores measure 4.5–6(–7) × (2.5–)3–3.5 µm. Occurs in troops on mummified fungal fruitbodies, mostly in heathland habitats. Common hosts include species of *Russula* and *Lactarius*.

The species is rather hard to identify unless found growing on heathland. The other collybias ◁◁ can be recognized by their yellow or red-brown sclerotia that developed in rotten fruitbodies during the previous season.

Widespread and rather common; mainly August–October.

Dendrocollybia racemosa is a small, yet spectacular, pale grey collybioid, with stems that form side-branches with a small knob at the end. The stem arises from a spherical, rather large, black sclerotium, which was formed in an agaric fruitbody the previous season. The side-branches produce asexual spores (conidia). The gills lack cystidia. The ellipsoid, inamyloid, smooth spores measure 4.5–6.5 × 2–2.5 µm. Mainly occurs on the fruitbodies of *Lactarius* and *Russula*, but, when fruiting, there are no traces left of the host.

This species is a unique and unmistakable agaric.

Rare and rather scattered distribution; mainly August–October.

Asterophora are small agarics that grow on the rotten fruitbodies of *Russula* and *Lactarius*. Both species form asexual, thick-walled resting spores (chlamydospores).

Asterophora lycoperdoides is a small, but rather thick-fleshed agaric that, with age, becomes brown-dusted. The gills are ± poorly developed. It produces highly characteristic, ± star-shaped, large, brown chlamydospores. The hyaline, smooth, inamyloid basidiospores measure 5–6 × 3–4 µm. Occurs on old fruitbodies of *Russula*.

Asterophora parasitica ▽ is less thick-fleshed, has more regular gills, and the chlamydospores are ± fusiform.

Occasional, locally quite common, nearly absent from the boreal zone; mainly August–November.

Asterophora parasitica is a ± greyish, rather small, radially fibrillose agaric. At maturity the gills become brownish and are filled with fusiform chlamydospores. The basidiospores measure 5–7 × 3–4 µm. Occurs on old fruitbodies of *Russula* and *Lactarius*.

Asterophora lycoperdoides △ is more powdery on the cap at maturity and has star-shaped chlamydospores. The gills are usually rather well developed in *A. parasitica* but rather reduced in *A. lycoperdoides*.

Occasional, locally quite common, nearly absent from the boreal zone; July–November.

Strobilurus is small genus of, long-stemmed collybioid agarics that grow on conifer cones. They have conspicuous cystidia in the hymenium and on the stem.

Strobilurus esculentus is a rather small, variable, white-spored, somewhat tough collybioid. The stem is characteristically bicoloured, pale above and yellow-brown below, and the top is covered by cystidia. The taste is mild. The ± thick-walled, fusiform cheilo- and pleurocystidia are are somewhat crystal-covered. The smooth, drop-shaped, inamyloid spores measure 4.5–6.5 × 2.5–4 µm. It grows on ± degraded *Picea* cones and is often deeply rooting.

This species can normally be recognized by substrate alone, but the more bitter-tasting *Pinus*-associated species, *S. tenacellus* ⋉, can also occur rarely on *Picea* cones but has pointed cheilocystidia. It may also recall a small *Gymnopus dryophilus* ▷▷ or possibly a *Mycena* (page 177). *Baeospora myosura* ▷, also found on cones, has crowded gills, amyloid spores and small, thin-walled cystidia.

Widespread and common; mostly in winter but can be found year-round.

Strobilurus stephanocystis is a mild-tasting *Strobilurus* that can only be recognized with certainty by its wide, capitate, thick-walled cheilocystidia with a ± evident crystalline covering. The spores are inamyloid and measure 4.5–7.5 × 2.5–4 µm. Exclusively on *Pinus* cones.

Strobilurus tenacellus ⋉ also grows on *Pinus* cones; it should be more bitter-tasting, but is better characterized by the pointed, less thick-walled cystidia. *Strobilurus esculentus* △ grows on *Picea* cones and has fusiform, less pointed cystidia.

Widespread and fairly common; mostly in the cooler months.

Baeospora is a genus of small, collybioid fungi with very crowded gills and amyloid spores.

Baeospora myosura is a rather small, thin-fleshed, white-spored collybioid with crowded, narrow, adnexed gills, a downy, ± unicoloured, somewhat brownish stem and almost free gills. The simple cheilocystidia are thin-walled and ± clavate. The small, smooth, amyloid spores measure 3–4.5 × 1.5–2 µm. Occurs on cones and cone-scales from a range of different conifers.

Species of *Strobilurus* ⊲⊲ have less floccose, distinctly bicoloured stems and more distant gills. The cystidia are also ± thick-walled and have crystals, and the spores are inamyloid.

Widespread and common; mostly September–December.

Baeospora myriadophylla is a rather slender, hygrophanous, collybioid agaric with greyish-lilac, crowded, adnexed gills. It has simple clavate cheilo- and caulocystidia. The small, smooth, amyloid spores measure 3.5–5 × 2–3 µm. Occurs on rotten trunks, mostly *Picea,* in forests of high nature conservation value.

Chromosera cyanophylla (page 148) occurs in similar places and has lilac to lavender gills, but these are more distant and deeply decurrent.

A rather rare species, mostly found in central and eastern Europe; very rare in Scandinavia and absent from western areas; mainly autumn.

Jean-Marc Moingeon

Thomas Læssøe

Pseudobaeospora are small to very small, fragile collybioids with bluish, greyish-violet or whitish colours and silky-fibrillose caps. The spores are small and may be thick-walled and dextrinoid. They are decomposers, usually deep in herbal vegetation and thickets; mostly calciphilous.

Pseudobaeospora pyrifera △ is a rather robust and thick-fleshed *Pseudobaeospora* with subdued greyish-violet colours and brownish-violet gills. The cap cuticle consists of radiating, inflated, short-celled hyphae, and stains blue-green to brownish-green with KOH. The cheilocystidia are ± broadly clavate. Basidia are 4-spored, with clamps, and with many thick-walled, so-called sclerobasidia. The almost spherical spores measure 2.8–3.7(–4.2) × 2.6–3.5(–3.8) µm, and are ± thick-walled and ± dextrinoid. Occurs on calcareous soil in thickets and forest edges.

Pseudobaeospora mutabillis ⋉ is similar but has a hymeniform cap cuticle.

Very rare and local; autumn.

Pseudobaeospora pillodii ▷ is a somewhat greyish-violet *Pseudobaeospora* that lacks clamps and cheilocystidia. The cap cuticle becomes ± red-brown with KOH. The spores measure (3.2–)3.7–4.5(–4.9) × (2.6–)2.9–3.6 µm, Q 1.15–1.25. Occurs in mull-rich, damp woodland and fens; probably calciphilous.

This species can recall *Cortinarius lilacinopusillus* (page 736) but this and a number of other similar species have clamps and cheilocystidia; the rather simple structure of the cap cuticle is also a good distinguishing feature.

Very rare and local; autumn.

Thomas Læssøe

Pseudobaeospora celluloderma is a small, violet, ± striate *Pseudobaeospora* with a cap cuticle of inflated, erect cells. Cheilocystidia are absent but clamps present. The spores measure (3–)3.5–4.5 × 2.5–3.5 µm, Q ca. 1.3, some are thick-walled and dextrinoid. Calciphilous, occurring in grassland and herb-rich forests, often deep in the vegetation.

Two other *Pseudobaeospora* species have inflated cells in the cap cuticle: *P. mutabilis* ⋉ is violet, more robust and has cheilocystidia; *P. basii* ⋉ is pale brown.

Very rare, but possibly somewhat overlooked as it has to be dug out of the turf; mainly August–November.

Thomas Læssøe/Jens H. Petersen

Pseudobaeospora calcarea is a rather slender, white, matt *Pseudobaeospora* with an acrid taste. It does not react with KOH. Has scattered clamps but no cheilocystidia. The upper cap cuticle consists of radiating, inflated hyphae. The basidia are 4-spored. The spores measure 3.1–4.8 × 2.5–3.4 µm. Calciphilous, mostly in mossy places, both in conifer forests and in coastal thickets.

There are many other species of white pseudobaeosporas, but *P. calcarea* is probably best distinguished by its acrid taste.

Rare and local; autumn.

Thomas Læssøe

Pseudobaeospora paulochroma is a pale, somewhat brownish-tinted, *Pseudobaeospora* with a radiating cap surface, the upper part forming a thin layer of narrow hyphae; stains yellow with KOH. The basidia are 4-spored and lack clamps. The stem apex has cystidia. Spores measure 4–4.5 × 3–3.5(–4) µm. On light, calcareous soils in thorny thickets.

Pseudobaeospora calcarea ⋉ has an acrid taste and is ± pure white; *P. albidula* ⋉ has more distant gills (11–17 compared with 19–24 in *P. paulochroma*) and lacks the thin layer of narrow hyphae in the upper cap cuticle.

Very rare and local; autumn.

Thomas Læssøe

Jan Vesterholt

Tephrocybe, ***Sphagnurus***, ***Myochromella*** and ***Sagaranella*** are all genera with relatively small collybioid fruitbodies that were formerly included in *Lyophyllum*. Many species have farinaceous smells and greyish to grey-brown colours.

Tephrocybe rancida is a rather small to medium-sized, dark, but pale pruinose, somewhat hygrophanous, umbonate, deeply rooting and rancid-farinaceous collybioid with adnexed, greyish gills. Spores ellipsoid, smooth to finely echinulate, and measure 7–8.5 × 4–4.5 µm. Typically occurs in rich deciduous forests, but also in scrub and conifer forests on better soils. It is not clear why it is rooting.

The combination of a rancid-farinaceous smell, a deeply rooting stem and a pruinose surface makes this species easily recognizable.

Widespread and fairly common, but can be nearly absent in some years; August–January.

Sphagnurus paluster is a rather anonymous, somewhat variable but mostly ± grey-brown, marginally striate, slightly farinaceous, slender collybioid. The gills are ± emarginate to somewhat decurrent and the cap usually has a small umbo or navel. Cheilocystidia can be present or absent. The inamyloid spores are smooth, ellipsoid and measure 5.5–8 × 3.5–5 µm. Occurs on *Sphagnum* in mires where it forms fairy rings, apparently killing the *Sphagnum*.

Species of *Mycena* (page 177) can grow in similar places but these have cystidia and amyloid spores.

Widespread and common; often early in the season, May–October.

Myochromella boudieri is a strongly rancid-farinaceous, pruinose, pale grey-brown, hygrophanous, somewhat umbonate, non-rooting collybioid with pale, adnexed gills and a dark stem with conspicuous pale floccules. The inamyloid spores are smooth and measure 6–8.5 × 3–4.5 µm. Occurs on rich, often disturbed, mostly calcareous soils in forests and scrub.

Myochromella boudieri belongs to a difficult species complex. *Lyophyllum coracinum* ⋉ is darker and the cap becomes depressed with age.

Widespread and occational, rarer towards the north; mainly September–November.

Sagaranella tylicolor is a very variable, but rather mycenoid collybioid with broad, adnexed, pale gills and a striate cap. The stem is floccose. Smell farinaceous or odourless. The finely echinulate, inamyloid spores measure 6–8 × 4–6.5 µm. Occurs in places where the soil or the moss have been enriched, *e.g.* where cows and other animals urinate, both in woodlands and in open habitats.

This species has a rather anonymous appearance, but can be easily identified if checked microscopically.

Widespread and probably rather common; June–December.

Lyophyllum anthracophilum is a rather small, grey-brown, ± marginally striate collybioid with a finely floccose stem and a ± evident rancid-farinaceous smell. The spherical, almost smooth spores measure 4.5–6.5 µm. It is associated with burnt ground.

The similar *L. ambustum* ⋉ also has spherical spores, but these are obviously verrucose.

Rare, but perhaps somewhat overlooked; mainly September–March.

Jan Vesterholt

Macrocystidia cucumis is a medium-sized (up to 5 cm wide), red-brown, convex, pruinose-downy, brown-spored collybioid that smells strongly of cucumber or pickled herring. The stem is downy and darker towards the base. The cap margin is mostly yellow-brown. The hymenium has large, fusiform cystidia, and similar cystidia cover the stem surface. The smooth, brownish spores measure 8–9.5 × 3–4 µm. Occurs on litter, wood chips *etc.*, mostly on very disturbed, rich soil.

A very characteristic species that is easy to recognize by the smell, pruinose cap and prominent cystidia. It has two varieties besides the main variety, both of which form small fruitbodies. Whilst the main variety *M. cucumis* var. *cucumis* is fleshy and non-striate with relatively narrow gills, *M. cucumis* var. *latifolia* ▽ is paler with a striate cap and broadly arched gills. It apparently also has a paler spore-deposit. *Macrocystidia cucumis* var. *leucospora* ⋉ is also very small, but with rather narrow gills and somewhat darker colours, but is still ± translucently striate.

Widespread and fairly common, decreasing towards the north; (July–) September–November (–January).

Thomas Kehlet

Macrocystidia cucumis var. ***latifolia*** is a small *Macrocystidia*, typically up to 1 cm wide, with broad, adnexed, very pale gills and a striate, pale yellow-brown cap, which may have a small central papilla. The smell is possibly weaker than in the main variety, *M. cucumis* var. *cucumis*, but the spores are the same. Typically found in grasslands.

See above for possible confusion with the other two varieties.

Rather rare or overlooked; mainly June–November.

Callistosporium is a small, collybioid genus with sordid yellow-brown colours and amber-yellow inclusions in the dead spores.

Callistosporium pinicola is a sordid yellow-brown to olive-brown collybioid that is bitter-tasting and white-spored with sordid curry-yellow, adnexed gills. Lacks cystidia. The small, smooth, inamyloid spores measure 2.5–4.5 × 2–3.5 µm, when dead, they have a large amber-yellow inclusion. Occurs on rather decomposed conifer wood, either solitary or in small clusters.

Other species, including *C. luteo-olivaceum* ▽ which reaches the Nordic countries, typically have larger spores. Species of *Gymnopilus* (page 602) have red-brown spore-deposits and no other collybioids have such yellow to olive brown colours combined with a bitter taste and occurrence on wood.

Rare, but apparently increasing; mainly September–December.

Callistosporium luteo-olivaceum is a dry, brownish-olive, white-spored collybioid with curry-yellow gills, which becomes a somewhat more sordid colour with age. The flesh stains violet with KOH. The spores measure 5–5.5(–6.5) × 3.5–4 µm, and, when dead, have an amber-coloured inclusion. Occurs solitary or in small clusters on rotten conifer wood.

Callistosporium pinicola △ has smaller spores and is less vividly coloured.

Recorded in Sweden, England, the Netherlands and farther south; autumn.

Federico Calledda

Collybiopsis*, *Connopus* and *Gymnopus are small to medium-sized, somewhat tough collybioids that mostly become applanate with age. The gills are adnexed. All are decomposers (see also page 306).

Gymnopus dryophilus is a yellow-brown collybioid with white, crowded gills. The stem base either lacks a bulb or, if present, it is ± insignificant; hyphal cords white. It is odourless. The cheilocystidia are very gnarled-branched and hard to find microscopically. The spores measure 4–7 × 2.5–4 µm. Occurs on leaf-litter in forests and carrs, occasionally in moss on living trunks in wet habitats.

Gymnopus aquosus ▽ has pinkish-tinted hyphal cords, a bulb at the stem base and inflated cheilocystidia. *Gymnopus ocior* ⋉ is typically darker, has mostly yellowish-tinted gills and usually occurs on wood chips and in other disturbed places. These species are very hard to identify, and material is often found that cannot be named with certainty.

Widespread and common; mainly mid-summer-late autumn.

Gymnopus aquosus is a pale yellow-brown, hygrophanous collybioid with crowded, pale gills, a smooth stem and a basal bulb with ± pink hyphal cords; no smell. The clavate cheilocystidia are ± swollen and easy to detect. The spores measure 5–8 × 3–4 µm. Grows on well-decayed leaf-litter, mostly in deciduous forests, but also in boggy habitats and mixed forests.

Gymnopilus dryophilus △ can often be recognized by its non-bulbous stem, whitish hyphal cords, and its inconspicuous, very irregularly gnarled-branched cheilocystidia.

Widespread and rather common; ± early in the season, May–October.

Gymnopus erythropus is typically a rather compact, odourless collybioid with a smooth, brownish-red stem that is dark-felty at the base. The pale gills contrast with the brownish-red cap and stem. The spores measure 5.5–9 × 3.5–4.5 µm. Usually occurs near and on dead wood of deciduous trees, mostly on rich soil.

Connopus acervatus ▽ has less vivid colours and is densely clustered on conifer wood. *Mycetinis scorodonius* (page 303) has a garlic-like smell, tougher fruitbodies with more distant gills, and grows in drier habitats.

Widespread but rather local, absent from the boreal zone; autumn.

Thomas Læssøe

Connopus acervatus is a distinctly clustered and very dark, conifer-associated collybioid that has a smooth stem with a felty base. It becomes pale when dry. Has crowded, pale gills and an insignificant odour. The spores measure 5.5–8 × 2.5–3.5 µm. Occurs on stumps of conifers.

Collybiopsis confluens ▷▷ grows on leaf-litter, is paler, and has a densely felty stem surface.

Widespread and common in the hemiboreal–boreal zone but rarer towards the south; July–December.

Collybiopsis confluens is an almost always distinctly clustered, rather slender collybioid with crowded, pale gills and a stem that is felty along its entire length. It lacks a distinctive smell or taste. The spores measure 6.5–8 × 3–4 µm. Grows mostly in large troops in fairy rings on thick leaf-litter in both deciduous and coniferous forests.

Connopus acervatus ◁◁ has darker, red-brown colours and a smooth stem except towards the base (but very old, water-soaked specimens of *Collybiopsis confluens* may look similar and cause confusion).

A very common and widespread species; July–November.

Thomas Læssøe

Collybiopsis obscuroides is a rather slender and very dark, odourless collybioid with a felty stem. The hyphae stain olive-brown in KOH. The relatively long spores measure 7.5–12 × 3.5–5 µm. Occurs in large troops, mostly in short grass in parkland.

Gymnopus fuscopurpureus ⋉ is even darker, occurs on leaf-litter and has smaller spores. *Gymnopus fagiphilus* ▽ occurs in the same habitat as *G. obscuroides* but does not react with KOH.

Locally common, at least in England, and also recorded in the Netherlands but generally rare or overlooked; late summer–autumn.

Gymnopus fagiphilus is a relatively pale-capped, but dark-stemmed, odourless collybioid with rather distant gills and dense pale felt on the lower part of the stem. The flesh does not stain green with KOH. The spores measure 6–9 × 3–4.5 µm. Occurs on thick *Fagus* leaf-litter on rich soil.

This species has been overlooked in the past, probably because it was confused with *G. fuscopurpureus* ⋉; however, the latter species reacts green with KOH.

Widespread and occasional, not present in the boreal zone; mostly September–November.

Collybiopsis peronatus is a rather tough, very variable collybioid with distant, somewhat yellow- to brown-tinted gills and a strong, pepper-like taste, an a sour odour. The stem base is woolly felty and mostly has a yellow shade that continues as fine floccules all the way to the top. The spores measure 8–10(–11) × 3–4 µm. Occurs on litter from deciduous and coniferous trees, mostly in fairy rings on acid soils; when picked, the litter adheres to the base.

The pepper-like taste, combined with a weak or sour smell, should make misidentifications unlikely. However, the very variable fruitbodies can cause confusion.

Widespread and very common; June–January.

Gymnopus hariolorum is a pale brown, rather large, early occuring collybioid with a cabbage-like smell. It has a very felty-hairy, basally clavate, almost white stem and crowded, almost white gills. The spores measure 6–7.5 × 3–4 µm. Mostly occurs on debris from deciduous trees on ± calcicolous clay soils.

Other collybioids or marasmioids (page 294-295) that have a cabbage-like smell are either smaller and/or darker, and have more distant gills.

A rather rare, local species, ± absent from the boreal zone; mostly April–June, but also August–October.

Gymnopus foetidus is a lignicolous, striate, red-brown, somewhat marasmioid collybioid with distant, rather dark gills, a dark stem that becomes almost black towards the base and a cabbage-like smell. The spores measure 7–10(–11) × 3.5–5(–5.5) µm. Occurs on branches and twigs of *Fagus* and *Quercus*, mostly in calcareous woodlands with a long continuity.

Despite having been placed in a series of different genera, *e.g.* *Micromphale* and *Marasmiellus* (page 302), this is an easy species to recognize. It is, however, not the only collybioid/marasmioid with a cabbage-like smell. *Gymnopus brassicolens* ▷ is also lignicolous, but has more crowded and paler gills. *Gymnopus impudicus* ▽ has a paler, typically longer stem, a ± non-striate cap and usually occurs on ± buried litter. *Gymnopus inodorus* ⋉ is very similar but, as the name implies, is odourless.

Widespread, and locally quite common, absent from the boreal zone; mid-summer–autumn.

Jan Vesterholt

Gymnopus impudicus is a rather small but robust, ± non-striate, red-brown collybioid with a cabbage-like smell. The gills are typically tinted the same colour as the cap and are ± distant. The stem surface is minutely downy and may have a violet tint. The spores measure 5.5–7.5 × 3–4.5 µm. It grows on soil in heath-like vegetation, *e.g.* with *Juniper*.

Similar species are either striate on the cap or/and lignicolous.

Widespread and occasional to rare, very rare in the boreal zone; July–October.

Gymnopus brassicolens is a red-brown, only slightly striate, lignicolous, somewhat marasmioid collybioid that has pale, rather crowded gills and a cabbage-like smell. The spores measure 5.5–7.5 × 2.5–4 µm. It mostly occurs on *Fagus* branches in ± calcareous woodlands.

Other collybioids or marasmioids (page 294) that have a cabbage-like smell typically have darker and, mostly, more distant gills.

A rather rare and locally occuring species, absent from the boreal zone; June–November.

Jan Vesterholt

Gymnopus fusipes is a very sturdy and fleshy collybioid with deeply rooting, mostly clustered stems and yellow- to red-brown colours; older specimens have red-brown spots on the entire fruitbody. Cheilocystidia short and ± gnarled-clavate. The spores measure 4.5–7(–7.5) × 3–4.5 µm. Ocurrs at the base of old *Quercus* trees (more rarely *Fagus*), which it presumably parasitizes; mainly in open park-like woodlands with many old trees.

An easily recognizable species that superficially recalls species of *Rhodocollybia* ▷▷, e.g. *R. maculata* ▷▷.

Widespread and locally quite common; June–November.

Jens Maarbjerg

Rhodocollybia is a small genus of rather large collybioids with pale vinaceous-brown spore-deposits, and with some dextrinoid spores (staining red-brown in iodine reagents). Decomposers, often in fairy rings.

Rhodocollybia maculata is a ± bitter-tasting, pale, red-brown-speckled, fleshy collybioid that almost resembles a species of *Tricholoma* (page 226), although the gills are adnexed, crowded and ± serrated. The stem is typically somewhat rooting. The almost spherical spores, of which some are thick-walled and dextrinoid, measure (5–)5.5–6.5(–7.5) × (3.5–)4.5–5.5(–7) µm. It occurs on acidic litter, mostly from conifers, but also in open heathland.

Whitish tricholomas (page 270) mostly have strong odours and ± emarginate gills. Other species in the genus *Rhodocollybia* are more red-brown.

Common and widespread; July–January.

Jens Maarbjerg

Rhodocollybia prolixa var. distorta is a large, red-brown, greasy collybioid with a stem that typically turns brown with age, pale, serrated gills and a mild taste. The almost spherical spores, of which some are thick-walled and dextrinoid, measure 4-6 × 3-4.5 (–5) µm. It is associated with debris from conifers and deciduous trees.

Rhodocollybia prolixa var. *prolixa* ⋉ is apparently somewhat darker. *Rhodocollybia filamentosa* ▷ has an innately fibrillose, almost dry cap and is somewhat smaller, but with larger spores.

Common in the hemiboreal–boreal zones but rare southwards; mainly September–October.

Jens Maarbjerg

Rhodocollybia filamentosa is an almost dry, radially fibrillose, red-brown, rather fleshy collybioid with a striate, clavate stem and crowded, almost free, white gills. The cheilocystidia are coralloid-branched, ± clavate and measure 19–40 × 4–9 µm. The spores, of which some are thick-walled and dextrinoid, measure 6.5–10(–11) × (3.5–)4–4.5 µm. Mostly associated with debris from conifers.

Rhodocollybia butyracea f. *butyracea* ⋉ is very similar, but lacks the obvious innate fibres on the cap.

Rarely reported, but perhaps also somewhat overlooked; mainly August–November.

Rhodocollybia butyracea* f. *asema is a distinctly hygrophanous, pale grey-brown collybioid with an umbonate, dark centre, particularly after drying. The stem is horn-grey and clavate and mostly smooth. The spores, of which some are thick-walled and dextrinoid, measure 6.5–10(–11) × (3.5–)4–4.5 µm. Occurs on leaf-litter, mostly from deciduous trees.

Rhodocollybia butyracea f. *butyracea* ⋉ has a more red-brown, unicoloured cap and does not have a dark centre after drying. It is mostly found on debris from conifers.

Widespread and very common; August–January.

Jan Vesterholt/Christian Lange

Xerula pudens is a white-spored, rather sturdy collybioid, with an orange-brown, felty cap and stem, pale, almost free gills and a deeply rooting stem. Has large cheilo- and pleurocystidia. The smooth, inamyloid spores measure 8.5–12.5 × 8–10.5 µm. Occurs on buried wood or perhaps living roots on ± calcareous, clay-rich soils in thermophilic, old deciduous woodlands.

Paraxerula caussei ▽ has a darker, more grey-brown colour, and a white, pruinose cap surface. *Xerula melanotricha* ⋉ is associated with conifers and has longer hairs on the cap (3 mm compared with 1 mm). Species of *Flammulina* ▷▷ have a smoother cap surface and more vivid colours, and the spores are ellipsoid to cylindrical rather than spherical.

Rare and local, not present in the boreal zone; mainly July–October.

Paraxerula caussei is a sturdy, rooting, white-spored collybioid with grey-brown, strongly pruinose cap and stem surfaces. The gills are white to cream, broad and narrowly adnate to emarginate. Pileo- and caulocystidia are hyaline. Has large cheilo- and pleurocystidia. The inamyloid, smooth spores measure 9–12 × 5.5–8 µm. Grows from buried wood or roots of deciduous trees, mostly *Fagus,* and mostly in calcareous woodlands.

Xerula pudens △ has a more orange-brown colour and is felty due to the coloured pileo- and caulocystidia. *Hymenopellis radicata* ▷ has a ± sticky cap surface. *Megacollybia platyphylla* (page 238) has a radially fibrillose cap surface.

Rare and local; mainly August–October.

Jan Vesterholt

Hymenopellis radicata is a tough-fleshed, white-spored collybioid with a greasy, yellow-brown, ± veined cap surface and, typically, a very deeply rooting, smooth, dry, brownish-tinted stem. The gill edge is mostly brownish, especially near the cap margin. The gills are narrowly adnate to emarginate and rather distant. Has large, clavate cheilo- and pleurocystidia. The smooth, large spores measure 12–17 × 9–14 µm. Occurs around living deciduous trees or stumps, from roots or buried wood.

Typical rooting specimens with greasy, somewhat veined caps are distinctive and unlikely to be misidentified. Small specimens may recall *Hydropus subalpinus* (page 182), but that species has a radially fibrillose cap, an undulating gill edge and allantoid spores. Very small specimens found directly on wood may cause confusion, but are easily recognized by the large, spherical spores.

Common and widespread; May–November.

Flammulina are mostly odourless collybioids with ± slimy caps, dark-felty stems and mainly yellow cap colours. They are decomposers on dead wood.

Flammulina velutipes is an orange-yellow *Flammulina* with a somewhat glutinous cap, moderately crowded, pale gills and a ± brown-felty, tough stem. It has rather long and narrow spores, measuring 6–9.5 × 3–4 µm, with a Q value at ca. 1.8–2.4, and with only a few spores longer than 10 µm. Grows on various deciduous trees, usually on dying trees, but sometimes roots into dead wood.

Flammulina elastica ▽ is very similar but usually has more distant gills and longer spores. *Flammulina finlandica* ⋉ also has longer spores. *Flammulina populicola* ⋉ has smaller spores (6–7.5 × 3.5–4.5 µm) and a preference for *Populus*. In central Europe, another intermediate species occurs, *F. rossica* ⋉, but differs in having a cap cuticle of clavate cells. The edible enokitake mushroom was until recently considered a cultivated form of *F. velutipes,* but is now recognized as an East Asian species, *F. filiformis* ⋉.

Common and widespread; mainly October–May.

Flammulina elastica is an orange-yellow *Flammulina* with ± distant gills, a ± brown-felty stem and long, narrow spores measuring 8–11.5 × 3–4 µm, a Q value above 2.3 and numerous spores that are longer than 10 µm. It is mostly associated with *Salix* in wet habitats.

Flammulina velutipes △ normally has more crowded gills, a lower spore Q, and only few spores longer than 10 µm; *F. finlandica* ⋉, currently only known from *Salix* in Finland, is very similar, but has crowded gills and even longer, and much broader spores (av. 9.88 × 4.05 µm, Qav. 2.46).

Apparently common and widespread; mainly October–April.

Flammulina fennae is a very pale-capped, clustered and rooting *Flammulina* with spores measuring 6-8 × 3-5 µm and a spore Q between 1.5 and 1.7. The centre of the cap may be tinted in yellow and brown, but the cap is always almost white at the margin. Roots into buried wood of deciduous trees, mostly with a very long and thin root-like extension.

Other species of *Flammulina* may also root into wood, *e.g.* *F. velutipes* ◁, but in that species the cap is coloured to the margin, and the spore Q is somewhat higher.

Rather rare to occasional and local; can be found most of the year.

Jaroslav Malý

Mucidula mucida is a rather spectacular large, slimy, clay-buff to porcelain-white, mostly clustered, white-spored, collybioid with a ± coloured ring on a tough and basally swollen stem. The large, spherical, inamyloid spores measure 14–18 x 11–16 µm. Found almost exclusively on *Fagus*, both on fallen trunks, large branches and on standing dead or weakened trees.

A very distinct species that should be very easy to recognize, although young and dark clay-buff specimens can be slightly confusing.

Widespread and common, absent from the boreal zone; May, August–December.

Jens Maarbjerg/Thomas Læssøe

Marasmioids

The marasmioid fungi have applanate to convex, rather tough caps and highly variable gill attachments, including a collarium around the stem (page 312). The stems are mostly dark, except for the apex, and some species have strong odours. The fruitbodies revive after desiccation. Microscopically many species have cystidia with finger-like protuberances.

The marasmioid fungi can be very difficult to distinguish from the collybioids, especially the genus *Gymnopus* (pages 290 & 306). Likewise, *Xeromphalina* (page 185) creates a bridge between marasmioids and mycenoids.

OTHER SIMILAR FUNGI:
– collybioids usually have adnexed gills and the stems are mostly not darker than the caps (page 274).
– marasmioid species without stem or with excentric stems are to be found amongst the pleurotoids (page 60).

FURTHER READING: 13, 156, 181, 202.

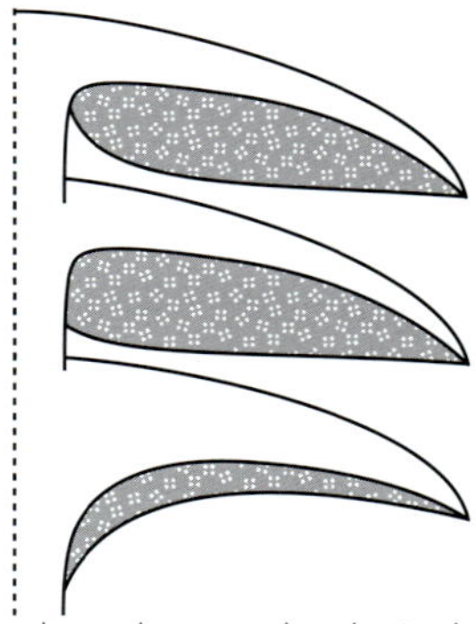

Mycetinis is recognized by the smell of stale garlic or old rubber-bands.

Mycetinis alliaceus is a large, tough, strongly garlic-smelling marasmioid with a very dark, finely felty-downy, stiff stem. The gills are pale and rather distant. The spores measure 7.5–11 × 6–8 µm. Almost exclusively on trunks and buried twigs of *Fagus*.

Mycetinis querceus ▽ is browner and occurs directly on leaf-litter. *Mycetinis alliaceus* var. *subtilis* ⋉ is a rather common dwarf form with a pale, marginally striate cap.

Widespread and very common in parts of temperate Europe, rare or absent from others; May–January.

Mycetinis querceus is a strongly garlic-smelling, large, pale marasmioid; the stem becomes felty towards the base. The spores measure 7–10 × 4–5 µm. Occurs mostly on *Quercus* leaf-litter late in the season (often Q. *rubra*), but also on *Fagus* leaf-litter.

Mycetinis alliaceus △ has darker stems, grows on wood and, in comparison to *M. querceus*, has cheilocystidia. *Mycetinis scorodonius* ▽ has a reddish, smooth stem. Similar species of *Gymnopus* (pages 290) are either odourless or have a cabbage-like smell.

Widespread, but mostly rather uncommon; mainly October–January.

Jean-Marc Moingeon

Mycetinis scorodonius is a medium-sized marasmioid with a shiny, reddish stem, pale cap and a strong garlic smell. Spores 7–10 × 3–5 µm. Occurs in open, dry places, *e.g.* in dunes, and can be found on *e.g.* stems of *Artemisia campestris* and similar coarse herbs, but occasionally also on wood, *e.g.* in *Syringa* hedgerows.

Crinipellis scabellus ▷▷ occurs in the same habitats but has projecting hairs on the stem and is odourless. *Marasmius oreades* ▷▷ is less red and has a different smell. Other *Mycetinis* species have ± felty stems, are larger and more bound to woodland.

Widespread and rather common; mainly July–November.

Rhizomarasmius includes odourless species with ± decurrent gills and downy stems that gradually become darker towards the base.

Rhizomarasmius undatus is a rather compact, completely pruinose-downy, odourless marasmioid with adnate to somewhat decurrent gills. The cap is pale, and the downy, rooting stem gradually becomes dark reddish-brown towards the base. The cheilocystidia are large and capitate. The lemon-shaped to fusiform spores measure 8–12 × 4–6 µm. Roots into the underground parts of *Pteridium aquilinum*, mostly in calcareous habitats.

This species is not easy to confuse, but may perhaps recall a small *Xerula* (page 298), which, amongst other things, may be recognized by its ± spherical spores and almost free gills. Species of *Mycenella* (page 178), may also look rather similar, but have ± spherical, mostly coarsely verrucose spores.

Widespread but local and very rare or absent from some areas; mainly September–November.

Rhizomarasmius setosus is a small, white marasmioid with a few, somewhat decurrent gills and a long, thin stem which gradually darkens towards the base and has scattered, very long hairs. The spores measure (9–)10–14.5 × 4–6 µm. Occurs on leaf-litter, mostly from *Fagus,* but also from other deciduous trees, *e.g. Salix*. May occur on twigs, but rarely.

The bicoloured, long-haired stem should prevent confusion with small, white-capped *Hemimycena* and *Mycena* species (pages 188 & 121). *Marasmius epiphyllus* ▷▷ lacks the long hairs, has a wider cap with vein-like gills, and prefers leaves from other trees.

Very common in most of temperate Europe, except the boreal zone; mainly July–December.

Collybiopsis includes small to medium sized, pale, smooth to downy-stemmed, marasmioid species with ± decurrent gills.

Collybiopsis ramealis is typically a rather pale marasmioid with a short, pale brown stem that becomes finely scaly towards a somewhat widened base. The gills are adnate to rather decurrent; they may may meet in a small, thickened ring (but not a true collarium). The clavate cheilocystidia have gnarled, short projections. Cap cuticle hyphae have numerous short protuberances. The ± fusiform spores measure 7.5–12 × 2.5–5 µm. Occurs in troops on twigs of both deciduous trees and conifers, including *Juniperus*.

Collybiopsis vaillantii ▽ has, typically, a longer stem, more ellipsoid spores and irregular, gnarled cheilocystidia.

Widespread and very common; mainly June–January.

Collybiopsis vaillantii is a pale brownish marasmioid with somewhat decurrent gills and a long, floccose-downy stem, which is brownish towards the base. The cheilocystidia are irregularly gnarled, and the cap cuticle has no small protuberances. The spores measure 7–10 × 3–5 µm. Occurs on various herbs, including grass culms in open habitats, mostly deep in the turf.

Collybiopsis ramealis △ has more obvious scales on the stem, the cheilocystidia are clavate and the cap cuticle cells have short, gnarled protuberances. The rather tough texture and bicoloured stem of *C. vaillantii* should prevent confusion with mycenoids (page 177).

Widespread and common; mainly June–November.

Thomas Læssøe

"Marasmiellus" tricolor is an easily recognizable marasmioid with almost pink, decurrent gills, a chalk-white cap and a stem that soon becomes grey-black towards the base. It has a variable number of cheilocystidia, and the spores measure 8.5–12.5 × 3.5–6 µm. On rhizomes/roots in herb-rich, open, dry grassland communities.

"Marasmiellus" trabutii ⋉ occurs in saline grasslands on, *e.g. Juncus maritimus*, but also on *Schoenus nigricans*, *etc.*, from the Netherlands southwards. It has a gelatinous cap cuticle (dry in *M. tricolor*).

Widespread, but nowhere common; mainly June–September.

Paragymnopus perforans is a rather small and slender marasmioid with a smell of rotten cabbage. The stem is felty and increasingly reddish-brown towards the base. The gills are pale and adnexed. The spores measure 6–9.5 × 3–5 µm. Occurs in troops on needle debris.

The combination of felty stem, smell and habitat is distinctive.

Widespread and very common; mainly April–January.

Gymnopus androsaceus is a small, slender marasmioid with a smooth, horse hair-like stem and adnexed, ± grey-brown-tinted, distant gills. It is odourless. The spores measure 6.5–9 × 3.5–5 µm. Occurs mostly on twigs in forests and heathland habitats.

Paragymnopus perforans △ smells of cabbage and has a felty stem (hand lens required!). *Collybiopsis quercophila* ▷ looks similar, but the stem is minutely downy and it always occurs on *Quercus* leaves.

Widespread and common; mostly in the autumn.

Collybiopsis quercophila is a small, slender, rather pale marasmioid with brownish colours and an almost smooth to minutely downy stem surface; the stem base lacks felt. It is odourless. The spores measure 7.5–9 × 4.5–5.5 µm. Occurs exclusively on *Quercus* leaf-litter.

Marasmius bulliardii ▷▷ has the gills attached to a collarilum, and *M. epiphyllus* ▷▷ has a chalk-white cap and veined, anastomosing gills.

Widespread and rather common, decreasing towards the north; mostly present in the autumn.

Crinipellis scabella is a brown, radially hairy-felty, white-spored marasmioid with a reddish-brown to dark brown, strongly felty-hairy, rigid, thin stem. The cheilocystidia are slender and gnarled. The inamyloid, smooth spores measure (6.5–)7.5–9.5(–10.5) × (4–)4.5–6(–8) µm. Grows on coarse, herbaceous stems, including grass culms, in dry, open habitats, *e.g.* dunes, but may also occur on twigs in dry scrub.

Other European species of *Crinipellis* are more southern. Species of *Collybiopsis* ◁◁ and *Mycetinis* ◁◁ do not have radially and coarsely felty cap surfaces or projecting brown hairs on the stem.

Widespread and common in suitable habitats, but rare in the hemiboreal–boreal zone; mainly June–January.

Marasmius is a rather variable genus of tough, drought-tolerant fungi with adnexed gills or gills connected to a collarium around the stem. They are typically decomposers of herbaceous stems and leaf-litter.

• ***Marasmius oreades*** is a rather atypical *Marasmius* with almost fleshy, ± unicoloured, pinkish-buff, hygrophanous, umbonate caps, rather distant, adnexed, wide, pale buff gills, a minutely hairy stem surface and a smell recalling bitter almonds or more pleasantly spicy. The spores measure 8–11 × 4–6 µm. It kills grass and occurs as fairy rings in all kinds of grassland.

Marasmius collinus ⋉ is a more elegant looking species with an almost smooth and rather fragile stem, and an unpleasant metallic or gas-like smell recalling that of *Lepiota cristata* (page 326); it occurs in similar habitats, but it is not known from lawns. May also recall some species of *Gymnopus* (page 290), but these do not produce fairy rings in grass turf.

Widespread and very common; May–November.

Marasmius torquescens is a rather large *Marasmius* with a finely, hairy-felty stem that has a thick-felty base, and rather distant gills that do not have a coloured edge. The cap is buff and the stem becomes darker towards the base. The spores measure 7–10 × 3–5.5 µm. Occurs on leaves and small twigs in leaf-litter in ± rich deciduous forests.

Marasmius cohaerens ▷ has a smooth, more shiny stem, a brown gill edge with projecting cystidia, more crowded gills and, typically, a somewhat darker cap colour.

Widespread and common, but very rare in the boreal zone; mainly July–November.

Marasmius cohaerens is a rather large, buff to clay-buff, finely pubescent, ± hygrophanous, odourless *Marasmius* with a ± brown gill edge. The stem is rigid, smooth, shiny, and black towards a felty base. The gills are covered with brown, projecting cystidia (hand lens required!). The spores measure (7–)8–10.5 × 4–5.5(–6) µm. Mostly found on leaf-litter in rich deciduous forests.

Marasmius torquescens ◁ has a felty stem, distant gills that lack brown cystidia and usually a somewhat paler cap. May initially recall some brown-spored agarics (LBMs – page 616) but has a tough texture and white spore-deposit.

Widespread and fairly common; mainly August-December.

Thomas Læssøe/Jens H. Petersen

Marasmius wynnei is a large *Marasmius* with a greyish to almost white, semiglobose to applanate cap and a very distinct bicoloured, finely floccose stem that is pale above and reddish-brown below. The spores measure 6–9 × 3–4.5 µm. Grows in big troops or clusters and binds thick layers of leaf-litter together with white sheets of mycelium. Occurs on rich soil in deciduous forests and scrub.

The distinctly bicoloured, ± floccose stem, the tough texture and grey-toned cap make this species hard to confuse. Large mycenas (page 177) are more soft-fleshed and do not bind the leaf-litter in sheets. Species of *Gymnopus* (pages 290 & 306), are less contrasty.

Widespread and fairly common, except in the hemiboreal–boreal zones; August–January.

Jens H. Petersen/Thomas Kehlet

Marasmius epiphyllus is a white *Marasmius* with distant, vein-like ± anastomosing gills and a minutely downy stem that darkens towards the base. Spores 8–12(–14) × 3–5 µm. Occurs typically on petioles of various deciduous trees in moist places, *e.g.* in *Fraxinus* woodland, but may also occur in drier, more open sites.

Marasmius epiphylloides ⋉ occurs on the leaves of *Hedera* and has setose cells in the cap cuticle, whereas *M. epiphyllus* only has projecting, smooth pileocystidia. *Rhizomarasmius setosus* ◁◁ is smaller and has long, projecting hairs on the stem.

Widespread and fairly common in most regions; mainly August–January.

Marasmius anomalus is a slender, smooth- and slim-stemmed *Marasmius* with pale brownish to pale brick-red colours on the radially deeply furrowed cap. The gills are distant and adnexed. The long, narrow spores measure (12–)16–22(–24) × (3–)3.5–4.5(–6) µm. Occurs in very dry grassland, on grass or rabbit or hare droppings.

May recall *Marasmiellus vaillantii* ◁◁, which has a paler, floccose stem, less brown in the cap and smaller spores. *Marasmius curreyi* ▷▷ may also look similar, but its gills are attached to a collarium.

Found mostly in coastal areas in northwestern parts, where it can be locally common; August–December.

Jacob Heilmann-Clausen

Marasmius siccus is a moderately large, brownish-orange to pale brick-red, beautifully plicate-sulcate *Marasmius* with white gills, which are not connected to a collarium. The very large spores measure 15–23.5 × 2.5–5 µm. Occurs on litter in rather dry habitats with *Betula*, *Salix* etc.

Marasmius curreyi ▷▷ is much smaller, and the gills are attached to a collarium.

Found mostly in warmer parts of the far north of temperate Europe; August–October.

Cryptomarasmius minutus is a very small and very slender, pale red-brown *Marasmius* with few gills, which are not joined to a collarium but are either almost free or adnexed. The stem is as thin as a thread, and smooth. The spores measure 6–9.5 × 2–4(–5) µm. Occurs in damp sites on leaves of *Salix* and *Populus*.

Marasmius limosus ▷▷ is just as small, but the gills are connected to a collarium and it grows on monocots (*e.g. Phragmites*, *Juncus* and *Carex*).

Widespread, but rarely reported, as easily overlooked; mainly June–October.

Marasmius hudsonii is a small, yet spectacular *Marasmius* with projecting, long, stiff, red-brown hairs on the white cap and reddish-brown stem. The spores measure 7.5–13.5 × 4–6.5 µm. Occurs exclusively on leaves of *Ilex aquifolium* in ± damp woodlands.

This distinctive species should cause not identification issues, but can be difficult to locate, although it sometimes occurs in mass fruitings.

Fairly widespread and common in the British Isles but rare elsewhere; mainly October–January.

Marasmius rotula is a small, ivory-white *Marasmius* with white, distant gills that connect to a collarium. The stem is smooth and rather short, and white above darkening to red-brown to purple-black towards the base, which is not felty. The spores measure 7–8.5 × 3.5–4.5 µm. Almost always occurs in troops on twigs in the litter layer, but may also root into buried twigs in lawns *etc.*

Marasmius bulliardii ▽, which occurs on the leaf-litter of deciduous trees, and *M. wettsteinii* ⋉ which occurs on needle debris, are also collariate but have brown-tinted caps and are somewhat more gracile.

Widespread and very common; peaks early in the season but may be found June–November.

Thomas Læssøe

Marasmius bulliardii is a pale brownish, furrowed-sulcate and umbilicate *Marasmius* with a dark 'eyespot' in the cap centre. The gills are connected to a collarium. The dark stem has a tendency to produce side branches. The spores measure 8–9.5 × 3.5–4.5 µm. Occurs on thick leaf-litter from deciduous trees, mostly in large troops.

Marasmius rotula △ is very similar but has an ivory-white cap and grows on twigs; *M. wettsteinii* ⋉ is also very similar, but is found on needle debris and is somewhat smaller.

Widespread and rather common, except in the boreal zone; mainly September–November.

Marasmius limosus is a very small, slender *Marasmius* with few white gills that are connected to a collarium, a very finely downy, brown-tinted cap, and a smooth stem that is white at the apex but black towards the base. The spores measure 8.5–9.5 × 5–6 µm. Occurs in fens and boggy areas on grasses, sedges and rushes, mostly deep down in damp vegetation.

Distinctive and unlikely to be misidentified once the characters have been checked with a hand lens. *Marasmius curreyi* ▽ is larger and the cap is orange to brick-red. Small white mycenas (pages 188 & 121) do not have dark, rigid, smooth stems.

Widespread and very common, but easily overlooked due to its habitat (rubber boots are needed!); mainly August–November.

Marasmius curreyi is a small, orange to brick-red *Marasmius* with a furrowed-sulcate cap that has a darker 'eyespot' at the centre; the few, whitish gills are connected to a collarium. The stem is thin, smooth and dark towards the base. The spores measure 8–14 × 4–6 µm. Occurs on rotten parts of grasses, sedges and rushes in hay fields, lawns, meadows, fens *etc.*

Can be told from *M. limosus* △ by its wider cap with more vivid colours and its preference for mostly drier habitat. Also recalls *M. rotula* ◁ and *M. bulliardii* ◁, but these species occur on woody debris or leaf-litter and have white to pale brown caps.

Widespread and common, but apparently rare towards the north; peaks during the summer after heavy rains, but can be found into the autumn.

Jens H. Petersen/Jan Vesterholt

Cystoderma and the like

This group is characterized by the granulose surface of the cap and stem that constitutes remnants of a universal veil. In some species a ring or ringzone is present on the stem. The gills are adnexed, adnate or emarginate and the spore-deposit is either whitish or brownish-yellow. The spores are amyloid or inamyloid and smooth. The flesh may contain asexual spores. Some species have hymenial cystidia.

OTHER SIMILAR FUNGI:
– the lepiotoids have completely free gills (page 320).

FURTHER READING: 156, 181, 324.

Left: young caps. Right: expanded caps with remnants of veil at the margin.

Phaeolepiota aurea is a large, fleshy, completely dry, orange-yellow agaric with a large, flaring to hanging ring, the upperside of which is ± coloured by the brownish-yellow spores. The entire fruitbody is mealy-grainy. The gills are adnexed. The odour may be somewhat like cyanide or bitter almonds. The spores are fusiform, smooth, and measure 11–12 × 4.5–5.5 µm. Occurs on rich, often disturbed, soils, *e.g.* in fertilized lawns or stands of nettles, typically in troops and fairy rings.

Could possibly be confused with the ± fibrillose *Gymnopilus spectabilis* (page 602) but that species has ornamented spores and grows on wood. It may also recall an enlarged *Cystoderma* ▷▷ or *Cystodermella* ▷▷ but the spore-deposit colour is distinctive. In Japan it is parasitized by a species of *Squamanita* (page 228), which may reflect its close relationship with *Cystoderma*, a common host for this parasitic genus of agarics.

Widespread and locally common, more or less absent from the boreal zone; mainly September–November.

Cystoderma **&** ***Cystodermella*** are white or yellow to orange agarics with a grainy universal veil that is left as a granular surface and as a ± evident ring. *Cystoderma* has amyloid spores, whilst the spores of *Cystodermella* ▷▷ are inamyloid.

Cystoderma amianthinum is a rather small, yellow-orange to cinnamon-buff *Cystoderma* with ± white gills, no violet on the stem apex, an ephemeral ring and a ± wrinkled and grainy cap surface. The smell is unpleasant, musty-earthy. The cap flesh is whitish and always comprises normal hyphae that do not break down to conidia. The spores are amyloid and measure 5–6(–7) × (2.5–)3–3.5 µm. Occurs mostly on acidic soils amongst mosses in open and semi-open habitats.

Cystoderma jasonis ▷ has gills that are tinted pale brown and cap flesh that consists mostly of brownish, squared, asexual spores (arthroconidia). *Cystoderma amianthinum* may be parasitized by *Squamanita paradoxa* (page 229).

Widespread and common; mainly August–November.

Cystoderma jasonis is a rather small, squat *Cystoderma* with yellowish-brown colours and mostly brown-tinted gills and an ephemeral ring. The smell is mostly faint. The cap flesh is pale yellowish or brownish to brownish-violet and filled with squared, asexual spores (arthroconidia). Spores amyloid, measure (5.5–)6–7.5(–9) × 3–4.5 µm. Mostly on rather acidic litter, on needle debris and amongst mosses.

Cystoderma amianthinum ◁ has paler gills, the flesh lacks arthroconidia, and it has smaller, sexual spores. The mostly northern *C. jasonis* var. *lilacipes* ⋉ has a ± violet-tinted stem apex and fewer arthroconidia; *C. jasonis* var. *niveum* ⋉ is a rare, white variety. *Cystoderma simulatum* ⋉ has even smaller spores and is mostly late-fruiting.

Widespread and rather common; mainly September–November.

Jan Vesterholt/Jens H. Petersen

Cystoderma carcharias is a medium-sized, pale rose-grey *Cystoderma* with a distinct flaring ring. The small, amyloid spores measure 4–5.5 × 3–4 µm. The smell is unpleasant, musty-earthy. It may also occur in an almost pure white form. Grows on soil amongst litter and mosses, mostly on rather rich soils – both in deciduous woodland and in conifer forests.

Cystoderma jasonis △ also occurs in a white form, but it has an indistinct ring and larger spores.

Widespread and fairly common, but seems to have declined in at least part of temperate Europe; mainly September–November.

Vello Liiv

Cystodermella are rather large *Cystoderma*-like species with inamyloid spores.

Cystodermella granulosa is a rather sturdy, ± brownish-red *Cystodermella* that lacks cystidia but has a faint smell. The inamyloid spores measure 3.5–5 × 2–3 µm. It also occurs in a rare white form (*Cystodermella granulosa* var. *ambrosii* ⋉). Typically grows on rather acidic litter amongst mosses in both ± open habitats and in forests and plantations.

Cystoderma jasonis ⊲⊲ is smaller, has duller colours and amyloid spores. Both *Cystodermella cinnabarina* ⊳ and *C. adnatifolia* ⊳ are a more vivid orange-red, and the former has conspicuous cystidia.

Widespread, and fairly common in northern parts of temperate Europe but rather rare and perhaps declining farther south; mainly August–October.

Vello Liiv

Jan Vesterholt

Cystodermella cinnabarina is a vivid, apricot-orange *Cystodermella* with prominent, urticiform cystidia. The gills are white and rather crowded. The smell is faint. The spores are inamyloid and measure (3.5–)4–5 × 2–2.5 µm. Occurs in damp, mossy, deciduous woodlands and conifer forests and plantations.

Cystodermella adnatifolia ▽ is similar, but lacks cystidia. *Cystodermella granulosa* ◁ is less vividly coloured and also lacks cystidia. Species of *Cystoderma* ◁◁ have amyloid spores and are never as brightly coloured.

Widespread and fairly common in the hemiboreal zone, but rather rare farther south; August–November.

Thomas Læssøe

Cystodermella adnatifolia is a vivid, apricot-orange *Cystodermella* that lacks cystidia. The spores are inamyloid and measure (3.5–)4–4.5(–5) × (2–)2.5–3 µm. Smell faint, somewhat sweetish. Often occurs in *Quercus*-dominated woodland, amongst mosses and leaf-litter.

Cystodermella cinnabarina △ is very similar, but has gills with urticiform cystidia. *Cystodermella granulosa* ◁ also lacks cystidia but is more brownish-red.

Mainly found in the hemiboreal–boreal zones, very rare farther south; mainly August–November.

Thomas Læssøe

Lepiotoids

The lepiotoids have whitish or greenish spore-deposits and free gills but the universal veil is less membranous in comparison to the amanitoids, p. 352. In some species the universal veil is powdery or granulose, sometimes in the form of small spines. Often have complex, muff-like rings formed by parts of both the universal and the partial veil. The spores are often dextrinoid.

The lepiotoids are decomposers and are mostly found on black, mull soils, *e.g.* in damp, rich woodlands, along road-sides with *Urtica*, in scrub and gardens, and on old compost. Some species of *Macrolepiota*, *Chlorophyllum* and *Lepiota* also grow in open dry grasslands, including dunes.

The smaller lepiotoids apparently need a specific climatic scenario in order to produce fruitbodies, and years may pass without significant fruiting. When this does happens it is often over in a week or two.

OTHER SIMILAR FUNGI:
– the amanitoids are similar but have a membranous universal veil (page 352).
– *Limacella* and *Chamaemyces* may look similar but have a greasy/slimy cap surface (page 348).
– *Cystoderma* and others have adnexed to emarginate gills (page 314).
– *Floccularia* has adnexed to emarginate gills (page 252).
– *Agaricus* (page 500) has black spores. Young fruitbodies of *e.g. A. sylvicola*, with immature spores, can be recognized by the sweet marzipan smells. Reddning species may recall species of *Chlorophyllum*, but these have more complex rings.

FURTHER READING: 6, 85, 156, 183, 226, 332.

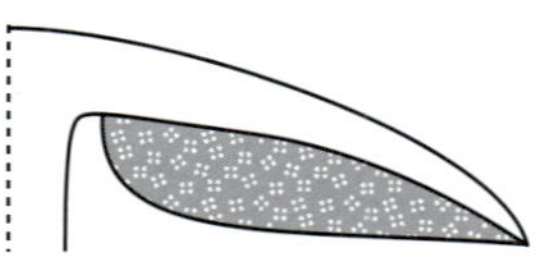

free gills

Micro-drawings: spores.

Approximate species number applies to temperate Europe.

Leucocoprinus
5+ species
page 336
Leucocagaricus
20+ species
page 337
rather small; thin-
ned; with ring, mostly
th ± plicate margin
small to medium-
sized; mostly with
well-defined, muff-
like ring
I+ (dextrinoid),
with a germ pore
with a hyaline
cover
I+/- (dextrinoid),
with or without a germ
pore; one species has a
hyaline cover
spore colour
medium-sized to large;
with well-defined,
muff-like, loose,
mostly double ring
lepiotoids
ave free gills with whitish
greenish spore-deposits,
and non-membranous
universal veils
I+ (dextrinoid),
thick-walled, with a germ
pore with a hyaline cover
Macrolepiota
7 species
page 345
other
similar fungi
I+/- (dextrinoid), thick-
walled, with an uncovered
germ pore
medium-sized to large;
with well-defined,
muff-like, loose,
mostly double ring
spore-deposit white;
gills free
spore-
deposit
black;
gills free
surface
greasy/slimy
with membranous
universal veil
Chlorophyllum
5 species
page 341
acella/Chamaemyces
page 348
amanitoids
page 352
Agaricus, page 500

Melanophyllum includes small, powdery and veiled species with slightly roughened spores, free, brightly coloured gills and, initially, a green spore-deposit.

Melanophyllum haematospermum is a rather small, grey-brown, powdery, marginally veiled lepiotoid with blood-red, free gills. When completely fresh the spore-deposit is green, but soon changes to a pale red-brown. The cap surface consists of rather large, ± spherical elements. The spores are rough and measure 5–6.5 × 2.5–4 µm. Occurs on mull-rich soils, old compost, in deciduous forests on rich soil, in old gardens, in coastal scrub, *etc.*

Distinctive and unlikely to be misidentified.

Widespread and fairly common in the nemoral zone but with scattered occurrences in the boreal zone; (June–) September–October (–November).

Jan Vesterholt

Melanophyllum eyrei is a rather small, very powdery, marginally veiled lepiotoid with unique turquoise gills. The cap is almost white, but usually has a slight cinnamon-buff to yellow-green tint. The spore-deposit is green. The finely verrucose spores measure 4–5.5 × 2.5–3.5 µm. Occurs on mull in rich and mostly calcareous sites in deciduous forests and scrub, typically together with other lepiotoids.

This species is unique and unlikely to be misidentified.

Rare to very rare and local in most of temperate Europe; mainly August–October.

Thomas Læssøe

Cystolepiota is a genus of ± small lepiotoids with the cap covered in a powdery layer from the universal veil. Often with ± powdery stems, sometimes dentate at the margin or with an annular zone. The powder typically consists of globose cells. The spores are non-dextrinoid. They are decomposers that prefer mull-rich soils in scrub and forests.

Cystolepiota hetieri is a medium-sized, very powdery, whitish to rose lepiotoid with slightly yellow-tinted, free gills. The flesh and surface redden when bruised and with age. The spores are iodine inert and measure 4.5–6 x 2–3 µm. Occurs on mull, mainly in deciduous forests and scrub, rarely in conifer forests.

Cystolepiota adulterina ⋉ is very similar, but somewhat larger and does not redden. It is less powdery with age and greyer on the cap. *Cystolepiota moelleri* ⋉ and *C. cystidiophora* ⋉ have vinaceous-brown granules, although these discolour in the latter.

Rather rare, but locally fairly common, decreasing towards the north; mainly August–November.

Cystolepiota seminuda is a small, slender, very powdery, white to somewhat pink lepiotoid. The gills are white to yellowish, and free. The stem powder disappears with age to reveal a reddish colour. Cap margin dentate from veil when young. The spores are smooth, iodine inert and measure 3.5–4.5 × 2–2.5 µm. On mull soils or heavily decomposed wood on rich soil in deciduous and mixed forests.

Small, white species of *Leucoagaricus* ▷▷ are not powdery and have rings on the stem. Other white *Cystolepiota* species are more fleshy.

Widespread and rather common, decreasing towards the north; July–November.

Cystolepiota pulverulenta is a medium-sized, very powdery, pale lepiotoid, the powder quickly staining brown, both on cap and stem. The smooth to finely ornamented spores do not stain in iodine, or slowly become dextrinoid, and measure (3.5–)4–5(–6) × 2.5–3.5 µm. The veil powder consists of irregularly formed elements. Occurs on rich soils and old compost in parks, scrub and forests.

Cystolepiota hetieri ◁ has spherical elements in the veil powder and reddens.

Nemoral, very rare; August–November.

Thomas Læssøe

Cystolepiota icterina is a rather small, pale sulphur-yellow, powdery lepiotoid with yellowish, free gills. The stem becomes reddish towards the base with age. The spores are dextrinoid and measure 3.5–5.5 × 2.5–3 µm. Grows on mull, mostly in thermophilic, ± calcareous deciduous forests.

Other yellow lepiotoids do not have a layer of powder or spherical cells on the cap.

Very rare and local; mainly August–November.

Thomas Læssøe

Cystolepiota bucknallii is a rather small to medium-sized, very powdery, ± lilac lepiotoid with pale sulphur-yellow, free gills. The odour is very strong and highly nauseous, gas-like. The spores are dextrinoid and measure 7–9 × 2.5–3.5 µm. Occurs on rich, often disturbed soil, mostly in deciduous forests and scrub, often with other lepiotoids.

Distinctive and unlikely to be misidentified.

Widespread and fairly common, but absent from the boreal zone and very rare in the hemiboreal zone; mainly August–November.

Thomas Læssøe

Lepiota is a genus of small to medium-sized agarics with free gills and white spore-deposits. The cap surface is ± cracking from the margin, but has a ± smooth centre. Many species have veil remnants in the form of a ring or girdles on the stem. The spores are dextrinoid. Many speices are toxic (amatoxin).

Lepiota cristata is a medium-sized, rather pale *Lepiota* with an orange-brown cap cuticle that splits to reveal the pale flesh. The stem has a ± evident, loose ring, but the veil may also adhere to the cap margin. The odour is unpleasant metallic or gas-like. The cap cuticle is a hymeniderm of short, pear-shaped cells. The dextrinoid spores are bullet-shaped and measure 5.5–8.5 × 2.5–3.5 µm. On ± disturbed soil along roads, both in rather open habitats and in forests and gardens.

Lepiota lilacea ▷ has a dark, greyish-violet ring and ellipsoid spores. Other similar species have a radial cap surface with long elements in the cuticle (*e.g. L. subalba* ▷) or fusiform to ellipsoid spores.

Common and widespread, decreasing towards the north; June–November.

Lepiota subalba is a rather small to medium-sized, pale cream *Lepiota* with a radial structure in the cap surface. The stem may be ± yellow-brown-tinted and has pale, floccose girdles, or a weakly developed ring. The cap cuticle has long elements. The dextrinoid spores are bullet-shaped and measure 7.5–10 × 2.5–3.5 µm. Occurs on mull soils in scrub and forests, often in the company of other lepiotoids.

Other very pale lepiotoids do not have bullet-shaped spores.

Rather rare, very rare in the boreal zone; July–October.

Thomas Læssøe

Lepiota erminea is a medium-sized *Lepiota* with whitish colours except for the ± brownish cap centre. The cap has pale remnants of a veil at the margin, while the stem is floccose below a ± evident, floccose ring. The ± fusiform, dextrinoid spores measure 10–20 × 5.5–7 µm. Occurs in sandy grassland and dunes.

Lepiota oreadiformis ⋉ is very similar, but is browner on the cap. *Lepiota subalba* △ grows in forests and scrub and has bullet-shaped spores. Species of *Leucoagaricus* ▷▷ have ± muff-like rings and different spores.

Rather common and widespread; July–November.

Jan Vesterholt/Jens Maarbjerg

Lepiota lilacea is a rather small *Lepiota* with brown cap scales and a muff-like ring with a greyish-violet underside. It has a sweetish smell. The dextrinoid, ellipsoid spores measure 4–6.5 × 2.5–4 µm. Occurs on disturbed soils, *e.g.* in mulched beds in city parks, gardens, *etc.*

The similar *L. cristata* ◁ has a pale underside to the ring, and the spores have a lateral apiculus (bullet-shaped). Both species have a hymeniform cap cuticle.

Rather rare and southerly, absent from the hemiboreal–boreal zones; mainly April–October.

Emilio Pini/Jan Vesterholt

Jan Vesterholt

Lepiota ignivolvata is a large, rather pale *Lepiota* with an orange-brown cap centre and a similarly coloured ring. The stem base stains orange-brown with age and after picking. The odour is unpleasant, ± like that of *Lepiota cristata* ◁◁. The cap cuticle has long, tapering elements. The dextrinoid, ± almond-shaped spores measure 8.5–11 × 5–6 µm. Occurs on mull soils in thermophilic sites in deciduous or mixed forests.

Lepiota clypeolaria ▷ does not discolour orange and has larger spores; *L. magnispora* ▷ also has much longer spores. The same is true for *L. cortinarius* ⋉, where, in addition, the spores have a lateral apiculus (bullet-shaped).

Very rare, absent from northern regions; mainly August–October.

††**Lepiota subincarnata** is a rather small, fleshy, flesh-pink to brownish-rose *Lepiota* with a ± scaly cap, and the stem has girdles below a narrow, brownish-red ring. The dextrinoid, ± almond-shaped spores measure 6–7.5 × 3–4 µm. The cap cuticle has up to about 200 µm long, ± tapering cells. Occurs on mull soils in gardens, scrub and forest edges – not especially demanding.

Lepiota brunneoincarnata ▷▷ is darker with larger spores. *Lepiota helveola* ⋉ also has larger spores and more brightly coloured girdles.

Widespread and occasional to rather common, absent from the north; August–November.

†?**Lepiota magnispora** is a rather large, ± apricot-orange *Lepiota* with yellow-brown-tinted, floccose girdles on the stem. The dextrinoid, fusiform, sway-backed spores measure 13–25 × 4–6 µm, and the apiculus is almost lateral. Occurs on mull soils amongst leaf-litter in both deciduous and coniferous forests.

This species is also known as *L. ventriosospora*. *Lepiota clypeolaria* ▽ has white, floccose girdles on the stem and relatively wider, less sway-backed spores. *Lepiota ignivolvata* ◁ is even more fleshy and typically has an almost basal, narrow, orange ring; the stem base discolours orange.

Rather common and widespread; July–November.

††**Lepiota clypeolaria** is a rather large, pale, scaly *Lepiota* with a somewhat brownish cap centre and white, floccose girdles on the stem. The dextrinoid spores are fusiform and measure 11–18 × 4.5–6 µm. Occurs in damp, rich woodlands amongst litter and mosses.

Lepiota magnispora △ has more yellow-brown colours, particularly on the lower stem girdles, and the spores are more sway-backed.

Widespread and fairly common but rarer than *L. magnispora* in the nemoral zone; July–November.

Thomas Læssøe

†?**Lepiota tomentella** is a small, pale brown, finely scaly *Lepiota* with rather indistinct, pale brown girdles on the stem and no ring. Has long, multicellular hairs in the cap cuticle. The dextrinoid spores are bullet-shaped and measure 7–9.5 × 3–4 µm. Occurs on ± calcareous, light soils in coastal thorny thickets, in deciduous forests, *etc.*

This species belongs to a taxonomically complex group of rather poorly defined species. *Lepiota pseudolilacea* ⋉ has a ring. *Lepiota cingulum* ▽ is larger with more distinct girdles on the stem. *Lepiota echinella* ⋉ has ellipsoid spores. *Lepiota pilodes* ⋉ stains orange and has unicellular hairs in the cap cuticle.

Rare (very rare in the north); mainly August–October.

Jan Vesterholt

†?**Lepiota cingulum** is a rather compact, medium-sized, red-brown to purple-brown *Lepiota* with minute scales on the cap and conspicuous brown girdles below an indistinct ring zone on the stem. The cap centre is almost tomentose and dark. Has long cells in the cap cuticle that measure 100–350 × 12–15 µm. The bullet-shaped, dextrinoid spores measure 8.5–11 × 3.5–4.5 µm.

Lepiota tomentella △ is smaller and has smaller spores.

Very rare, absent from the hemiboreal–boreal zones; mainly September–November.

††Lepiota felina is a rather small to medium-sized *Lepiota* with dark scales that contrast with the white flesh. It has a well-defined ring that is dark on the underside and margin. The cap centre is almost black. The ellipsoid spores measure 6–8.5 × 4–5 µm. Occurs on needle debris on rich soil, but occasionally in deciduous forests, *e.g.* on very decomposed wood.

Lepiota lilacea ◁◁ has a hymeniderm cap cuticle; *L. felina* has two kinds of longer elements in the cap cuticle. Similar species of *Leucoagaricus* ▷▷ have thick-walled spores and muff-like rings.

Widespread and fairly common; mainly August–November.

†?Lepiota boudieri is a ± medium-sized, rather vivid orange-brown to more matt rust-brown, felty-scaly *Lepiota* with whitish gills. Odour faint. The cap cuticle elements are 3–6-celled and clamped. The bullet-shaped, dextrinoid spores measure 7–10 × 3–4 µm. Occurs on rich soils in parks, scrub and deciduous forests – rarely with conifers.

Lepiota castanea ▽ has smaller, darker fruitbodies, smells of cedar wood and lacks clamps in the cap cuticle. *Lepiota pilodes* ⋉ discolours more orange with age.

Widespread and fairly common, except in the north; July–November.

Thomas Læssøe

††Lepiota castanea is a rather slender, ± red-brown, appressed-scaly *Lepiota* that lacks an evident ring but has ± well developed red-brown girdles below the ring zone. The gills discolour orange with age. Smells of cedar wood. The cap cuticle has long elements that are 2–4-celled. The bullet-shaped spores measure 7–14 × 3–5.5 µm. Occurs on rich, mostly somewhat disturbed soils along roadsides or in deciduous and coniferous forests, *etc.*

Lepiota ignicolor ⋉ has 1-celled hairs in the cuticle, but otherwise is very similar.

Rather common and widespread; July–December.

††***Lepiota fuscovinacea*** is a medium-sized, rather fleshy, grey-rose *Lepiota* with a very woolly, floccose stem covering and no well-defined ring. The cap has dark, concentric, scaly zones. There are no clamps, and the cap cuticle consists of uniform, long cells. The dextrinoid, ± ellipsoid spores measure 4–6.5 × 2–3.5 µm. Occurs on rich soils, mostly disturbed, in parks, deciduous and coniferous forests, *etc.*

Lepiota brunneoincarnata ▽ has more distinct, dark girdles at the base of the stem and no woolly covering.

Rather rare, absent from northern regions; mainly September–October.

Jacob Heilmann-Clausen

††***Lepiota brunneoincarnata*** is a rather fleshy, but not very large, short-stemmed, rather dark red-brown *Lepiota* with dark, red-brown to wine-red girdles on the lower half of the stem, but no true ring. It has clamps and the cap cuticle cells are of different lengths. The dextrinoid, ± almond-shaped spores measure 7–10 × 4–5 µm. Occurs on mull soils in gardens, scrub and deciduous forests.

Lepiota fuscovinacea △ has a woolly-felty stem surface. Other similar species are smaller or paler, or have an evident ring.

Rather rare, absent from the boreal zone; July–November.

Thomas Læssøe

††*Lepiota griseovirens* is a medium-sized, grey-green, minutely scaly *Lepiota* with a ring zone and girdles below. The flesh does not discolour with age. The cheilocystidia are rather variable in shape, from ± cylindrical to clavate. The bullet-shaped spores measure 6–10 × 3.5–4 µm. Occurs on light, calcareous soils, *e.g.* in *Syringa* hedgerows, scrub and in various types of deciduous forest.

Lepiota grangei ▽ is more vivid green, and the flesh discolours orange. *Lepiota poliochloodes* ⋉ is more felty than scaly and has somewhat smaller spores that are only slightly bullet-shaped.

Widespread, rather rare, very rare in the boreal zone; mainly August–October.

†?*Lepiota grangei* is a medium-sized *Lepiota* with dark, ± blue-green scales on the cap and stem, and flesh that discolours brownish-orange with age and when bruised. The bullet-shaped spores measure 8–13 × 3.5–4.5 µm. Occurs mostly on disturbed soils along forest roads, often in ± calcareous deciduous forests.

Lepiota griseovirens △ does not stain orange. *Lepiota poliochloodes* ⋉ is less scaly and has smaller spores.

Rather rare, absent from the boreal zone; mainly August–October.

Thomas Kehlet

Echinoderma is recognized by the conical, prominent cap scales that are remnants of the universal veil, and by the brown colours. The spores are dextrinoid or inamyloid. With or without prominent cheilocystidia.

† ***Echinoderma asperum*** is a large, rather fleshy, brown, spiny-scaly *Echinoderma* with very crowded, bifurcate, whitish gills that have a pale edge. The stem has a large membranous, ± hanging ring with brown scales on the underside. The odour is almost metallic, somewhat like that of *Lepiota cristata* ◁◁. The cheilocystidia are broadly clavate The dextrinoid spores measure 6–10 × 2.5–3.5 µm. Occurs on rich soils or on old compost in deciduous forests, parks, gardens, and sometimes in coniferous forests.

Echinoderma perplexum ▽ lacks an obvious ring and does not have distinctly bifurcate gills; the spores are also smaller. *Echinoderma hystrix* ▷ has guttulate spines, a dark gill edge and a different smell.

Widespread and rather common, decreasing towards the north; August–November.

Jacob Heilmann-Clausen

†? ***Echinoderma perplexum*** is a rather large *Echinoderma* with ± grey-brown, somewhat woolly cap scales that leave a silk-like surface when lost. The stem has an ephemeral ring and somewhat woolly, grey-brown scales. The white gills are barely bifurcate. Has numerous cheilocystidia. The dextrinoid spores measure 4.5–6.5 × 2.5–3 µm and have a somewhat lateral apiculus. Occurs on ± rich, calcareous soils in deciduous forests and very rarely in coniferous forests; it is mostly found in ancient woodlands.

Echinoderma calcicola ⋉ has significantly longer cap scales (1–5 mm compared with 1–2 mm).

Rare and local; August–November.

†?***Echinoderma jacobi*** is a rather small to medium-sized *Echinoderma* with brown, woolly scales on the cap and stem. The stem has a ring zone with woolly girdles below. The gill edge is fertile, and lacks cystidia. The dextrinoid spores measure 3.5–5.5 × 2.5–3 µm. Occurs on rich soils in forests, scrub, gardens and parks.

Echinoderma echinacea ⋉ has pointed scales that are up to 2 mm long, and it is not woolly.

Widespread but rather rare; mainly August–October.

†?***Echinoderma hystrix*** is a large, fleshy *Echinoderma* with guttulate, brown, pointed cap scales and a dark gill edge. The stem has strongly guttulate scales and a prominent ring. The gills are whitish, free and crowded. The odour is very unusual, somewhat like *Sambucus* flowers or cat urine. The dextrinoid spores measure 5.5–7 × 2.5–3 µm. Occurs on rich, ± disturbed soils in deciduous forests.

Echinoderma asperum ◁ has a pale gill edge, lacks guttulate scales and smells metallic.

Rare and local; August–November.

Jan Vesterholt

Jens Maarbjerg/Christian Lange

Leucocoprinus & **Leucoagaricus** are small to large agarics, which mostly have a thin-fleshed and ± plicate cap, and a muff-like ring on the stem. The spores are mostly dextrinoid and may have a germ pore that is sometimes covered by a hyaline cap. Mostly on rich, disturbed sites. The two genera are not morphologically well delimited.

† ***Leucocoprinus birnbaumii*** is a bright yellow *Leucocoprinus* with a grainy-scaly cap surface and a somewhat clavate, widened stem and a muff-like ring. The cap cuticle is complex with elements of numerous dimensions and shapes. The dextrinoid, thick-walled spores measure 8–12 × 5–9 µm and have a covered germ pore. A ± tropical species, mostly known from greenhouses and potted plants; very rare in the wild.

The other tropical, ± yellow species of *Leucocoprinus* are less frequent as introductions: *L. straminellus* ⋉ is paler yellow and has smaller spores; *L. fragillimus* ⋉ is extremely thin-fleshed, paler and has more spherical elements in the cuticle.

Occasional; all year indoors.

Leucocoprinus brebissonii is a pale *Leucocoprinus* with a dark brown cap centre surrounded by concentric rings of brown scales. The margin is furrowed-floccose. The stem ring is ± white and ± muff-like. The dextrinoid spores measure 8.5–13 × 5–8 µm and have a covered germ pore. Occurs on rich, disturbed soils in forests and parks, but also in greenhouses, *etc.*

Small, dark, scaly species of *Lepiota* ◁◁ are not furrowed at the cap margin, the ring is less muff-like and the spores do not have a germ pore. *Leucoagaricus melanotrichus* ▷▷ is finely grey-hairy on the cap, and the spores are much smaller.

Rare in the north, but quite common farther south; mainly June–November.

Jens Maarbjreg

Leucoagaricus badhamii is a large and fleshy, almost white to pale brownish, indistinctly scaly *Leucoagaricus*, which, when bruised, immediately stains blood-red and then quickly black. The surface stains green with KOH or ammonia. The rather thick-walled, dextrinoid spores measure 5.5–7.5 × 3–4.5 µm and have no germ pore. Occurs on ± calcareous, rich soils in scrub, parks and deciduous forests.

A few other species also have a strong, reddening stain. *Leucoagaricus georginae* ⋉ is similar but much smaller. *Leucoagaricus croceovelutinus* ⋉ has more obvious scales, stains red-brown with KOH; and the flesh stains orange-brown when bruised. *Leucoagaricus americanus* ▽ and *L. meleagris* ⋉ also redden, but less strongly.

Very rare to rare, southerly, absent from the hemiboreal–boreal zones; September–November.

Leucoagaricus meleagris is a rather large, clustered *Leucoagaricus* with small, dark red brown scales that create a dark brown umbo; the flesh stains yellow and then red. Cap surface red with KOH and green with ammonia vapours. The stem has small, dark scales and is cylindrical to fusiform and may have a small, dark ring. Spores thick-walled, ellipsoid, (7.5–)8–10 (–11.5) × (5.5–)6–8 µm with a small germ pore. Cap surface with very narrow, pointy elements. On composting wood chips, etc.

Leucoagaricus americanus ⋉ has bigger, paler scales on the cap and a more swollen, almost smooth stem. The spores have a more distinct germ pore; *L. badhamii* △ reddens strongly.

Very rare and meteoric; mainly July–October.

Ebbe Kristensen

Leucoagaricus melanotrichus is a rather small, thin-fleshed, finely black-haired *Leucoagaricus*. The stem has a small, rather ephemerous, dark-edged muff-like ring with dark hairs below. The flesh stains reddish-brown when bruised. Lacks clamps. The somewhat thick-walled, non-dextrinoid spores do not have a germ pore and measure 5.5–7.0 × 3.5–4.5 µm. Occurs in scrub, deciduous and coniferous forests, on litter or very decomposed wood.

Lepiota felina ◁◁ has scales rather than hairs and clamps. *Leucocoprinus brebisonii* ◁◁ has much larger spores and no hairs on the stem.

Very rare and southerly; mainly September–October.

Helmut Pidlich-Aigner

Leucoagaricus ionidicolor is a vinaceous, rather fleshy *Leucoagaricus* with a very prominent, muff-like, vinaceous ring with a flaring upper rim. Below the ring the stem is somewhat swollen-clavate and also vinaceous. The dextrinoid spores measure 6–7 × 3.5–4.5 µm and lack a germ pore. Occurs on rich, calcareous soils, sometimes near decaying tree trunks, both in deciduous and coniferous forests.

Distinctive and unlikely to be misidentified.

A very rare species with only a scattering of records; mainly September–October.

Thomas Læssøe

Leucoagaricus sericifer is a small, almost completely white *Leucoagaricus* with a rather narrow, ± muff-like ring situated below the middle of the stem. The cap centre may be somewhat buff. The cheilocystidia are very variable, ± lageniform and lack crystals. The ± narrowly almond-shaped to fusiform, dextrinoid spores measure 6–9.5 × 3–5 µm (from 4-spored basidia). Mostly on light, calcareous soils in scrub and forests.

Leucoagaricus serenus ⋉ has clavate cystidia; *L. crystallifer* ⋉ has fine crystals on the cystidia; and *L. cygneus* ⋉ has spherical elements in the cap cuticle.

Rather rare and local; autumn.

Leucoagaricus sublittoralis is a medium-sized, delicately pinkish to brownish-red-tinged, umbonate, very finely scaly *Leucoagaricus* with a white stem and white muff-like ring that has a flaring upper rim. The cheilocystidia are clavate. Spores acute almond-shaped, dextrinoid, measure 6.5–10 × 3.5–5 µm, without germ pore. On light, calcareous soils, including sand in coastal scrub or forests.

Leucoagaricus wichanskyi ⋉ has white veil remnants as a patch on the cap and more narrow cheilocystidia (4–11.5 compared with 6–15 µm). *Leucoagaricus tener* ⋉ has more cylindrical cheilocystidia, more ellipsoid spores and more contrasty cap surface.

Rare and southern; autumn.

Thomas Læssøe

Leucoagaricus leucothites is a pale, fleshy, *Agaricus*-like *Leucoagaricus*, which, with age, acquires pinkish-tinted gills. The cap surface is very finely felty-scaly and ± yellowing when bruised. The ring is muff-like with a flaring upper rim. The ± almond-shaped, dextrinoid spores measure 7–11 × 5–7 µm. Occurs mostly in lawns in gardens, parks, open forests, *etc.*

Leucoagaricus leucothites var. *carneifolius* ⋉ has a distinct grey cap surface, and perhaps a more intense yellowing. *Macrolepiota excoriata* ▷▷ is more scaly on the cap and also has a more obvious stem covering.

Widespread and common in some years, decreasing towards the north; May–November.

Thomas Læssøe

Leucoagaricus nympharum is a magnificent, large *Leucoagaricus* with recurved, pale grey to somewhat darker grey-brown scales and a complex, loose, muff-like ring. The stem base is ± widened. The cap centre is wide, smooth and ± greyish. The flesh stains somewhat yellow, and later brown. The cheilocystidia are large and rather variable. The dextrinoid spores measure 8–12 × 6–8 µm and have a germ pore with a hyaline cap. Occurs on needle debris on ± rich soils, including on anthills and under *Juniperus*.

Species of *Macrolepiota* ▷▷ usually have a more evident, cracking stem covering and less recurved scales on the cap. Species of *Chlorophyllum* ▷ have distinctly reddening flesh and lack a hyaline cap over the germ pore.

Rather common and widespread; May–December.

Chlorophyllum is a genus of large lepiotoids with complex, loose, muff-like rings and reddening flesh. The spores are ± dextrinoid, thick-walled and the germ pore lacks a cap. One species occurs in dry grasslands and is secotioid (page 1247).

• ***Chlorophyllum olivieri*** is a rather monochrome, grey-brown, scaly *Chlorophyllum* with an orange-red flesh reaction, and a distinct, ± loose, muff-like, double ring. The cheilocystidia are ± clavate. The dextrinoid spores have a somewhat truncate end with an indistinct germ pore, and measure 8–11 × 5.5–7 µm. Occurs on thick litter, mostly from conifers on richer soils.

The other species of *Chlorophyllum* have a more marked contrast between the flesh and the cap scales. *Leuco-agaricus nympharum* ◁ has smaller, recurved scales and stains yellow.

Common and widespread; May–December.

• ***Chlorophyllum rhacodes*** is a very contrasty *Chlorophyllum* with large, coarse cap scales. The ring is loose, muff-like and distinctly double with a ± greyish, felty underside. The stem base is onion-shaped. The flesh reddens, as is typical for the genus. The cheilocystidia are almost balloon-shaped and measure 10–35 × 8–25 µm. The dextrinoid spores measure 9–12.5 × 5.5–8 µm. Occurs on disturbed soil in gardens, scrub, parks and forests.

The spelling of *rhacodes* has recently been settled, but *racodes* is often seen in the literature. *Chlorophyllum brunneum* ▷ has a simpler, one-layered ring, and less spherical cheilocystidia. *Chlorophyllum olivieri* ◁◁ has a double ring but the cap is not nearly as contrasty.

Rather common and widespread; June–December.

† *Chlorophyllum brunneum* is a contrasty *Chlorophyllum* with a very brown cap centre and scattered, coarse scales that radiate outwards, revealing the white flesh underneath. The ring is loose, muff-like, thick and brown underneath, but not distinctly double. The flesh is reddening, as is typical for the genus. The cheilocystidia are clavate and measure 25–50 × 8–20 µm. The dextrinoid spores measure 10–12 × 7–8.5 µm. Occurs on disturbed, rich soil, *e.g.* old compost in gardens and parks, rarely in forests.

Chlorophyllum rhacodes ◁ is very similar, but has a distinct double ring; *C. olivieri* ◁◁ also has a double ring but is not nearly as contrasty.

Fairly widepread and common, but rare or absent from northern parts of temperate Europe; mainly June–November.

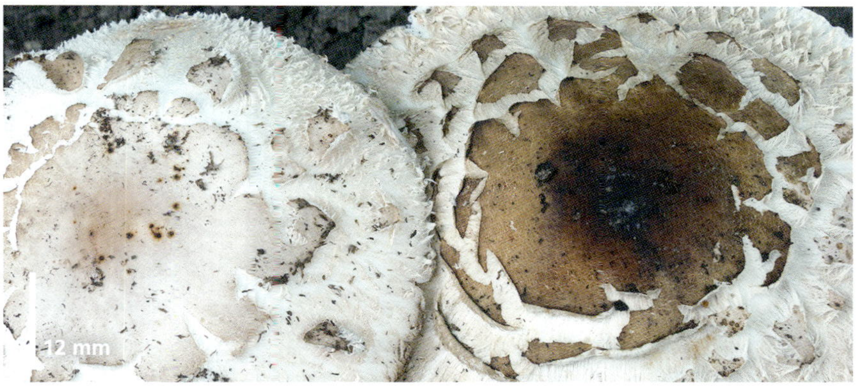

Maurice Pelissier

† ***Chlorophyllum molybdites*** is a *Chlorophyllum* with olive-green gills and spore-deposit. The ring is loose and double. The spores measure 9–11 × 7–8 µm. In temperate Europe only known indoors from waterparks and similar places.

Other species of *Chlorophyllum* have white spore-deposits, apart from the almost puffball-like (secotioid), rare steppe species, *C. agaricoides* (see Vol. 2, p. 1247), which has passive spore dispersal and a ± olive spore mass. There are three similar tropical/subtropical species.

Only indoors – occurs naturally in southern parts of North America and other tropical/subtropical regions; summer–autumn.

Macrolepiota species form very large fruitbodies with a loose ring. The flesh is non-staining or somewhat yellowing or reddening. The stems often have a ± distinct snakeskin pattern. The spores are dextrinoid, thick-walled and have a covered germ pore.

• **Macrolepiota excoriata** is one of the smaller, very pale *Macrolepiota* species. The cap is fibrillose-scaly, somewhat brownish on the umbo and becomes almost white towards the edge. The ring is white, loose and muff-like. The stem surface is finely felty, but lacks a distinct snakeskin pattern. The spores measure 11–17 × 8–11 µm. Occurs in dry grassland and is not very demanding.

Leucoagaricus leucothites ◁◁ has a less fibrillose cap surface, a smoother stem and stains ± yellow. *Macrolepiota mastoidea* ▷▷ is taller and has patchy remnants of a veil on the cap.

Rather common and widespread; June–November.

Jan Vesterholt

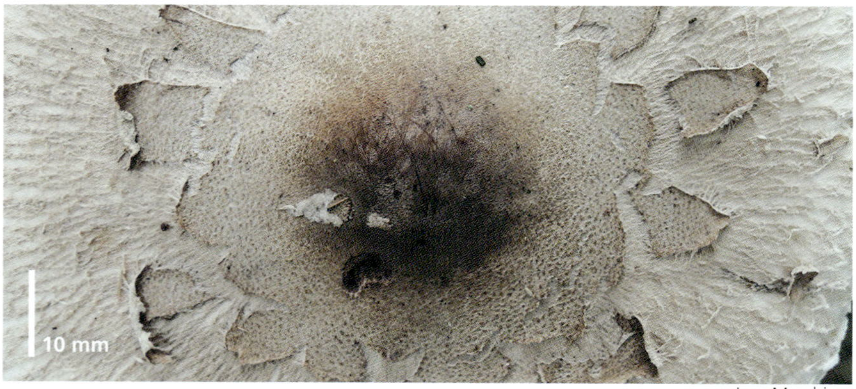

Jens Maarbjerg

• ***Macrolepiota mastoidea*** is a large, elegant, distinctly umbonate, rather pale brown *Macrolepiota* with a ± smooth, brown cap centre and flaky/patchy scales radiating outwards. The stem has a pale snakeskin pattern on an even paler background and carries a loose, muff-like, double ring. The flesh stains slightly yellow or red. The dextrinoid spores measure 11–18 × 7.5–12 µm. Occurs in open forests or in dry, open habitats, but rarely in sand dunes.

Some authors accept more than one species with a similar suite of characters, and the name *M. konradii* is also frequently used. *Macrolepiota procera* s.l. ▷ is larger, has a more distinct snakeskin pattern on the stem and somewhat darker colours on the cap.

Widespread and fairly common; June–November.

• ***Macrolepiota procera*** is a very large, stately *Macrolepiota* with a distinct brown snakeskin pattern on the stem and a very large, muff-like, complex double ring. The umbonate cap has concentric, brown ± recurved scales. The gills are free and cream or have a pink tinge. The flesh only reddens slightly in the stem cortex. The dextrinoid spores have a small hyaline cap covering the germ pore, and measure 5–18 × 8–12 µm. Found mainly in sandy, open places, *e.g.* in coastal grasslands, including old dunes.

This species belongs to a difficult and not fully understood species complex. *Macrolepiota olivascens* ⋉ discolours green on the cap and gills, but the delimitation against *M. procera* is not completely clear. The colour of the spore-deposit may be of importance. The presumably rare *M. fuliginosa* ⋉ apparently has looser, larger cap scales on a radially fibrillose base. Other species in the genus either redden strongly, are paler, or have less distinct snakeskin patterns on the stem.

Widespread and rather common; June–November.

Thomas Læssøe

Chamaemyces and *Limacella*

These genera are characterized by ± free gills, white spore-deposits and the absence of a membranous universal veil, although some species may have a slimy universal veil. They have membranous or thread-like partial veils left as a ring or ring zone on the stem. The cap surfaces are smooth, do not crack, and are mostly greasy to slimy. They often have an farinaceous smell.

Chamaemyces and *Limacella* are decomposers, which occur mostly in exceptionally mull-rich habitats in scrub and forests. Most, perhaps all, species are associated with calcareous soils.

Chamaemyces is closely related to the lepiotoids (page 320), while *Limacella*, *Limacellopsis* and *Zhuliangomyces* is relatively close to the amanitoids (page 352).

OTHER SIMILAR FUNGI:

– the amanitoids have ± free gills and a membranous universal veil that covers the greasy cap (page 352).

– volvariellas have free gills, a membranous universal veil seen as a volva, and a brownish-rose spore-deposit (page 487).

– the lepiotoids have free gills and powdered, plicate, scaly or cracking, dry, cap surfaces (page 320).

FURTHER READING: 90, 156, 181.

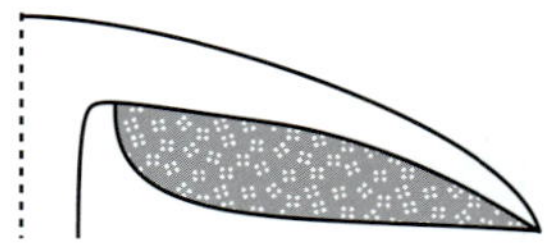

free gills

Thomas Læssøe

Chamaemyces fracidus recalls something between an *Amanita* and a *Limacella*, but it is easily recognized by the grainy-scaly covering below the ring-zone and by the wet, mostly guttulate, convex, pale buff cap, which may have remnants of a veil at the margin. The stem is also ± covered in yellow-brown drops. The whitish gills are free to almost free. The odour is strong, ± gas-like or sour. Has clavate cheilo- and pleurocystidia. The spores are smooth, inert to iodine, measure 4.5–5.5 × 2.5–4 µm, and lack a germ pore. Occurs on calcareous soils in deciduous forests (typically *Fagus*), mostly in sites with numerous other rare species of fungi.

This species also goes by the name *Lepiotella irrorata*. It may recall *Echinoderma hystrix* (page 335), but that species has pyramidal, brown scales on the cap. Farther south a darker type, *Chamaemyces fracidus* f. *pseudocastanea*, can be found.

Rather rare and local, absent from the boreal zone; mainly July–October.

Martin Kříž

Jens Maarbjerg/Jens H. Petersen

Limacella*, *Limacellopsis and ***Zhuliangomyces*** are white-spored agarics with ± free gills, a ± greasy to slimy universal veil and a thread-like to membranous partial veil. The spores are smooth to finely verrucose, inamyloid or dextrinoid.

Limacellopsis guttata is an only slightly greasy *Limacella* with a large, projecting to hanging white ring, which initially is covered in drops. The stem is ± clavate. The spores are smooth to finely rough, inamyloid, non-dextrinoid and measure 4.5–6.5 (–8) × (3.5–)4–5.5(–6) µm. Occurs on rich soils, mostly calcareous, in deciduous and coniferous forests.

Distinctive and unlikely to be misidentified. Species of *Amanita* (page 352) always have a membranous universal veil and a gill edge fimbriated by cystidia – species of *Limacella* lack cystidia. Most lepiotoids (page 320) have a powdery-scaly cap and cheilocystidia.

Widespread, but uncommon; (July–) September–October (–November).

Jan Vesterholt

Zhuliangomyces illinitus is a slimy, ivory-white species that lacks a membranous ring but does have a ± distinct slimy ring zone. The stem is smooth and slimy below the ring zone. The spores are ± dextrinoid, smooth to finely verrucose, and measure 5–6 × 4–5 µm. Typically occurs in calcareous conifer forests.

Tricholomella constricta (page 244) is not slimy and has a narrow, membranous ring.

Rather rare, absent from most of the northwestern lowlands; late summer–autumn.

Limacella vinosorubescens is a spectacular, brownish-red, farinaceous *Limacella* with reddening flesh and reddish belts below the ring zone. The ± dextrinoid, ± verrucose spores measure 3.5–5 × 3–5 µm. Occurs on rich, mostly calcareous clay in parks and deciduous forests, mostly with other rare fungi, *e.g.* species of *Echinoderma* (page 334) and *Leucoagaricus* (page 337).

The rather similar *Limcella glioderma* ▽ does not redden in the flesh.

Very rare and local; mainly August–September.

Thomas Læssøe/Jens H. Petersen

Limacella glioderma is a rather large, vividly coloured, brownish-orange, somewhat slimy, farinaceous, non-reddening *Limacella* with a rather evident floccose ring. The ± dextrinoid and ± verrucose spores measure 3.5–5 × 3–5 µm. Occurs on rich, mostly calcareous soils in deciduous and coniferous forests.

Limacella vinosorubescens △ has a reddening flesh and is also redder overall.

Rather rare and local, with most records from the hemiboreal zone; mainly August–September.

Thomas Læssøe

Limacella delicata is a rather small, relatively dry, almost odourless *Limacella* with rather subdued, vinaceous-brown colours. Has vinaceous-brown veil remnants below the ring zone. The spores are inamyloid and smooth to very finely rough, and measure 4–5 × 3.5–5 µm. Occurs on calcareous soils in forests and parks.

Other *Limacella* species are either paler, more slimy or smell strongly farinaceous, but some authors treat all three species on this page as varieties of *L. delicata.*

Rather rare and local, mostly found in the hemiboreal zone; autumn.

Christian Lange

Amanitoids

The amanitoids are recognized by free, or more rarely, adnexed gills, whitish or somewhat greenish spore-deposits and, almost always, a membranous universal veil (velum), which, at maturity, may be seen as a volva at the base of the stem and/or as loose patches on the cap. The spores may be amyloid or inamyloid.

Most of the amanitoids form ectomycorrhiza and are, therefore, usually found in forests and parks. The genus *Saproamanita* is saprotrophic.

Amanita is a species-rich genus that includes some of the most notorious poisonous mushrooms, not least the ringless species (*Amanita* sect. *Vaginatae*) which form a very confusing group where specific identification can be difficult.

OTHER SIMILAR FUNGI:
– the lepiotoids have either a powdery universal veil, or it is absent (page 320).
– limacelloids are rather similar, but have a slimy universal veil (page 348).
– volvariellas are separated by their brownish-rose spores (page 486).

FURTHER READING: 149, 156, 183, 218, 263, 281.

Emilio Pini

Saproamanita are decomposing *Amanita*-like agarics. They usually occur in open land and mostly form fairy rings.

Saproamanita vittadinii is a magnificent, ± white to pale buff amanitoid with highly projecting scales on the cap and stem. The stem also has white to pale buff belts along almost its entire length; it lacks a bulb but does have have a complex ring. The gills are tinted buff or green. The spores are ellipsoid, amyloid and measure (9–)10–13 (–15) × (6.5–)7.5–10(–11) µm. Occurs in fairy rings without connections to mycorrhizal trees or scrub.

The mycorrhizal *Amanita solitaria* ▷▷ is very similar, but has echinulate scales on the lower part of the stem.

Mainly a southern species but reaches as far north as the Netherlands and central England; autumn.

Emilio Pini

José Ignacio Fernández Simón

Saproamanita singeri is a small, rather slender, ± scaly, ivory-white amanitoid with cream-buff gills and an indistinct, finely scaly, adhering volva. There is no ring, and the stem is ± rooting and lacks a bulb. The spores are broadly ellipsoid, amyloid and measure 7.5–11× 5–8 µm. It is a decomposer that, *e.g.,* occurs on rich soil in stands of *Taxus*.

Distinctive and unlikely to be misidentified.

It reaches southern England in the north and also occurs in France – a very rare species; summer–autumn.

Carel Breukink

Saproamanita inopinata is a large, floccose-fibrillose amanitoid with dull, grey-brown colours, but with, at first, white, later buffish-tinted gills. The stem has a ± evident ring, but no volva. It is mostly somewhat rooting. The flesh is white, but discolours slowly buffish. The amyloid, ± ovoid spores measure (7.5–)8–10(–12) × 6–7(–8) µm. A decomposer on disturbed soils in parks and gardens with both deciduous trees and conifers.

Distinctive and unlikely to be misidentified.

Considered to be an introduced alien in temperate Europe, rare; autumn.

Nico Dam

Jan Vesterholt

The genus ***Amanita*** has free to, rarely, adnexed gills, ± white spore-deposits and a ± membranous universal veil. The species are ectomycorrhizal.

Amanita solitaria is a rather small, but robust and fleshy, almost white *Amanita* with greenish to pale buff gills and permanent, upright, ± conical veil scales on the cap. The stem is clavate with belts of pointed scales at the base and lacks a volva. The spores are ellipsoid, amyloid and measure 9–12 × 6–8 µm. Occurs with *Fagus* on calcareous soils in thermophilic forests and parks.

Drought-damaged specimens of *Amanita excelsa* ▷▷ and *A. pantherina* ▷▷ may look similar, but the cap scales are less pointed and the gills white. *Amanita strobiliformis* ▷▷ has more flaky scales and, typically, larger fruitbodies. The saprotrophic *Saproamanita vittadini* ◁◁ has scaly belts along the entire length of the stem.

Widespread and common in the south, reaching as far north as Denmark and England; August–September (October).

Amanita ovoidea is a very fleshy, large (up to 35 cm wide), ivory-white *Amanita* with a smooth cap that has ± obvious, floccose remnants of a veil at the margin. It also has a very large, white to cinnamon-buff, robust volva at the stem base, but only rarely a well-defined ring. The ellipsoid, amyloid spores measure (6.3–)7.5–10.5(–15) × (4.9–)5.2–7(–8.4) µm. Occurs in thermophilic forests and parks on ± calcareous soils, mostly mycorrhizal with *Quercus*.

The very large fruitbodies should make a correct identification relatively straightforward.

Reaches the south of England and southern parts of Germany, very rare; autumn.

Emilio Pini

Amanita strobiliformis is a large, fleshy, white to grey-tinged *Amanita* with floccose, ± angular, mealy, white to greyish scales on the cap and stem, and almost pure white gills. The widened stem base is bulb-shaped and lacks a volva. The spores are ellipsoid, amyloid and measure 10–13.5 × 7–8.5 µm. Occurs with various deciduous trees on ± calcareous clay soils in deciduous forests and parks.

Amanita solitaria ◁◁ is smaller, with more pointed scales and greenish to pale buff gills. *Amanita virosa* ▷ has a ± smooth, conical cap and an evident volva at the base. The white form of *A. citrina,* var. *alba* ▷▷, smells of raw potato and does not have floccose veil remnants.

Rather rare to locally quite common, absent from the boreal zone and nearly so from the hemiboreal zone; (June–)July–October.

Amanita strobiliformis and *A. solitaria* ◁◁ are thought to be parasitized by *Squamanita schreieri* (page 229).

Jens Maarbjerg

Jens Maarbjerg

††***Amanita virosa*** is a tall, slender, ± conical, porcelain-white *Amanita* with a floccose ring on the stem and floccose belts below; the stem has a large volva at the base. The cap either lacks or has only a few flakes of the universal veil. The gills are ± free and white to pale cream. The odour is faintly nauseous. The spores are amyloid, spherical and measure 7–8 µm. Mycorrhizal with both conifers and deciduous trees on ± poor soils.

Amanita citrina ▷▷ has a more obvious ring and no floccose belts towards the base, a round bulb at the base, no volva and smells of raw potato. The spores are also amyloid, but somewhat larger. *Amanita strobiliformis* ◁ is more fleshy, has a different cap shape and permanent remnants of the veil on the cap. The white form of *A. phalloides* ▷▷ has no floccose belts on the stem and a more flattened cap.

Rather common and widespread, less so in the nemoral lowlands; (July–) August–October (–November).

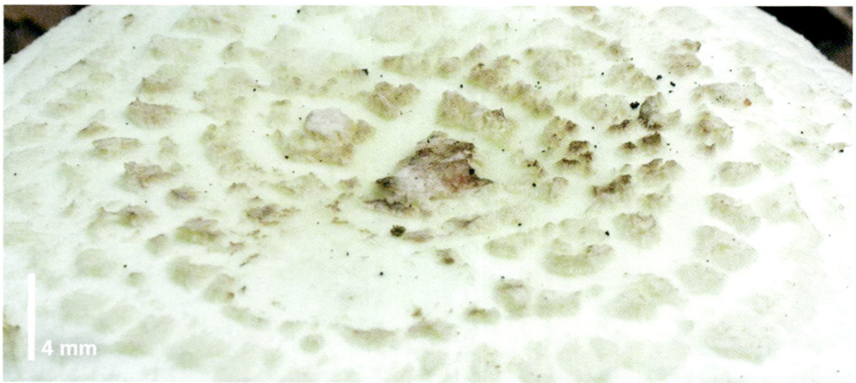

Thomas Læssøe

Amanita citrina has a raw potato-like odour; a round, sharply delimited bulb; an evident, striate, white or yellowish ring; and pale yellow-green (*A. citrina* var. *citrina*, lower pictures) or white (*A. citrina* var. *alba,* top picture) colours. It has more or less obvious veil remnants on the cap. The spores are amyloid, ± spherical, and measure 8–10 × 7–9 µm. Mycorrhizal with both deciduous and coniferous trees on various soil types.

The white variety can be told from *A. virosa* ◁◁ by the ± smooth stem, lack of a volva and the potato-like smell. *Amanita citrina* var. *intermedia* ⋉ recalls something between *A. porphyria* ▷▷ and *A. citrina* var. *citrina*: it has a rather grey cap, but no grey on the ring and a distinct smell of raw potato. This variety possibly prefers open habitats with *Helianthemum* as the mycorrhizal partner. *Amanita phalloides* ▷ can be very pale, but has an evident volva at the stem base and a different smell.

Very common and widespread; (July–) August–October (November).

††*Amanita phalloides* is a rather variable, grey-green, yellowish, olive-green to greyish, very poisonous *Amanita* with a large white volva at the stem base and a large, hanging ring on the almost white stem. The cap mostly has no veil remnants. The odour is somewhat nauseous. The spores are amyloid, almost spherical to broadly ellipsoid and measure 7.5–9 × 6–7.5 µm. Occurs on rich soils with deciduous trees, *e.g. Fagus* and *Corylus*, but also with *Abies*.

Typical specimens should be obvious, but washed-out, greyish specimens, particularly if ringless, could be misidentified. *Volvopluteus gloiocephalus* (page 487), has a more slimy cap cuticle that completely lacks greenish shades, red-tinted gills, a brownish-rose spore-deposit and no ring. *Amanita phalloides* occurs in a pure white form, *A. phalloides* var. *alba*, which is separated from *A. virosa* ⪡ by its smooth stem.

Rather common to locally very common, decreasing towards the north; mainly August–October.

Morten Christensen

Amanita gemmata is a rather small, ochre-yellow *Amanita*, which recalls *A. pantherina* ▷ in shape, with a marginate bulb and non-striate ring. The cap has a variable amount of the white universal veil. The spores are inamyloid and measure 8–12 × 6–9 µm. Occurs mostly on very sandy soils with conifers, but may also grow with *Abies* on richer soils.

Distinctive and unlikely to be misidentified.

Widespread and fairly common in central Europe, but almost absent from the hemiboreal–boreal zones; (May–) July–November.

Thomas Kehlet

Amanita franchetii is a rather slender, pale *Amanita* with yellowish veil remnants on the cap in the shape of small scales, and also in scaly belts on the clavate stem base. The striate ring is also yellow and yellow squamulose at the edge. The flesh does not stain red. The spores are amyloid and measure 8–10 × 6–7 µm. Occurs on rich soils in thermophilic forests and parks with deciduous trees.

Amanita rubescens ▽ may have yellowish colours, but the flesh reddens.

Rare, absent from the boreal zone; (June–) July–September (–October).

• ***Amanita rubescens*** is a fleshy, ± yellow-brown *Amanita* with evident reddening, particularly in worm holes *etc.* Ring striate; universal veil brownish-grey, seen as flat scales on the cap and as belts near the clavate stem base. Odour rather neutral. Spores amyloid, broadly ellipsoid, 8–9 × 6–7 µm. Occurs mostly with deciduous and coniferous trees on poor soils.

May occur in a pure white form *A. rubescens* var. *alba* ⋉ and in a form with yellowish veil remnants, var. *annulosulphurea* ⋉, which may be confused with *A. franchetii* △. No other temperate European species of *Amanita* reddens.

Very common and widespread; (June–) July–October.

Amanita excelsa is a rather variable fleshy, grey-brown *Amanita* with a striate, hanging ring, which is usually dentate and grey at the margin. The stem surface is somewhat floccose, and the stem base is normally rather brownish, but there is no true reddening. The cap typically has whitish to pale grey, somewhat mealy veil remnants. The stem base widens and lacks a volva. The spores are amyloid and measure 7–10 × 6–8 µm. Occurs with both deciduous and coniferous trees, mostly on somewhat rich soil.

Amanita rubescens ◁ is very similar but more yellow-brown with brown-grey veil remnants and stains reddish, particularly in worm holes. *Amanita pantherina* ▽ has smaller, more contrasting, pure white cap scales, a non-striate ring and a marginate bulb. *Amanita excelsa* is sometimes split into at least two species (*A. excelsa* and *A. spissa*), and a white form is named *A. excelsa* f. *alba* ⋉.

Widespread and common; July–October (–November).

† ***Amanita pantherina*** is a rather small, ± grey-brown *Amanita*, typically with small, contrasting, white veil remnants on the cap, a non-striate ring, a marginate bulb and no distinct smell. The spores are inamyloid, ellipsoid and measure 10–12 × 7–9 µm. Occurs mostly on ± rich soil in thermophilic forests and parks, or even in more open landscapes, *e.g.* dune slacks with *Quercus*.

Amanita excelsa △ has a ± similar coloration, but the ring is striate and the stem base clavate. It is normally more fleshy and it has amyloid spores.

Widespread and fairly common, but rare in the boreal zone; (June–) July–October (–November).

Amanita porphyria is a rather slender but fleshy, grey-brownish *Amanita* with a striate ring that has a greyish underside, and a ± rounded stem base without a volva. The cap has no or only a few remnants of the veil. Smells of raw potato. Spores amyloid, spherical and measure 8–10 µm. Mostly mycorrhizal with *Picea*, but rarely may occur with deciduous trees.

Amanita excelsa ◁◁ has grey teeth only on the ring, lots of small remnants of veil on the cap and does not normally smell of raw potato (although some reports suggest that it can).

Rather common and widespread; mainly July–October.

††**Amanita regalis** is a magnificent, large, yellow-brown to orange-brown *Amanita* with a yellowish universal veil and a large, yellowish, hanging, striate ring. The stem base has belts but no volva. The inamyloid spores measure 9–11(–12) × 6.5–8 µm. With *Picea* in moist, mossy forests and plantations, rarely with *Betula* in the uplands.

Amanita muscaria ▷ is very similar, but never has a brown cap surface. Some older molecular phylogenies indicate that *A. regalis* could be a variety of *A. muscaria*, but the species is still widely accepted and has a very well-defined distribution. *Amanita excelsa* ◁◁ is greyer, has greyish veil remnants and amyloid spores.

Amanita rubescens ◁◁ is less intensely coloured and has reddening flesh and amyloid spores.

Fairly common in the hemiboreal–boreal zone but rare to very rare farther south; summer–autumn.

Jens Maarbjerg

††***Amanita muscaria*** is a large, fleshy *Amanita* with a red to yellow-orange cap, ± free, white gills, a striate, hanging ring on the white stem and a white to yellow universal veil, which is left as white scales on the cap surface and at the bulbous stem base. There is no volva. The broadly ellipsoid, inamyloid spores measure 8–10 × 6–7 µm. Mycorrhizal with both deciduous trees and conifers on mostly poor soils.

Typical specimens are hard to confuse with any other species. However, old specimens, where the veil has disappeared from the cap, may recall a red *Russula* species (page 372), but these have crumbly flesh and no ring or basal bulb. See also *A. regalis* ◁ and *A. caesarea* ▷▷.

Very common and widespread; (July–) September–October (–November).

Jaroslav Malý

Jaroslav Malý

●● ***Amanita caesarea*** is an impressive, large, fleshy *Amanita* with a huge, thick, white volva at the stem base. The ± free gills are yellow. The beautifully golden-orange cap is typically smooth, lacking veil remnants, but has a furrowed-striate margin. The ± yellow stem has a large, hanging, yellow ring. The spores are inamyloid and meaure 8.5–13 × 6–8.5 µm. Occurs in thermophilic, deciduous or evergreen forests, *e.g.* with evergreen *Quercus*.

Unmistakable in a European context, even though washed-out specimens of *A. muscaria* ◁◁ can sometimes cause confusion, but that species has no volva and the gills are white. In other parts of the world several very similar species can be found, *e.g. A. hemibapha* s.l. ⋉ in Asia and *A. jacksonii* ⋉ in North America, both of which are prized edibles.

Absent from the northern part, but reaches southern Germany – not present in the British Isles, very rare; autumn.

Amanita* sect. *Vaginatae is recognized by the absence of a partial veil, and thus absence of a stem ring. They have a characteristic sulcate cap margin, but this is also the case for some of the ringed amanitoid species. The spores are inamyloid.

Amanita crocea is a vivid orange-yellow to apricot-orange, non-ringed *Amanita* with a paler cap margin, floccose brownish-yellow belts down the long stem, and a white, cohesive volva at the base. The spherical to almost spherical spores measure 9.5–12 × 8.5–11 µm. Occurs with various deciduous trees, mostly *Betula* and *Fagus*, and perhaps also with conifers.

Amanita contui ▽ is shorter stemmed and has a less vividly coloured cap, but making an accurate identification is difficult. The spores are ± alike.

Widespread, fairly common to uncommon; (June–) July–October.

Amanita contui is a monochrome ± buff to pale orange-yellow, rather short-stemmed, fleshy, non-ringed *Amanita* without conspicuous belts on the almost white stem, and a white volva. The almost spherical to broadly ellipsoid spores measure 9–12.5 × 8–10.5 µm. Apparently mycorrhizal with both deciduous trees and conifers.

Formerly often called *A. flavescens*. *Amanita crocea* △ is very similar but typically has a darker cap centre and distinct yellowish to orange belts on the stem.

Seemingly rare, but distribution is unclear at present; autumn.

Amanita fulva is a relatively vividly coloured, yellow-brown to orange-brown, non-ringed *Amanita* with rust-coloured spots on the volva, no obvious belts on the stem and, normally, no veil on the cap. The mostly spherical, inamyloid spores measure 10–12 µm. Occurs on poor, acidic soils, mostly with *Betula* and *Quercus*, but also with *Fagus*.

Amanita betulae ⋉ can usually be separated by its duller colours, a more evident umbo, lack of rust-coloured spots on the volva, and by its rather obvious, coloured belts on the stem; the volva should also have more spherical cells than in *A. fulva*. *Amanita crocea* ⧏⧏ and *A. contui* ⧏⧏ are both a more vivid orange-yellow. All four species can be ectomycorrhizal with *Betula*.

Very common to common; (June–) July–October.

Amanita battarrae is a ± grey-brown to grey-olive, usually obviously concentrically zonate non-ringed *Amanita*, which normally has a brown gill edge and a pale, brown-spotted, rather short volva that mostly comprises thread-like hyphae rather than globose cells. The stem is finely grey-brown floccose. The spores are spherical and measure 10–13 µm. Occurs in both deciduous and coniferous forests, typically on better, clay-rich mull soils.

There are several very similar species, including *A. pachyvolvata* ⋉, which apparently has a more robust volva and a less zoned cap.

Rare in the north, but more common farther south; mainly August–October.

Amanita submembranacea is an ± olive-brown, medium-sized, non-ringed *Amanita* with a grey universal veil that may persist as large patches on the cap. The stem surface may stain brownish (but not greyish) when handled. The volva is rather fragile and greyish and consists of both thread-like and spherical elements. Spores spherical, 10–14 µm. Mostly on acidic soils with *Fagus*, *Betula*, *Quercus etc.*, rarely on more calcareous soils.

The more fleshy *A. ceciliae* ▷▷ has more obvious scales on the cap, and the stem stains dark grey when handled; it occurs on rich soils.

Widespread and fairly common to rather rare in some areas; (June–) August–October (–November).

Jens H. Petersen/Michael Krikorev

Thomas Læssøe

Amanita ceciliae is a very large, fleshy, non-ringed *Amanita* with a grey universal veil, which typically persists as greyish flakes on the yellow-brown to olive-brown cap, and as flakes above the somewhat widened stem base. The stem surface discolours dark grey when handled. The spherical, inamyloid spores measure 11.5–14 µm. Occurs on rich soils with deciduous trees, mostly in rather open, park-like deciduous forests.

Very washed-out specimens may resemble *A. lividopallescens* ▷. *Amanita submembranacea* ◁◁ is smaller, has a non-staining or brown-staining stem surface and normally few or no veil remnants on the cap; it typically occurs on less rich soils.

Widespread, and locally quite common, decreasing towards the north; (June–) July–October.

Jan Vesterholt

Michael Krikorev

Yves Deneyer

Amanita lividopallescens is a large, long-stemmed, ± grey to pale cinnamon-buff, non-ringed *Amanita* with a greyish snakeskin pattern on the non-staining stem, and a universal veil that comprises predominantly spherical cells. The veil typically breaks so that a few fairly small patches are left on the cap, while other parts adhere to the stem. The spores are spherical and measure 10–13 µm. Mostly found in thermophilic parks with old *Quercus*.

The veils of *A. submembranacea* ◁◁ and *A. ceciliae* ◁ are greyer than the veil of *A. lividopallescens*.

Rare and local, absent from the boreal zone; (June–) July–October.

Guillaume Eyssartier

Amanita vaginata is an almost monochrome, grey-capped, non-ringed *Amanita* that either lacks veil on the cap or has a single large patch. The volva is white to somewhat greyish and non-spotty, and consists mainly of thread-like elements. The stem surface is pale and quite finely floccose. The spores are spherical, inamyloid and measure 9–12 µm. Occurs both with deciduous and coniferous trees, but mostly found with *Fagus*.

This species may rarely occur in a pure white form, which is very difficult to identify. *Amanita argentea* ▽ may look very similar but has broadly ellipsoid spores. Material with pinkish gills and some rust-coloured spots on the volva is currently accepted as *A. vaginata*.

Widespread and common; (June–) July–October (–November).

Jan Vesterholt

Amanita argentea is a silvery-grey, non-ringed *Amanita*, which normally lacks a veil on the cap and has a rather small, whitish volva that consists mostly of thread-like hyphae. The spores are broadly ellipsoid and measure 9.5–13.5 × 8.5–10.5 µm. Occurs on thermophilic, mostly calcareous sites such as open old parks, deer parks, *etc.*, mostly with *Quercus*.

Amanita vaginata △ may look very similar but has almost spherical spores.

Very rare in the north, but more common in southern parts of temperate Europe; (June–) July–September.

Amanita olivaceogrisea is a small, grey-olive, non-ringed *Amanita* with pale greyish-brown veil remnants. The universal veil is made of both thread-like hyphae and spherical elements in about equal amounts. The stem has fine, grey-olive floccules. The almost spherical spores measure 9–13 µm. Occurs in damp places with *Betula*, *Alnus*, etc.

Amanita friabilis ▽ also occurs in very damp woodlands with *Alnus* but has darker cap colours, more obvious grey scales on the cap and a ± disintegrating volva.

Distribution and frequency poorly known, but it is widespread and probably more common than generally assumed; (July–) August–October.

Thomas Kehlet

Amanita friabilis is a small, grey-brown, non-ringed *Amanita* with a universal veil that quickly breaks up (no volva present at maturity), and conspicuous, greyish belts on the stem. The cap surface has irregular, greyish remnants of the universal veil. The broadly ellipsoid spores measure 10–12.5 × 8–10 µm. Occurs on damp soil with *Alnus*.

Amanita olivaceogrisea △ has less grey on the universal veil and a well-defined volva.

Rare and local; late summer–autumn.

Jean-Marc Moingeon

Russula

More than 200 species in temperate Europe.

Russula is a genus of fleshy agarics with fragile flesh, and spores that have amyloid ornaments. The spore-deposit is white to orange-yellow.

A difficult genus. Among the most important macroscopical characters are the cap colour, the colour of the spore-deposit, the taste and the smell. However, the cap colours can be variable, since the pigments are water-soluble and easily washes out. The spore colour must be judged by obtaining a thick spore-deposit. The taste is fairly constant; some species are burning hot like strong chilli but for the milder species it is necessary to chew on a suitably large sample for some time. There exist also truffle-like species of *Russula* (see page 1266).

Russulas form ectomycorrhiza. Some species may grow with many different tree species, while others are ± limited to partners of certain genera. It is therefore important to note which ectomycorrhizal trees and bushes are present in the vicinity of a fruiting *Russula*. Mild-tasting russulas are considered edible but only a few are really choice edibles.

Other similar fungi:
– lactarioids have a similar fragile flesh and spores with similar ornaments, but the flesh exudes a white or coloured latex when cut (page 414).
– *Paxillus* and *Gomphidius* can have fragile flesh, but the spores are brown to black (pages 596 & 756).

Further reading: 76, 92, 151, 156, 165, 191, 228, 276, 287, 288.

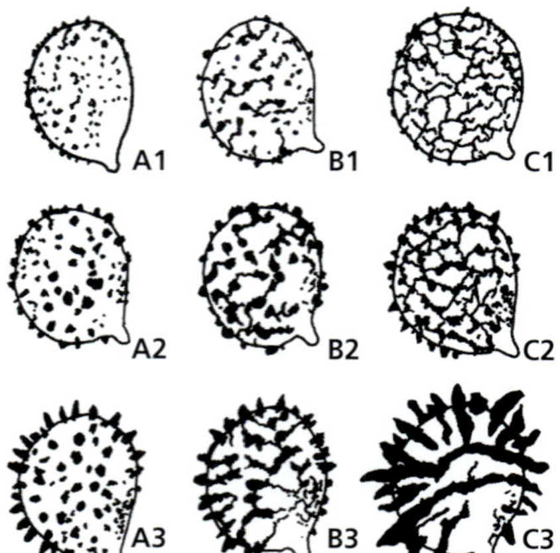

Spore types (ornament stained with iodine reagents):
column A with isolated warts or spines.
column B with somewhat connected warts or spines.
column C with reticulum (C1–C2) or crests (C3).

mild, pale-spored russulas
page 386
acrid, pale-spored russulas
page 390
taste mild
to moderately acrid; with
yellowish to reddish colours
taste
distinctly acrid
russulas
are fragile agarics
without milk
SPORES WHITE TO CREAM
SPORES PALE BUFF-YELLOW
taste mild; smell of
shellfish; stem browning
(green with $FeSO_4$)
SPORES PALE BUFF-YELLOW
SPORES BUFF- TO ORANGE-YELLOW
shrimp-russulas
subgen. *Xerampelinae*
page 396
WITH SHORT AND LONG GILLS
WITH LONG GILLS ONLY
ORANGE-YELLOW
SPORES BUFF- TO ORANGE-YELLOW
taste mild
to moderately acrid
taste
distinctly acrid
acrid, dark-spored russulas
page 408
mild, dark-spored russulas
page 398

The **funnel-russulas** are large, pale, funnel-shaped russulas with numerous short gills between the long gills. They are very similar to the fleecy lactifuii but these have milk (page 417).

Russula chloroides is a large, whitish, funnel-russula with crowded gills (more than 10 gills/cm at the margin), green shades on either the gills or stem apex and with a pale cream spore-deposit. The spores measure 8–11 × 7–9 µm and have 1–1.5 µm long spines with thin connecting lines (type B3). The taste is acrid, the odour often fish-like and the texture firm. Occurs mostly with *Fagus* on both calcareous and/or clay-rich soils but also on more acidic soils.

Other funnel-russulas are similar, but have mostly more distant gills, lack green colours and have spores with a lower ornament.

Widespread and common but rare in the boreal zone; mostly July–October.

Jan Vesterholt

Russula delica is a large, whitish, acrid-tasting, funnel-russula with medium-crowded gills (fewer than 10 gills/cm at the margin) and with a pale cream spore-deposit. The spores measure 8–11 × 6.5–8.5 µm and are verrucose with connecting lines (type B2). Mycorrhizal mostly with *Fagus*, but also with other deciduous trees and with conifers.

Russula chloroides △ has more crowded gills, greenish shades on gills or stem apex and spores with spines up to 1.5 µm long. Other, rarer, funnel-russulas have darker spore-deposits.

Widespread and common; mostly July–October.

Russula pseudodelica is a large, whitish, sharp-tasting funnel-russula with medium-crowded, cream gills (fewer than 10 gills/cm at the margin) and a pale buff-yellow spore-deposit. The spores measure 8–11 × 7–9 µm and are echinulate with thin connecting lines (type B3). Mycorrhizal with *Fagus*, probably predominantly on better soils.

The two common funnel-russulas, R. *delica* ◁ and *R. chloroides* ◁, have paler spores and gills. *Russula pallidospora* ⋉ is possibly identical, but there is currently no consensus; according to the literature, *R. pseudodelica* should have spores with isolated warts in comparison to the reticulate spores of *R. pallidospora*.

A ± southern, rare but possibly somewhat overlooked species; summer–autumn.

Russula flavispora is a large, cream to buffish, acrid-tasting, funnel-russula with a deeply buff-yellow to orange-yellow spore-deposit. The spores measure 6.5–8.5 × 5.5–7 µm and have isolated spines (type A3). Mycorrhizal with deciduous trees in thermophilic sites on clay-rich soil.

Recalls a faintly coloured version of *R. delica* ◁, but with buff-yellow gills. It has a much darker spore-deposit than all other members of the funnel-russula group.

Southern, but reaches as far north as southern Denmark, very rare; mainly September–October.

Tobias Bøllingtoft

The **charred russulas** are large, ± dark russulas with numerous, short gills between the long gills. The flesh either reddens and then blackens or blackens immediately. The mummified black fruitbodies stay put for months, often into the following spring. Fresh fruitbodies can be parasitized by species of Asterophora (page 281).

Russula nigricans is a large to very large, mild, but unpleasant-tasting charred russula, which at first reddens and later blackens after bruising. The entire fruitbody becomes brownish-black to black and rots very slowly. The gills are very distant, thick and very fragile. The spores measure 6.5–8 × 6–7 µm and are verrucose with connecting lines (type B2). Mainly mycorrhizal with *Fagus*, but may also be found with conifers.

This species is the only charred russula with distant, thick gills.

Widespread, except in the boreal zone, very common in the other zones; mostly July–October.

Russula adusta is a large, mild, but unpleasant-tasting charred russula, which, when bruised, at first reddens and later blackens. It has medium-crowded, not especially thick, mostly somewhat pinkish-tinted gills and a striking smell of old wine barrels or a damp cellar. The spores measure 7.5–9 × 6–7.5 µm and have a low reticulate ornament (type C1). Mycorrhizal with conifers, mostly *Pinus*.

This species is the only charred russula that smells of a damp cellar, and the only one found predominantly in coniferous forests.

Widespread and rather common; mainly June–October.

Russula densifolia is a medium-sized charred russula, which at first reddens and later blackens when bruised. It has crowded, mild to slightly hot-tasting, whitish gills. The spores measure 6.5–8 × 5.5–7 µm and are verrucose with connections (type B2). The odour is insignificant. Mycorrhizal with deciduous trees, mostly *Fagus* and possibly also with *Picea*.

The very similar *R. acrifolia* ⋉, has distinctly hot-tasting gills.

Widespread and common, but absent from the boreal zone; mainly July–October.

Russula albonigra is a medium-sized, rather pale charred russula that blackens (without any red transition) when bruised. The whitish gills are crowded and have at first a slightly bitter, but then a distinctive menthol-like taste. The spores measure 7.5–9.5 × 6–7.5 µm and have small, ± connected warts that almost form a reticulum (type B1–C1). Mycorrhizal with deciduous trees, mostly *Fagus*, but according to some sources also with conifers.

A number of similar blackening, mild tasting species occur in Europe. *Russula atramentosa* ⋉ has pleurocystidia that stain dark in sulfovanilin. *Russula ustulata* ⋉ and *R. nigrifarcta* ⋉ have pleurocystidia without oil drops and the flesh stains green with $FeSO_4$.

Russula anthracina ▽ has acrid-tasting gills.

Rather local and rare to occasional; mainly July–October.

Jan Vesterholt

Russula anthracina is a medium-sized charred russula, which normally blackens after bruising although in rare cases it may have a short, reddish transition. The gills are crowded and acrid-tasting. The spores measure 7.5–10 × 6–8 µm and have a low reticulum (type C1). Mycorrhizal with deciduous trees, mostly *Fagus*, on better soils

The very similar *R. albonigra* △ is recognized by its mild, menthol-like taste.

Rather local, occasional but very rare in the boreal zone; mainly July–October.

The **fetid russulas** are medium-sized to large, ± smelly russulas with yellow-brown to brownish colours and a strongly furrowed cap margin. The stems are ± hollow or chambered, and there are no short gills between the long gills.

Russula recondita is a rather small, short-stemmed, pale grey-brown fetid russula with a smell of rubber boots or fish and a mild to bitter (not acrid) taste. It usually has rust coloured spots at the stem base. The spores measure 7–8.5 × 6–7 µm and have ± isolated spines (type A2–B2). Mycorrhizal with a number of deciduous trees, particularly *Quercus* and *Fagus*. Occurs mostly in urban habitats.

Other fetid russulas typically have a different smell (*e.g.* of marzipan, fruit, oil or camembert cheese) and generally a more unpleasant taste. *Russula insignis* ⋉ tastes mild, but is darker grey-brown and have a strong red reaction with KOH at the stem base.

Widespread and rather common, but absent from the boreal zone; mainly June–October.

Russula amoenolens is a medium-sized, grey-brown fetid russula with a smooth, shiny cap centre, a smell of camembert cheese or fruit, and a very acrid taste. The spores measure 7–8.5 × 5–6.5 µm and have warts up to 0.8 µm in height (type A2). Mainly mycorrhizal with *Quercus*.

Mostly confused with *R. sororia* ⋉, but that species is somewhat larger, has a more wrinkled cap centre and the warts on the spores are only 0.4 µm in height. *Russula pectinatoides* △ and *R. insignis* ⋉ taste mild, while other similar species have more vivid colours and different odours.

Widespread and common, except in the boreal zone; mainly June–October.

Russula foetens is a large, yellow-brown, sticky, fetid russula with a rancid, oily, unpleasant smell and an acrid taste. The gills become somewhat spotty with age and do not have a darkened edge. The spores measure 7.5–10 × 7–8.5 µm and have isolated warts up to 1.5 µm in height (type A3). Mycorrhizal with both deciduous and coniferous trees, *e.g. Fagus*, *Quercus* and *Picea*.

Russula subfoetens ⋉ is very similar, but the warts on the spores are only up to 0.8 µm in height and the flesh stains distinctly yellow with KOH. *Russula fragrantissima* ⋉, *R. grata* ▽ and *R. illota* ▷ are similar, but smell ± like marzipan.

Widespread and common; mainly July–October.

Russula grata is a medium-sized, yellow-brown fetid russula with a smell of marzipan and a mild to acrid taste. The gills do not have a darkened edge. The spores measure 7.5–10 × 8–8.5 µm and have crests up to 2 µm in height (type C3). Mostly mycorrhizal with *Fagus*.

Russula illota ▷ is similar, but larger, and the gill edge has dark dots. *Russula fragantissima* ⋉ is also larger, smells more like aniseed, and the spores have crests up to 1 µm in height. The other fetid russulas smell *e.g.* oily, rubbery or like camembert cheese.

Widespread and rather common, except in the boreal zone; mainly July–October.

Russula illota is a large to very large, yellow-brown fetid russula with more or less obvious, dark dots on the gill edge. The odour is distinctly like marzipan before bruising, but more complex after. The taste is acrid. The spores measure 7–9 × 6–8 µm and have high, ± connected warts (type B3). Mycorrhizal mostly with *Quercus* and *Fagus*, usually in thermophilic sites on rich soil.

The equally large *R. fragrantissima* ⋉ and the smaller *R. grata* ◁ are similar, but do not have dark dots along the gill edge and the spores have large crests (type C3). The other fetid russulas have a different smell, *e.g.* oily, rubbery or like camembert cheese.

Widespread but absent from the boreal zone, occasional; July–October.

Russula farinipes is a rather small, relatively long-stemmed and elegant, pale apricot-orange fetid russula with a somewhat fruity odour and a very acrid taste. The stem is very pale and the apex mostly finely floccose. The spores measure 6–8 × 5–7 µm and have isolated warts (type B2). Mycorrhizal with *Fagus* and *Quercus*, often in open, park-like forests.

Other fetid russulas are mostly less elegant, browner and usually larger or have different odours.

Widespread but absent from the boreal zone, occasional; mainly June–October.

The **jade-russulas** are small to large, mild-tasting russulas with greenish, bluish or grey-rose colours and white to cream spore-deposits.

• ***Russula cyanoxantha*** is a medium-sized to large, variable, mixed green to violet, mild, nutty-tasting jade-russula with soft, pliable gills, a long white stem and white spore-deposit. The spores measure 7–9 × 6–7.5 µm and have rather small, isolated warts (type A2). The reaction with $FeSO_4$ on the flesh is almost negative or grey-green. An unusual form is completely green and has a cracking cap cuticle. Mainly mycorrhizal with *Fagus*, but is claimed to have a wider range of hosts.

This species is recognized by the green and violet cap colours and gills that are so soft they can cope with being stroked. *Russula vesca* ⊳⊳ also has moderately soft gills but a more brownish-rose cap colour and 'shows teeth' in the margin; it also tends to be smaller and have a shorter stem.

Widespread and common; mostly peaks in late summer, June–October.

• ***Russula aeruginea*** is a medium-sized, almost monochrome, green, mild-tasting jade-russula with fragile gills and a cream spore-deposit. It usually develops small, rust-coloured spots on the cap and gills. The spores measure 7–8.5 × 5.5–7 µm and are verrucose with connecting lines (type B2). Mainly mycorrhizal with *Betula*.

This species recalls the russulas around *R. grisea* ⊳⊳, but is recognized by the monochrome greenish colour on the cap and its association with *Betula*.

Common and widespread; mainly June–October.

• ***Russula heterophylla*** is a medium-sized to large, green or violet to grey-rose, mild-tasting jade-russula with rather fragile, bifurcated gills and a white spore-deposit. The spores measure 5.5–7 × 4.5–6 µm and have small, somewhat connected warts (type B1). The flesh reacts grey-rose with $FeSO_4$. Mycorrhizal with *Fagus*, *Quercus* and other deciduous trees.

The bifurcated gills are distinctive. Microscopically, this species can be told from other green-violet russulas by the hyaline to pale brown hairs in the cap cuticle and the small spores. *Russula aeruginea* ◁ has a cream spore-deposit.

Widespread and fairly common; mainly July–October.

Russula innocua is a very small, mild-tasting jade-russula with a pale green cap centre, pale cap margin, strikingly distant, fragile, whitish gills and an almost white spore-deposit. The spores measure 7.5–10 × 6.5–9 µm and have completely isolated small spines (type A2). With *Fagus* on clay-rich soils.

Recalls a tiny *R. aeruginea* ◁ but that species has cream spores and occurs with *Betula*, mostly on more sandy soils. Other similar green russulas are mostly larger and the spores have at least some warts that are connected by lines.

Very rare, or perhaps overlooked, with only scattered occurrences; mainly July–October.

Jan Vesterholt

Jan Vesterholt

Russula grisea is a medium-sized, grey-green, mild-tasting jade-russula with fragile gills, firm flesh and a deep cream spore-deposit. The stem may have a violet tinge. The spores measure 6.5–8.5 × 5–6.5 µm and have a low ornament of connected warts (type A2). Mycorrhizal with *Fagus*, *Quercus* and other deciduous trees.

Belongs to a difficult species complex but is recognized by its dominating green colour that is usually supplemented by faint violet tinges. *Russula ionochlora* ⋉ is very similar, but can possibly be separated by its pale cream spore-deposit.

Rather common and widespread; mainly June–October.

Russula parazurea is a medium-sized, more or less blue-green, mild-tasting jade-russula with fragile gills, a striking pruinose cap surface and a cream spore-deposit. The spores measure 6–8 × 5–6.5 µm and have a low reticulate ornament (type C1). Mainly mycorrhizal with *Fagus* and *Quercus*.

This species is mainly told from other similar species, *e.g. R. grisea* △, by the rather monochrome, pruinose cap surface.

Rather common and widespread; mainly July–October.

• **_Russula virescens_** is a medium-sized to large, mild-tasting, firm-fleshed jade-russula with fragile gills and a waxy, grey-green cap surface that quickly cracks into squamules. The spore-deposit is white. The spores measure 6.5–8.5 × 6–7 µm and have ± reticulate warts (type B1). Mainly mycorrhizal with *Fagus* on dry, clay-rich soils.

The green, cracking cap is very distinctive. Other similar species that may crack are *R. cyanoxantha* ⊲⊲ (with pliable gills), *R. heterophylla* ⊲⊲ (with bifurcated gills) and *R. vesca* ▷ (with brownish-red colours).

Widespread and fairly common, absent from the boreal zone; mainly July–October.

• ***Russula vesca*** is a medium-sized, mild and nutty-tasting jade-russula with a greyish-red cap and somewhat pliable gills. With age the cap cuticle withdraws from the margin to reveal a narrow, white zone of gills – which looks like it is 'showing its teeth'! The spore-deposit is white. The spores measure 6.5–8 × 5–6 µm and have small, isolated warts (type A1). Mycorrhizal with deciduous and coniferous trees on nutrient-poor, often sandy soils.

Russula cyanoxantha ◁◁ has even softer gills and green and violet cap colours. Other similar species have fragile gills and no other species of *Russula* in temperate European 'shows teeth'.

Widespread and common; June–October, peaking early in the season.

Russula mustelina is a medium-sized, mild-tasting, almost odourless *Russula* with a uniform, orange-brown cap, fragile, pale cream, somewhat forked gills and a cream spore-deposit. The spores measure 7–10.5 × 5.5–7.5 µm and have ± connected warts (type B2). Mycorrhizal with conifers.

This is the only uniform orange-brown, mild-tasting, pale spored *Russula* that occurs with conifers. The fetid russulas ◁◁ have brown colours, but they also have ± strong odours and tastes.

Fairly common in the hemiboreal zone but decreasing to the north and south; mainly July–October.

The **mild, pale-spored russulas** are small to large, ± mild-tasting russulas with yellowish to reddish colours and pale (white to cream) spore-deposits.

Russula velutipes is a medium-sized, mild-tasting, white spored *Russula* with a coral-red to grey-rose cap and white gills. The white stem is widened and flocculose at the apex. Spores 6–8 × 5–6.5 µm, with connected warts (type B2). Mainly mycorrhizal with *Fagus*.

The widened, floccose stem apex is unique; *R. velenovskyi* ▷▷ has darker spores; *R. lepida* ▽ is firmer, tastes of pencil wood and is ± red on the stem.

Widespread and fairly common; absent from boreal zone; July–October.

Russula lepida is a medium-sized, matt-red *Russula* with white gills, a white-and-red stem with particularly hard flesh, a taste of pencil wood (cedar) and a pale cream spore-deposit. Spores 8–9 × 7–8 µm, with net-like ornament (type C2). Mainly mycorrhizal with *Fagus* on mor soils.

The taste is unique. A few other species, *e.g. R. violeipes* ▷, also have very hard flesh, but are different colours. *Russula lepida* occurs in a completely white or cream form; it also goes by the name *R. rosea*.

Common in the nemoral zone but absent from the boreal zone and rare in the hemiboreal zone; mainly July–October.

Russula emeticicolor is a small, white-spored, mild-tasting, rather firm-fleshed, scarlet *Russula* with a white, grainy covering on the cap (hand lens required!). The spores measure 6–8.5 × 6–7 µm and have isolated spines (type A3). Mycorrhizal with *Fagus*, mostly on clay soils.

Russula lilacea ⋉ and *R. roseo-aurantia* ▷ also have pale granules on the cap, but the former has a ± grey-rose cap, while the latter is ± cream with rose tinges.

A local, rarely reported species, mainly known from the nemoral zones; July–September.

Russula violeipes is a medium-sized, mild-tasting, fishy-smelling, pale yellow, rarely pale violet, *Russula* with a hard-fleshed, tapering stem that usually has violet tinges. Spore-deposit cream. Spores 7–9 × 6–7.5 µm and have a low reticulate ornament (type C2). Mainly mycorrhizal with *Fagus*.

The rarer *R. amoenicolor* ⋉ has a wine-red/yellow-spotted cap and wine-red colours on the stem, which is not especially tapered or firm-fleshed. *Russula amoena* ⋉ is as wine-red as *R. amoenicolor*, but spore ornament is coarser than that species or *R. violeipes*.

Widespread and fairly common, absent from the boeal zone; mainly July–October.

Russula roseoaurantia is a rather small, white-spored, mild-tasting, rather soft-fleshed, pale yellow and somewhat rose-tinged *Russula* with a pale, grainy covering on the cap (hand lens required!). The spores measure 7–9 × 6.5–7 µm and have warts with numerous connecting lines (type B2). Mycorrhizal with *Fagus*, probably mostly on clay-rich soils.

Two other small russulas have grainy caps, *R. lilacea* ⋉ and *R. emeticicolor* ◁, but the caps are a darker grey-rose to scarlet.

Rare and local, but probably also somewhat overlooked; July–October.

Jan Vesterholt

Russula puellaris is a small, rather soft-fleshed, mild-tasting, cream-spored *Russula* with brownish-violet cap colours. The entire fruitbody stains yellow-brown with age. The spores measure 6.5–9 × 5.5–7 µm and have type A2 ornament. Mycorrhizal with both deciduous and coniferous trees, mostly on damp, acidic soils.

Russula alnetorum ▷▷ is acrid-tasting, occurs with *Alnus* and is greying on the stem. Both *Russula melliolens* ▷▷ and *R. viscida* ▷▷ are brown-staining, but larger and firmer. The somewhat acrid-tasting *R. versicolor* ▷▷ occurs with *Betula* and is yellowing on the stem.

Widespread and rather common; mostly July–October.

Russula ochroleuca is a medium-sized, slightly acrid-tasting *Russula* with an ochre-yellow cap, whitish gills and a white, but with time greying stem. The spore-deposit is white. The spores measure 8–10 × 7–8 µm and have type B3 ornament. Mycorrhizal with both deciduous and coniferous trees, and may occur on many soil types.

Russula fellea ▷▷ may look similar, but it is uniformly pale cinnamon-buff on cap, gills and stem, and is distinctly acrid-tasting. *Russula claroflava* ▷▷ is vivid yellow on the cap, very mild-tasting and has an orange-yellow spore-deposit.

Widespread and very common; mostly August–November.

Russula viscida is a medium-sized, cream-spored, at first moderately acrid-, later mild-tasting *Russula* with a shiny, rather variable, wine-red or pale yellowish cap and characteristic arched gills. The stem stains brown from the base and stains red with KOH. Spores 8–11 × 6.5–9.5 µm with type C2 ornament. With deciduous trees and *Abies*, mostly in parks or old deciduous forests on rich soil.

Russula melliolens ▽ is very mild-tasting and smells of honey at the stem base when dry. *Russula undulata* ▷ has similar colours and taste but the stem stains grey, not brown, with age.

Rather rare and local, absent from the boreal zone; mainly July–October.

Russula melliolens is a medium-sized, cream-spored, mild-tasting, firm-fleshed *Russula* with a shiny, wine-red cap, and a stem that, with age, discolours brown from below. The flesh of the stem has a faint honey-like smell upon drying. Spores 8.5–11 × 8–9.5 µm with type C1 ornament. Mainly mycorrhizal with *Quercus* and *Fagus*.

Russula viscida △ is at first moderately acrid-, later mild-tasting, mostly odourless and has somewhat more variable cap colours. *Russula undulata* ▷ has very similar colours but the taste is moderately acrid and the stem stains grey with age.

Rather rare and local, absent from the hemiboreal–boreal zones; July–October.

Russula undulata is a medium-sized, white-spored, mild- to moderately acrid-tasting *Russula* with a shiny, dark wine-red cap with yellowish spots. The white stem stains grey with age. The spores measure 6.5–9 × 6–7 µm and have type B3 ornament. Mainly mycorrhizal with *Quercus*, but also with *Fagus* and other deciduous trees.

Russula viscida ◁ and *R. mellio-lens* ◁ are similar but the stems do not stain grey, discolouring brown instead, and the spore-deposits are cream. *Russula undulata* is also known as *R. atropurpurea*.

Widespread and common, except in the boreal zone; June–October.

Russula aquosa is a medium-sized, moderately acrid- or almost mild-tasting, white-spored *Russula* with a greyish-red, dark-centred cap and a mostly greying, sometimes red-tinged stem. Smells ± like horseradish. The spores measure 7–8.5 × 6–7 µm and have type C2 ornament. Mainly in *Sphagnum* with conifers or *Betula*.

Russula atrorubens ▷▷ and *Russula fragilis* ▷▷ are distinctly acrid-tasting and smell fruit-like; *R. fragilis* also has a serrate gill edge. *Russula gracillima* ▽ has a cream spore-deposit and the spores have isolated spines rather than a net-like ornament as in *R. aquosa*.

Mainly recorded in the hemiboreal–boreal zones; mainly July–October.

Jan Vesterholt

Russula gracillima is a rather small *Russula* with greenish-reddish pastel cap colours and a white- to red-tinged stem. The gills and spore-deposit are cream. Tastes moderately acrid to almost mild and smells ± fruity. The spores measure 7–8.5 × 5.5–6.5 µm and have type A3 ornament. Mycorrhizal with *Betula* in ± damp habitats.

Russula versicolor ▷▷ is yellowing on the stem and has darker spores. A number of similar, white-spored species, *e.g. R. atrorubens* ▷▷ and *R. fragilis* ▷▷, are distinctly acrid-tasting and occur with other hosts.

Widespread and rather common; mainly July–October.

Jan Vesterholt

The **acrid, pale-spored russulas** are small to medium-sized, very acrid-tasting russulas, with white to cream spore-deposits. Many species have faint, fruity odours.

Russula emetica is a small to medium-sized *Russula* with a scarlet cap, white stem, white gills, white spore-deposit and a very acrid taste. The spores measure 8–11 × 7.5–8.5 µm and have up to 1 µm long spines with net-like connections (type C2). Perhaps exclusively mycorrhizal with conifers, mostly in rather damp, mossy habitats on poor soils.

May be difficult to separate from *R. mairei* ▽ but that species typically has a green tinge on the gills and occurs with *Fagus*. Many species have been described in this group, *e.g.* *R. silvestris* ▷ and *R. griseascens* ⋉, which are rather similar macroscopically to *R. emetica*, but occur mostly in deciduous forests and have somewhat narrower spores (7–10 × 6–8 µm); the latter also has a markedly greying stem and grows in wet habitats.

Common and widespread; mainly July–October.

Russula mairei is a small, acrid-tasting, white-spored *Russula* with a scarlet cap, white stem, and white gills that generally have a faint greenish tinge. The spores measure 6.5–8 × 5.5–6.5 µm and have short (up to 0.5 µm), ± connected warts (type B2–C1). Mycorrhizal with *Fagus* on mull and mor soils.

Russula silvestris ▷ and *R. emetica* △ never have a greenish tinge on the gills and have more reticulate spores (type C2). *Russula luteotacta* ▷ stains yellow.

Widespread and common in the nemoral zones; mainly July–October.

Russula silvestris is a small to medium-sized, scarlet, acrid-tasting, white-spored *Russula* with no green tinge on the gills. The spores measure 7–10 × 6–8 µm and have spines up to 1 µm long with net-like connections (type C2). Mycorrhizal with numerous hosts on a range of soil types.

The spore ornament of *R. emetica* ◁ recalls that of *R. silvestris* but the spores are larger and it is probably only associated with conifers. *Russula mairei* ◁ has smaller spores with warts up to 0.5 µm in height.

Widespread and common; mainly July–October.

Russula luteotacta is a small to medium-sized, acrid-tasting *Russula* with a ± scarlet cap that has a paler, furrowed margin, a white stem, and rather distant, pale cream gills that slowly turn bright yellow when bruised. The spore-deposit is white. The spores measure 7–9 × 6–7.5 µm and have short, mostly isolated warts (type B2). Mycorrhizal with *Fagus* and *Quercus* on clay-rich soils, mostly together with other uncommon fungi in open park-like habitats.

The very acrid taste, scarlet colour and yellow stain is a unique combination of characters.

Occasional, absent from the boreal zone; mainly July–October.

Russula betularum is a small, acrid-tasting, white-spored *Russula* with a pale rose to completely white cap, white gills and a white stem. The spores measure 8.5–11.5 × 7.5–8.5 µm and have spines with net-like connections (type C2). Mycorrhizal with *Betula* on damp, poor soils.

Russula raoultii ▷▷ can also be small, pale and acrid-tasting, but occurs on rich soils with *Quercus* and *Fagus*. The other acrid-tasting, pale-spored russulas are more brightly coloured.

Common and widespread; mostly July–October.

Russula fragilis is a small to medium-sized, very acrid-tasting, white-spored *Russula* with a grey-rose, wine-red or greenish cap, white stem, white, serrated gills and a fruity smell. The flesh is very fragile. The spores measure 7.5–9 × 6–7.5 µm and have warts that are reticulately connected (type C2). It is probably mycorrhizal with both deciduous and coniferous trees on different soil types, but is found mainly with *Quercus* on poor soils.

The serrated gill edge (use hand lens) is unique amongst the acrid-tasting, wine-red russulas, but this is not always obvious. *Russula pelargonia* ▽ is similar but occurs with *Populus* and the gill edge is entire. *Russula laccata* ▷ and *R. atrorubens* ▷ also have entire gill edge. *Russula queletii* ▷▷ has cream spores and wine-red colours on the stem. Other russulas with coloured stems have even darker spores and occur with conifers (see page 412). There are several varieties and forms of *R. fragilis*, *e.g.* a yellow variety, *R. fragilis* var. *gilva*.

Widespread and common; mainly July–November.

Jan Vesterholt

Russula pelargonia is a rather small, acrid-tasting, white-spored *Russula* with a grey-rose, pale lilac, wine-red or greenish cap, white stem, and white, entire gill edge. The smell is flower-like, ± like *Pelargonium*, but not very strong. The spores measure 7–9 × 6–7.5 µm and have rather connected spines (type B3). Mycorrhizal with *Populus* (mostly *Populus tremula*), both on damp and drier soils

The similar species on this page-spread do not normally occur with *Populus*.

Widespread, occasional; mainly July–October.

Russula laccata is a small, very acrid-tasting, white-spored, wine-red *Russula* with whitish gills and stem. It also occurs in a pale greenish form. The spores measure 6.5–9 × 6–7.5 µm and have rather connected warts (type B2). Occurs in dune slacks and older *Salix* carrs with various species of *Salix*.

It is the only acrid-tasting *Russula* on damp soil with *Salix*; *R. subrubens* ▷▷, from the same habitat, is a shrimp-russula with a mild taste, a shellfish-like smell and darker spores. The similar *R. fragilis* ◁, *R. atrorubens* ▽ and *R. pelargonia* ◁ grow with other hosts.

Widespread, rather common to scarce; mostly August–November.

Russula atrorubens is a small to medium-sized, very acrid-tasting, white-spored *Russula* with either a rather pale, grey-rose to greenish cap or a more pure red and shiny cap with a darker, sometimes almost blackish, centre. The stem and gills are white. The odour is fruity or like coconut. The spores measure 6–8 × 5–6.5 µm and are verrucose with connecting lines (type B2). Occurs with conifers and possibly also with *Salix*.

A dark form is rather similar to *R. aquosa* ◁◁, but that species is only mildly acrid-tasting and occurs in wet habitats. The pale version is very similar to *R. pelargonia* ◁, which occurs with *Populus*, and to *R. fragilis* ◁, which has a serrated gill edge. *Russula gracillima* ◁◁ may also look similar but is only moderately acrid-tasting and has cream spores.

Widespread and rather common; mainly July–October.

In some small russulas the gills are entirely replaced by a layer of green perithecia from *Hypomyces luteovirens* (Vol. 2, page 1543).

Jens Maarbjerg

Russula queletii is a small to medium-sized, acrid-tasting *Russula* with a greyish wine-red, shiny cap, a greyish wine-red stem, white to cream gills, a cream spore-deposit and a fruity smell (like gooseberry compote). The spores measure 7.5–9 × 6.5–7.5 µm and have type A3 ornament. Mycorrhizal with *Picea*.

Russula sardonia ▷▷, *R. sanguinea* ▷▷, *R. torulosa* ⋉ and *R. rhodopus* ▷▷ are, like *R. queletii*, all acrid-tasting, conifer-associated species with red colours on the stem, but they have darker, buff-yellow spore-deposits and do not smell fruity.

Widespread and common; mostly July–November.

Jens Maarbjerg

Russula alnetorum is a small, rather soft-fleshed, acrid-tasting, white-spored *Russula* with brownish-violet, wine-red to greenish-yellow cap colours and a tendency to stain grey and yellow on the stem with age. The gills are not serrated. The spores measure 8.5–10.5 × 7–8 µm and have warts with numerous connections (type B2). Mycorrhizal with *Alnus* on damp, even wet, soils.

Russula puellaris ◁◁ is rather similar, but mild-tasting, does not stain grey, and occurs on mor soils with other deciduous and coniferous trees.

Rather rare, but easily overlooked; rather late-occurring, August–November.

Russula consobrina is a medium-sized, extremely acrid-tasting, pale cream-spored, greyish to brownish, shiny *Russula*; the stem has a tendency to stain grey and red with age. The spores measure 8.5–10 × 7.5–9 µm and have warts that are connected in a reticulum (type C2). Mycorrhizal with *Picea*, mostly in the boreal zone.

Fairly easy to recognize based on its habitat, very acrid taste and grey-brown colours.

Fairly common in the hemiboreal–boreal zones but very rare in the nemoral zones; mainly July–October.

Russula fellea is a rather small to medium-sized, acrid and very unpleasant-tasting, sweetish-smelling, white-spored *Russula* with a rather uniform buff to cinnamon-buff colour on cap, gills and stem. The spores measure 7.5–9 × 6–7 µm and have warts that are connected in a long-meshed reticulum (type C2). Mainly mycorrhizal with *Fagus*.

The moderately acrid-tasting *R. ochroleuca* ⊲⊲ has a somewhat clearer yellow cap and obviously paler gills and stem. The burning-acrid-tasting *R. raoultii* ▽ is pale yellow on the cap and has white gills. *Russula farinipes* ⊲⊲ is more furrowed-sulcate at the cap margin and has more distant gills, which are obviously paler than the cap. Other species of fetid russulas may have similar colours to *R. fellea* but have marzipan-like, oily, rubbery or camembert cheese-like odours.

Widespread and common, rare in the hemiboreal zone and absent from the boreal zone; mostly August–October.

Russula raoultii is a rather small, acrid-tasting, white-spored *Russula* with a pale yellow cap, a white stem, white gills and a white spore-deposit. The spores measure 7–8.5 × 5.5–7 µm and have warts in a reticulum with oblong meshes (type C2). Mainly mycorrhizal with *Fagus*.

Russula fellea △ is similar, but is a darker buff on cap, gills and stem. The yellow form of *R. fragilis* ⊲⊲ has serrated gills. *Russula solaris* ⊳⊳ has more vivid colours and darker spores with isolated warts.

Widespread, occasional, absent from the boreal zone; June–October.

The **shrimp-russulas** are small to large, brown-staining, mild-tasting russulas with pale buff-yellow spore-deposits and a smell of shellfish. This group of russulas can be difficult to recognize, but the smell of older specimens is revealing, and $FeSO_4$ stains the flesh and surfaces green (see inset image above).

•• ***Russula xerampelina*** is a medium-sized, vividly coloured shrimp-russula with ± saturated carmine colours on the cap and stem. The spores measure 8–10 × 7–8.5 µm and have slightly connected warts (type B2). Occurs with conifers, particularly *Pinus*, but also *Picea* and *Abies*, typically on calcareous soils.

Other shrimp-russulas are less intensely red and typically occur with deciduous trees. Other red russulas with carmine- red stems are mostly sharp-tasting and do not stain brown.

Widespread and common to more local; mainly July–October.

Thomas Læssøe

• ***Russula subrubens*** is a rather small shrimp-russula with a red to brownish-red cap and sometimes also with red tinges on the stem. The spores measure 8–10 × 6–7.5 µm and have somewhat connected warts (type B2). Mycorrhizal with *Salix*, including *Salix repens* in dune slacks.

Other shrimp-russulas are larger or differently coloured, and typically occur with other partner trees. It is related to *R. pascua* ⋉, which is mostly considered to be an alpine-subarctic species.

Widespread and fairly common to rather rare; July–November.

●● ***Russula faginea*** is a medium-sized to large shrimp-russula with a brownish-red to carmine cap and, normally, a white stem. The spores measure 8.5–11 × 7.5–8.5 µm and have isolated spines (type A3). Mycorrhizal with *Fagus* on better soils.

Russula graveolens ▽ also occurs on rich soils but is typically mycorrhizal with *Quercus* and is usually has more blotched and brownish cap colours. Other shrimp-russulas generally occur with other hosts on lighter soils.

Widespread and fairly common in the nemoral zones; July–October.

●● ***Russula graveolens*** is a medium-sized shrimp-russula with a rather uneven coloration and a brownish-red to carmine cap. The spores measure 7.5–10 × 6–8 µm and have isolated spines (type A3). Mycorrhizal with *Quercus* on better soils, *e.g.* in open parks with older trees.

Russula faginea △ is also found on rich soil but typically occurs with *Fagus* and is a more monochrome brownish-red to carmine. Other shrimp-russulas usually occur with other hosts on lighter soils.

Widespread and fairly common to occasional, except in the boreal zone; June–October.

Jan Vesterholt

● ***Russula clavipes*** is a medium-sized shrimp-russula with olive-green to brownish cap colours. The stem may have a pink tinge. The spores measure 8–9 × 6.5–7 µm and have ± isolated spines (type A3–B3). Predominantly mycorrhizal with *Betula*.

Other shrimp-russulas have redder cap colours and typically occur with other hosts. *Russula elaeodes* is a synonym.

Widespread and occasional, apparently more common in the northern parts of temperate Europe; July–October (–November).

The **mild, dark-spored russulas** are small to large and mild-tasting, with buff-yellow to orange-yellow spore-deposits.

• ***Russula claroflava*** is a medium-sized to large, mild-tasting *Russula* with a bright yellow cap, which mostly stays intensely coloured. It has whitish to buff-yellow, greying gills, a white stem that becomes distinctly grey with age, and a buff-yellow spore-deposit. The spores measure 8–9.5 × 6.5–7.5 μm and have rather connected spines (type B3). Mycorrhizal with *Betula*, mostly in ± damp places, and may occur in very large amounts in boggy *Betula* forests, even as early as midsummer.

Russula ochroleuca ◁◁ is rather similar but has more sordid colours and white spores. *Russula decolorans* ▷ and *R. vinosa* ▷▷ also stain grey, but have orange and wine-red caps respectively.

Widespread and common; mainly June–October.

Russula risigallina is a small, apricot-orange to rose, very soft-fleshed, mild-tasting *Russula* with a white stem and an orange-yellow spore-deposit. The spores measure 7.5–9 × 6–7.5 μm and usually have isolated spines (type A3) but, rarely, connected warts (type B2). Mycorrhizal with *Fagus*.

Russula nauseosa ▷▷, *R. odorata* ▷ and the pure yellow *R. vitellina* ⋉ are similar in stature and spore colour but occur with *Picea*, *Quercus* and *Betula* respectively. The rare *Fagus*-associated *R. melzeri* ⋉ has a more scarlet cap, which is distinctly matt and somewhat cracking.

Common in the nemoral zones; mainly June–October.

• **Russula decolorans** is a rather large, orange to brownish-orange, mild-tasting *Russula* with a cream to buff-yellow spore-deposit and strongly grey-staining stem and gills. The spores measure 9–14 × 8–11 µm and have mostly isolated spines (type A3). Mycorrhizal with *Pinus* and *Picea* on sandy soils, including coastal plantations on dunes.

Russula vinosa ▷▷ and *R. claroflava* ◁ also stain grey on the gills and stem but their caps are wine-red and bright yellow respectively. *Russula paludosa* ▷▷ may have almost the same colours as *R. decolorans* and occurs in the same habitat, but does not stain grey and is mostly slightly acrid-tasting.

Widespread and common in the north, much rarer in the nemoral zones; June–October.

Russula odorata is a small, pale pinkish, pallescent, very soft-fleshed, mild-tasting *Russula* with a white stem, an orange-yellow spore-deposit and a distinct fruity/flowery odour. The spores measure 6.5–8.5 × 5.5–7 µm and have rather connected warts (type B2). Mycorrhizal with *Quercus*, often in park landscapes.

There are three similar species in stature and spore colour: *R. risigallina* ◁, *R. melzeri* ⋉ and *R. vitellina* ⋉, but these occur with *Fagus*, *Fagus*, and *Betula* respectively and do not have pale pinkish caps.

Widespread and locally common, absent from the boreal zone; June–October.

Russula velenovskyi is a medium-sized, ± shiny, mild-tasting orange-red *Russula* with a white stem that may rarely have faint orange-red tinges. The cap cuticle typically continues as red lines on the gill edges. It has a pale buff-yellow spore-deposit and smooth hyphae in the cap cuticle. The spores measure 6.5–9 × 6–7 µm and usually have isolated spines (type A3). Mainly mycorrhizal with *Betula* and *Fagus*.

The similar *R. laeta* ▷, which is a more coral-red/brick-red, has a completely white stem and a slightly darker spore-deposit; it also has hyphae with incrusted pigment in the cap cuticle. *Russula font-queri* ▽ is very similar, but can be separated by its redder stem, darker, orange-yellow spore-deposit, and by the spores having some connected warts (type B2). *Russula velutipes* ◁◁ has a white spore-deposit and a floccose, widened stem apex.

Widespread and rather common; mainly July–October.

Jan Vesterholt

Russula font-queri is a medium-sized, orange-red *Russula* with a white stem that has a red tinge, an orange-yellow spore-deposit, a mild taste, and smooth hyphae in the cap cuticle. The spores measure 8–9 × 6.5–7.5 µm and have partly connected warts (type B2). Mycorrhizal with *Betula* and possibly with *Fagus*.

The very similar *R. velenovskyi* △ has paler, buff-yellow spores and, usually, a completely white stem and spores with ± isolated spines. *Russula laeta* ▷ is also similar, but has a white stem and occurs with *Fagus* and *Quercus*.

Widespread but rather rare; July–October.

Russula pseudointegra is a medium-sized to large, scarlet *Russula* with a white stem and buff-yellow gills. The cap colour is typically saturated and does not change much over time. The cap cuticle is thin and cannot be peeled off. It is not acrid, but has a very characteristic astringent taste, which is evident only when exposed to the back of the mouth. The odour may recall a mixture of menthol and *Pelargonium*, but is, as in most russulas, rather weak. The spores measure 7–8.5 × 6.5–8 µm and have somewhat connected warts (type B2). Mainly mycorrhizal with *Quercus*, quite often with older trees in park-like settings.

With its three-coloured appearance and distinctive taste, this is one of the more easily recognized species of *Russula*.

Widespread and rather common, except in the boreal zone; mainly July–October.

Russula laeta is a medium-sized, ± coral-red to rose, mild-tasting *Russula* with a white stem, a deep buff-yellow spore-deposit and hyphae that have incrusted pigment in the cap cuticle. The spores measure 8–9 × 7–8 µm and mostly have isolated spines (type A3). Mycorrhizal with *Fagus* and *Quercus*.

The similar *R. velenovskyi* ◁ is more orange-red with slightly paler spores and may have red tinges on the stem. There are several other similar mild-tasting, yellow-spored species, and identification is difficult.

A rather rarely reported species, absent from the boreal zone; mostly July–October.

Jan Vesterholt

• ***Russula paludosa*** is a large, long-stemmed, shiny, orange-red to red *Russula* with buff-yellow spores, a mild to slightly acrid taste and a white stem with red shades. The spores measure 8–11 × 7–9 µm and are echinulate with connecting lines (type B3). Mycorrhizal with *Pinus* and other conifers on poor soils – may be a dominant fungus during the summer in dry *Pinus* plantations.

May recall *R. vinosa* ▷▷ or *R. decolorans* ◁◁ which can occur in similar places, but both stain grey in the flesh and on the gills. *Russula integra* ▷▷ is also large and red and occurs with conifers, but has a white stem and distinctly darker gills that are coloured by the orange-yellow spores. Orange specimens of *Russula paludosa* are very similar to *R. intermedia* ▷▷ but that species has a somewhat more acrid taste and, typically, a white stem and darker, orange-yellow spore-deposit; it is mycorrhizal with *Betula*.

Widespread and very common in the hemiboreal–boreal zone and the northern part of the nemoral zones but rarer farther south; mainly June–October.

Russula aurea is a medium-sized, orange-red *Russula* with striking golden gills (especially the gill edges), a buff-yellow to orange-yellow spore-deposit and a mild taste. The spores measure 7–10 × 6–8 µm and have rather distant spines with distinct connections (type B3). Mycorrhizal with *e.g. Fagus* and *Picea* on ± calcareous soils.

Provided the golden gill edge is observed there should be no doubt as to the identification of this species.

Widespread but local; mainly July–October.

Russula carpini is a medium-sized, mild-tasting *Russula* with a greenish-violet, shiny cap, a white, mostly somewhat excentric stem, and an orange-yellow spore-deposit. The spores measure 7.5–10 × 7–8 µm and have isolated spines (type A3). Mycorrhizal with *Carpinus*.

Recalls a small version of *R. romellii* ▷▷ but that species occurs with *Fagus* and has a ± central stem. *Russula sericatula* ⋉ may grow with *Carpinus*, and is very similar, but mostly lacks greenish cap colours and cystidia in the cap cuticle.

A rather southern but fairly widespread species; June–September.

Jan Vesterholt

Russula turci is a medium-sized, mild-tasting, dark wine-red *Russula* with a buff-yellow to orange-yellow spore-deposit and an iodoform-like smell from the stem base. The spores measure 7.5–9 × 6.5–7.5 µm and have very connected warts (type C2). Mycorrhizal with *Pinus*.

Russula amethystina ⋉ is similar, but has only partly connected warts on the spores. The two species may be conspecific.

Widespread and occasional; mainly August–November.

Jan Vesterholt

Russula nitida is a rather small, soft-fleshed, mild-tasting *Russula* with wine-red to grey-rose cap colours, which mostly continue on the gill edges. It has a sulcate cap margin, red tinges on the stem and a buff-yellow spore-deposit. The spores measure 7.5–10 × 6.5–8.5 µm and have isolated spines (type A3). Mycorrhizal with *Betula*, on damp poor soils.

This species often appears more pale-spored than the deposit reveals. *Russula sphagnophila* ⋉ is very similar but has cream spores with numerous connected spines.

Widespread and rather common; mostly July–October.

• ***Russula olivacea*** is a large, mild-tasting *Russula* with a wine-red to olive-green, non-shiny (matt) cap. The stem has a red tinge and usually a red collar at the apex. The spore-deposit is orange-yellow and the smell is indistinct. The spores measure 8–11 × 7–9 µm and have isolated spines (type A3). Mainly mycorrhizal with *Fagus* on rich, mostly clay soils.

Russula romellii ▷ and *R. curtipes* ▷ are rather similar and both may occur with *Fagus*. However, they have completely white stems and *R. romellii* has a shiny, somewhat greasy cap; *R. curtipes* has a duller cap with a somewhat cracking margin and a strikingly short stem. *Russula sericatula* ⋉ and *R. carpini* ◁◁ are more fragile and have white stems.

Widespread in the nemoral zones and common to occasional, rare or absent from other zones; mainly July–October.

• ***Russula romellii*** is a large, mild-tasting *Russula* with an unusual mixture of ± wine-red to olive-green, rarely completely pale green, cap colours, a shiny cap surface and a white stem. The spore-deposit is orange-yellow. The spores measure 6.5–9 × 5.5–7 µm and have connected spines (type B3). Mycorrhizal with *Fagus* and *Quercus*, mostly on ± rich soil types.

Russula olivacea ◁ is rather similar but has a duller cap and the stem has a red tinge and usually a red collar. *Russula curtipes* ▽ is duller and has a somewhat cracking margin; the cap is typically significantly paler (mostly almost pale brown at the centre) and the stem strikingly short. Both may occur with *Fagus*. *Russula viscida* ◁◁ and *R. undulata* ◁◁ have shiny caps and are a very similar colour, but both have paler, buff-yellow spores and a somewhat acrid taste.

Widespread and rather common, except in the boreal zone; mainly June–October.

• ***Russula curtipes*** is a large, mild-tasting *Russula*, typically with a pale brownish-rose cap centre and a dull, more orange-red to vinaceous-brown and usually cracking margin. The stem is white and strikingly short. The spore-deposit is orange-yellow. The spores measure 7–9.5 × 6–7.5 µm and have type B3 ornament. Mycorrhizal with *Fagus* on rather rich soils.

Russula olivacea ◁ has a matt cap, and a red tinge and usually a red collar on a relatively long stem. *Russula romelii* △ has a shiny, somewhat greasy and more blotched wine-red cap and the stem is longer. Both may occur with *Fagus*.

Widespread but uncommon; absent from the boreal zone; June–October.

Jan Vesterholt

Jan Vesterholt

• ***Russula vinosa*** is a large, wine-red, mild-tasting *Russula* with a buff-yellow spore-deposit and strongly greying stem and gills. The spores measure 9–11 × 7–9 µm and have isolated spines (type A3). Mycorrhizal with *Pinus* and *Picea* on sandy soils, including in coastal plantations.

Russula decolorans ◁◁ and *R. claroflava* ◁◁ also stain grey but the caps are brownish-orange and bright yellow respectively. *Russula paludosa* ◁◁ is a brighter red and does not stain grey.

Widespread and common in the north of temperate Europe, but much rarer farther south; mostly July–October.

Russula cessans is a medium-sized, vinaceous, mild-tasting *Russula*, which washes out to a more brownish colour. The stem is white and rather soft-fleshed. The spore-deposit is orange-yellow. The spores measure 7.5–9 × 6.5–7.5 µm and have moderately connected warts up to 1 µm in height (type B2). Mycorrhizal with *Pinus* on many different soil types.

Russula nauseosa ▷ is very similar but smaller and occurs with *Picea*. *Russula integra* ▷ is also similar but larger and the stem is 1.5–3 cm wide.

Widespread and fairly common to rather rare; mostly July–November.

●● ***Russula integra*** is a large, variable, but typically wine-red to brownish-red, mild and pleasant-tasting, shiny *Russula* with an orange-yellow spore-deposit. The stem is white and rather firm-fleshed. The spores measure 8–11 × 7–9 µm and have isolated, long (up to 1.7 µm) spines (type A3). Occurs with various conifers, also in plantations and parks.

May recall *R. paludosa* ◁◁ but that species has paler gills and spores, more scarlet colours and a longer, reddish-tinged stem. *Russula nauseosa* ▽ and *R. cessans* ◁ have similar colours and habitat, but they are much smaller (their stems are 0.5–1.5 cm wide, compared with 1.5–3 cm in *R. integra*) and they have a lower ornament on the spores.

Widespread and common in the north, but much rarer in the nemoral zones; mainly July–October.

Russula nauseosa is a small, vinaceous, mild-tasting, soft-fleshed *Russula* that washes out to an olive-brown colour. The spore-deposit is orange-yellow. Spores 7–10 × 6–8 µm and have somewhat connected warts up to 1 µm high (type B2). Mycorrhizal with *Picea*, mostly on rather rich soils.

Russula cessans ◁ is very similar but somewhat larger and occurs with *Pinus*; *R. puellaris* ◁◁ has paler spores and stains yellow-brown on the stem and gills; *R. versicolor* ▷▷ has paler spores and is moderately acrid-tasting. *Russula risigallina* ◁◁ occurs with deciduous trees.

Common and widespread, but decreasing towards the south; June–November.

The **acrid, dark-spored russulas** are small to large, with pale buff-yellow to orange-yellow spore-deposits.

Russula cuprea is a small to medium-sized, distinctly acrid-tasting *Russula* with a monochrome or mottled brownish-orange cap that washes out to brownish-rose or greenish and a striate margin. The stem is whitish, gills buff-yellow, and spore-deposit orange-yellow. The spores measure 8.5–11 × 6.5–8.5 μm and have isolated spines 1–1.5 μm long (type A3). Occurs with *Fagus*, *Tilia*, *Quercus* and *Carpinus* on clay-rich, thermophilic sites.

Russula cuprea belongs to a group of acrid-tasting, yellow-spored russulas, found with deciduous trees, that lack rust-brown spots and have a white stem. It is best separated from *R. veternosa* ▷ and *R. decipiens* ⋉ by microscopy but the latter should have a greying stem and the former often has a distinctive cap and honey-like smell.

Uncommon, absent from the boreal zone; mostly July–October.

Russula versicolor is a rather small, moderately but definitely acrid-tasting *Russula* with a pale wine-red to yellow-green cap and whitish, somewhat yellowing stem. The gills are cream and the spore-deposit a deep cream to buff-yellow. The spores measure 6–9 × 5.5–7 μm and have ± connected warts (type B2). Mycorrhizal with *Betula*, often in rather open habitats.

Shares habitat, spore colour and taste with *R. depallens* ▷▷, but that species has a ± greying stem and distinctly crowded gills. *Russula puellaris* ◁◁ is very mild-tasting and stains stronger.

Rather common and widespread; mostly July–October.

Russula solaris is a small to medium-sized, vivid yellow, rather soft-fleshed, acrid-tasting *Russula* with a white stem and creamy-yellow gills and spore-deposit. The colour washes out from the margin. The spores measure 7–8 × 5–7 µm and have isolated spines up to 1 µm long (type A3). Mainly mycorrhizal with *Fagus* and *Quercus*.

Russula claroflava ⊲⊲, *R. risigallina* ⊲⊲ and *R. vitellina* ⋉ may be the same colour, but are mild-tasting. *Russula fellea* ⊲⊲ is acrid-tasting, but white-spored and more buff coloured. *Russula ochroleuca* ⊲⊲ has white spores and is only slightly acrid-tasting.

Widespread and fairly common in the nemoral zones; July–October.

Russula firmula is a medium-sized, acrid-tasting *Russula* with a violet, greyish-violet to purple-brown, shiny cap that has a dark centre, a white stem, dark buff-yellow gills and an orange-yellow spore-deposit. Spores 8–10.5 × 7–8.5 µm, with isolated, spines up to 1 mm long (type A3). Smells of *Pelargonium*. Mycorrhizal with conifers.

Recalls a small, acrid-tasting version of *R. romellii* ⊲⊲ but that species occurs with decidous trees. *Russula badia* ⊳⊳ occurs with *Pinus* and only slowly becomes very acrid-tasting.

Fairly common in the hemiboreal–boreal zone but very rare farther south; mainly July–September.

Russula veternosa is a small to medium-sized, moderately acrid-tasting *Russula* with a grey-rose to brownish-rose, matt cap with a pallescent centre. The stem is white, gills buff-yellow and spore-deposit orange-yellow. The spores measure 6.5–8.5 × 5.5–7 µm and have isolated spines 1 µm long (type A3). With *Fagus* on rich soils.

Similar species are best separated by microscopy (see discussion under *R. cuprea* ⊲), but *R. veternosa* can often be recognized by its two-coloured cap and somewhat honey-like smell. *Russula zonatula* ⊳⊳ is smaller.

Widespread and fairly common in the nemoral zones; mostly July–October.

Michael Krikorev

Russula intermedia is a large, reddish-orange, moderately but definitely acrid-tasting *Russula* with a white, rarely somewhat reddish-tinged stem, buff-yellow gills and orange-yellow spore-deposit. The spores measure 7–8 × 6.5–7 µm and have isolated spines (type A3). Occurs with *Betula*.

Like a dark-spored *R. paludosa* ◁◁, which occurs with conifers. *Russula pseudointegra* ◁◁ occurs with *Quercus,* is scarlet and has a different taste. Other russulas with deciduous trees have smaller fruitbodies and less uniform caps, *e.g. R. maculata* ▷, *R. cuprea* ◁◁) and *R.* veternosa ◁◁.

A mostly northern species; August–October.

Jan Vesterholt

Russula rubra is a small, acrid-tasting *Russula* with a matt, scarlet, pallescent cap, white stem, cream gills, buff-yellow spore-deposit and a farinaceous to honey-like smell. The spores measure 6.5–8.5 × 6–7.5 µm and have somewhat connected warts (type B2). Occurs with *Quercus* and *Fagus*.

Recalls a small, pale *R. pseudointegra* ◁◁ but has an acrid taste. *Russula rutila* ⋉ has a shiny cap, orange-yellow spores and only a faint smell. Other similar species with deciduous partners have less uniform cap colours, *e.g. R. maculata* ▷, *R. cuprea* ◁◁ and *R. veternosa* ◁◁.

Rare and southern; mainly August–October.

Jan Vesterholt

Russula zonatula is a small, acrid-tasting *Russula* with a brownish-red cap centre, becoming a paler red towards the edge. It has a white stem, cream gills and a buff-yellow spore-deposit. The spores measure 7–9 × 7–8 µm, are almost spherical, and have ± isolated warts up to 1 µm in height (type B2–A2). Mycorrhizal with *Fagus* on clay-rich, ± calcareous soils.

May recall a small version of *R. veternosa* ◁◁, but that species has narrower spores (6.5–8.5 × 5.5–7 µm).

Rare and southern; mainly July–September.

Russula maculata is a small to medium-sized, acrid-tasting *Russula* with a rather subdued, brownish-red to brownish-orange cap, which may wash out to rose and acquire small, brown spots. The stem is whitish or rose-tinged, very hard and stains brown with age. The gills are buff-yellow and the spore-deposit orange-yellow. Spores 8–10 × 7–9 µm, with partially connected warts 1 µm high (type B2). Mycorrhizal with *Fagus*, *Quercus* and *Tilia*.

Russula globispora ⋉ is very similar but can be separated by the larger spores (9.5–14 × 8–12 µm) that have 1–1.5 µm long, isolated spines (type A3). *Russula cuprea* ◁◁ does not normally develop brown spots or stain brown; the spores have completely isolated spines.

Distribution not fully understood due to confusion with *R. globispora*; mainly August–October.

Russula depallens is a medium-sized, moderately but definitely acrid-tasting *Russula* with a pale rose to grey-rose cap that has a ± radiating, translucent marbled appearance that washes out to a yellowish or whitish colour. Some specimens can be completely cream. The stem is white or has red tinges, and stains grey with age. The gills are strikingly crowded and cream. The spore-deposit is deep cream to buff-yellow. Spores 8–9.5 × 6–7 µm, verrucose, partially with numerous connections (type B2). Mycorrhizal with *Betula*, often in open sites.

Shares habitat, spore colour and taste with *R. versicolor* ◁◁, but that species is typically smaller and has a yellowing stem and less crowded gills.

Widespread and common; May–November.

Jan Vesterholt

Russula sanguinea is a medium-sized, moderately but definitely acrid-tasting *Russula* with a greyish-red, ± matt to somewhat shinycap, a red stem, cream gills and a buff-yellow spore-deposit. The spores measure 7–8.5 × 6.5–7.5 µm and have somewhat connected warts (type B2). Mycorrhizal with *Pinus*.

Russula helodes ⋉ is similar but occurs in *Sphagnum* with *Picea* and has larger spores (8.5–10.5 × 7.5–8.5 µm) that have a reticulate ornament (type C2). *Russula rhodopus* ▽ also occurs with *Picea* and has a distinctly shiny cap; *R. queletii* ◁◁ has more vinaceous colours. *Russula sardonia* ▷ and *R. torulosa* ⋉, both micorrhizal with *Pinus*, have darker, more vinaceous colours.

Widespread and locally common to rare; July–November.

Russula rhodopus is a medium-sized, very acrid-tasting *Russula* with a shiny red cap, red stem, cream gills and a buff-yellow spore-deposit. The spores measure 7.5–9 × 6.5–7.5 µm and have a reticulate ornament (type C2). Mycorrhizal with *Picea*, mostly on acidic soils.

Russula queletii ◁◁ also occurs with *Picea* but has vinaceous colours and paler spores; *R. helodes* ⋉ has larger spores (8.5–10.5 × 7.5–8.5 µm). *Russula sanguinea* △ occurs with *Pinus* and is more matt red. *Russula sardonia* ▷ and *R. torulosa* ⋉, both found with *Pinus*, have more vinaceous colours.

A mostly northern species with a few southern records; August–October.

Russula sardonia is a medium-sized, very acrid-tasting, umbonate *Russula* with a dark wine-red, shiny cap, wine-red stem, buff-yellow gills that stain rose with ammonia, and a buff-yellow spore-deposit. The spores measure 7–9 × 6–7 µm and are verrucose with some connecting lines (type B2). Mycorrhizal with *Pinus*, mostly on sandy soils.

The combinaton of a wine-red, umbonate cap, and both yellow gill flesh and spores is distinctive. *Russula torulosa* ⋉ is very similar, but the gill flesh is not yellow, it is barely umbonate, tastes only moderately acrid and smells of apple; the gills do not stain with ammonia. *Russula xerampelina* ◁◁ is also rather similar and occurs mostly with *Pinus* but is mild-tasting and non-umbonate.

Widespread and rather common; July–November.

Russula badia is a medium-sized, at first mild but later distinctly acrid-tasting, cedar wood-smelling *Russula* with a brownish-red cap, a white stem with red tints, buff-yellow gills and a buff-yellow to orange-yellow spore-deposit. Spores 8–11 × 6.5–8.5 µm, with connected, medium-length warts (type B2). Mycorrhizal with *Pinus*.

Recalls *R. cessans* ◁◁ but the initially mild and later very acrid taste is characteristic. The many other acrid-tasting, red-stemmed russulas (*R. sardonia* △, *R. queletii* ◁◁, etc.) typically have pale buff-yellow spores and are immediately acrid-tasting.

Mostly northern but rare in the nemoral zones; mainly September–October.

Jan Vesterholt

Lactarioids

The genera *Lactarius* and *Lactifluus* are recognized by their fleshy fruitbodies with milk. They have the same fragile flesh as the genus *Russula*, and similar spores with a very amyloid ornament of spines, reticulae or crests. Most species also have large cystidia on the gills. Some are mild-tasting but many are very acrid.

The most important characters for species identification are the colour and stain of the milk, surface texture of the cap (including the degree of sliminess), overall colour, taste and mycorrhizal partner.

All lactarioids are mycorrhizal with trees and shrubs or, in the alpine zone, with *e.g. Dryas* and *Salix herbacea*.

The lactarioids are very closely related to *Russula*, and in the tropics the genera are less easy to differentiate.

With the exception of *Lactarius helvus*, which smells like curry and is slightly toxic, all mild-tasting European lactarioids are edible. However, the acrid lactarioids are toxic and some contain strong mutagens.

Other similar fungi:
– some *Mycena* species have milk, but the stems are less than 2 mm wide (page 203).
– in the subtropics/tropics other agarics, *e.g. Lactocollybia* ⋉, may have latex.

Further reading: 92, 105, 152, 156, 165, 228, 327.

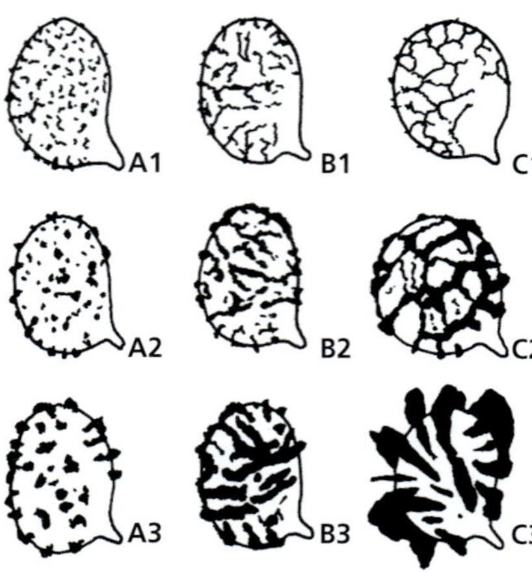

Spore types (ornament stained with iodine-reagents):
Column A with isolated warts or spines.
Column B with somewhat connected warts or spines.
Column C with reticulum or crests (C3).

More than 100 species in temperate Europe.

fishy lactifluii
(Lactifluus p.p.)
page 416
lilac-staining lactarii
page 434
yellow-staining lactarii
page 437
pink-staining lactarii
page 440
orange-milked lactarii
page 444
Lactarius helvus
page 447
fleshy, with very ample and sticky milk, and fishy smell
milk and flesh greyish-violet on drying
milk at first white but quickly yellow on the gill flesh
milk at first white, later rose or only on the gill flesh
milk rather sparse, orange from the start
with smell of curry or stock cubes
brown milk on drying
violet milk on drying
yellowing milk
reddening mælk
orange milk
clear milk
lactarioids
have fleshy fruitbodies with milk (latex) and spores with an amyloid ornament

The genus ***Lactifluus*** includes lactarioids that are more closely related to *Russula* than to species of *Lactarius*. There are two distinct groups with rather large, fleshy fruitbodies and plentiful white milk.

The **fishy lactifluii** has sticky latex that stains brown.

•• ***Lactifluus volemus*** is a very firm-fleshed, medium-sized, ± orange-brown lactarioid with copious white, sticky milk that stains brown. It smells of shellfish, and has thick-walled cystidia. The spores measure 9–11 × 8.5–10.5 µm and have a rather low reticulate ornament (type C2). Occurs in older deciduous forests and older conifer plantations; mycorrhizal with *Fagus*, *Quercus* and *Picea*.

The southern *L. oedematopus* ⋉ is similar but has 60 µm long hairs on the cap (other species of fishy lactifluii have hairs 85–125 µm long). *Lactifluus subvolemus* ▽ is also very similar but a paler orange-yellow.

Widespread, usually not very common but can be common locally, absent from the boreal zone; mainly July–October.

•• ***Lactifluus subvolemus*** is a smooth, matt, ± yellow to pale orange-brown, brown-staining, shellfish-like smelling lactarioid with sticky, white milk. The spores measure (7.7–)7.9–11.6 × (7.2–)7.3–11.1 µm and have a ± reticulate ornament up to 1.6(–1.8) µm in height (type C2). Occurs on various soil types in both deciduous and coniferous forests, often with *Fagus*.

Lactifluus volemus △ can ± only be separated by its darker, ± reddish-brown colours. The Mediterranean *L. oedematopus* ⋉ has shorter hairs in the cap cuticle (less than 60 µm long).

Widespread but as only recently recognized, records may relate to *L. volemus*; mainly August–October.

The **fleecy lactifluii** is a group of lactarioids with large, fleshy, whitish, funnel-shaped fruitbodies and copious white milk. They generally have a somewhat felty cap surface.

Lactifluus vellereus is a large, chalk-white to yellow-brown speckled, funnel-shaped, thick- and firm-fleshed *Lactifluus* with a rather felty cap surface and medium-crowded, cream gills. The milk is mild when isolated from the flesh, but the flesh is acrid. The spores measure 7.5–12 × 6.5–9.5 µm, longer than 9 µm on average, and have a low, reticulate ornament (type C1). Mycorrhizal with both deciduous and coniferous trees, but mainly with *Fagus*.

Lactifluus bertillonii ⋉ is similar but has a burning, acrid-tasting milk and smaller spores (7–10 × 5–7.5 µm).

Widespread and common, very rare in the boreal zone; (July–) August–October (–November).

Lactifluus piperatus is a large, fleshy, somewhat funnel-shaped, pale cream *Lactifluus* with very crowded gills and a very acrid taste. Milk/flesh do not change colour upon drying and usually do not react with KOH. The cap cuticle is a thin felt of hyphae (10–30 µm) above the cellular flesh. The spores measure 7–10.5 × 5–7.5 µm and have a low ornament (type B1). Mainly mycorrhizal with *Fagus* on ± rich soils.

The similar but rarer *L. glaucescens* ⋉ has a thicker cap covering (80–120 µm), the milk dries greenish and reacts orange with KOH.

Widespread and rather common, decreasing towards the north; mainly July–October.

Lactarius is a large genus with small and gracile to large, fleshy fruitbodies, and milk (latex) that may be white, coloured or watery-clear.

The **bearded lactarii** (this page spread) are acrid lactarioids with white milk and very bearded cap margins. They are not directly edible, but can be made so after salting.

Lactarius torminosus is a ± large, fleshy, somewhat funnel-shaped, flesh-pink, ± zonate *Lactarius* with a distinctly hairy-bearded cap margin and a very acrid taste. The milk is white and does not change colour. The spores measure 7.5–9.5 × 5.5–7 µm and have a rather low ornament (type B2). Mycorrhizal with *Betula* on peaty soils.

Lactarius pubescens ▷ is paler and somewhat less bearded and occurs in drier habitats. *Lactarius torminosulus* ⋉, perhaps just a form of *L. torminosus*, is more northern and smaller; specimens with these characters have been found as far south as Denmark.

Widespread and rather common; (July–) August–October (–November).

Hypomyces spadiceus (Vol. 2, page 1542) can attack and disfigure the gills of this and other related species.

Thomas Læssøe

Lactarius spinosulus is a rather small, pinkish-tinted *Lactarius* with strigose hairs, particularly at the cap margin. The milk is white, does not change colour, and tastes bitter to acrid. The spores measure 6–8.5 × 5–6.5 µm and have a striate-crested ornament (type C3). Occurs mainly with *Betula* in dry woodlands.

Other bearded lactarii are typically larger and do not have a striate spore ornament.

Rather common in the hemiboreal–boreal zones, rarer farther south; (July)–September–October.

Lactarius pubescens is a very pale, rather creamy, somewhat downy-bearded *Lactarius* with a white, unchanging, sharply tasting milk. The spores measure 6–8.5 × 4–6 µm and have very connected warts (type B2). Mycorrhiza with *Betula*, mostly in rather dry habitats, often with solitary trees in parks and churchyards.

Lactarius torminosus ◁ is tinted pinkish, more bearded and prefers wetter habitats; *L. scoticus* ▽ is smaller and occurs in very wet boggy habitats – both grow with *Betula*.

Widespread and common; (August–) September–October (–November).

Lactarius scoticus is a small, funnel-shaped, very pale *Lactarius*. The cap margin is initially downy-bearded but later becomes smooth. The milk is white, does not change colour and tastes acrid. The spores measure 5.5–7.5 × 4.5–5.5 µm and have very connected warts (type B2). Mycorrhizal with *Betula* in very wet *Betula-Salix* carrs.

Lactarius pubescens △ is larger and occurs in drier habitats.

Mainly recorded in the Nordic countries and northern parts of the British Isles; (July–) August–October (–November).

Jens Maarbjerg

Lactarius mairei is a yellow-brownish, somewhat zonate, medium-sized *Lactarius* with a strigose beard at the cap margin. The milk is white, does not change colour and tastes acrid. The spores measure 6–9 × 5–7 µm and have rather reticulate, connected spines (type B3–C2). Occurs on rich soils where it is mycorrhizal with *Quercus*.

Lactarius torminosus ◁ is more flesh-pink, mycorrhizal with *Betula* on peaty soil and is often larger.

Rare, thermophilic and local, absent from the boreal zone; mainly August–October.

Jacob Heilmann-Clausen/Carsten Brandt

Jens H. Petersen/Jens Maarbjerg

The **sticky lactarii** have white milk and sticky to slimy caps. Most species are distinctly acrid-tasting.

Lactarius zonarius is a large, orange-brown *Lactarius* with dense, concentric zones on the depressed cap, which is ± felty at the margin. The milk is white, unchanging and very acrid-tasting. The flesh reddens somewhat and the stem may have pits. Has 4-spored basidia. The spores measure 6.5–9 × 5–7.5 µm and have somewhat connected, high warts (type B2–B3). Mycorrhizal with *Quercus* on ± calcareous soils in thermophilic forests and parks.

Lactarius evosmus ▽ is mycorrhizal with *Salix*, *Populus* and *Quercus*, predominantly zonate at the ± smooth margin, lacks pits in the stem and has a lower spore ornament. *Lactarius acerrimus* ▷ also has a smooth cap margin, and the gills near the stem are strongly branching. The smooth-capped *L. porninsis* ▷▷ is less zonate and grows with *Larix*. *Lactarius zonarioides* ⋉ is also similar but is found with *Picea* and *Abies*.

Rare, absent from the boreal zone; mainly August–October.

Thomas Læssøe

Lactarius evosmus is a large, orange-brown *Lactarius* with a ± zonate, ± smooth cap margin, but without zones towards the centre. Stem pale, short and lacks pits. The flesh sometimes reddens. The smell is strongly fruity. Milk white and very acrid-tasting. The spores measure 6–9 × 5–7 µm and have low, connected warts (type B1). Mycorrhizal with *Populus*, *Quercus* and *Salix* on ± calcareous soils.

Lactarius acerrimus ▷ has branching (anastomosing) gills and 2-spored basidia. *Lactarius zonarius* △ is zonate to the centre, has pits on the stem, occurs only with *Quercus* and has a higher spore ornament.

Rare, absent from the boreal zone; (July–) August–October.

Lactarius acerrimus is a medium-sized to large, squat and mostly irregularly shaped *Lactarius* with a zonate, ochre-yellow cap. The gills are distinctly branching (anastomosing) near the stem and the basidia are 2-spored. The spores measure 9.5–14 × 8–11.5 µm and have connected spines (type B3). Mycorrhizal with *Quercus* in thermophilic woodlands, including parks and cemeteries, mostly on clay soils.

Lactarius evosmus ◁ and *L. zonarius* ◁ occur in similar habitats and can look rather similar but have less distinctly branching gills and 4-spored basidia.

Occasional, absent from the boreal zone; (July–) August–September (–October).

Lactarius controversus is a mostly large, rather short-stemmed and pale *Lactarius* with a somewhat depressed cap that has indistinct, somewhat reddish zones. The gills are pinkish and crowded. The milk is white, does not change colour and is very acrid-tasting. The spores measure 6–8.5 × 4–6 µm and have crests (type B3–C3). Mycorrhizal with *Salix* and *Populus*, often in open habitats including dune slacks and semi-urban areas.

Distinctive and unlikely to be misidentified.

Locally common to occasional, absent from the boreal zone; August–November.

Lactarius pallidus is a medium-sized, very sticky-slimy, pale brownish to somewhat flesh-pink *Lactarius* with white milk that does not change colour and is almost mild to somewhat acrid-tasting. The spores measure 6.5–8.5 × 5–6.5 µm and have connected spines (type B3). Mycorrhizal with *Fagus* on rich mull soils.

Other very similar species are mycorrhizal with conifers. *Lactarius vietus* ▷ has greyish milk on drying and is mycorrhizal with *Betula*.

Rather common in southern Scandinavia and in the British Isles but seemingly less so farther south; (July–) September–October (–November).

Jan Vesterholt

Lactarius musteus is a pale, rather small, slimy, non-zonate *Lactarius* with white, almost mild-tasting milk that does not change colour. The spores measure 7–9.5 × 5.5–7.5 µm and have crests (type B3). Mycorrhizal with *Pinus* on sandy, nutrient-poor soils.

Lactarius albocarneus ▷ is similar but is mycorrhizal with *Abies* on rich soils. *Lactarius pallidus* △ is also similar but occurs with *Fagus* on rich soils.

Rather common in the hemiboreal–boreal zones but very rare or absent from the nemoral zones; mainly August–October.

Lactarius albocarneus is a rather small to medium-sized *Lactarius* with a pale, very slimy cap and a sticky stem. The milk is white, does not change colour and and tastes acrid. The spores measure 7–9.5 × 6.5–8 µm and have connected spines (type B3). Mycorrhizal with *Abies* on ± calcareous soils.

Lactarius pallidus ◁ and *L. musteus* ◁ look similar but are associated with other partner trees; in addition, the latter grows on poor, sandy soils. Both species have somewhat smaller spores than *L. albocarneus*.

Mostly recorded in *Abies* plantations in Denmark and near the Alps; August–November.

Lactarius vietus is a rather pale violet-grey, sticky, ± zonate, convex to depressed *Lactarius* with pale gills. The white milk dries lead-grey to olive-grey and the taste is very acrid. The spores measure 7–9.5 × 5.5–7.5 µm and have connected warts (type B2). Mycorrhizal with *Betula* in damp habitats.

The more northern *L. pilatii* ⋉ is darker, ± distinctly umbonate and usually occurs in mires with *Sphagnum* and *Betula*.

Widespread and rather common, August–November.

Lactarius blennius is a medium-sized, ± zonate, ± olive-grey, sticky, very variable *Lactarius* with a concolourous cap margin, crowded, whitish gills and white, somewhat browning, very acrid-tasting milk. The spores measure 5.5–8 × 4.5–6.5 µm and have warts connected in a somewhat striate ornament (type B2–B3). Mycorrhizal with *Fagus* on ± rich mull soils.

Lactarilus fluens ▷ is less sticky, has a very narrow, pale cap margin and the gills are more cream-coloured. The rather similar *L. circellatus* ▽, which is mycorrhizal with *Carpinus,* and *L. pyrogalus* ▷, which is mycorrhizal with *Corylus,* have darker gills.

Common to very common in the nemoral zones, absent from the boreal zone; (July–) August–October (–December).

Lactarius circellatus is a somewhat sticky, greyish to grey-brown, distinctly zonate *Lactarius* with cream gills. The milk is acrid-tasting, white and dries olive-brown. The spores measure 5.5–8 × 4.5–6.5 µm and have warts connected in a striate ornament (type B2–B3). Mycorrhizal with *Carpinus* on rich soils.

Lactarius pyrogalus ▷ is mycorrhizal with *Corylus,* less zonate, more greenish, and has more distant gills. *Lactarius blennius* △ and *L. fluens* ▷ are mycorrhizal with *Fagus* and less zonate on the cap.

Fairly common, absent from the hemiboreal–boreal zones; (June–) July–October (–November).

Lactarius fluens is a medium-sized to large, ± zonate, slimy-sticky to greasy or almost dry, olive-grey *Lactarius* with cream gills and a narrow, pale cap margin. The milk is white and acrid-tasting but discolours brownish on the gills. The spores measure 6–8 × 5–6.5 µm and have warts connected in a somewhat striate ornament (type B2–B3). Mycorrhizal with *Fagus* on rich soils.

Lactarius blennius ◁ has whiter gills and the cap margin is not pale.

Rather common, but absent from the hemiboreal–boreal zones; (July–) August–October (–December).

Lactarius pyrogalus is a rather dry, grey-green, medium-sized *Lactarius* with rather distant, pinkish-buff to buff gills and a burning acrid-tasting, white milk that dries grey-green. The spores measure 5–7.5 × 4.5–6 µm and have warts connected in a striate ornament (type B2). Mycorrhizal with *Corylus*.

Lactarius circellatus ◁ is distinctly zonate, has paler and more crowded gills and is mycorrhizal with *Carpinus*. *Lactarius blennius* ◁ and *L. fluens* △ are mycorrhizal with *Fagus* and have paler gills.

Rather common and widespread, decreasing towards the north; (July–) August–October (–November).

†***Lactarius necator*** is a sticky to almost dry, dull olive *Lactarius*, with a centrally almost almost black, somewhat depressed cap. The stem is a similar colour but the gills are pale. The milk is white and burning acrid-tasting. The fruitbody becomes purple with KOH. The spores measure 6.5–9 × 5–6.5 µm and have warts connected in a reticulum (type C2). Mostly mycorrhizal with *Picea* and *Betula* on various soil types and in a range of woodland types.

Unmistakable due to the dark olive colours. This species is also known as *L. plumbeus* and *L. turpis*.

Very common and widespread; (July–) August–October (–November).

Apparently carcinogenic and should therefore not be used for salting and subsequent culinary purposes.

Thomas Læssøe

Lactarius hysginus is a medium-sized, cinnamon to red-brown, mostly somewhat pale-edged, ± sticky *Lactarius*, with a smell recalling that of *Hygrophorus discoxanthus* (page 167) or the larvae of *Cossus cossus*. The milk is white, does not change colour and is very acrid-tasting. The spores measure 6–8 × 5–6.5 µm and have connected warts (type B2). Mycorrhizal with *Picea* and *Pinus,* but also known to form mycorrhiza with deciduous trees.

Lactarius hysginoides ⨯ has a mild taste and occurs in damp woodland with *Picea*, *Betula* and *Salix*.

Rare in the south but quite common in the hemiboreal–boreal zones; mainly September–November.

Jan Vesterholt

Lactarius trivialis is a large, somewhat sticky, mostly long-stemmed, greyish-violet-brown *Lactarius*, usually with not very obvious zonation but sometimes with zones of spots. The stem apex is narrow and the gills are crowded and cream. The milk is white, drying grey on the gills, yellowish on paper and is burning acrid-tasting. Stains orange-yellow with KOH. The spores measure 7.5–10 × 6–8 µm and have connected warts (type B2). Mycorrhizal with *Picea* and *Betula* on damp, boggy soils.

Lactarius vietus ◁◁ is smaller, lacks a narrow stem apex and does not react orange-yellow with KOH. *Lactarius flexuosus* ⋉ has more distant gills and shorter spores (less than 8 µm long).

Common in the hemiboreal–boreal zones, much rarer farther south; July–October (–November).

Lactarius porninsis is a vivid orange, zonate, smooth, somewhat sticky *Lactarius*. The milk is white, does not change colour, and tastes mild to bitter. The spores measure 6.5–9.5 × 5–7.5 µm and have connected warts (type B2). Mycorrhizal with *Larix*.

A fairly easily recognizable species. Other somewhat similar species have acrid-tasting milk and different mycorrhizal partners.

Very rare and local in temperate Europe except in some regions of the Alps; (August–) September–October (–November).

Jean-Marc Moingeon

The **foxy lactarii** is a confusing group of white-milked, dry lactarioids. The milk either does not change colour or is somewhat yellowing. They are, in general, brown to brownish-orange or, more rarely, ± brownish-violet and are usually ± mild tasting, although *L. rufus* is very acrid-tasting.

Lactarius rufus is a medium-sized, dry, red-brown, monochrome *Lactarius* with a small umbo in a ± marked depression. The gills are rather pale and the milk does not change colour and is very acrid-tasting. The spores measure 6.5–10 × 5.5–7.5 µm and have reticulately connected warts (type B1–C1). Mostly mycorrhizal with conifers, more rarely with *Betula*, *Fagus etc.*

No similar species are as acrid-tasting. The *Fagus*-associated *L. subdulcis* ▷ is not as reddish. *Lactarius tabidus* ▷ is more slender and has ± yellowing milk. *Lactarius fulvissimus* ▷▷ occurs on rich soils, has a strong smell and is mild to bitter-tasting.

Very common and widespread; (June–) July–October (–November).

Jan Vesterholt/Jacob Heilmann-Clausen

Lactarius obscuratus is a small, rather vivid red-brown *Lactarius*, which may have olive shades at the cap centre. The milk is white and mild-tasting. The spores measure 6.5–9 × 5–7 µm (less than 6.5 µm wide on average) and have connected spines (type B3). Ectomycorrhizal with *Alnus* on ± damp soils.

There are two other small, brownish to olive-coloured *Lactarius* associated with *Alnus*: *L. cyathuliformis* ⋉, which has spores that on average are wider than 7 µm; and *L. omphaliformis* ⋉, which has a cracking/scaly cap cuticle.

Widespread and rather common; mainly June–October.

Lactarius subdulcis is a small to medium-sized, dry, matt, red-brown *Lactarius* with a rubber-like smell and white, unchanging, mild to bitter-tasting milk. The spores measure 7–9.5 × 5.5–7.5 µm and have moderately connected warts (type A3). Mycorrhizal with *Fagus* on ± mull-type soils, often in the same sites as *L. blennius* ◁◁.

Lactarius tabidus ▽ is ± wrinkled and the milk dries yellowish on paper/cloth. *Lactarius decipiens* ⋉ has a pink tinge, smells of *Pelargonium* and has ± yellowing milk.

Very common and widespread, absent from the boreal zone; (June–) August–October (–January).

Thomas Læssøe

Lactarius tabidus is a rather slender, small to medium-sized, orange-brownish, almost odourless *Lactarius*. If often has a small umbo and a ± wrinkled cap centre. The white, mild to slightly acrid-tasting milk dries yellowish on paper/cloth. The spores measure 6–8.5 × 5–7 µm and have almost isolated pointed warts (type A3). Occurs on ± acidic soils with both deciduous and coniferous trees.

The *Fagus*-associated *L. subdulcis* △ is less wrinkled, smells rubbery, and the milk does not change colour. *Lactarius lacunarum* ⋉ is less wrinkled and the spore ornament is ± reticulate.

Very common and widespread; (June–) August–October (–November).

Lactarius camphoratus is a very dark, ± medium-sized, red-brown, ± wrinkled *Lactarius* with a strong, spicy, camphorous to curry-like smell. The stem is darker than the cap. The milk is watery-white and mild-tasting. The spores measure 6–8.5 × 5–7 µm and have an echinulate ornament (type A3) with thin connecting lines. Occurs on acidic soil but also on mossy stumps, in both coniferous and deciduous forests.

The somewhat smaller *L. rostratus* ⋉ is more wrinkled and has very pointed pleurocystidia; it tends to occur on clay with *Fagus*.

Common and widespread; mainly July–October (–November).

Jens H. Petersen/Jan Vesterholt

Lactarius rubrocinctus is a rather large, orange-brown, mostly radially wrinkled *Lactarius* usually with a darker, red-brown stem that typically has an even darker zone at the apex. The white milk does not change colour and the taste is mild to slightly acrid. The gills have a decurrent tooth and discolour ± violet when bruised. The spores measure 6.5–9.5 × 5.5–7.5 µm and have connected warts (type B2). Mycorrhizal with *Fagus* on ± calcareous soils.

Lactarius fulvissimus ▷ is more vividly coloured, has a stronger smell and the stem apex is not dark.

Apparently more common in northern parts of the nemoral zones but absent from the hemiboreal–boreal zones and rare farther south; mainly (June–) August–October.

Lactarius aurantiacus is a rather small *Lactarius* with adnate gills, vivid brownish-orange to orange-brown colours and a shiny, monochrome cap. The milk is white, does not change colour and is mild-tasting. The spores measure 6.5–9.5 × 5.5–7.5 µm and have connected warts (type B2). Mycorrhizal with both deciduous and coniferous trees.

Lactarius fulvissimus ▷ is usually more fleshy and has gills with a decurrent tooth. The two species can also be separated by differences in cap cuticle structures and spore ornament. *Lactarius porninsis* ◁◁ has a zonate cap.

Rather common and widespread; (July–) August–October (–November).

Lactarius fulvissimus is a medium-sized, vivid orange-brown *Lactarius* with a relatively pale stem. The gills are decurrent and have a tooth. The milk is white, ± unchanging in colour and mild to bitter-tasting. The odour is pungent, recalling that of *Lepiota cristata* (page 326). The spores measure 6–9 × 5.5–7.5 µm and have somewhat connected spines (type B3). Occurs on clay-rich soils with various deciduous trees.

Lactarius aurantiacus ◁ is smaller, has adnexed gills and is almost odourless.

Occasional to locally common, absent from the boreal zone; (June–) August–October (–January).

Lactarius hepaticus is a rather small to medium-sized, matt, grey-brown, smooth to finely rugulose, somewhat stinkbug-like smelling *Lactarius* with white milk that dries yellowish. The spores measure 6.5–9.5 × 5.5–7.5 µm and have warts connected in a reticulum (type C2). Mycorrhizal with *Pinus*, mostly on poor, sandy soils, *e.g.* in plantations in dunes.

Other species of *Lactarius* that occur in the same environment are typically more vividly coloured.

Rather common but somewhat local and rare in the north – seems to prefer an Atlantic climate; mainly August–November.

Lactarius quietus is a medium-sized, rather pale, red-brown, zonate *Lactarius* with a stinkbug-like smell and somewhat watery, pale cream milk. The stem is typically rather dark, especially towards the base. The taste is mild at first but becomes somewhat acrid. The spores measure 6.5–9 × 5.5–7 µm and have oblong, connected warts (type B2). Mycorrhizal with *Quercus* on a range of different soil types.

Lactarius chrysorrheus ▷▷ also occurs with *Quercus* but is typically somewhat paler and the milk quickly becomes sulphur-yellow.

Very common and widespread, except in the boreal zone; (July–) August–October (–November).

Lactarius serifluus is a rather small, dark grey-brown, rather monochrome, stinkbug- or lovage-like smelling *Lactarius* with a watery-whitish, mild-tasting milk. The spores measure 6.5–8.5 × 5.5–8 µm and have a high reticulum that almost forms crests (type C2–C3). On rich, clay soils, mostly with *Quercus* in thermophilic sites.

Other similar species have white (not watery-white) milk. *Lactarius hepaticus* ◁◁ occurs on poor soils with *Pinus*. *Lactarius rostratus* ⋉ has a smaller, rugose-wrinkled cap. *Lactarius camphoratus* ◁◁ typically has a very dark stem and a more curry-like smell.

Rather local but widespread south of the boreal zone; July–October (–November).

Lactarius mammosus is a medium-sized, dark brown, strongly sweet, coconut-smelling *Lactarius* with rather scanty, white, mild-tasting milk. The spores measure 6–9 × 5–6.5 µm and have warts that are connected in a reticulum (type C2). Occurs mostly on ± sandy, poor, bare soil, mostly with *Pinus* or *Betula*.

Lactarius glyciosmus ▽ has a similar smell but is paler and occurs with *Betula* in damp places.

Common in the hemiboreal–boreal zones, much rarer farther south; (June–) August–October.

Lactarius glyciosmus is a slender, rather small, pale grey-brown *Lactarius* with cream gills and a pleasant smell of cakes made using coconut. The white milk is rather scanty and mild-tasting. The spores measure 5.5–8.5 × 4.5–7 µm and have rather connected warts (type B2). Mycorrhizal with *Betula*, mostly on damp, peaty soils.

Lactarius mammosus △ is darker and typically occurs on bare (sandy) mineral soils with *Pinus*.

Widespread and common; August–November.

Lactarius lilacinus is a small to rather small, rose to grey-rose, rather pale, dry *Lactarius* with cream gills and a velutinous cap surface. The milk is whitish and somewhat acrid-tasting. The spores measure 6.5–9 × 5.5–7 µm and have a rather reticulate ornament that almost forms crests (type C2–C3). Mycorrhizal with *Alnus*.

Lactarius spinosulus ◁◁, one of the bearded lactarii, has striate-winged spores (type C3) and a strigose-hairy cap margin.

Widespread but uncommon in most regions; (August–) September–October (–November).

The **lilac-staining lactarii** have milk that is at first whitish, but both milk and flesh quickly become lilac to violet after exposure. Some of the sticky lactarii (*Lactarius vietus* and the species with speckles, ◁◁) have white milk that dries greyish or grey-green.

Lactarius uvidus is a rather large, slimy-sticky, greyish to reddish-grey *Lactarius* with milk and flesh that discolour to lilac. The taste is bitter. The spores measure 8.5–11.5 × 6.5–8.5 µm and have an incomplete reticulum of confluent warts (type B2). Occurs on damp soils with *Betula* or *Picea*.

There are only a few possible confusion species in temperate Europe. *Lactarius luridus* ⋉ has a zonate, greasy cap and the spores have a more obvious reticulate ornament. *Lactarius violascens* ▽ is dry to sticky and the spores have a more winged-crested ornament. *Lactarius vietus* ◁◁ has milk that dries greyish.

Rather common in the hemiboreal–boreal zones, but rarer farther south; (July–) August–October.

Jan Vesterholt

Lactarius violascens is a rather compact, grey-brownish, dry to sticky *Lactarius* that discolours ± dark violet. The cap may be somewhat zonate. The taste is mild to rather bitter. The spores measure 7.5–11.5 × 6.5–8.5 µm and have an almost crested reticulum (type C2–C3). Mycorrhizal with *Quercus*, *Carpinus* and *Tilia* on ± calcareous soils in thermophilic deciduous forests.

Lactarius luridus ⋉ does not discolour so darkly and the spore ornament is shorter. *Lactarius uvidus* △ is more slimy and grows with other hosts.

Rare to local, absent from the boreal zone; mainly August–October.

Jan Vesterholt

Lactarius flavidus is a medium-sized, yellowish-cream, ± zonate-speckled, sticky *Lactarius* that discolours violet. The milk is white at first but later discolours reddish-violet, and it is ± acrid-tasting. The spores measure 8.5–11.5 × 6.5–8.5 µm and have connected warts (type B2). Occurs with various deciduous trees on ± calcareous soils.

Lactarius flavopalustris ⋉ occurs in damp sites with *Betula*. It has a pitted stem and the milk only changes colour when in contact with the flesh. *Lactarius aspideus* ▽ typically has no, or only a few, creamy drop-like blotches or zones on the cap; it grows with *Salix* in damp places and has a distinctly reticulate spore ornament (type C1–C2).

Rare and local; autumn.

Lactarius aspideus is a rather small to medium-sized *Lactarius* with milk that discolours violet and a slimy, pale yellowish cap. The spores measure 6.5–9.5 × 5.5–8 µm and have a low reticulate ornament (type C1–C2). Mycorrhizal with *Salix* in damp or wet habitats.

The much rarer *L. flavidus* △ has a more zonate-spotty cap and the spores lack a reticulum. *Lactarius flavopalustris* ⋉ grows in damp sites with *Betula*, has a pitted stem and milk that only changes colour when in contact with the flesh. *Lactarius uvidus* ◁ is larger and has more greyish to violet-grey cap colours.

Widespread and locally common; (June–) August–October (–November).

Jan Vesterholt

Thomas Læssøe

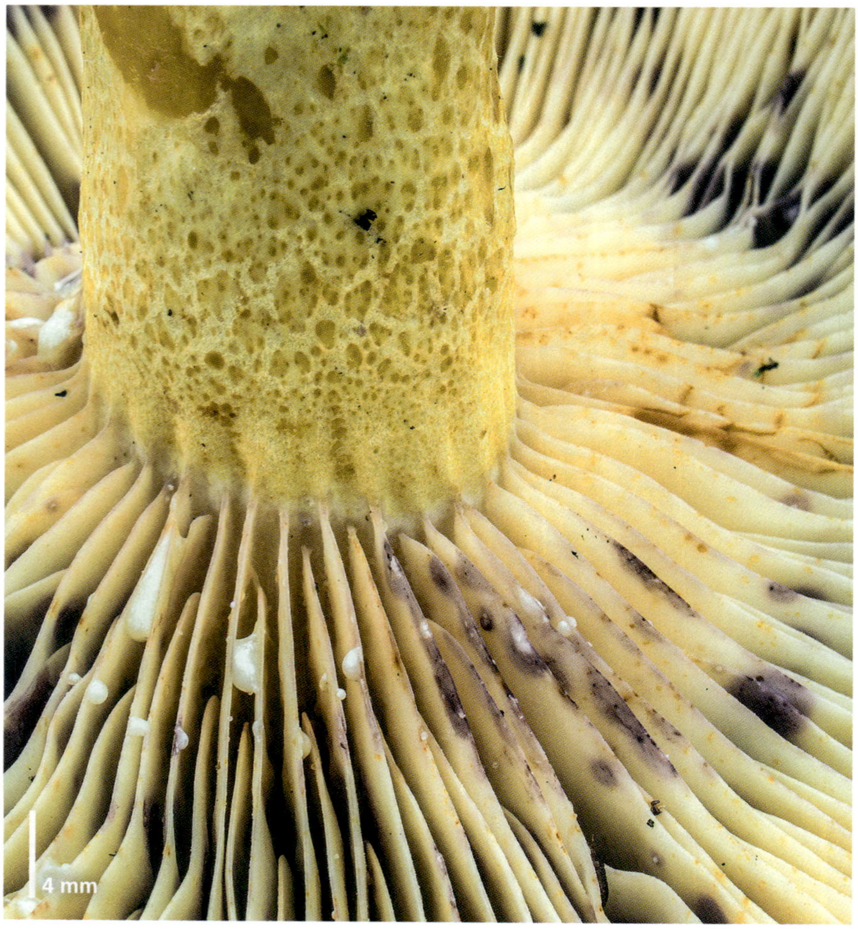

Lactarius repraesentaneus is a large, spectacular, bearded, orange-brown, felty-hairy *Lactarius* that discolours violet. The stem is pitted, the odour recalls hyacinth flowers, and the taste is mild to bitter. The spores measure 8.5–10.5 × 7–9 µm and have connected warts (type B2). Occurs on damp soil with *Betula* or *Picea*.

In the temperate zone of Europe this is the only yellow-bearded *Lactarius* with milk that discolours violet. In the arctic, the similar but paler *L. dryadophilus* ⋉. can be found with *Dryas* as its mycorrhizal partner. *Lactarius scrobiculatus* ▷ and similar species have milk that discolours yellow.

Fairly common in the hemiboreal–boreal zone but rare or absent farther south; (June–) August–October.

The **yellow-staining lactarii** have white milk that quickly changes to yellow.

Lactarius tuomikoskii is a medium-sized to large, bright yellow, monochrome *Lactarius* with milk that discolours sulphur-yellow, a sulcate margin and a somewhat pitted stem. The spores measure 7.5–10 × 6–7.5 µm and have connected warts that form a reticulum (type C2). Occurs with *Picea* in damp, mossy forests.

Lactarius olivinus ⋉ is duller and has a somewhat hairy-scaled cap; *L. leonis* ⋉ has a pitted stem and smaller spore; and *L. auriolla* ⋉ is smaller and has smaller spores.

Rather common in the boreal zone and is also known from central Europe; late summer–autumn.

Lactarius scrobiculatus is a large, orange-brown, strongly zonate *Lactarius* with a felty-hairy cap margin and a distinctly pitted stem. The milk discolours sulphur-yellow and the taste is bitter to acrid. The spores measure 7–10 × 5.5–7.5 µm and have a ± spiralling pattern of confluent warts (type B2). Occurs with *Picea* in calcareous damp woodlands.

Lactarius toumikoskii △ is sulcate at the cap margin and not as zonate. *Lactarius leonis* ⋉ is also less zonate and has a complete reticulum on the spores.

Quite common in the hemiboreal–boreal zones and in the subalpine zone of central Europe, but rare or absent from the nemoral zones; mainly August–October.

Guillaume Eyssartier/Jan Vesterholt

Lactarius aquizonatus is a medium-sized to large *Lactarius* with milk that discolours yellow, and a pale yellow-brown, very zonate cap, which has a slimy beard at the margin. The spores measure 6–8.5 × 4–6 µm and have a rather low, reticulate ornament (type C1–C2). Occurs on dry or rather dry calcareous soils, mostly with *Betula*, but also with *Pinus*, *Salix* and *Populus*.

Lactarius citriolens ▽ differs in having a dry beard and narrower spores; in addition, the lower part of the stem is mostly swollen and very hard and felty. *Lactarius resimus* ▷ has a downy cap margin and the spores have a lower Q value.

Uncommon, mostly in the hemiboreal–boreal zones; August–November.

Jan Vesterholt

Lactarius citriolens is a ± yellow, dry-bearded, very firm-fleshed *Lactarius* with milk that discolours yellow. The stem base is covered in long, rigid hairs. The spores measure 6.5–9 × 4.5–6.5 µm and have rather low, connected warts (type B2). Mycorrhizal with *Betula*, *Corylus* or *Quercus* on ± calcareous soils.

Lactarius aquizonatus △ is very similar, but has a slimy beard at the cap margin and narrower spores. *Lactarius resimus* ▷ has a downy, not bearded, cap margin. *Lactarius scrobiculatus* ◁◁ is darker yellow and has a pitted stem.

Uncommon in the hemiboreal–boreal zone, rare in the nemoral zones; late summer–autumn.

Lactarius resimus is a large to very large, very pale *Lactarius* with milk that discolours yellow. The cap is, at most, somewhat zonate near the downy margin. The spores measure 7–9.5 × 5.5–7 µm and have connected warts (type B1–B2). Occurs on various soil types, apparently predominantly with *Betula* and *Pinus*.

Lactarius aquizonatus ◁ has a slimy beard at the cap margin, and *L. citriolens* ◁ has a dry beard.

Uncommon, mainly in the hemiboreal–boreal zone, very rare in the nemoral zones; late summer–autumn.

Jacob Heilmann-Clausen

Lactarius chrysorrheus is a rather pale yellow-brown, ± zonate, dry, smooth *Lactarius* with milk that discolours sulphur-yellow. The odour is weak and the taste bitter to somewhat acrid. The spores measure 5.5–9 × 5–7 µm and have striate to reticulate connected warts (type B2–C2). Typically occurs with *Quercus* on poor soils

Lactarius quietus ◁◁ is normally darker and has milk that does not discolour. Other yellow-milked species are ± slimy or ± bearded.

Rather common in the nemoral zones, rare in the hemiboreal zone; (July–) August–October (–January).

The **pink-staining lactarii** have white milk that in most cases changes slowly to rose. In most species the change only happens when the milk is in contact with the flesh, but the milk of *Lactarius acris* can also change colour when isolated from the flesh.

Lactarius acris is a medium-sized *Lactarius* with a pale grey-brown, sticky cap. The milk is at first mild-tasting but later very acrid, and quickly discolours rose even when isolated from the flesh. The spores measure 7–9.5 × 6.5–8.5 µm and have crests up to 1.5 µm in height (type C3). Occurs on rich soil with various deciduous trees, mostly *Fagus*.

There are several similar **lactarii**, but none have the combination of a sticky cap and milk that quickly stains rose when isolated from the flesh.

Uncommon to rare, absent from the boreal zone; July–October.

Lactarius azonites is a rather small, monochrome, pale grey-brown, dry, white-stemmed *Lactarius* with milk that slowly changes to rose when in contact with the flesh. The spores measure 7.5–9.5 × 7–8.5 µm and have narrow wrinkles and warts that form an incomplete reticulum (type B3). Mycorrhizal with *Quercus* on rich soil.

Similar species have crested ornaments on the spores and typically have darker stems. *Lactarius azonites* can occur as an almost white form, *L. azonites* f. *virgineus* ⋉.

Uncommon to rather rare, absent from the boreal zone; July–October.

Lactarius pterosporus is a medium-sized rather pale, almost clay-buff, dry *Lactarius* with milk that slowly stains rose when in contact with the flesh. The cap surface is mostly distinctly radially wrinkled, and the stem is tinted the same colour as the cap. The gills quickly become almost ochre. The spores measure 7–8.5 × 6–8 µm and have a high, wing-like ornament, up to 2.5 (–3) µm, that forms a striate pattern (type C3). Occurs on rich soil, mostly with *Fagus*.

Lactarius pterosporus is among the paler species of pink-staining lactarii. *Lactarius azonites* ◁ has a faintly wrinkled cap and spores that are not winged. *Lactarius acris* ◁ has a sticky cap and milk that quickly stains rose without flesh contact.

Widespread and fairly common to rather local, except in the boreal zone; mainly July–October.

Lactarius fuliginosus is a grey-brown to darker brown, dry *Lactarius* with white milk that stains faintly rose when in contact with the flesh. The stem is pale brown to grey-brown. The spores measure 7.5–9 × 6.5–8.5 µm and have an ornament of low wrinkles approx. 1 µm in height (type B3). On rich soil, possibly only mycorrhizal with deciduous trees.

Lactarius picinus ⋉ is ± identical but possibly restricted to conifers; *L. azonites* ◁ is typically paler and has a white stem.

Widespread but the distribution is poorly known due to confusion with other species; mainly July–October.

Jan Vesterholt

Lactarius romagnesii is a dark brown, mostly sulcate-furrowed, dry, somewhat felty *Lactarius* with milk that stains rose on contact with the flesh. The stem is tinted the same colour as the cap, and the ± ochre gills are very distant. The spores measure 7.5–10 × 6.5–8.5 µm and have an irregular, crested ornament up to 2(–2.5) µm in height, (type C3). Occurs on rich soils, mostly with *Fagus*.

Similar species with the same ecology, *e.g. L. pterosporus* ◁◁, are paler, have a less sulcate cap margin and more crowded gills.

Mostly recorded in southern Scandinavia and southern England; mainly July–October.

Lactarius ruginosus is a grey-brown, ± smooth, dry *Lactarius* with a sulcate-furrowed margin. The stem is rather pale, and the rather distant gills are ochre. The white milk slowly stains rose on contact with the flesh. The spores measure 6.5–9 × 6–8 µm and have a somewhat irregular ornament of wrinkles or crests up to 2 µm in height (type C3). Occurs with *Fagus* and *Carpinus* on ± calcareous soils.

Lactarius pterosporus ◁◁ has more crowded gills and a wrinkled cap surface; the spore ornament is more striate.

Rather rare or somewhat overlooked, absent from the boreal zone; mainly July–October.

Lactarius lignyotus is a very dark, mostly umbonate/papillate *Lactarius* with a felty to almost smooth and pruinose surface on the cap and stem, and whitish, decurrent, rather distant gills that may have a dark edge. The white milk slowly stains rose when in contact with the flesh. The spores measure 8.5–10 × 7.5–9 µm and have a somewhat echinulate-crested ornament up to 1.5 µm in height (type B3–C3). Occurs on rich soil in old, mossy *Picea* forests and plantations.

Almost recalls a dark *Hygrophorus* (page 175) but the texture and milk are diagnostic. *Lactarius picinus* ⋉ also occurs with conifers but is non-umbonate/papillate and has more crowded, less decurrent gills that lack a dark edge.

Rather common in the hemiboreal–boreal zones, very rare in the nemoral zones; July–October (–November).

The **orange-milked lactarii** are characterized by their red to orange, mostly rather scanty milk and ± orange gills. They are all conifer-associated.

•• ***Lactarius deliciosus*** is a ± orange to pale greyish-orange, zonate, dry *Lactarius* with a distinctly orange, pitted stem. The orange milk either does not change colour or only becomes reddish after more than 30 minutes. The taste is mild and the texture firm. The spores measure 8–10.5 × 6.5–8 µm and have a reticulate ornament up to 0.5 µm in height (type C2). Occurs with *Pinus*, mostly on calcareous sand.

Lactarius deterrimus ▷ always has a smooth stem, flesh that quickly becomes greenish, and is mycorrhizal with *Picea*. *Lactarius quieticolor* ▷ has a darker, more greyish cap, almost blue-green flesh and often a smoother stem; the milk always becomes red. *Lactarius salmonicolor* ▽ is associated with *Abies* and has milk that always discolours wine-red. *Lactarius sanguifluus* ▷▷ has wine-red milk.

Widespread and rather common; (July–) August–November.

•• ***Lactarius salmonicolor*** is a large, orange, ± zonate *Lactarius* with orange milk that slowly discolours more wine-reddish. The stem has ± obvious pits. The flesh either does not discolour or only becomes slightly green. The spores measure 9.5–12 × 7–9.5 µm and have connected warts (type B2). Mycorrhizal with *Abies* on ± calcareous soils.

Lactarius deliciosus △ is mycorrhizal with *Pinus* but has shorter spores (below 9.3 µm on average, compared to 10.5 µm or more in *L. salmonicolor*).

Mostly recorded in the alpine region in ± native *Abies* forests; late summer–autumn.

• ***Lactarius deterrimus*** is an orange, soon discolouring greenish *Lactarius* with a mostly azonate cap. It has orange milk that also discolours green. The taste is mild. The stem is orange, hollow, lacks pits and discolours ± green. The spores measure 8–11.5 × 6.5–8.5 µm and have somewhat connected warts up to 0.5 µm in height (type B2). Mycorrhizal with *Picea*.

The *Pinus*-associated *L. deliciosus* ◁ has distinct pits on the stem and does not stain as green. *Lactarius quieticolor* ▽ has a somewhat more greyish cap, almost blue-green flesh and also occurs with *Pinus*. The southern *L. semisanguifluus* ▷▷ is *Pinus*-associated and has milk that after some minutes stains brownish-red. *Lactarius fennoscandicus* ⋉ occurs with *Picea* but has a more sombre, densely zonate cap and only grows in northern parts of temperate Europe.

Widespread and common; July–October (–November).

•• ***Lactarius quieticolor*** is a greyish-orange to bluish-tinted, zonate, dry *Lactarius* with orange milk that stains red after approx. 20 minutes and then dries greenish. The flesh under the cap cuticle is mostly blue-green. The ± greyish stem can be smooth or pitted. The spores measure 8–10.5 × 6.5–8.5 µm and have oblong ridges that almost form a reticulum (type B2–C2). Mycorrhizal with *Pinus*.

Similar species have more orange cap colours. *Lactarius fennoscandicus* ⋉ has somewhat similar colours but occurs with *Picea* and has isolated warts on the spores.

Widespread, but ± uncommon; (July–) August–October (–November).

Jens H. Petersen/Jens Maarbjerg

Jan Vesterholt

•• ***Lactarius sanguifluus*** is a medium-sized to large, dry, somewhat zonate *Lactarius* with dull orange colours and wine-red flesh and milk. The stem is rather pale and white pruinose, and either lacks pits or has just a few. A faint green discolouration can be seen in old bruises. The spores measure 8–9.5 × 6.5–8 µm and have warts connected in a somewhat reticulate ornament (type B2). Thermophilic and mycorrhizal with *Pinus* on calcareous soils.

Lactarius semisanguifluus ▽ initially has orange milk that only stains wine-red after about 5 minutes.

A mainly southern species with just a few records from the north; autumn.

Jukka Vauras

•• ***Lactarius semisanguifluus*** is a medium-sized, sticky to dry, somewhat zonate *Lactarius* with milk that is at first reddish-orange but after a few minutes stains coral-red and later brownish-red. The cap is initially pinkish-buff to dull orange but soon becomes jade-green. The stem is ± pale orange and usually has some orange or jade-green pits. The spores measure 7.5–10 × 6–8 µm and have low, connected warts (type B1). Mycorrhizal with *Pinus* in open woodland on calcareous soil.

Lactarius quieticolor ◁◁ has milk and cap colours that change less distinctly. *Lactarius sanguifluus* △ always has wine-red milk.

Very rare in the north but increasing farther south; autumn.

† ***Lactarius helvus*** is a medium-sized to large, dry, felty, mild-tasting *Lactarius* with clear milk and a strong smell of curry, stock cubes or lovage. The cap is finely felty to minutely scaly and clay-buff to cinnamon-buff or vinaceous-brown. The stem is smooth and mostly a deeper brown. The spores measure 6–8.5 × 4.5–6.5 µm and have warts that are connected in an almost complete reticulum (type B2–C2). Occurs on peaty soil or directly in *Sphagnum*; mycorrhizal with *Betula*, *Picea* and *Pinus*.

No other species of *Lactarius* has completely clear milk, and the smell of stock cubes is also characteristic. *Lactarius serifluus* ◁◁ has watery-white milk (like skimmmed milk) but is smaller, darker, smooth-capped and smells more like stinkbugs.

Common and widespread; mainly July–October (–November).

Entolomatoids

This group includes the genus *Entoloma* in a broad sense and the closely related genera *Clitopilus*, *Clitopilopsis*, *Clitocella*, *Rhodophana* and *Rhodocybe*. They are characterized by adnexed-adnate, emarginate to decurrent gills and ± brownish-rose spore-deposits. The spores are either verrucose, striate-ridged or three-dimentionally angular (faceted). An important character of entolomatoids is whether the pigment in the cap cuticle is incrusted or intracellular. Species with excentric or lateral stems are treated under the pleurotoid fungi (page 88).

Some species apparently have a parasite-like association with plants in the *Rosaceae,* while other species around *Entoloma rhodopolium* are mycorrhizal. The ecology of many species is still not known.

All genera belong in the same family, the *Entolomataceae*.

Other similar fungi:
– *Pluteus*, *Volvariella* & *Volvopluteus* have the same spore colour but free gills (page 486).
– some species of *Lepista* have brownish-rose spores and tricholomatoid fruitbodies. They are recognized by their minutely warty spores and a characteristic sweetish smell (page 249).
– some clitocyboids and the omphalinoid *Arrhenia discorosea* have ± brownish-rose, smooth spores (pages 98 & 138).
– the large collybioids in the genus *Rhodocollybia* have somewhat brownish -rose, smooth spores (page 296).

Further reading: 45, 89, 156, 158, 181, 182, 204, 223, 224, 335, 336, 337, 338.

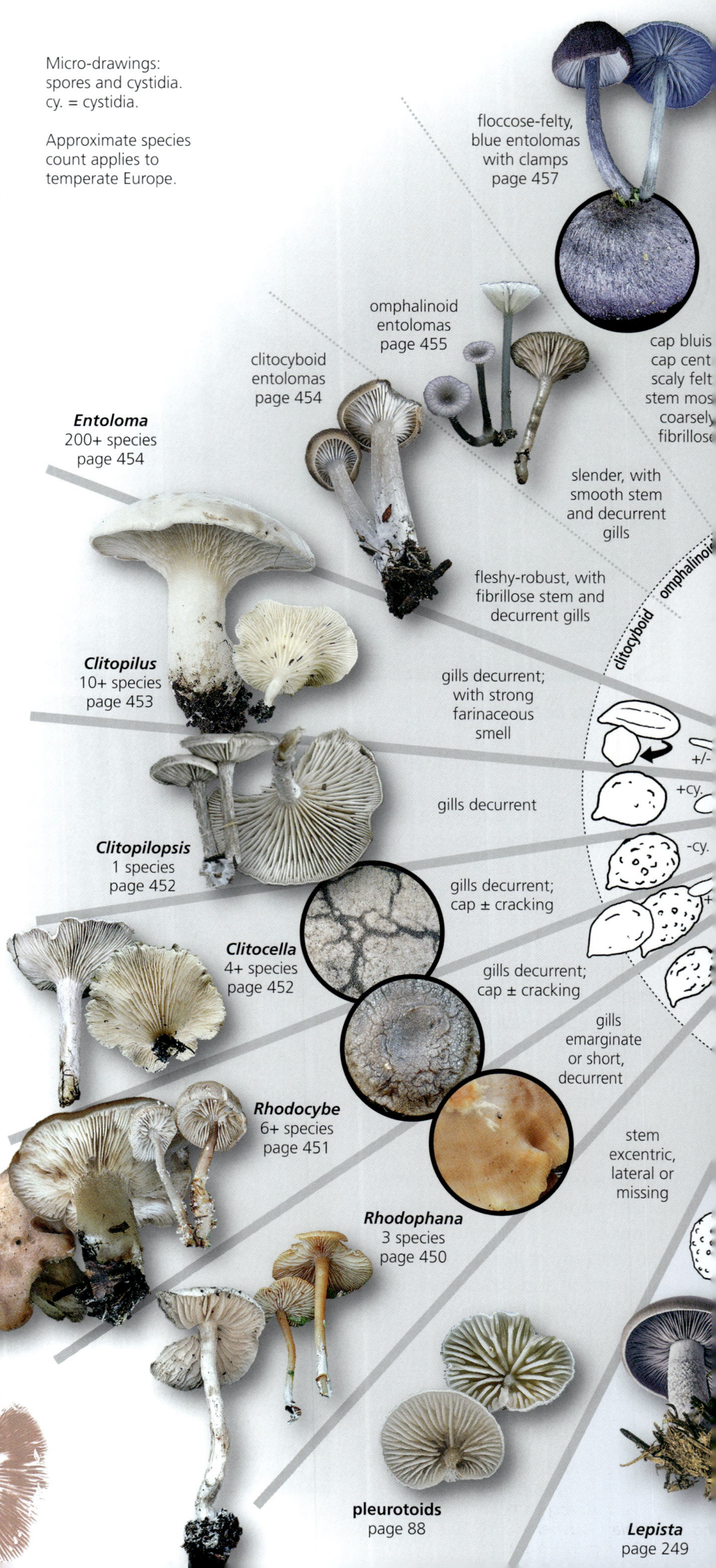

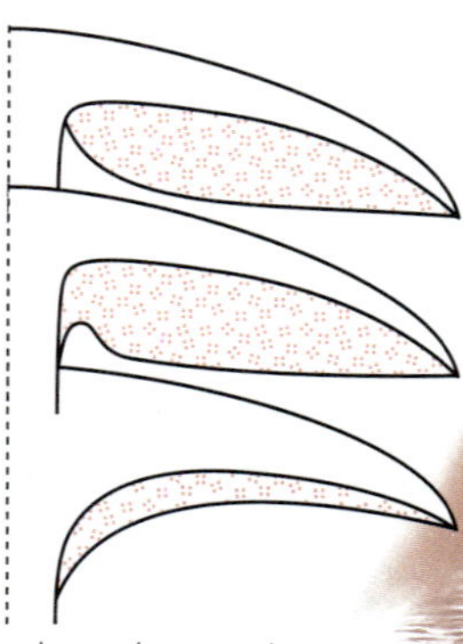

adnexed, emarginate and decurrent gills

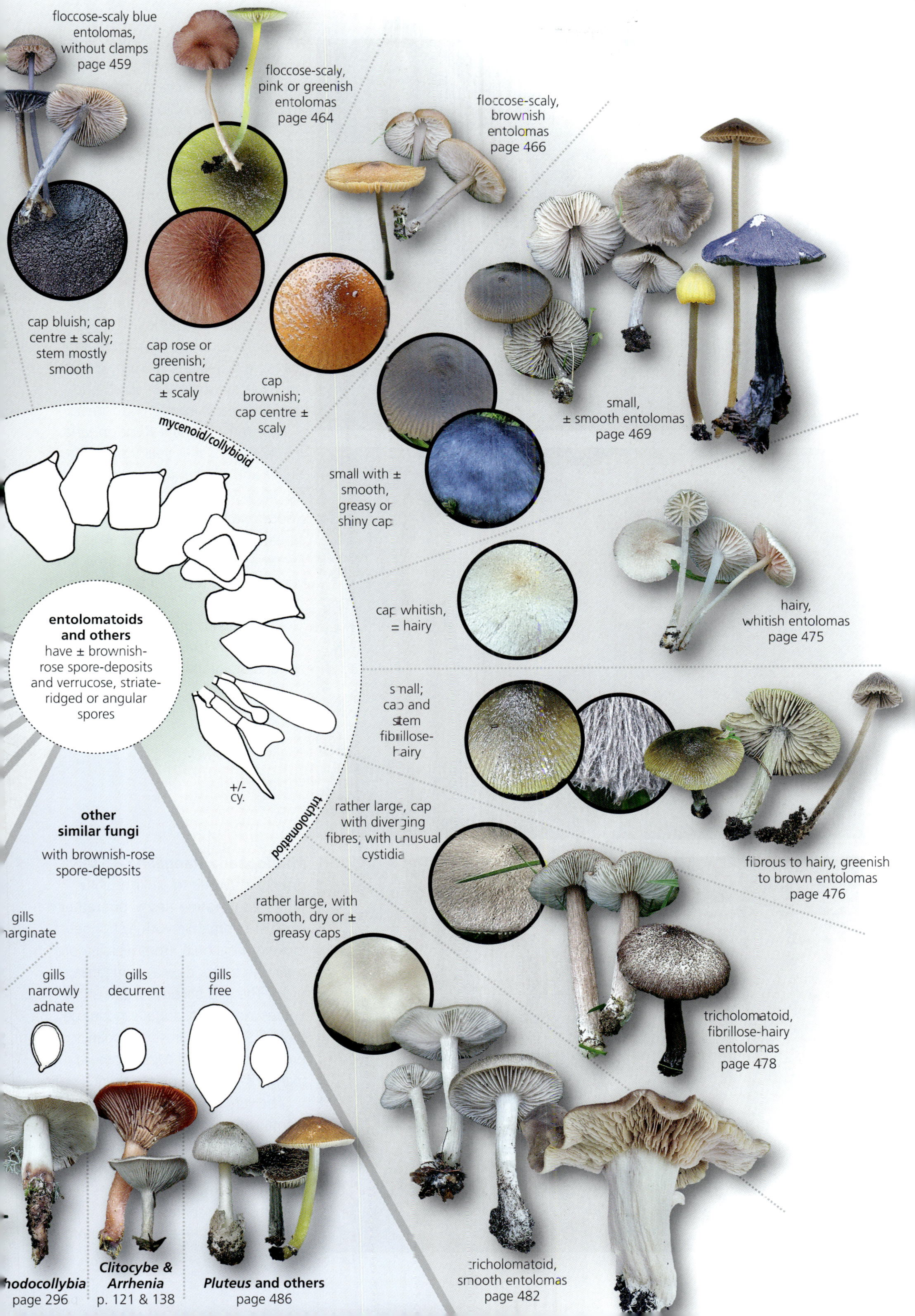
floccose-scaly blue entolomas, without clamps page 459
floccose-scaly, pink or greenish entolomas page 464
floccose-scaly, brownish entolomas page 466
cap bluish; cap centre ± scaly; stem mostly smooth
cap rose or greenish; cap centre ± scaly
cap brownish; cap centre ± scaly
small, ± smooth entolomas page 469
mycenoid/collybioid
small with ± smooth, greasy or shiny cap
cap whitish, ± hairy
hairy, whitish entolomas page 475
entolomatoids and others
have ± brownish-rose spore-deposits and verrucose, striate-ridged or angular spores
small; cap and stem fibrillose-hairy
+/- cy.
tricholomatoid
rather large, cap with diverging fibres, with unusual cystidia
fibrous to hairy, greenish to brown entolomas page 476
other similar fungi
with brownish-rose spore-deposits
gills marginate
rather large, with smooth, dry or ± greasy caps
tricholomatoid, fibrillose-hairy entolomas page 478
gills narrowly adnate
gills decurrent
gills free
hodocollybia page 296
Clitocybe & Arrhenia p. 121 & 138
Pluteus and others page 486
tricholomatoid, smooth entolomas page 482

Jan Vesterholt

Rhodophana includes rather small white to orange-brown entolomatoids with a brownish-rose spore-deposit. The spores have oblong warts. Typically on calcareous soils. One species parasitizes other agarics.

Rhodophana nitellina is a rather vivid, slender, small, brownish-orange, rancid-farinaceous entolomatoid with paler, adnate to slightly decurrent gills. There are no cystidia on the gill edge, and the hyphae lack clamps. The spores are finely nodulose-verrucose and measure 6.5–8.5 × 4–5 µm. Occurs on ± calcareous soil in scrub and deciduous and coniferous forests.

The rather similar *Rhodophana melleopallens* ⋉ is yellower and duller and has smaller spores ((4–)4.5–7 × 3–4(–4.5) µm). *Rhodophana nitellina* also recalls a small, brownish-orange species of *Lactarius* (page 428), but these have very different amyloid spores and the flesh has milk.

Widespread but rather rare with most records from the hemiboreal zone in Scandinavia; mainly August–October.

Thomas Læssøe

Rhodophana stangliana is a very distinctive, white to pinkish-white, farinaceous, rather fleshy, but rather small entolomatoid with a strange-looking basal bulb, from which more than one fruitbody may sometime grow. The gills become ± pinkish. Spores indistinctly angular-nodulose, 5–7 × 3.5–5 µm. Parasitizes agarics, probably species of *Myochromella* (page 287), in woodlands on rich soils.

The very unusual bulb makes identification straightforward. This species has been accepted in the genus *Squamanita* (page 228), but these have greyish, squamulose fruitbodies and smooth, white spores.

Very rare and meteoric; autumn.

Rhodocybe includes species with rather decurrent gills, a somewhat cracking cap cuticle and nodulose-verrucose, brownish-rose spores. Occurs on soil in forests and in more open grassland, often in fairy rings.

Rhodocybe gemina is a large, fleshy, flesh-pink, ± red-spotted, perfumed, rather tricholomatoid *Rhodocybe*. The spore-deposit is a rather dark brownish-rose. The cheilocystidia are simple. The spores are ± bumpy and measure 4.5–7 × 3–4.5 µm. Typically in fairy rings on rich soil in parks, gardens, deciduous and coniferous forests, and in scrub and hedgerows.

Gerhardtia incarnatobrunnea (page 244), is very similar, but has a white spore-deposit, a less characteristic smell, somewhat more obvious warts on the more fusiform spores and lacks cystidia. Both species lack clamps.

Widespread and rather common to occasional, absent from the boreal zone; late summer–autumn.

Rhodocybe caelata is an odourless, brown-grey, rather slender, somewhat umbilicate, not or only indistinctly zonate, untidy-looking *Rhodocybe* with a pale, somewhat splitting coating and adnate to somewhat decurrent gills. Cystidia have yellow contents in KOH. Spores finely nodulose, ellipsoid, (5.5–)6–9 × (3–)3.5–4(–4.5) µm. On sandy soils, mostly in open, often coastal habitats.

Clitocella popinalis ▷▷ lacks cystidia, blackens somewhat and is more fleshy. Species of *Clitocybe* (page 98) with coloured spore-deposits also lack cystidia and have completely smooth spores.

Widespread; (May–) August–November.

Clitocella has a cracking cap cuticle as in some species of *Clitocybe* (page 98), decurrent gills, a farinaceous smell and nodulose-verrucose, brownish-rose spore-deposit.

Clitocella popinalis is a brown-grey, ± blackening, farinaceous, ± bitter-tasting, rather fleshy *Clitocella*. Lacks clamps and cystidia. The spores measure 5–6.5 × 4.5–5 µm. Occurs on peat and soil, mostly in open habitats – may occur in fairy rings.

Rhodocybe caelata ◁◁ is thinner-fleshed and cystidia stain yellow in KOH.

Widespread and rather common, absent from the boreal zone; August–November (–January).

Clitocella fallax is a chalk-white, thin-fleshed clitocyboid *Clitocella* with ± yellowish-pinkish gills, a farinaceous smell and a bitter taste. The spores are finely roughened and slightly angular and measure 5–7 × 3–4 µm. Lacks cystidia. On light, calcareous soil, mostly in *Syringa* hedgerows and *Crataegus* thickets, but also in dunes.

Species of *Clitocybe* (page 98) have smooth spores, and species of *Entoloma* ▷▷ have distinctly angular spores. *Clitopilus prunulus* ▷ has striate-ridged spores.

Rare and rather scattered, but probably somewhat overlooked; July–January.

Leif Stridvall

Clitopilopsis hirneola is a slender, grey, finely felty, mild-tasting entolomatoid. Has rather large, multicellular cystidia that do not stain yellow in KOH, and lack clamps. The ± angular-nodulose spores measure 5.5–7.5(–8) × 4–5.5(–6) µm. Occurs on soil and litter on sandy soils in conifer woodland and in dunes.

Rhodocybe parilis ⋉ is very similar but lacks cystidia. Small, grey species of *Clitocybe* (page 98) have no cystidia, paler spore-deposits and smooth spores.

Mostly a northern species with only a few records from the nemoral zones; August–October.

Clitopilus is a genus of farinaceous, whitish entolomatoids with deeply decurrent gills, brownish-rose spore-deposits and longitudinally striate-ridged spores. For pleurotoid species of *Clitopilus*, see page 88.

Clitopilus prunulus is an almost white to pale greyish, fleshy, strongly farinaceous *Clitopilus* with decurrent, rather crowded gills. The cap margin is pale and initially inrolled. Lacks prominent cheilocystidia. The spores have longitudinal ridges and measure 9–12 × 4–6 µm. Occurs on soil near deciduous or coniferous trees where it grows in a very similar manner to a mycorrhizal fungus and is in some sources stated to be one.

Clitopilus cystidiatus ⋉ is described with ± distinct, hyphoid cheilocystidia, but this character seems to be dubious. Various species of *Clitocybe* (page 98) may also have brownish-rose spore-deposits, but lack a strong farinaceous smell and have smooth spores. *Pseudoclitopilus rhodoleucus* (page 107), has verrucose, amyloid spores.

Common and widespread; (June–) July–October (–November).

Clitopilus scyphoides is a small, slender, mostly excentrically stemmed, minutely downy, white *Clitopilus*. The spores have longitudinal ridges and measure 5.5–8.5 × 3–4.5 µm. Occurs in dry, mostly calcareous grassland and seems more tolerant than many other grassland fungi of limited fertilizer applications.

Omphalina mutila (p. 140), does not have a red spore-deposit and have white spores without ridges. Small white species of *Entoloma* ▷▷ have angular spores. Similar-looking specimens from woodlands often prove to be species of *Clitocybe* (page 98).

Widespread, occasional; (June–) July–October (–November).

Thomas Læssøe

Entoloma is a large genus of agarics with adnexed, adnate, emarginate to decurrent gills and pleurotoid, omphalinoid, clitocyboid, collybioid, mycenoid or tricholomatoid fruitbodies. The brownish-rose spore-deposit and the very distinctive, angular (faceted) spores are the main distinguishing characters. They are mostly thought to be decomposers, but at least some species are mycorrhizal and others probably parasitic on plants in the *Rosaceae* (see also the pleurotoid species of *Entoloma*, page 88).

Entoloma fridolfingense is a medium-sized, dark, non-striate, finely pruinose to downy, umbilicate, clitocyboid *Entoloma* with pale, decurrent gills. The odour when bruised is ± farinaceous. The pigment is intracellular. Has clamps. Cheilocystidia 40–90 × 12–25 µm. Spores 5–9 angled, measure 7.5–12.5 × 5.5–8.5(–9) µm. Occurs in open, ± calcareous habitats.

Entoloma nausiosme ⋉ is similar and also rare, but has a very strong, unpleasant smell.

Rare, *e.g.* Austria, Germany and Norway; summer–autumn.

Thomas Læssøe

Entoloma rhodocylix is a small, slender, very thin-fleshed, omphalinoid *Entoloma* with distant and deeply decurrent, at first white, later pink gills. Has cheilocystidia, clamps and relatively wide spores that measure 8–10 × 7–9 µm. Occurs in acid, damp, mossy places, mostly directly on decomposed, brown-rotted wood.

The brownish-rose spore-deposit should prevent confusion with omphalinoids and mycenas. *Entoloma iodiolens* ⋉ lacks cystidia and clamps and smells of iodoform.

Widespread and rather frequent; June–November.

Entoloma undatum is a rather variable, grey, zonate, omphalinoid, short-stemmed, somewhat silvery, fibrillose, farinaceous *Entoloma*. The decurrent gills are at first greyish, later grey-rose. The hyphae have clamps; lacks cystidia; the pigment is incrusted. The spores measure (7–)7.5–10 × 6–7 µm. Typically occurs on light, calcareous soils in scrub, along roads in deciduous forests *etc.*

A complex with many very similar species. *Entoloma lanicum* ⋉ apparently lacks zones on the cap and the spores are slightly smaller (7–9 × 5.5–7 µm). *Clitopilopsis hirneola* ◁◁ does not have angular-faceted spores.

Widespread and rather common; June–November.

Entoloma rusticoides is an omphalinoid, ± translucently striate *Entoloma* with grey-brown to red-brown colours and paler, decurrent, distant gills. The stem is finely floccose to smooth. Lacks cheilocystidia and clamps but has incrusted pigment in the cap cuticle. The spores measure 8–10.7 × 6.9–10.4 µm. On sandy, mostly disturbed soil in open habitats.

Entoloma graphitipes ⋉ is completely smooth-stemmed and paler. Species of *Omphalina* (page 139) and *Arrhenia* (page 135) are very similar but have ± white spore-deposits and the spores are not angular-faceted.

Widespread, but rare or overlooked; mainly June–October.

Thomas Kehlet

Entoloma parkensis is a ± omphalinoid, translucently striate, ± dark brown *Entoloma* with rather pale, decurrent gills. The pigment is mostly intracellular, but also very finely incrusted. Has clamps, but no cystidia; basidia 4-spored. The spores measure 9–11 × 7–8 µm. Occurs in boggy woodlands, but also in dry grassland.

Entoloma politum ▽ is usually paler, smells mostly nitrous and has no incrusted pigment.

Rare or perhaps overlooked; mainly August–November.

Thomas Læssøe

Entoloma politum is a grey-brown to yellow-brownish, translucently striate, hygrophanous, omphalinoid *Entoloma* with a smooth, shiny stem. The gills are pale and ± decurrent. It has a ± distinct nitrous smell. Has clamps, but no cystidia. The pigment is intracellular. The spores measure (7.5–)8–10(–10.5) × (6.5–)7–8(–8.5) µm. Mostly in troops on damp soil, typically in *Salix* carrs.

The similar *E. pernitrosum* ⋉ has incrusted pigment; *E. caccabus* ⋉ (incl. *bisporigerum*) has a strong farinaceous smell; and *E. bryorum* ⋉ occurs in dry habitats.

Widespread and rather common; June–December.

Thomas Læssøe

Entoloma incarnatofuscescens is a small, slender, omphalinoid *Entoloma* with a bluish-grey, smooth stem and a greyish-lilac, translucently striate and umbilicate cap. It lacks cystidia and clamps. The spores measure 8–10.5(–11) × 6–8(–9) µm. Occurs on clay-rich or mull soils in scrub, parks, gardens and deciduous forests.

The combination of a deeply umbilicate cap and rather vivid colours should make confusion with other species unlikely.

Widespread, but absent from the boreal zone and nearly so from the hemiboreal zone; June–November.

Entoloma euchroum is a smallish, felty, violet-blue *Entoloma* with blue gills that have a ± distinctly darker edge. The stem is less intensely blue and felty-scaly fibrillose. The odour is aromatic and the taste unpleasant. With cheilocystidia and clamps. The spores measure (8–)9–11.5 × 6–8 µm. Occurs on wood of deciduous trees, mostly *Alnus*; often found, for example, in cavities of standing, broken *Alnus* trees.

Should be unmistakable. Other species from the same habitat with paler blue gills do not have darker gill edges; *e.g. E. lepidissimum* ⋉ and *E. syringaecolor* ⋉.

Widespread, occasional, ± absent from the boreal zone; mainly August –November.

The species on this and the following page form a phylogenetic group – subgenus *Leptonia*. They are recognized by their felty-floccose-scaly caps and their bluish-violet colours, either on the cap or stem, or both, and, unlike many other blue entolomas, they have clamps.

Entoloma chytrophilum is a rather small, very dark blue *Entoloma* with white gills that have a concolorous edge and lack cystidia. The cap is finely floccose-scaly at the centre, and the stem is finely felty-scaly. The hyphae have clamps. The spores are rather thin-walled, more nodulose than angular, and measure 8.5–11.5 × 5.5–7 µm. Occurs on rich soils and sometimes on rotten wood, typically in coniferous forests, amongst woody and needle debris.

Other similar species have smaller and/or more angular spores, mostly with thicker walls.

Widespread but apparently rare; June–October.

Thomas Læssøe

Jacob Heilmann-Clausen

Entoloma allochroum is a rather compact *Entoloma* with lilac-tinged, floccose-scaly cap and a violet, very fibrillose-scaly stem. The gills are pale, and lack blue colours. Has clamps and few, or no, rather poorly differentiated cheilocystidia. The thick-walled spores have 5–7 sharp angles and measure 8.3–12.5 × 6.2–9.5 µm. Occurs on soil or on very rotten wood in damp, deciduous forests on mull soils.

The stem of *E. dichroum* ⋉ is even more scaly and a purer dark blue. *Entoloma tjallingiorum* ▽ lacks bluish colours in the cap.

Rare, apparently absent from the boreal zone; mainly July–October.

Thomas Kehlet

Entoloma lampropus is a slender *Entoloma* with, at most, somewhat violet tinges at the margin of the brown, scaly cap. The stem is blue and fibrillose. The hyphae in the cap cuticle have both incrusted and intracellular brown pigment. Has clamps; with or without cheilocystidia. The spores have 6–9 soft angles and measure 8.8–11.5 × 6–7.5 µm. Occurs on soil or decomposed wood in rich, mainly coniferous forests.

The farinaceous *E. placidum* ⋉ is found primarily in deciduous woodlands and does not have incrusted pigment; *E. tjallingiorum* ▽ has a more scaly stem.

Rare to occasional; autumn.

Jens H. Petersen/Jan Vesterholt

Entoloma tjallingiorum is a scaly, rather compact, brown-capped *Entoloma* with a fibrous-scaly, violet-blue stem. The gills are pale and lack blue shades. Has clamps and ± septate, tramal cheilocystidia. The thin-walled spores have 6–9 soft angles and measure 8–11(–12) × 5.8–7.2(–7.6) µm. Occurs on soil or rotten hardwood in rich forests.

Entoloma tjallingiorum var. *alnetorum* ⋉ has a very pale brown cap; var. *laricina* ⋉ has multicellular pseudocystidia and is known from *Larix* woods. *Entoloma placidum* ⋉ and *E. lampropus* △ have fibrillose stems.

Rare, but widespread; July–November.

Entoloma serrulatum is a rather variable *Entoloma*, but always has dark, serrated gill edges (coloured cystidia grouped in small teeth). The cap is dark grey-brown to almost black and scaly but usually not translucently striate. The gills are ± blue. Also rarely found in red forms (see 464). Lacks clamps. The sharp-angled spores measure 9–12 × 6.5–8 µm. Occurs in dry, mostly calcareous, unfertilized grasslands and along roadsides, more rarely in open woodlands and sometimes even on rotten wood.

Entoloma caesiotinctum ⋉ is a browner and more translucently striate very similar soecies. *Entoloma chalybaeum* ▷▷ has a non-serrated, brownish gill edge.

Widespread and rather common; June–November.

The floccose-scaly ± blue entolomas on the following pages (sect. *Cyanula*) are a large and difficult group in which all the species are unclamped, unlike the previous ± blue species in subgenus *Leptonia*. Most species occur in unfertilized grasslands.

Entoloma mougeotii is a violet-grey, scaly *Entoloma* with whitish gills that can either have or lack a brown edge. The stem is felty along its entire length. Lacks clamps but has cheilocystidia. The sharp-angled spores measure 9–11.5(–12.5) × 6–8(–9) µm. Occurs in ± calcareous, unfertilized grasslands and fens.

A form with brown gill edges is called *E. mougeotii* var. *fuscomarginatum* ⋉. *Entoloma alvarense* ⋉ is recognized by its smaller spores and perhaps browner colours. *Entolomas griseocyaneum* ▷▷ has less pronounced violet colours.

Widespread, fairly common to rare; June–October.

Jan Vesterholt

Thomas Kehlet

Entoloma porphyrogriseum is a brown to very dark, blue-black to black *Entoloma* with similar colours on its somewhat fibrillose stem, but with pure white young gills and gill edges. Lacks clamps. The cheilocystidia measure 30–60 × 6–15 µm, and the spores 7,5-10 x 6-9 µm. Occurs in unfertilized grasslands, typically with many other entolomatoids, hygrocyboids, clavarioids, *Geoglossaceae, etc.*

Entoloma caeruleum ⋉ is similar when young, but becomes paler with age, unlike *E. porphyrogriseum* which usually remains dark. *Entoloma aethiops* ⋉ has larger spores (11.5–14.5 × 8–11 µm).

Widespread, fairly common to rare; mainly June–October.

Thomas Læssøe/Thomas Kehlet

Entoloma atrocoeruleum is a rather dark, minutely scaly, not translucently striate *Entoloma* with a blackish-blue to washed-out brown-grey cap that has blue or violet tinges. The gills are white and a finely fibrillose, blue-black to violet-tinted stem. Lacks cheilocystidia and clamps. The spores measure 8–12 × 6–8 µm. Occurs in unfertilized, ± calcareous grasslands.

Similar to *E. corvinum* △ and *E. caeruleum* ⋉ but both have cheilocystidia.

Widespread, fairly frequent to rare; mainly June–October.

Jan Vesterholt/Thomas Læssøe

Entoloma griseocyaneum is an almost medium-sized, fibrillose-scaly *Entoloma* with pale brown cap colours and a pale, blue-grey-tinted, longitudinally very fibrillose stem. Lacks cheilocystidia and clamps. The spores measure 9–13.5(–14) × 6.5–8(–9) µm. Occurs in unfertilized, mostly calcareous grasslands.

Entoloma scabropellis ⋉ is probably just a colour form without blue stem. *Entoloma coeruleofloccosum* ⋉ has cheilocystidia. *Entoloma anatinum* ▽ has a darker brown cap and darker fibres on the stem. *Entoloma viaregale* ⋉ has a floccose-pruinose stem and smells farinaceous. *Entoloma scabrosum* ⋉ is darker brown and smells ± aromatic.

Widespread, fairly common to rare; mainly June–October.

Entoloma anatinum is a dark brown, floccose-scaly *Entoloma* with ± depressed cap centre. The stem is matt, tinged in blue and fibrillose. Lacks a distinct smell or taste. Clamps and cheilocystidia absent. The 6–9-edged spores measure 9–14 × 6.5–8 µm. Occurs in unfertilized, species-rich grasslands. This species is a very good indicator of grasslands of high conservation value.

Entoloma viaregale ⋉ has a more pruinose rather than fibrillose stem. *Entolomas asprellum* ⋉ and *E. nigroviolaceum* ⋉ have broader spores and smell ± rancid-farinaceous.

Widespread but rare; mainly June–October

Jan Vesterholt

Entoloma chalybaeum has greyish-blue colours on the cap and stem and, when young, bluish gills. The gill edge is ± brownish and entire, and lacks clusters of cystidia. The stem is ± smooth and has a white base. The odour is indistinct. Lacks clamps; cheilocystidia measure 35–75 × 5–25 µm. The sharp-angled spores measure 8.5–12.5 × 6–8.5 µm. Occurs in unfertilized grasslands, including in open forests.

Entoloma chalybaeum var. *lazulinum* ⋉ has a translucently striate cap. *Entoloma serrulatum* ◁◁ has a serrated, usually darker gill edges that have cystidia in clusters. *Entoloma corvinum* ◁◁ has pure white gills when young.

Widespread, common to occasional; July–October (–December).

Jan Vesterholt

Entoloma poliopus s.l. is a ± dark brown, floccose-scaly, somewhat marginally translucently striate *Entoloma* that lacks blue colours on the cap. The stem is ± greyish-blue and has a white base. The gills are pale grey and may have a brownish edge. Has cheilocystidia but no clamps. The spores measure 9.5–13.5(–14.5) × 6–10 µm. Occurs in unfertilized grasslands.

Very variable and presumably constitutes a species complex. Split into three varieties: var. *poliopus* ⋉ has brown gill edges, spores longer than 10 µm and long cheilocystidia; var. *parvisporigerum* ⋉ has spores up to 10 µm and large cheilocystidia; and var. *discolor* ◁ has spores of intermediate length (9–11.5 µm) and cheilocystidia up to 40 µm long. *Entoloma sodale* ▷ is yellow-brown and translucently striate.

Widespread, rather common to occasional; mainly June–October.

Entoloma sodale is a floccose-scaly, yellow-brown, typically rather translucently striate *Entoloma* with a bluish, ± smooth stem. The gills are ± cream with concolourous edges. The smell and taste are indistinct. The cheilocystidia are broadly clavate and measure 20–80 × 10–30 µm. Lacks clamps. The spores measure 9.5–14 × 7–9 µm. Occurs in unfertilized grasslands, mostly with many other species of *Entoloma*, *Hygrocybe, etc.*

Entoloma poliopus ◁ has a darker brown, not translucently striate cap and narrower cheilocystidia. *Entoloma lividocyanulum* ▽ has no cystidia.

Widespread, uncommon to rare; mainly June–October.

Tobias Bøllingtoft

Entoloma lividocyanulum is a pale yellow-brown, floccose-scaly *Entoloma* with a darker spot in the middle of the translucently striate cap. The ± smooth stem quickly loses its bluish colour. The gills are pale cream with concolourous edges. Cystidia and clamps are absent. The spores measure 7–10 × 6.5–7.5 µm. Occurs in unfertilized grasslands, mostly on species-rich sites.

Entoloma lividocyanulum var. *lilacinum* ⋉ has a sordid lilac cap when young. *Entoloma sodale* △ has large, wide cheilocystidia. *Entoloma incarnatofuscescens* ◁◁ has deeply decurrent gills.

Widespread; uncommon to rare; mainly June–October.

Entoloma incanum is a very distinctive, rather slender *Entoloma* with a bright green stem that discolours bluish when bruised. The cap is floccose-scaly at the centre and a delicate grass-green at first, later becoming more brownish-green with obvious translucent striation. The odour is very distinct, recalling burnt rubber or mouse urine. Lacks clamps and cystidia. Spores 10–14 × 7.5–10 µm. Occurs on calcareous, open habitats, occasionally on almost pure mineral soils and may be found in large troops, *e.g.* with *Geoglossum cookeanum* (Vol 2, p. 1352).

Very young specimens may recall a species of *Hygrocybe* (page 142), but identification should be straightforward.

Widespread and locally common; mainly June–October.

Rasmus Ejrnæs

Entoloma sarcitulum var. majusculum is a pink form of a normally yellow brown *Entoloma* with a translucently striate, finely floccose-scaly cap and a finely floccose to smooth stem that may be more vividly coloured near the apex (see page 468). The gills are pale pinkish buff, pink to almost white, adnexed to emarginate and have smooth, concolorous edges. Lacks cystidia and clamps. The spores measure 9-11 x 6-7 µm.

Other ± pink, floccose-scaly species of *Entoloma* have cheilocystidia.

The pinkish form is very rare; summer–autumn.

Entoloma serrulatum in a beautifully pinkish version with lilac to brown, serrated gill edge (page 459). In unfertilized, species-rich grasslands.

Entoloma sarcitulum var. majusculum ◁ lacks cystidia and has a non-serrated gill edge; *E. roseum* ▽ has cystidia but a non-serrated edge.

The pinkish colour-form is very rare; summer–autumn.

Thomas Læssøe

***Entoloma roseum* ss. auct.** is a pinkish, floccose-scaly *Entoloma* with white gills that have white to pink, smooth edges. The stem is ± smooth and paler than the cap. The cheilocystidia measure 20–60 × 5–20 µm. Lacks clamps. The spores measure 8.5–10.5(–11) × 7–8 µm. Occurs on unfertilized, ± calcareous grassland.

May be a colour form of an otherwise darker species. *Entoloma serrulatum* △ has cystidia but the gill edges are serrated and distinctly coloured. *Entoloma sarcitulum* var. *majusculum* ◁ lacks cystidia and has smooth gill edges.

Widespread, but rare; summer–autumn.

Thomas Læssøe

Entoloma catalaunicum is a rose-red, not translucently striate, minutely scaly *Entoloma*, which usually has a somewhat violet cap margin. The pale gills are serrated with a concolorous or blue-black edge. The fibrillose stem is reddish above and violet-blue towards the base. The cheilocystidia measure 25–75(–100) × 7–20 µm. Lacks clamps. The spores measure 9–10.5(–11) × 6.5–8 µm. Occurs in ± calcareous, unfertilized, species-rich grasslands.

May be a colour form of an otherwise darker species; the pink form of *E. serrulatum* △ has no violet colours on the cap margin; *E. roseum* △ does not have serrated gill edges.

Widespread but rare; summer–autumn.

Jean-Marc Moingeon

Thomas Kehlet

Entoloma queletii is a pale, yellow-brown *Entoloma* with an unstable lilac-pink tint, especially towards the floccose-scaly cap centre. The gills are white and the floccose stem is ± tinted the same colour as the cap. It may be almost odourless or distinctly perfumed, recalling that of *Lepista irina* (page 249). Cheilocystidia are present and measure 20–80 x 10–15 µm. Lacks clamps. The spores measure 10–13 × 6.5–9(–9.5) µm.

Occurs in calcareous fens, on steep coastal slopes, in scrub *etc.*; rarely on rotten wood.

Entoloma ochromicaceum ⋉ and *E. kervernii* ⋉ both have smaller spores, and the former has a smooth stem.

Widespread, but rather rare to very rare; mainly July–September.

Thomas Læssøe/Thomas Kehlet

Entoloma exile is a floccose-scaly, translucently striate *Entoloma* with a dark cap centre. The cap margin can be slightly bluish. The gills are white. The smooth, grey, yellowish-grey to olive-grey stem is typically brownish-orange at the base, particularly in older specimens. Lacks clamps. The many-celled cheilocystidia measure 20–100(–120) × 5–20 µm. The spores measure 9–12(–12.5) × 6–8(–9) µm. Occurs in unfertilized grasslands.

Entoloma chloropolium ▷▷ and *E. lividocyanulum* ◁◁ lack cystidia. *Entoloma poliopus* ◁◁ has a darker cap. *Entoloma sodale* ◁◁ is, at most translucently striate only at the margin and is a more bluish colour. *Entoloma longistriatum* ▷ never has a reddish stem base and is less 'dark-eyed'.

Widespread, and fairly common to occasional; June–November.

Entoloma longistriatum has a floccose-scaly, matt, ± yellow-brown, translucently striate cap. The cheilocystidia measure 20–55 × 6–20 µm. Lacks clamps. The spores measure (9–)10–14.5(–15) × 6–9(–10) µm (var. *longistriatum*). Occurs in mostly species-rich, unfertilized grasslands.

Entoloma longistriatum var. *microsporum* ⋉ has spores that measure 6.5–10 × 5.5–9 µm. *Entoloma exile* ◁ has a very distinct dark 'eye' in the cap centre and typically has a brownish-orange stem base.

Widespread, occasional to rather rare; mainly July–October.

Jan Vesterholt

Entoloma pseudoturci is a rather dark, brown, ± not translucently striate, floccose-scaly *Entoloma* with pale brown gills. The ± smooth stem is grey-brown to red-brown. The flesh and stem base do not discolour red. The cheilocystidia measure 20–60 × 5–15 µm. Lacks clamps. The spores measure 8.5–10 × 6.5–7 µm. Occurs in unfertilized grasslands.

Entoloma turci ⋉ has reddening flesh; *E. scabropellis* ⋉ has a fibrillose-floccose stem.

Widespread, but rare; mainly June–October.

Jan Vesterholt

Thomas Kehlet

Entoloma sarcitulum* var. *majusculum is a very distinctly translucently striate, floccose-scaly, yellow-brown *Entoloma* with a smooth, somewhat blue-green-tinted stem that does not have brownish-orange colours at the base. The gills are whitish with concolorous edges. Does not have a distinctive smell or taste. Lacks clamps and cystidia. The spores measure 9–11 × 6-7 µm. Occurs in unfertilized, ± calcareous, grass-dominated habitats.

Entoloma exile ◁◁ has cheilocystidia and a ± brownish-orange stem base.

Widespread, occasional to rare; mainly June–October.

See also the pink form p. 464.

Jens H. Petersen/Thomas Læssøe

Entoloma formosum is a vivid, apricot-orange, ± translucently striate, floccose-scaly *Entoloma* with whitish gills that have concolorous edges. Lacks clamps. The cheilocystidia measure 20–50 × 5–15 µm, and the spores 9–14 × 6–9 µm. Occurs in rather damp, ± calcareous, unfertilized grasslands.

Entoloma ochromicaceum ⋉ is apparently a paler, less vivid yellow, not translucently striate and normally has brown gill edges. *Entoloma xanthochroum* ⋉ has cinnamon-buff gills, typically with a brown edge. *Entoloma sphagnorum* ⋉ has white gills with a brown edge.

Widespread, occasional to rare; mainly July–October.

Entoloma pleopodium is a vivid greenish-yellow to apricot-orange, bell-shaped *Entoloma* with a somewhat darker, finely floccose cap centre. The stem is tinted in red-brown to yellow-brown and floccose to almost smooth. It usually has a distinctly sweetish-nauseous odour. Lacks cheilocystidia, but has clamps. The spores measure 8–11.5 × 6.5–8 µm. Occurs on ± rich, mostly disturbed soil in gardens, parks, path edges in forests *etc.*

Entoloma olivaceostipitatum ⋉ does not have clamps and the stem is tinted olive-green. *Entoloma chlorophyllum* ⋉ has cheilocystidia and usually an olive tinge to the gills.

Widespread, occasional, absent from the boreal zone; July–November.

Tobias Bøllingtoft

Entoloma hebes is a rather slender, ± grey-brown, translucently striate, smooth *Entoloma* with a small dark papilla. The stem is ± distinctly floccose-fibrillose. The odour is rancid. Has clamps and cheilocystidia that have a ± capitate apex and measure 25–60 × 4–12 µm. The spores measure 8–12 × 5.5–8 µm. Occurs on ± calcareous, rich soil in gardens, scrub and deciduous forests.

Entoloma hirtipes ⋉ is usually more robust and has wider spores. *Entoloma tibiicystidiatum* ⋉ has basidia mixed with the cheilocystidia, and a lower spore Q.

Widespread and fairly common except in the northern parts, where it is mostly absent; June–December.

Thomas Læssøe

Jan Vesterholt

Entoloma fernandae is an often umbonate, ± translucently striate, grey-brown *Entoloma* with a slightly felty cap centre. The grey-brown stem is ± smooth. The gills are also pale grey-brown. Lacks clamps and cystidia. The pigment is incrusted and can also be seen as intracellular lumps. The spores measure 7.5–10 × 5.5–7.5 µm. Occurs in sandy, open, mossy sites.

Entoloma xanthocaulon ⋉ has a yellowish stem. Other similar species either have clamps or more fibrillose stems.

Widespread, occasional; mainly July–November.

Jan Vesterholt

Entoloma juncinum is a relatively slender, ± translucently striate, smooth, grey-brown to red-brown, hygrophanous *Entoloma* with an almost smooth stem and rather pale gills. Smell and taste ± farinaceous. Has clamps, but lacks cystidia. The pigment is finely incrusted. Spores 8–10.5 × 7–9.5 µm. On soil in open woodland, wooded bogs and parks.

Can be told from the many similar brown *Entoloma* species by its ± smooth stem, rather slender stature and relatively wide spores. *Entoloma sericeum* ▷ has a fibrillose stem and coarse incrustation on the hyphae.

Widespread and common; July–December.

Entoloma infula is a ± concentrically undulating, translucently striate, grey-brown, hygrophanous *Entoloma* with a ± evident umbo or papilla and almost pure white gills. The rather tough stem is ± yellow-brown and smooth. Smell and taste somewhat farinaceous. Has clamps, but lacks cystidia. The pigment in the cap cuticle is both incrusted and intracellular. The spores measure 7–10 × 5–7 µm. Occur in unfertilized grasslands.

Entoloma infula var. *chlorinosum* ⋉ smells like chlorine and has a yellow-brown stem.

Widespread and probably rather common; June–November.

† ***Entoloma sericeum*** is a robust, translucently striate, hygrophanous, dark brown *Entoloma* that soon becomes rather applanate. The gills are brownish, and the stem is ± silvery fibrillose. Smells and tastes rancid-farinaceous. Has clamps, lacks cystidia. The pigment is coarsely incrusted. The spores measure 7–10.5 × 6.5–9.5 µm. Occurs in open habitats, tolerant of some eutrophication with fertilizers.

Entoloma juncinum ◁ is more slender and has a smoother stem. There are many similar species but they have different combinations of pigments, clamps, cystidia and smells.

Very common and widespread; (June–) August–January.

Entoloma minutum is a small, slender, deeply translucently striate *Entoloma* with a small dark spot in the applanate cap centre. The stem is ± smooth and pale. Has clamps, lacks cystidia. The pigment is finely incrusted. The rather wide spores measure 8–11(–11.5) × 6.5–8.5 µm. Occurs mostly in damp, herb-rich places, in wooded bogs and meadows.

Entoloma pygmaeopapillatum ⋉ is even smaller and has an evident papilla on the cap, but separation of the two species can be problematic.

Widespread but much overlooked; July–November.

Thomas Kehlet

Entoloma ameides is a somewhat shiny, brown-grey, hygrophanous, ± medium-sized *Entoloma* with a very distinct, sweetish, *Prunus padus*-like smell. The gills are grey. Has clamps, lacks cystidia. The pigment is finely incrusted. The spores measure 9.5–11 × 7–8 µm. Occurs on soil in calcareous scrub and in more open, unfertilized grassy habitats.

Entoloma sacchariolens ⋉ has a similar smell, but it has cheilocystidia (often mixed with basidia).

Widespread but rare; July–September.

Jacob Heilmann-Clausen

Entoloma cetratum is a long-stemmed, yellow-brown, translucently striate, smooth, hygrophanous *Entoloma* with a somewhat paler, floccose stem. The gills are broad and adnexed to emarginate. The smell and taste are indistinct. The basidia are 2-spored. Lacks clamps and cheilocystidia. The spores measure 9.5–14 × 7–9.5 µm. Occurs on or amongst litter in acidic habitats, mostly in mossy conifer plantations.

Several 4-spored species from the same habitat can look similar, *e.g. E. lanuginosipes* ⋉ and *E. pallescens* ⋉. *Entoloma turbidum* ▷ is a similar shape and occurs in the same habitat, but is more fleshy, a duller brown and has a greyish stem. The spores are also completely different (much smaller with very weak angles).

Widespread and common; April–January.

† ***Entoloma vernum*** is a dark, ± translucently striate, hygrophanous, mostly smooth, conical, ± umbonate *Entoloma* with a fibrillose, dark stem. The smell and taste are insignificant. Has incrusted and intracellular pigment. Lacks cheilocystidia, but has clamps. The spores measure 9–12 × 7–9 µm. Grows amongst mosses and needle debris in conifer plantations.

Entoloma papillatum ⋉ typically occurs later in the year and has an almost smooth stem and a farinaceous smell.

Widespread, occasional; March–May.

† *Entoloma turbidum* is a compact, somewhat tricholomatoid, dull brown to red-brown *Entoloma* with a smooth, greasy cap surface, which, at most, is only translucently striate at the margin. The gills are whitish, and the stem white to tinted silver-grey, but sometimes yellowish at the base. Smell and taste insignificant. Lacks cystidia, has clamps. The spores are very indistinctly angular and small, 6.5–8.5(–9) × 6–7(–7.5) µm. Occurs on ± peaty soils in heathlands and in coniferous woodland.

Entoloma pseudoturbidum ⋉ has a darker stem, a farinaceous smell and occurs in deciduous forests. *Entoloma cetratum* ◁ is more elegant, yellow-brown and deeply translucently striate.

Widespread and rather common; June–December.

† *Entoloma nitidum* is a ± conical, smooth, beautifully dark blue tricholomatoid *Entoloma* with a blue, somewhat shiny, longitudinally fibrillose stem. The gills are crowded, at first white but later sordid reddish-grey. Smells and tastes somewhat farinaceous. Has clamps, lacks cheilocystidia. The spores measure (6.5–)7–8.5 (–9) × (5.5–)6–7.5(–8) µm. Occurs mostly in mossy coniferous forests, less often in deciduous forests and then only on mor soils.

Many of the other bluish entolomas have floccose-scaly rather than smooth caps ◁◁. Similar, but even more fleshy entolomas occur in unfertilized grasslands or on very calcareous soils, see *E. madidum* and *E. bloxamii* ▷▷.

Widespread, occasional to rare, absent from northern parts of temperate Europe, August–October.

Jens Maarbjerg

Thomas Læssøe

Entoloma rhombisporum is a rather anonymous, brown, marginally striate, ± smooth *Entoloma* with a striate, grey-brownish stem (very floccose in var. *floccipes*). Smells and tastes farinaceous. Has cheilocystidia that measure 35–80 × 10–25 × 3–9 µm, and clamps. The pigment is incrusted (but not in var. *floccipes*) and intracellular. The spores are 4-edged (cuboid/rhombic) and measure 8–10.5(–11) × 7–10.5 µm. Occurs in unfertilized grasslands or in mires amongst *Sphagnum*.

Apparently a species complex. *Entoloma conferendum* ▽ has star-shaped spores and lacks cystidia.

Widespread, occasional to rare; mainly August–October.

Entoloma conferendum is a brown-grey, hygrophanous, ± bell-shaped *Entoloma* with a silvery, longitudinally fibrillose stem. The gills are strikingly broad and arched/hanging. The smell and taste are insignificant to mealy. Lacks cheilocystidia, but has clamps. The pigment is intracellular. The spores are ± star-shaped and measure 7.5–13(–14) × 7–13 µm. Occurs in a range of habitats but found mostly in damp, ± acidic plantations, wooded bogs, in *Sphagnum*, *etc.*

Umistakable microscopically. *Entoloma conferendum* var. *incrustatum* ⋉ has incrusted pigment; var. *pusillum* ⋉ is usually very small and lignicolous (probably just a synonym). *Sphagnurus paluster* (page 286) is less striate on the stem and has completely different, white spores.

Widespread and common; May–December.

Jan Vesterholt

***Entoloma sericellum* s.l.** is a rather small, very variable, almost chalk-white to ochre-tinged, fibrillose *Entoloma* with white, later pink gills. The stem is floccose to almost smooth. Has scattered cheilocystidia and clamps. The spores measure 8–12 × 6–9 µm. Occurs on soil in unfertilized grasslands, in grassy verges along forest roads, on dark soil in wooded bogs, *etc.*

Entoloma cuboidoalbum ⋉ is another white species with more cuboid spores. *Entoloma rugosum* ⋉ – if a genuine species – should be recognizable by its well-differentiated caulocystidia. *Entoloma cephalotrichum* ▽ is chalk-white and has capitate cystidia on the cap and stem.

Rather common and widespread; June–November.

Entoloma cephalotrichum is a small to very small, chalk-white, floccose-hairy *Entoloma* with capitate hairs on both cap and stem. The gills are white to pink and emarginate. Lacks cheilocystidia and clamps. The spores measure 8–12.5 × 6–7(–7.5) µm. Occurs on dark, damp soil in scrub, wooded bogs and damp forests.

Other small white entolomas lack capitate hairs.

Rare or possibly overlooked, absent from the boreal zone; July–October.

Entoloma versatile is a rather small, fairly fleshy, fibrillose, hairy and ± umbonate *Entoloma* with metallic green cap colours and a silvery-grey, fibrillose stem. The gills are rather thick and grey. The cheilocystidia have a long, pointed tip and measure (40–)60–110 × 10–25 µm. The nodulose-angular spores measure (9–)9.5–11.5(–12.5) × 7–8(–9) µm. Occurs on calcareous mull soils in scrub or in open deciduous forests and parks.

Entoloma araneosum ▽ is very similar in shape and hairiness, and even very similar microscopically. However, it is grey to grey-brown and lacks the greenish metallic lustre.

Rare and local; mainly September–October.

Entoloma araneosum is a somewhat silvery, very hairy, convex to umbonate *Entoloma* with grey, adnexed, broad, thickish gills. It may recall an *Inocybe* (page 652). The stem base may have a tuft of reddish hairs. The cheilocystidia have a pointed tip and measure 45–75(–100) × 10–30 µm. The nodulose-angular spores measure (9.5–)10–14(–15) × 7–8(–9) µm. Occurs on ± calcareous mull soils in scrub and open deciduous forests, damp woodlands, parks, *etc.*

Other small, grey, hairy *Entoloma* species have ± clavate cystidia.

Widespread and occasional; May–December.

Entoloma dysthaloides is a small, ± convex, dark brown, hairy *Entoloma* with distant, dark gills and a felty-hairy stem that has septate hairs with coarsely incrusted pigment and more than three cells. The cheilocystidia are clavate and measure 25–60 × 15–30 µm. The spores are nodulose-angular and measure 10.5–13.5(–15) × 7–8(–8.5) µm. Occurs on ± calcareous mull soils in scrub, borders of boggy forests, *etc*.

Entoloma hirtum ⋉ occurs in drier, more open habitats, is paler, and the stem hairs are, at most, 2-celled. *Entoloma dysthales* ⋉ is very similar, but has larger spores (13.5–19.5(–21.5) × (7.5–)8–10.5(–11) µm).

Widespread and rather rare, but probably also overlooked; May–November.

Thomas Læssøe

Entoloma strigosissimum is a small, strigose-hairy *Entoloma* with ± red-brown, thick-walled, one-celled, pointed hairs (setae), mostly on the cap. The cheilocystidia are ± clavate and measure (25–)40–80 ×
12–22 µm. Lacks clamps. The spores are nodulose-angular and measure (12.5–)14–19(–22) × 7–11 µm. Occurs on ± calcareous mull soils in damp forests, coastal thickets, *etc*.

Other small, hairy *Entoloma* species lack setae.

Widespread and very rare, but probably also somewhat overlooked; mainly July–September.

This species has been reported as a host for the *Cordyceps*-like *Hypomyces porphyreus* (Vol 2, p. 1544).

Thomas Læssøe

Jan Vesterholt

Entoloma porphyrophaeum is a sturdy, fleshy, tricholomatiod, ± umbonate *Entoloma* with grey-brown colours on the rather coarsely fibrous cap surface. The stem is long, somewhat clavate, fibrillose-floccose and tinted the same colours as the cap. The gills are pale and almost free. Lacks a distinct smell or taste. The cheilocystidia are large, 20–60 × 7–20 µm, capitate and have a 3–15 µm wide head at the top of an undulating neck. Has clamps. The spores measure 8–12 × (5.5–)6–8 µm. Occurs in unfertilized, ± calcareous grasslands.

Entolomas jubatum ▷ and *E. fuscomarginatum* ⋉ have darker stems and the necks of the cystidia are not undulating.

Widespread, occasional to rare; July–November.

Thomas Læssøe

Entoloma scabiosum is a ± tricholomatiod *Entoloma* with a very fibrillose, splitting, dark grey-brown cap surface. The stem is fibrillose in the colours of the cap, and the emarginate gills are white to sordid pink. The cheilocystidia have an undulating, ± capitate top with a small, pointed extension and measure 25–90 × 5–20 µm. Lacks clamps. Spores five-angled, 7–8(–9) × 5.5–7 µm. On rich soil in scrub and deciduous forests, rarely in more open, grassy habitats.

May, at first glance, look similar to a species of *Inocybe* (page 652).

Rather rare to very rare, absent from parts of the hemiboreal–boreal zones; July–October.

Entoloma jubatum is a ± tricholomatiod, conical, grey-brown, radially fibrillose to minutely scaly *Entoloma* with a rather dark, fibrillose stem. The gills are grey-brown. Smells and tastes ± rancid-farinaceous. The cheilocystidia are capitate, have ± conspicuous heads and measure 20–50 × 8–20 × 2–7 µm. Has clamps. The spores measure 7–10(–11) × 5.5–7.5(–8) µm. Occurs mostly on sandy soils, both acidic and basic, in open habitats.

Entoloma fuscotomentosum ⋉ – if a genuine species – should have more obvious, pointed cap scales and paler gills. *Entoloma elodes* ▽ is somewhat hygrophanous.

Widespread, occasional to rare; mainly August–November.

Jean-Marc Moingeon

Entoloma elodes is a grey-brown, somewhat shiny, ± hygrophanous tricholomatiod, umbonate *Entoloma*. The stem is pale brown and longitudinally fibrillose. The gills are at first very pale then greyer. Smells and tastes farinaceous. The cheilocystidia are 25–70 × 7–20 µm and a mixture of clavate and capitate. Has clamps. The spores measure (8–)9–11.5(–12.5) × (6–)7–9(–9.5) µm. Occurs on damp, peat soils or directly in *Sphagnum*.

The rather similar *E. porphyrophaeum* and *E. jubatum* do not have hygrophanous caps and have only capitate, more pronounced, cystidia. Brown-edged specimens are sometimes given the name *E. fuscomarginatum*, but are regarded here as conspecific. Other similar species have distinctly capitate cystidia and are mostly more felty.

Widespread, but northern, rather local; mainly August–October.

Jean-Marc Moingeon

Entoloma bloxamii* var. *bloxamii is a compact, tricholomatoid, umbonate, beautifully blue, later violet-brown *Entoloma* with whitish, serrated gills. The stem is ± steel-blue with a yellow, tapering base. The smell and taste are indistinct to farinaceous. Lacks cystidia, but has clamps. The spores measure 7.4–9.4(–10.8) × 6.7–9.7 µm (av. 8.2 × 8.3 µm). Occurs in unfertilized grasslands of high conservation value.

Entoloma bloxamii var. *rubellum* ▽ has red, not blue, pigments; *E. atromadidum* ⋉ has a darker, blackish-blue cap; and *E. madidum* ▷ is smaller-spored. *Entoloma ochreoprunuloides* f. *hyacinthinum* ⋉ has smaller spores. *Entoloma nitidum* ◁◁ is darker blue, has smaller spores and is a woodland species.

Widespread but rare; mainly August–October.

Jürgen Schnieber

Entoloma bloxamii* var. *rubellum is a ± pinkish, sometimes ochraceous, tricholomatoid *Entoloma* with a broad umbo and smooth cap surface. The gills are at first white, later pinkish-tinted, and have a ± uneven, serrated edge. The stem is fleshy, whitish and ± smooth. The smell and taste are indistinct to farinaceous. Lacks cystidia, has clamps. The spores measure 6.5–8 × 5.5–8 µm. Occurs on soil in open thermophilic forests.

Although a very distinctive taxon, molecular analyses have shown it to be a variety of *E. bloxamii* △ rather than a separate species.

Very rare, mostly recorded in central Europe; autumn.

Machiel Noordeloos

Entoloma madidum is a ± steel-blue tricholomatoid *Entoloma* with innately radiating fibres on it non-hygrophanous, umbonate cap. The whitish gills are adnate to decurrent with a ± serrated edge. The stem is blue with a pale base. Smells and tastes farinaceous. Lacks cystidia, has clamps. The thin-walled, 6–8-angled spores measure 5.9–7.6 × 5.8–7.5 μm, av. 6.8 × 6.7 μm, Q 1–1.2. Occurs on clay-rich soils in unfertilized grasslands or in open forests.

Entoloma bloxamii var. *bloxamii* ◁ is almost identical and best separated by its spore measurements. *Entoloma atromadidum* ⋉ has a darker, blackish-blue cap and larger spores (6.2–8.3 × 6.4–8.0 μm).

Widespread, but rare; autumn.

Entoloma prunuloides is a compact, rather pale, grey, ± smooth *Entoloma* with distinct farinaceous smell and taste. The gills are white, broad and crowded. The spores are indistinctly edged, thin-walled and measure 6.5–8(–9) × 6.5–8 μm. Occurs in unfertilized, mostly species-rich, poor or calcareous grasslands.

Entoloma ochreoprunuloides ⋉ (including *E. ochreoprunuloides* f. *hyacinthinum* ⋉) has shorter spores and is either darker or has violet shades at the cap margin.

Widespread, fairly common to rather rare; June–October.

Thomas Læssøe

† ***Entoloma rhodopolium*** is a medium-sized to rather large, pale yellow-brown, hygrophanous *Entoloma* with no, or only a weak, nitrous smell. The ± smooth stem and gills are the same pale colour. The pigment is intracellular. Lacks cystidia, but has clamps. The spores measure 7.5–10 × 7–8.5 µm. Occurs on black, rich soils, mostly in deciduous forests.

The species belongs to a difficult, not yet fully understood complex. *Entoloma sericatum* ▽ has incrusted pigment in the cap cuticle.

Widespread and common; July–November.

†? ***Entoloma sericatum*** is a highly variable, rather large, ± yellow- to grey-brown *Entoloma* with an initially minutely downy-scaly, later glabrescent, cap surface. The stem is pale and ±floccose. Smell faint, ± mealy. The pigment is ± incrusted and ± intracellular. Spores 7.5–9.3 × 6.5–7.8 µm. In ± damp habitats with *e.g. Betula* and *Salix*.

Proposed synonyms include *E. myrmecophilum*, *E. bipelle*, *E. rimulosum* and *E. subarcticum*. *Entoloma rhodopolium* △ has intracellular pigments but usually a nitrous smell; *E. majoloides* ⋉ has ± olive tinges. There are several other similar species.

Widespread and very common to common; mainly August–November.

Jan Vesterholt

Entoloma sordidulum is a rather small, robust, thick-fleshed, dark grey-brown, mostly non-striate tricholomatoid *Entoloma* with pale gills and a strong mealy smell and taste. Lacks cystidia, has clamps. The pigment is incrusted and ± intracellular. The rather wide spores measure 8–11.5 × 7–8.5 µm. Occurs on clay-rich soils, mostly with *Quercus*, often in parks and larger gardens.

Entoloma subradiatum ⋉ is more striate and the pigment is only intracellular. *Entoloma sordidulum* belongs to a difficult species complex.

Widespread and rather common, except in the boreal zone; May–December.

Entoloma porphyrocephalum is a medium-sized to large, somewhat *Gymnopus*- to *Melanoleuca*-like *Entoloma* with a convex, matt, smooth, greyish white to pale grey-brownish cap. The gills are pale, crowded and ± emarginate. The stem is mostly ± excentric, ± fibrillose and tinted the same colours as the cap. The cheilocystidia are very large, 40–110 × 10–25 µm, and ± fusiform. The spores measure (10–)11–12.5(–14) × 7–8.5(–9.5) µm. Occurs in open habitats on calcareous soils.

Entoloma excentricum ⋉ is similar. Microscopically, the very long cystidia make this complex easy to recognize.

Widespread but rare and local; late summer–autumn.

Thomas Læssøe

††***Entoloma sinuatum*** is a very large, rather pale, tricholomatoid *Entoloma* with emarginate gills that are at first yellow, but later somewhat rose, and an uneven, yellow-brown, smooth cap. Smell rancid-farinaceous. Lacks cheilocystidia. Has intracellular pigment and clamps. The sharp-edged, rather wide spores measure 8–11 × 7–9.5 µm. Occurs on ± clay-rich and/or calcareous soils, typically with *Quercus*, in old parks and open thermophilic, deciduous forests.

Other large fleshy entolomas do not have yellow gills.

Widespread, but rather local, absent from the boreal zone; mainly August–October.

†***Entoloma lividoalbum*** is a fleshy, rather short, pale-stemmed *Entoloma* with a rather pale, grey-brown to yellow-brown cap, pale, emarginate gills and a mealy smell and taste. It has intracellular pigment and clamps. The rather wide spores measure 8–10(–11) × 8–9 µm. Occurs in open, sunny places on clay-rich soils, typically with *Quercus* in parkland.

Similar species typically have darker colours on the gills and stem or a different pigmentation in the cap cuticle. *Entoloma majoloides* ⋉ tends to have faint olive tinges.

Widespread south of the hemiboreal–boreal zone and rather common; June–November.

Entoloma clypeatum is a thick-fleshed, tricholomatoid, ± grey-brown, smooth, ± hygrophanous, farinaceous, early-summer *Entoloma*. The gills are pale and emarginate. Lacks cystidia, but has clamps. The spores measure (8.5–)9–11(–12) × (7–)7.5–10 µm. Occurs on rich soils with *Crataegus* and other woody members of the *Rosaceae*.

Entoloma saundersii ⋉ has a finely fibrillose cap surface and occurs with *Ulmus*. *Entoloma aprile* ⋉ is smooth, hollow-stemmed, more slender and also occurs with *Ulmus*.

Widespread and rather common, decreasing towards the north; April–June.

Thomas Læssøe/Jan Vesterholt

Entoloma sepium is a fleshy, pale, not hygrophanous, tricholomatoid early-summer *Entoloma*, which develops yellow to orange-red spots when bruised and with age. The gills are pale. Smells and tastes mealy. Lacks cystidia, but has clamps and intracellular pigment. Spores (7.5–)8–11(–12) × (6.5–)7–10(–11) µm. On ± calcareous soils with *Prunus*-species.

Entoloma clypeatum △ occurs in the same habitat but is typically darker and does not redden. *Entoloma niphoides* ▽ is entirely white and never reddens.

Widespread south of the hemiboreal–boreal zones, but rather local; May–June.

Jürgen Marqua

† ***Entoloma niphoides*** is an almost pure white early-summer *Entoloma*. The cap is somewhat hygrophanous and smooth. The stem is ± longitudinally fibrillose and the flesh does not redden. Smells and tastes farinaceous. Lacks cystidia, but has clamps. The 5–8-edged spores measure 8–10(–11.5) × 7.5–10(–11) µm. Occurs on soil, with *Crataegus*, *Prunus*, etc.

Entoloma sepium △ is somewhat darker and the flesh reddens.

Local and rare; May–June.

Pluteoids

Volvopluteus, *Volvariella* and *Pluteus* are characterized by their free gills and brownish-rose spore-deposits. They are decomposers either of wood or organic material in the soil. One species, *Volvariella surrecta*, is a parasite on fruitbodies of *Clitocybe nebularis*.

OTHER SIMILAR FUNGI:
– *Amanita* has white spore-deposits and the poisonous species have a partial veil that is left as a ring on the stem (page 352).
– *Limacella* is ± sticky and has white spore-deposits and no volva at the base (page 348).
– *Leucoagaricus leucothites* has pinkish gills, white spores, and a ring but no volva (page 339).
– entolomatoids have attached gills and ridged or angular spores (page 448).

FURTHER READING: 6, 89,140, 141, 156, 182.

Volvopluteus gloiocephalus is a large agaric with a smooth, greasy to slimy, greyish cap surface, a long, greyish stem and a rather large, white volva (the remnants of the universal veil) at the base. The spore-deposit is brownish-rose. Cheilocystidia clavate. The spores are smooth and measure 10.5–17.5 × 6.5–9 µm. Mostly found in disturbed places, including stubble fields, compost heaps, bales of damp straw, *etc.* It can occur by the thousands in stubble and in newly sown fields late in the season.

The amanitoids (page 352) usually have the same greasy-smooth cap surface but they are generally ectomycorrhizal and have white spores; the most similar species, *Amanita phalloides* (page 359), has a ring on the stem. Species of *Volvariella* ▷▷ are dry to somewhat sticky on the cap and always distinctly radially fibrillose. Species of *Pluteus* ▷▷ lack a universal veil and thus a volva.

Widespread and common, decreasing towards the north; May–December.

Jens H. Petersen/Jan Vesterholt

Volvariella has a brownish-rose spore-deposit, free gills, a fibrillose-hairy cap surface and a membranous universal veil that is left as a volva at the stem base. All are decomposers.

Volvariella bombycina is a large to very large *Volvariella* with the cap covered in long, rigid, whitish to yellowish hairs. The volva is very large, grey-brown and scaly. The free gills are very crowded. The cheilocystidia are flask-shaped to fusiform, 40–110 × 10–25 µm. Spores 7–10 × 4.5–6 µm. Typically occurs on standing, decaying deciduous trees, including *Fagus* and *Aesculus*, in forests and urban settings, rarely in houses.

Volvopluteus gloiocephalus ⊲⊲ has a smooth, greasy cap and occurs in disturbed places such as stubble fields. Other species of *Volvariella* are less fleshy and have smaller volvas.

Widespread, occasional, absent from the boreal zone; June–October.

Volvariella caesiotincta is a rather large, grey to blue-grey-tinted, silky-hairy *Volvariella* that smells of *Pelargonium*. The volva is greyish. The spores measure 5.5–7.5 × 3.5–5 µm. Occurs on very rotten deciduous wood, mainly *Fagus*, and often in sites with a long continuity and numerous other rare fungi.

Volvariella bombycina ◁ is even larger and has a yellowish tinge on the hairs. *Volvariella murinella* ▽ is smaller and has simple clavate cystidia, whereas the cystidia in *V. caesiotincta* have a pointed tip.

Rare to very rare, absent from the boreal zone; mainly July–October.

Thomas Læssøe

Volvariella murinella is a greyish, rather small *Volvariella* with a ± dry, finely fibrillose cap surface. The volva is whitish to grey-brown and rather small. The odour recalls *Pelargonium*. The cheilocystidia are flask-shaped to clavate. The spores measure 5.5–7.5 × 3–4.5 µm and do not darken significantly in KOH. Occurs on soil and very rotten wood, mostly on calcareous soil.

The cultivated *V. volvacea* ⋉, which occurs on compost in some parts of Europe, is larger, darker and smells of radish. *Volvariella caesiotincta* △ is also larger and has pointed cystidia. *Volvariella taylori* ⋉ and *V. pusilla* ▷▷ have spores that stain darker in KOH.

Widespread, except in the boreal zone, occasional; mainly July–November.

Lucie Zíbarová

Jens Maarbjerg

Volvariella hypopithys is a white, rather small to medium-sized, silky-hairy and downy-stemmed *Volvariella* with a ± lobed, white to grey-tinged volva. The spores only darken slightly in KOH and measure 5.5–8 × 4–5.5 μm. Occurs on ± calcareous soils along roadsides in deciduous forests, *etc*.

Volvariella murinella ◁◁ has a more greyish cap and smells of *Pelargonium*; the spores do not darken in KOH. *Volvariella pusilla* ▽ has a smooth stem and spores that stain yellow in KOH.

Widespread, occasional, but absent from the boreal zone; (July–) August–October.

Thomas Læssøe

Volvariella pusilla is a small, white to pale greyish, fibrillose, somewhat sticky *Volvariella* with a ± brown, very lobed volva. The stem is ± smooth and white to somewhat grey-brown-tinted. The spores stain rust-brown in KOH and measure 5.5–8 × 4–5.5 μm. Typically in gardens and parks, but also in greenhouses, and on mull soils or compost.

Volvariella taylori ⋉ is not sticky and has a more coloured, typically larger cap. *Volvariella murinella* ◁◁ has a dry cap, a *Pelargonium*-like smell and spores that are KOH inert.

Widespread but nearly absent from the hemiboreal–boreal zones, occasional; mainly July–October.

Jens H. Petersen/Rasmus Ejrnæs

Volvariella surrecta is a medium-sized, silky-fibrillose, white, convex *Volvariella* with a thick, ± lobed, white volva. The stem is pale and finely floccose. The cheilo- and pleurocystidia are large and measure 50–90 × 10–30 μm. The somewhat thick-walled spores measure 6.5–8.5 × 3.5–4.5 μm. Occurs on the ± rotten fruitbodies of *Clitocybe nebularis* (page 126).

The somewhat similar *V. hypopithys* △ fruits directly on soil and has a more downy-hairy stem surface.

Mainly in the nemoral zones, rather rare; September–December.

Pluteus is a genus of agarics with free gills, smooth spores and brownish-rose spore-deposits. All are decomposers and found predominantly on very rotten wood.

Pluteus cervinus is a large to very large *Pluteus* with a grey-brown, innately radially fibrillose cap (which may be more felty at the centre), a mostly brown-tinged, scaly stem and a smell of radish. The cystidia are thick-walled and hooked. The cap cuticle comprises horizontal, elongated, brown cells that lack clamps. The spores measure (6–)6.5–8–9.5) × (4–)4.5–6.5 µm. Occurs mostly on deciduous wood, including wood chips and sawdust.

Pluteus primus ⋉ and *P. pouzarianus* ▽ are very similar, but both occur on conifer wood and have clamps in the cap cuticle. *Pluteus pouzarianus* also differs in having a paler grey-brown and very shiny cap.

Very common and widespread; April–January.

Pluteus pouzarianus is a rather large grey-brown, silvery-shiny, innately radially fibrillose *Pluteus* that is ± felty at the cap centre. The stem is tinted the same colours as the cap. Has a faint smell. The cystidia are hooked and there are clamps on 10–35% of the septa in the cap cuticle. The spores measure (5.5–)6–8 × 4–5.5–6) µm. Occurs on conifer wood including wood chips.

Pluteus primus ⋉ has clamps on all septa in the cap cuticle and is less shiny; *P. alniphilus* ⋉ have clamps at some septa and grows on deciduous wood.

Widespread and probably rather common; June–November.

Sten Larris

Pluteus atromarginatus is a rather large to large *Pluteus* with a dark grey-brown, radially fibrillose, smooth and ± silky cap surface and gills with distinct, dark brown, felty-serrated edges. The cystidia are thick-walled and hooked. The spores measure 6–8 × 4–5.5 µm. Occcurs on very rotten conifer wood, including wood chips and sawdust.

Pluteus umbrosus ▷▷ has dark gill edges, and may look similar, but can be separated by its more velvety and yellow-brown cap, and unhooked cystidia.

Widespread, occasional to rare, decreasing towards the south; mainly July–October.

Pavel Moran

Pluteus petasatus is a large, pale, ± clustered, rather sticky *Pluteus* with a brownish, felty cap centre that often cracks into a scaly pattern. It is almost odourless. The cystidia are thick-walled, hooked and lack clamps. The spores measure 5–7.5 (–8) × 3.5–5(–6) µm. Occurs mostly on sawdust, but also on deciduous and coniferous wood.

Can be difficult to separate from other pale, clampless plutei, but the brownish cap centre is a good character when compared to pale forms of *P. cervinus* ◁◁ and *P. pellitus* ⋉.

Rather rare; mainly June–October.

Thomas Læssøe

Pluteus salicinus is a rather large to large *Pluteus* with ± bluish-grey to grey-green shades on the cap and also sometimes on the stem, which may also be more pure grey. Young specimens may be very felty at the cap centre. The cystidia are thick-walled and hooked. Spores ± ellipsoid, somewhat thick-walled, (6.5–)7–9.5(–12) × 4.5–6.5(–9) µm. On deciduous wood, mainly in damp or wet woodlands such as *Alnus* carrs.

Species in the *P. cervinus* ◁◁ complex never have bluish or greenish tinges to the cap or stem.

Widespread and rather common; May–December.

Pluteus umbrosus is a medium-sized to large *Pluteus* with a yellow- to reddish-brown, velvet-like, ± veined cap surface and similar, brown-velvety, gill edges; even the stem is brownish floccose. The cystidia are simple clavate. The spores measure (5–)5.5–6.5(–7) × (3.5–)4–5 (–5.5) µm. Occurs on large dead wood of deciduous trees, mostly *Fagus*.

A pure white to slightly yellowish variety is named *P. umbrosus* var. *albus* ⋉. *Pluteus pusillulus* ⋉ is much smaller and does not have pleurocystidia. *Pluteus atromarginatus* ◁ has a smooth cap and hooked cystidia.

Widespread, occasional to rare, very rare in the boreal zone; June–November.

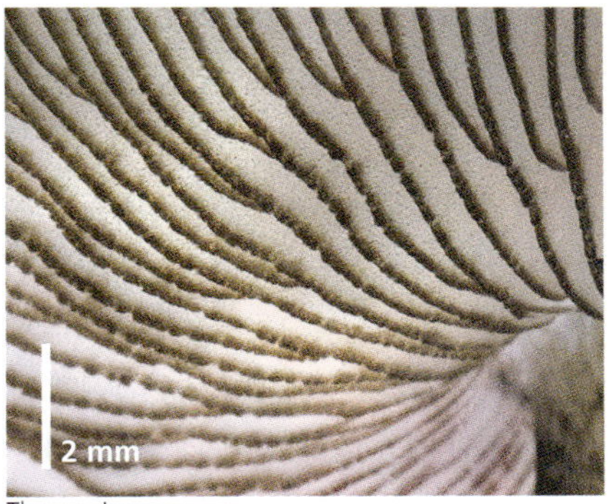

Thomas Læssøe

Pluteus ephebeus is a medium-sized to large *Pluteus* with a dark grey-brown, radially scaly to fibrillose cap, the surface of which cracks to reveal the white flesh between the fibres. The stem has some grey-brown hairs at the base. The cystidia are simple clavate. The cap cuticle has hair-like, brown, ± erect cells. Spores 5.5–8.5 × 4.5–6 (–7) µm. Typically on ± disturbed, mull soils or wood chips mixed with soil.

Drought-damaged specimens of *P. cervinus* ◁◁ and *P. salicinus* ◁ can look similar but are easily recognized by their hooked cystidia.

Widespread, but rather rare, decreasing towards the north; May–October.

Thomas Læssøe

Pluteus leoninus is a medium-sized *Pluteus* with a yellow to apricot-orange, finely velvety-felty cap and a yellowish, longitudinally fibrillose stem. The cap cuticle consists of upright, hair-like cells and the cystidia are simple clavate. The spores measure (5.5–)6–8 × (4.5–)5–6.5(–7) µm. Occurs on deciduous wood, both in old-growth *Fagus* forests and in old wooded bogs with many dead *Alnus* or *Betula* trunks.

May be confused with *P. chrysophlebius* ▷▷ but that species typically has green tinges to the yellow cap, and a cap cuticle comprised of pear-shaped cells in a hymeniform layer (and, with a hand lens, it does not appear hairy).

Widespread, occasional to rare; mainly June–October.

Thomas Læssøe/Jens Christian Schou

Pluteus roseipes is a medium-sized, typically rather long-stemmed, greyish-brown, felty *Pluteus* with a similarly felty, rose-tinged stem. The cystidia are simple clavate. The spores measure (6–)6.5–8.5(–9) × 5.5–7 µm. Occurs on wood (mostly buried) in calcareous *Abies* forests and plantations.

This species is closely related to *P. leoninus* △ but has completely different colours and a different ecology.

Very rare in the Nordic countries, more common near the Alps, not recorded in the Netherlands or the British Isles; mainly July–October.

Pluteus hispidulus is a very small *Pluteus* with conspicuous, grey-brown hair-like scales on a paler background. The stem is white and finely floccose towards the base. The cap cuticle has horizontal to upright chains of elongated cells. The cheilocystidia are ± clavate, sometimes slightly capitate. The spores measure (5.5–)6–8(–8.5) × (4–)5–6 µm. Occurs on mull soils or very decomposed deciduous wood.

The very similar *P. exiguus* ⋉ has strikingly long end-cells in the cap cuticle hairs and ± pointed cystidia.

Widespread, but rather rare or perhaps somewhat overlooked; mainly June–October.

Thomas Læssøe

Pluteus semibulbosus is a rather small, cream to pale grey-brown, convex *Pluteus* with a felty cap centre and a distinct, pale, downy bulb at the base of the stem. The cap is either non-striate or striate halfway towards the centre. The cap cuticle has horizontal to erect, elongated elements. The cystidia are simple clavate. The spores measure 5–8.5 (–9.5) × 4–7(– 7.5) µm. Occurs on decomposing wood from deciduous trees, mostly in sites with many other, rarer species.

This species belongs to an extremely difficult species complex, which is sometimes treated as one very variable species. *Pluteus plautus* ⋉ is a larger, darker member of the same group; it may recall *P. umbrosus* ◁◁ but does not have a coloured gill edge. The white form of *P. umbrosus* may recall both *P. plautus* and *P. semibulbous*. *Pluteus hiatulus* ⋉ is translucently striate to the cap centre.

Widespread, occasional, decreasing towards the north and absent from the boreal zone; mainly June–October.

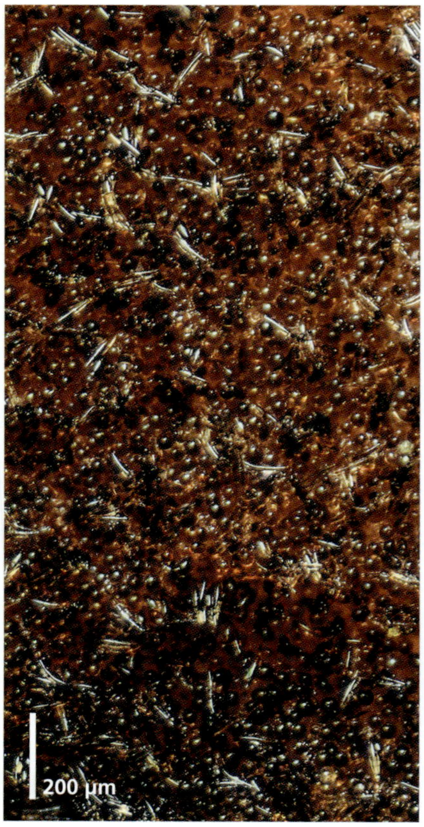

Pluteus podospileus is a very dark brown, small to rather small, slightly velvety-felty *Pluteus* with similar brown tufts on the pale stem; even the gill edge can have a brown tinge. The cap cuticle is both cellular and ± erect hairy, and, microscopically, both banana- and pear-shaped cells with brown contents are evident (picture to the right). The cystidia are simple clavate. The spores measure 5.5–7.5(–8) × (4–)4.5–6 µm. Occurs on mull soils, mostly in disturbed sites along roads *etc.* or on rotten wood.

Specimens with brown gill edges could be confused with *P. pusillulus* ⋉, but the cap cuticle of that species lacks pear-shaped elements and the long elements are longer.

Widespread, rather common to ± absent from the boreal zone; June–November.

Pluteus thomsonii is a small to rather small, grey-stemmed *Pluteus* that has a very ridged-veined, dark brown cap surface with a cellular cuticle of pear-shaped cells. The free gills are white to pink with concolorous edges. The cheilocystidia are clavate, but have a long thread-like extension. The spores measure (5.5–)6–8(–9) × 5–7(–7.5) µm. Occurs on very rotten wood from deciduous trees or, rarely, directly on soil.

Pluteus insidiosus ⋉ has the same thin extension to the cystidia but does not have a veined pattern on the cap surface. *Pluteus phlebophorus* ▷ is paler and the cystidia lack an extension.

Widespread, fairly common to ± absent from the boreal zone; June–November.

Thomas Læssøe

Pluteus phlebophorus is a medium-sized, yellow-brown to pale reddish-brown, faintly striate, finely pruinose *Pluteus* with a ± distinct pattern of veins on the cap surface. The crowded gills become pinkish at maturity and may have a brown-tinged edge. The cystidia are clavate and the cap cuticle consists of a palisade of crowded, pear-shaped cells. The spores measure (5.5–)7–8(–9.5) × (4.5–)5–7 µm. Occurs typically on large, very rotten trunks, often of *Fagus*.

Pluteus pallescens ⋉ has a distinctly translucently striate cap. *Pluteus luctuosus* ⋉ can, apparently, be recognized by its brown gill edge, but *P. phlebophorus* would appear to have the same character, at least in some fruitbodies.

Widespread and rather common to ± absent from the boreal zone; May–December.

Pluteus nanus is a small to medium-sized *Pluteus* with a smooth, matt, dark grey-brown, occasionally slightly veined cap and a whitish to greyish stem. The cap cuticle consists of pear-shaped cells. The spores measure (6.5–)7–9(–10) × 5.5–7 µm. Typically occurs on soil or wood chips, sometimes in very damp or wet sites.

Can be distinguished from similar species by the rather greyish, dark cap colours and by its more terrestrial habits; the cap surface changes colour when stroked. *Pluteus dietrichii* ⋉ has a distinctly cracking cap surface. *Pluteus cyanopus* ⋉ has faint bluish colours on the stem, but is perhaps just an aberrant form of *P. nanus*.

Widespread, fairly common to occasional; June–November.

Thomas Læssøe

Pluteus romellii is a small to medium-sized *Pluteus* with a pale yellow-brown, finely pruinose cap surface and a bright yellow stem, especially towards the base. The cellular cap cuticle consists of crowded, pear-shaped cells. The spores measure 6–7.5(–8) × (4.5–)5.5–6.5(–7) µm. Typically occurs on wood chips or soil mixed with wood chips, rarely on trunks or branches, mostly in mull-rich places or on calcareous clay.

Pluteus chrysophlebius ▽ has a golden to greenish-yellow cap and the stem is less bright yellow.

Widespread and fairly common but rather rare in the boreal zone; mainly April–November.

Pluteus chrysophlebius is a rather small to medium-sized *Pluteus*. The cap colour is initially a vivid red-brown but soon becomes golden to greenish-yellow. The stem is tinted the same colour as the cap. The cap cuticle is cellular. The spores measure (5.5–) 6.5–7(–7.5) × (4.5–)5–6(–6.5) µm. Occurs on wood from deciduous trees, often very decomposed large boles in sites with a long continuity.

Also known as *P. chrysophaeus*. *Pluteus romellii* △ has a shiny yellow stem and a brown cap. *Pluteus leoninus* ◁◁ has a felty cap with long elements in the cuticle.

Widespread and rare, and absent from the boreal zone; June–November.

Claes Ingvert

Pluteus aurantiorugosus is a small to medium-sized *Pluteus* with a smooth to finely wrinkled, very vivid orange-red cap and crowded, at first white, later pinkish, gills. The stem is yellow with small ± orange-red dots. The cap turns a duller yellow-brown in older specimens. The cellular cap cuticle consists of pear-shaped cells in a palisade. The spores measure (5–)5.5–7(–8) × (3.5–)4–5.5(–6) µm. Occurs on deciduous wood, *e.g.* of *Fraxinus*, *Ulmus*, *Salix* and *Populus*, mostly in cavities in large fallen or standing dead trees, mainly on rich soils.

Could be mistaken for a *Hygrocybe* (page 142) growing in an unusual habitat but if picked is unmistakable due to its free, crowded gills that soon turn pinkish.

Found mainly in the nemoral zones with a few scattered records from farther north, absent from the boreal zone; mainly August–October.

Thomas Læssøe

Agaricus and *Allopsalliota*

The genera *Agaricus* and *Allopsalliota* are characterized by free, permanent, non-dissolving gills and dark chocolate-brown spore-deposits. Ink caps (*Coprinus*, *Coprinopsis* and *Coprinellus*, page 520) also have ± free gills and ± black spores, but their gills normally dissolve with maturiy.

Molecular phylogenetic studies have led to descriptions of a large number of *Agaricus* species that cannot be recognized by morphological characters. Using traditional morphology, only 40+ broadly defined species can be keyed out in temperate Europe.

Agaricus and *Allopsalliota* are decomposers that often occur in fairy rings.

Many species are edible but the slowly yellowing species (the royals) may contain high levels of cadmium, and the yellow stainers can cause stomach upsets. In addition, most species contain potentially carcinogenic hydrazines (agaritin). Species of *Agaricus* should therefore only be consumed in fairly small amounts and then only when cooked.

OTHER SIMILAR FUNGI:
- *Stropharia coronilla* recalls a small, yellowish *Agaricus* but has narrowly adnate gills (page 590).
- annulate species of *Psathyrella* have narrowly adnate gills (page 546).
- similar lepiotoids have white spore-deposits (page 320).
- the very poisonous *Amanita virosa* has a membranous universal veil and a white spore-deposit, as do other species of *Amanita* (page 357).

FURTHER READING: 91, 147, 148, 156, 182, 226, 236, 237.

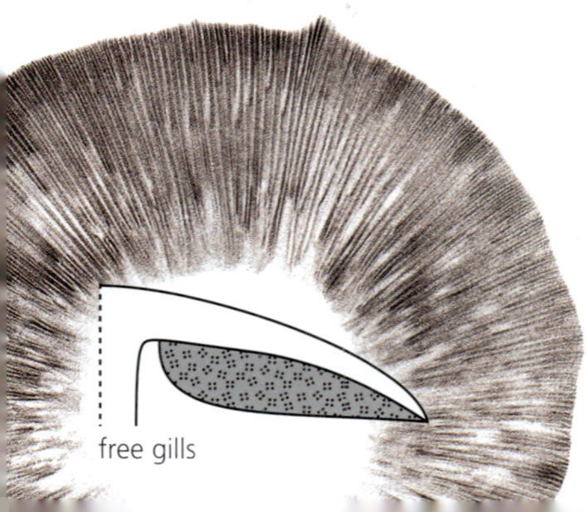

royals
4+ species
page 503
Micro-drawings:
spores and
cheilocystidia.
Approximate species
number applies to
temperate Europe.
ring thick, double;
fruitbodies rather large;
spores 5–12 µm long
dwarfs
4+ species, page 506
ring slender;
fruitbodies small;
spores 4.5–6.5 µm long
yellow stainers
3+ species, page 508
slowly yellowing;
smell of aniseed or
almonds
Agaricus **and**
Allopsalliota
have free gills that
are pale grey or pink
at first but quickly
become dark brown
immediately yellowing;
smell of carbol
(ink, phenols)
± brown or red-staining;
smell faint, sour or
unpleasant
red stainers
18+ species, page 510

Thomas Læssøe

Allopsalliota geesterani is a large, clustered agaric, which push the black soil up like a molehill. When bruised, the surface instantly turns yellow, but soon changes to red and then black. The smell is insignificant. The cheilocystidia are clavate. The brown spores measure 7–10 × 4.5–6 µm and have a lens-shaped thickening at the end but no germ pore. Occurs in open habitats, mostly near *Populus* on ± sandy soils.

The colour reactions and the very heavy, clustered fruitbodies are unique to this species.

Originally described from land reclaimed from the sea (polders) in the Netherlands, but since then has been reported from a number of other countries, including Great Britain and Germany; late summer–autumn.

Thomas Læssøe

Agaricus has free gills, a partial veil as stem rings and chocolate-brown spore-deposits. Genera such as *Coprinus*, *Coprinopsis* and *Coprinellus* share some of the characters but their gills ± dissolve at maturity (see page 520).

The royals, including "the prince" (this page), are recognized by their fleshy fruitbodies, aniseed-like smells and a slowly yellowing reaction when bruised.

•• ***Agaricus augustus*** is a very large, brown, scaly, slowly yellowing royal with an aniseed-like smell, a hanging and upwards-sheathing cogwheel ring and a ± white, clavate, scaly stem. The cheilocystidia are septate. The spores measure 7–10 × 5–6 µm and lack a germ pore. Occurs on rich soil in gardens, parks, deciduous and coniferous forests.

May superficially recall a red staining species ▷▷ but these are smaller and sour-smelling.

Rather common and widespread, becoming scarcer towards the north; mainly June–November.

Jens Maarbjerg

•• ***Agaricus crocodilinus*** is a very large, whitish, yellowing royal with an aniseed-like smell, an upwards-sheathing cogwheel ring and a white, clavate, scaly stem. The cheilocystidia are simple clavate. The spores measure 9–12 × 5.5–7 µm and lack a germ pore. Occurs on rich soil in scrub, forests and in ± open habitats.

May be confused with the somewhat smaller *A. arvensis* ▷, which has smaller spores, and with the much smaller, long-stemmed and small-spored *A. sylvicola* ▽ and *A. essetii* ⋉.

Widespread and fairly common, becoming scarcer towards the north; May–November.

•• ***Agaricus sylvicola*** is a medium-sized, whitish, yellowing royal with an aniseed-like smell, an upwards-sheathing cogwheel ring and a relatively long stem with a small bulb at the base. The spores measure 5–6 × 3–4 µm and lack a germ pore. Occurs on litter, particularly in coniferous forests.

The very similar *A. essettii* ⋉ has a marginate bulb with a flattened base and larger spores (6-8 × 4–5 µm), and *A. arvensis* ▷ has a relatively shorter stem that lacks a bulb, and even larger spores.

Widespread and common; May–December.

●● ***Agaricus arvensis*** is a large, whitish, slowly yellowing royal with an aniseed-like smell, an upwards-sheathing cogwheel ring and a mostly clavate, somewhat scaly, whitish stem. The cheilocystidia are broadly clavate. The spores measure 5–9 × 4.5–5.5 µm and lack a germ pore.

Molecular data have shown that a complex of species is involved that cannot be identified using classic, morphological characters. *Agaricus arvensis* may, in a broad sense, be confused with the very fleshy and large-spored *A. crocodilinus* ◁ and the longer-stemmed, small-spored *A. sylvicola* ◁ and *A. essettii* ⋉. *Agaricus osecanus* is treated here as a synonym of *A. arvensis*. The poisonous *A. xanthoderma* ▷▷ is rather similar, but stains yellow instantly when the cap margin is bruised or if the stem base is cut; it also has an ink-like smell.

Widespread and common; May–December.

The dwarfs are small, slowly yellowing *Agaricus* species with aniseed-like smells and often purplish cap colours.

• ***Agaricus dulcidulus*** is a dwarf, with yellowing flesh and an aniseed-like smell. The cap fibres are ± distinctly tinged wine-red, especially near the cap centre. The stem ring is rather narrow, thin and hanging. The spores measure 5–6 × 3.5–4 µm. Occurs on soil and litter, mostly on disturbed mull soils along road verges in forests, but also as fairy rings on needle debris and in lawns.

Here treated as a broadly defined species that is known to include a number of molecularly defined species. It has also been named *A. semotus*. *Agaricus brunneolus* ▷ has similar colours but is more fleshy.

Widespread and rather common, becoming scarcer towards the north; June–December.

Jan Vesterholt

Jens H. Petersen/Christian Lange

•• ***Agaricus brunneolus*** is a giant among the dwarfs, with a relatively compact and fleshy fruitbody, purple, hair-like scales on the cap and a thin, hanging, ± white stem ring. The flesh smells aniseed-like and yellows, particularly at the stem base, which is clavate and typically about 10 mm wide. The spores measure 5–6 × 3.5–4 µm. Occurs on rich, often clay soils in both deciduous and coniferous forests.

Agaricus dulcidulus ◁ is more slender, and usually lacks a distinctly clavate stem.

Widespread but occasional to rather rare; June–November.

• ***Agaricus lutosus*** is a small and slender, rather short-stemmed, yellowing, aniseed-scented dwarf with yellow-brown colours, mostly towards the cap centre, and a small stem ring. It normally has distinctly differentiated cheilocystidia that may be mixed with basidia. The spores measure 4.5–6.5 × 4–5 µm. Found in a range of open and forested habitats.

Agaricus dulcidulus ◁ has distinct purple fibres on the cap surface. *Agaricus comtulus* ⋉ normally has no cheilocystidia and is typically even paler.

Widespread but rather rare or perhaps overlooked, absent from the boreal zone; mainly July–October.

Thomas Læssøe

The yellow stainers have an unpleasant, ink-like (carbolic) smell. They typically stain immediately yellow and then greyish.

† ***Agaricus xanthodermus*** is a whitish to pale brown-grey, rather large *Agaricus* (a yellow stainer) with a double ring and a smooth stem that is widened at the base. When young, the cap is chalk-white but stains grey and cracks with age and when exposed to the sun (top picture). When bruised at the cap margin or cut at the stem base it instantly turns bright yellow. The smell is ink-like, especially evident when heated, and the taste is similar. The spores measure 4.5–6.5 × 3.5–5.5 µm. Occurs on rich, ± disturbed soils in gardens, parks and churchyards, rarely in more natural settings.

The instant yellow discolouration is the best separating character from the royals ◁◁. Other yellow stainers have darker, distinctly scaly caps.

Widespread, locally very common, absent from the boreal zone; May–January.

Jens Maarbjerg

Rasmus Ejrnæs

† ***Agaricus moelleri*** is a medium-sized *Agaricus* (a yellow stainer) with dark greyish to brown-grey, adpressed, fine scales on the cap surface. The smooth, white stem is widened at the base and has a double ring. The surface and the flesh stains yellow immediately when bruised. The odour is ink-like. The spores measure 4.5–6.5 × 3–4 µm. Occurs on ± calcareous, light soils in forest edges, scrub and parks.

Agaricus phaeolepidotus ▽ also has a somewhat ink-like smell, but the cap scales are brown and it yellows less strongly. *Agaricus xanthodermus* ◁ is initially chalk-white but usually turns grey and cracks when exposed to the sun; it never has minute coloured scales on the cap surface.

Widespread but rather local in the nemora-hemiboreall zones, absent from the boreal zone; June–November.

† ***Agaricus phaeolepidotus*** is a medium-sized *Agaricus* (a yellow stainer) with small, adpressed, hair-like, pinkish-buff to grey-brown cap scales. The smooth stem has a large double ring. The stem base especially stains somewhat yellow when bruised or cut. The stem surface may turn brown or redden somewhat. The odour is ink-like. The spores measure 4.5–6 × 3–4 µm. Occurs on calcareous, light soils in gardens, scrub and forests.

Agaricus moelleri △ has grey cap scales and the yellow stain is more pronounced. *Agaricus impudicus* ▷▷ does not stain yellow.

Local and rare, absent from the hemiboreal–boreal zones; July–November.

The red stainers are a group of *Agaricus* species recognized by their ± red-staining flesh and neutral, sour to fishy smells.

• ***Agaricus campestris*** is a rather small to medium-sized, typically white to pinkish-tinted, finely scaly *Agaricus* with free gills that are at first rose but later dark red-brown. The stem tapers towards the base and lacks a hyphal cord. The stem ring is thin and narrow and disappears relatively quickly; it is upwards-sheathing. There may be distinct girdles of scales below the ring. The flesh reddens somewhat and smells ± sour (like commercial mushrooms). Lacks cheilocystidia. The spores measure 6–9.5 × 4.5–7 µm. Occurs mostly in grass on open land.

Agaricus altipes ▷ has a non-tapering, relatively longer stem and somewhat smaller spores. *Agaricus moellerianus* ⋉ is very similar, but has somewhat smaller spores and yellows slightly. *Agaricus devoniensis* ▷ has a downwards-sheathing ring.

Common and widespread; May–December.

• ***Agaricus altipes*** is medium-sized, whitish, faint-smelling *Agaricus* with somewhat reddening/browning flesh, an upwards-sheathing ring and a rather long, clavate stem. Lacks cheilocystidia. The spores measure 7.5–8 × 5–5.5 µm. Occurs on mull soils in scrub and forests.

The very similar *A. campestris* ◁ also lacks cheilocystidia but normally has a more tapering stem, and fruiting peaks later in the season.

Widespread but rather rare, probably absent from the boreal zone; May–October, with an early peak.

Jens Maarbjerg

Christian Lange

• ***Agaricus devoniensis*** is a rather small, whitish *Agaricus* with free gills that are at first rose but later dark red-brown. The stem is ± cylindrical and lacks a hyphal cord. The ring is thin and narrow, disappears relatively quickly, and is downwards-sheathing. Has distinct girdles below the ring. The flesh has a faint smell and reddens somewhat. Has cheilocystidia. The spores measure 6–7 × 5–5.5 µm. More or less confined to coastal dunes.

Agaricus campestris ◁ is normally more fleshy and has an upwards-sheathing ring.

Widespread but uncommon; mainly September–December.

Thomas Læssøe

• ***Agaricus bitorquis*** is a medium-sized, fleshy, sordid whitish *Agaricus* with an early expanding cap, a distinct double ring that has an upwards-sheathing and a downwards-sheathing part, and somewhat reddening/browning flesh with a ± sour, urine-like smell. The young gills are grey-rose but turn purple-brown at maturity. The cap surface may stain sordid yellow in spots. Has 4-spored basidia and broadly clavate cheilocystidia. The spores measure 5.5–6.5 × 4.5–5 µm. Mostly found at town roadsides and may even expand below asphalt and break through.

The double ring and early expanding cap is also a character of the brown-scaled *A. subperonatus* ▷▷.

Widespread and common, absent or nearly so from the boreal zone; May–December.

Jens Maarbjerg

• ***Agaricus bernardii*** is a large to very large, compact, short-stemmed *Agaricus* with a cracking cap surface, strongly reddening flesh, a downwards-sheathing ring and an unpleasant sour smell. The young gills are grey-rose. Has irregular, narrowly clavate cheilocystidia. The spores measure 5.5–7.5 × 4.5–6.5 µm. Occurs on salt-exposed soils in coastal grasslands or along salted roads and tracks, also in churchyards.

The white cap, very thick, hard, reddening flesh, downwards-sheathing ring and habitat make this species unmistakable. *Agaricus subfloccosus* ▷▷ is a somewhat more gracile, brown-grey version of *A. bernardii*; it occurs with conifers.

Widespread and rather common to occasional, apparently nearly absent from the boreal zone; May–December.

• ***Agaricus benesii*** is a medium-sized, squamulose, white but very reddening *Agaricus* with an upwards-sheathing ring, a rather long, clavate stem and a faint smell. The cheilocystidia are narrowly clavate. The spores measure 5–7 × 3.5–5 µm. Occurs on rich, often calcareous soils in deciduous and coniferous forests.

Other whitish but reddening *Agaricus* species are not minutely scaly, are more squat, smell mostly sour-unpleasant and usually occur in open land. Other s of similar stature are brown-scaled ▷▷.

Widespread but rare, absent from the boreal zone; August–November.

Jens Maarbjerg

• ***Agaricus subfloccosus*** is a medium-sized, non-rooting, brown, appressed-scaly *Agricus* with ± reddening, sour smelling flesh and a thick double ring with an upwards-sheathing and a downwards-sheathing part. The basidia are 4-spored. The spores measure 5.5–7.5 × 4–6 µm. Occurs in parks, gardens and deciduous forests, on rich, ± disturbed soils.

Agaricus subperonatus ▷▷ is typically more fleshy and has a ± rooting stem.

Found mostly in the nemoral zones and rather rare; mainly May–October.

• ***Agaricus bisporus*** is a rather small to medium-sized, whitish or pale brown, minutely scaly, faint-smelling *Agaricus* with ± reddening flesh, a clavate stem and a complex ring that may be sheathing both upwards and downwards. The basidia are 2-spored. The spores measure 6.5–7.5 × 5–6 µm. Occurs on disturbed soil along roads, on compost, *etc.* Found almost exclusively in highly disturbed urban sites.

The complex stem ring separates this species from many other red staining species. If in doubt, the 2-spored basidia are diagnostic.

Widespread and fairly common; April–December.

• ***Agaricus litoralis*** is a whitish to brown, scaly, rather small and compact *Agaricus* with a thick, white hyphal cord at the stem base and a somewhat reddening reaction in the stem flesh. The ring is upwards-sheathing. Does not have a distinct smell. Has cheilocystidia but these are mixed with basidia. The spores measure 6–8.5 × 4.5–6 µm and lack a germ pore. Occurs in meadows and dunes, mostly on somewhat salt-influenced sandy soils near the coast.

Other relatively pale species of *Agaricus* with an upwards-sheathing ring do not have a distinct hyphal cord. *Agaricus lanipes* ▷▷ has an obvious hyphal cord but is darker brown and occurs in woodlands. *Agaricus devoniensis* ◁◁ is white, lacks a hyphal cord and has a downwards-sheathing ring.

Widespread and occasional, absent from the boreal zone; May–November.

• ***Agaricus cupreobrunneus*** is a ± red-brown, appressed scaly/hairy, rather small, compact *Agaricus* with a floccose, pale cap margin and a somewhat tapering stem that lacks a hyphal cord. The stem ring is white, floccose and upwards-sheathing and the flesh is slightly reddening. The odour is ± sour. The spores measure 7–10 × (4–)5–6 µm and have a germ pore. Occurs in dry grassland, including dunes.

Agaricus lividonitidus ⋉ has smaller spores.

Widespread, but mainly coastal, occasional to rather rare; August–November.

Jan Vesterholt

Thomas Læssøe

• ***Agaricus subperonatus*** is a solitary or ± clustered, fleshy *Agaricus* with brown, adpressed scales over the pale flesh and an early expanding cap. The wide, mostly ± buried, cylindrical or tapering stem has a thick double ring that is striate above, and both upwards- and a downwards-sheathing. The stem has ± distinct girdles below the ring. The thick flesh stains grey-rose when cut and has a sour smell. Has cheilocystidia and 4-spored basidia. The spores measure 6.5–7.5 × 5–6.5 µm. The fruitbodies mostly break through from deep down on rich, disturbed, ± calcareous soils along roadsides, in parks and rarely in more natural settings.

Agaricus bohusii ▷ is very similar but has recurved, pointed cap scales. *Agaricus subfloccosus* ◁◁ is less fleshy, non-rooting and typically occurs with conifers. *Agaricus bisporus* ◁◁ is smaller and has 2-spored basidia. *Agaricus bitorquis* ◁◁ occurs in the same environments but is much paler.

Widespread, occasional, absent from the boreal zone; April–January.

• ***Agaricus bohusii*** is a distinctly clustered and rooting species with a brown, densely scaly cap surface. The scales are ± pointed and recurved. The tapering and strongly rooting stem has brownish tinges and a thick, upwards-sheathing ring. The flesh reddens to some extent but may also stain somewhat yellow. Has cheilocystidia and 4-spored basidia. The spores measure 5.5–7 × 4.5–6 µm. Occurs on disturbed soil in parks, churchyards, *etc.*

Agaricus subperonatus ◁ is similar, but the the cap scales are appressed and not recurved.

Local and rare, absent from the boreal zone; mainly June–November.

Thomas Læssøe/Christian Lange

• **Agaricus lanipes** is a medium-sized, compact, short-stemmed *Agaricus* with a thick, white, strong hyphal cord at the base of the stem. The brown cap is appressed-scaly and the stem has brown, woolly girdles below the rather narrow, upwards-sheathing ring. The flesh is moderately reddening and lacks a distinct smell. The cheilocystida are globose to pear-shaped. The spores measure 6–7.5 × 4–5 µm. Occurs on rich mull soils, mostly in deciduous or mixed forests.

The brown colours combined with the thick hyphal cord make identification relatively straightforward. *Agaricus impudicus* ▽ may have hyphal cords but they are less prominent and it typically has a long, whitish stem that lacks distinct woolly girdles.

Widespread in the nemoral zones, but rather rare; mainly August–October.

• **Agaricus impudicus** is a slender, appressed, brown-scaled *Agaricus*, which does not, or only slightly, reddens in the flesh. The floccose stem has a thin, hanging cogwheel ring with a ± brownish underside and is widened at the base. The odour is somewhat metallic. Has cheilocystidia. The spores measure 5–7 × 3.5–4.5 µm. Occurs on litter in scrub and forests on mull, often calcareous, soils.

Agaricus sylvaticus ▷ is similar in habit, but reddens strongly. *Agaricus lanipes* △ has a shorter, woolly stem and the flesh reddens somewhat. *Agaricus phaeolepidotus* ◁◁ stains ± yellow in the stem base and smells ink-like.

Widespread and ± common in the nemoral zones; June–January.

Jens Maarbjerg

• ***Agaricus langei*** is a rather large, appressed, buffish, scaly *Agaricus* with strongly reddening flesh and cap surface when bruised. The stem is ± smooth and bears a large, thin ring; the base is widened. The cheilocystidia are almost balloon-shaped, one-celled, and 10–30 µm wide. The spores measure 5–9 × 3.5–6 µm (average 7 × 4 µm). Occurs on mull soils, mainly in deciduous forests.

Agaricus sylvaticus ▽ has a floccose stem beneath the ring, has on average smaller spores and has cheilocystidia with a different shape.

Widespread and common; April, June–January.

• ***Agaricus sylvaticus*** is a rather slender-stemmed, buffish, scaly, strongly reddening *Agaricus*. The cheilocystidia are narrowly clavate, several-celled and 7–12(–15) µm wide. The spores measure 4.5–6.5 × 3–4.5 µm (average 5.5 × 3.5 µm). Occurs on litter, mostly in conifer plantations on mull soils.

Agaricus langei △ has a ± smooth sten below the ring, on average larger spores and one-celled, ± balloon-shaped cheilocystidia. *Agaricus impudicus* ◁ is very similar, but does not redden or only reddens slightly.

Widespread and common; July–December.

Coprinoids

The coprinoids include the genera *Coprinus*, *Coprinellus*, *Coprinopsis* and *Parasola*, of which there are are more than 150 species in temperate Europe. They are recognized by the black spores and the almost always free gills. Many coprinoids also have a unique maturation process. The spores mature from the gill edges inwards and, as soon as the spores are released, that part of the gill disintegrates and becomes liquid, and so the process continues. This allows for very closely spaced gills, since the risk of spores being caught between them is reduced. Nevertheless, many spores are still caught and produce the inky liquid associated with the vernacular name ink cap. The coprinoids can be grouped according to the appearance of the surface of the cap and veil, *e.g.* whether the veil is grainy or thread-like. These characters are borderline between macro and micro, and a good hand lens is required. The coprinoids may produce many different spore types, from smooth and ellipsoid to lemon-shaped, heart-shaped or truncate and they may have prominent warts. They always have a germ pore and occasionally a loosening, colourless outer wall. All have cystidia.

The phylogenetic and morphological differences between psathyrelloid (page 546) and coprinoid fungi are not straightforward. Some members of *e.g. Coprinopsis* do not have gills that dissolve and are treated among the psathyrelloids.

All coprinoids are decomposers and some coprophilous species have a very short life-cycle, fruiting just a few weeks after spore germination. All fruitbodies are very short-lived. At least some of the coprilous species are able to kill competing mycelia in the substrate.

Other similar fungi:
- *Psathyrella* and the like mostly have adnate gills that never dissolve during maturation (page 546).
- *Bolbitius* are slimy-capped and have brown spore-deposits (page 618).

Further reading: 156, 182, 207, 208, 227, 290, 322.

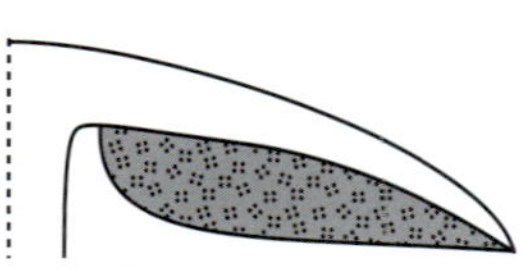

Coprinopsis nivea group
page 532
small to medium; cap powdery from globose veil cells
Coprinellus micaceus group,
page 537
Micro-drawings: spores first, thereafter cystidia. ve. = hyphae of the universal veil.
Approximate species number applies to temperate Europe.
Coprinellus
50+ species
page 537
cap ± orange-brown, with thread-like and spherical remnants of veil
Coprinellus disseminatus group
page 540
cap ± orange-brown, hairy from projecting cystidia (pileocystidia)
Coprinopsis
Coprinellus
Coprinellus
Parasola
ve.
coprinoids
have free gills
hat dissolve with
ge; the spores are
black
Parasola
13+ species
page 544
other similar fungi
cap smooth, mostly ± plicate; gills very free and very distant
spores
dark brown to black
spores
brown
Psathyrella
and the like
page 546
little brown mushrooms (LBMs)
page 616

Coprinus is a small genus of large coprinoids with floccose-scaly, whitish caps and dissolving gills. The spores are large and have an excentric germ pore. The closely related *Montagnea* (Vol. 2, p. 1247) has powdery spores.

• **Coprinus comatus** △ is a large, almost fleshy, coarsely fibrillose, recurved scaly, almost white *Coprinus* with an evident ring and a long, white, hollow stem. The cap centre may be somewhat cream. The gills are very crowded and may be somewhat reddish when young, but later dissolve into a black inky mass. The spores mature from the margins of the gills. The spores measure 10–13 × 6.5–8 µm; germ pore ± excentric. Typically in troops on new lawns, at roadsides, *etc.;* thrives only on disturbed soil.

The tall, white, coarsely fibrillose-scaly cap is unique. *Coprinus sterquilinus* ▽ is much smaller and occurs on dung. Coprinoids in the genera *Coprinopsis* ▷ and *Coprinellus* ▷▷ typically have smooth caps, although mostly covered with remnants of veil. There is some variation in *C. comatus*, *e.g.* in cap shape, with some forms almost egg-shaped. This form, or perhaps a separate species, tends to occur early in the season.

Widespread and common; mostly August–October, but may also fruit May–July and into November.

Coprinus sterquilinus ◁ is a large coprinoid, but a small *Coprinus*, with a ± pointed-scaly, almost white cap and an evident, but easily lost, narrow ring. The spores are very large, 17–26 × 10–15 µm; germ pore ± excentric. The flesh and gills redden in young specimens. Occurs on horse dung and middens, *e.g.* in grassland.

It should not be possible to confuse mature specimens of this species with *C. comatus* △ or other coprinoids, but when young it may recall *e.g. Coprinopsis cothurnata* ▷▷.

Widespread, but rare, probably absent from the boreal zone; mainly May–September.

Jens Maarbjerg

Coprinopsis includes white, greyish to grey-brown, smooth-capped coprinoids with thread-like or granular remnants of veil.

Coprinopsis picacea is a large, spectacular, mottled coprinoid with persistent adpressed, pale brown, grey-rose to almost white veil scales on a grey-brown to grey cap surface. The gills are very crowded and dissolve completely with age. Young fruitbodies in particular smell strongly fetid. The stem is white, finely floccose and lacks a ring. The spores are ellipsoid and measure 14–19 × 9.5–13 µm. Occurs mostly individually or in small loose groups on litter and wood chips in rich *Fagus* forests.

Other ± mottled coprinoids, *e.g. C. stangliana* ⋉, lack the fetid smell and typically occur in different habitats or grow in clusters. *Coprinopsis episcopalis* ⋉ is similar in ecology and size but has almost square spores with a protruding germ pore.

Widespread and common in the nemoral zones; mostly September–November.

Coprinopsis strossmayeri is a large, clustered, trooping coprinoid that has a narrow cap with prominent, pale, rather large veil scales. The spores measure 7–9 × 4.5–6 µm. The fruitbodies are connected to thick, brown hyphal cords in the substrate, which mostly consists of wood chips.

Coprinopsis atramentaria ▷▷ has less evident veil and larger fruitbodies. Species around *Coprinellus domesticus* ▷▷ have smaller and more coloured remnants of veil.

Very rare and local; mostly found during the summer.

David Boertmann

Jens Maarbjerg

† ***Coprinopsis atramentaria*** is a very fleshy, clustered, rounded-conical coprinoid that typically is almost pure grey. The veil remnants at the cap centre may be tinted dark grey-brown, and similar remnants may be seen on the lower part of the stem. The gills are greyish and extremely crowded. The spores are smooth and measure 8–11 × 5–6 µm. Occurs mostly in clusters on soil, but always grows from buried wood; often seen at the base of damaged trees in urban environments, but also in forests and parks.

Coprinopsis romagnesiana ▷ is a very similar sister species with orange-brown to red-brown veil scales. *Coprinopsis acuminata* ▷ has a more acute, slender cap with a ± smooth centre and narrower spores. Also *C. insignis* ▷▷ can be surprisingly similar

Widespread and common; April–December.

Coprinopsis atramentaria and allies contain coprin, a compound similar to disulfiram (antabus), which, if consumed with alcohol, causes reddening, nausea and raised blood pressure.

†? *Coprinopsis acuminata* is a grey, fleshy, large *Coprinopsis*, which typically has a rounded, ± smooth umbo on top of the rather narrow, conical cap. The thread-like veil is almost invisible. The spores measure 8–10.5 × 4–5 µm. Occurs mostly in troops on damp, clay-rich or mull soils in forests and scrub.

Coprinopsis atramentaria ◁ and *C. romagnesiana* ▽ are very similar but both have spores that are wider than 5 µm and typically somewhat wider caps without a distinct, smooth umbo.

Widespread, occasional; May–November.

† *Coprinopsis romagnesiana* is a ± clustered, large, fleshy, orange-brown, scaly *Coprinopsis* with greyish, very crowded gills. The stem has a distinctly defined lower part that has similar orange-brown scales. The spores measure 8–9.5 × 4.5–5.5 µm. Occurs on and around stumps or dying trees, or growing from buried wood.

The very similar *C. atramentaria* ◁ lacks the orange-brown colours. *Coprinopsis romagnesiana* was formerly treated as a variety of *C. atramentaria*.

Widespread, occasional, becoming scarcer towards the north; May–November.

Thomas Læssøe

Martin Kříž

Coprinopsis insignis is a large, grey, solitary or clustered, lignicolous *Coprinopsis* with a thread-like veil. The amygdaliform spores are distinctly verrucose, measure 11–14.5 × 7–8 µm, and have a rather small, central germ pore. Occurs on deciduous wood, mostly *Fagus*.

Also known as *C. alopecia*. *Coprinopsis atramentaria* ◁◁ can be surprisingly similar but lacks the thread-like veil and has smooth spores. The verrucose spores of *C. insignis* are a good separating character from other large lignicolous coprinoids.

Widespread, but rare; April to November.

Coprinopsis jonesii is a sturdy, typically trooping coprinoid. The cap has a very obvious, grey-white thread-like veil of unbranched elements. The smooth spores measure 6.5–9.5 × 6–8 µm. Found predominantly on burnt ground but may also occur on disturbed soil, even in the basements of buildings.

Coprinopsis geesterani ⋉ has much smaller fruitbodies and narrower spores. *Coprinopsis lagopus* ▷ has longer spores. Other similar species ▷▷ are mostly associated with dung.

Widespread, occasional; April–October.

Coprinopsis lagopus forms rather ephemeral, thin-fleshed, but rather large coprinoid fruitbodies with plenty of whitish thread-like veil on the cap. The stem is downy-hairy and very fragile. The spores measure 10–14 × 6–8.5 µm. Occurs on litter, wood chips and in lawns.

This species belongs to a very difficult complex that is still not fully understood. The very gracile and short-lived *C. lagopus* var. *vacillans* ⋉ that fruits at night seems to be genetically similar to the more fleshy versions of *C. lagopus*. See also *C. jonesii* ◁, which has shorter and wider spores, and other similar species on the following page spread.

Widespread and common, becoming scarcer towards the north; May–November.

Coprinopsis radiata is a rather small, rooting coprinoid with a similar thread-like veil to *C. lagopus* ◁◁. The spores measure 13.5–15 × 7.5–8.5 µm. Occurs on dung and manure.

The similar *C. pseudoradiata* ⋉ has smaller spores that measure 7.5–9.5 × 5–5.5 µm. *Coprinopsis macrocephala* ▽ and *C. cinerea* ▷ are typically more robust and have spores that measure 12–17 × 8–10 µm and 8.5–12 × 6–8 µm respectively.

Widespread, occasional, becoming scarcer towards the north; April–October.

Coprinopsis macrocephala is a rather compact, ± rooting coprinoid. The veil is, as in other species in the *C. lagopus* group, mostly unbranched with constricted septa separating allantoid cells. The spores measure 12–17 × 8–10 µm, have rounded ends and a central germ pore that is 2 µm wide. Occurs on old straw bales, middens, *etc.*

Coprinopsis lagopus ◁◁ and *C. jonesii* ◁◁ do not occur on old straw or dung and have shorter spores. *Coprinopsis cinerea* ▷ has smaller spores.

Widespread, probably fairly common, becoming scarcer towards the north; may occur year-round except during frosty conditions.

Coprinopsis cinerea is a rather compact, ± rooting *Coprinopsis* with a thread-like, unbranched, whitish veil. The spores measure 8.5–12 × 6–8 µm and the germ pore is central. Occurs on compost and middens and can survive the high temperatures that arise during composting. It can complete its entire life-cycle in just two weeks and for this reason (and others) is used as a popular model organism by *e.g.* geneticists.

Coprinopsis macrocephala ◁ is very similar, both in appearance and ecology, but the spores are larger.

Widespread and probably common; may be found all year, except during periods of very low temperatures.

Coprinopsis erythrocephala is a medium-sized coprinoid with orange-red, thread-like remnants of veil on the cap and stem. The egg-shaped cap remains closed for a long time. The veil consists of simple, unbranched cells. The spores measure 8–13 × 5–7.5 µm and the germ pore is central. Typically found growing among small woody fragments on disturbed damp soils, but may also occur on compost or wood chips.

No other species of *Coprinopsis* have similar orange-red velar remnants.

Widespread but rather rare, absent from the boreal zone; June–November.

Coprinopsis urticicola is one of several small coprinoids that grow on damp, rotting herb stems. It is characterized by the colourless, branched, thin-walled veil and somewhat fusiform, not flattened, rather pale spores that measure 5.5–9 × 4.5–6.5 µm (Q 1.25–1.45). Occurs in damp places, swamps *etc*, on rotting, mostly monocot vegetation.

Similar species have either different spore shapes, occur on other substrates or have cells in the veil with walls thicker than 1 µm; some have distinctly brown-coloured veils.

Common and widespread, but infrequently recorded; a typical summer species, found mainly June–September.

Erik Arnfred Thomsen

Coprinopsis kubickae is a small, pale coprinoid with a brownish, branched, thin-walled veil. The almost spherical to broadly ellipsoid spores measure 7–11.5 × 6–10.5 µm and the germ pore is central to slightly excentric. Occurs mostly on very wet, rotten vegetation, often half buried in mud, *e.g.* on coarse grasses.

Similar species with thin-walled veils are either pure white, grow in other habitats or have a different spore shape.

Probably widespread, but poorly known; summer-fruiting, May–August.

Coprinopsis phaeosporus is a small coprinoid with small ± brownish veil scales. The branched veil has 1–2 µm thick walls. The flattened spores are almost spherical, as seen in front view, but narrower in side view, measure 5.5–8 × 4.5–7 × 4.5–6 µm and the germ pore is excentric. Occurs on stems in dense, damp herbaceous vegetation after summer rains.

Coprinopsis xantholepis ⋉ does not have flattened spores. *Coprinopsis urticicola* ◁ has thin-walled veil cells. *Coprinopsis friesii* ⋉ has thicker veil walls, more than 2 µm.

Widespread but infrequently recorded, probably overlooked; a summer species fruiting mainly June–September.

Jens Maarbjerg

Coprinopsis echinospora is a medium-sized, rather floccose coprinoid with a pale veil on a dark grey background. The veil consists of rather narrow, somewhat branching hyphae. The amygdaliform spores are strongly ornamented with irregular warts and measure 8.5–12 × 6–8 µm; the germ pore is wide and truncate. Occurs on ± urea-covered branches underneath bird colonies or at bird roosting sites.

Coprinopsis phlyctidosporus ⋉ and *C. rugosobisporus* ⋉ have similar spore ornaments and veils, but the spores are ovoid rather than amygdaliform. Other species of *Coprinopsis* with ornamented spores have different types of veil or are much larger, *e.g.* *C. insignis* ◁◁.

Local and uncommon, but probably also overlooked; May–November.

Thomas Læssøe

Coprinopsis stercorea is a small, greyish-white, odourless coprinoid with projecting spines and tufts of veil. The stem is minutely downy. The veil comprises spherical, finely echinulate cells connected by thin branching hyphae and appears thread-like in very young specimens, but more powdery in mature ones. The basidia are 4-spored. The cheilo- and pleurocystidia are ± broadly lageniform. The spores have a central germ pore, an indistinct loosening outer wall and measure 5.6–7.7 × 3.2–4.2 µm. Occurs on dung, particulary deer dung.

An untidy appearance and narrow spores are particulary characteristic of this species. *Coprinopsis foetidella* ⋉ has wider spores and greyer fruitbodies. The slightly larger *C. trispora* ▷▷ has, as the name implies, 3-spored basidia.

Widespread and common; most of the year, although mainly in the autumn.

Coprinopsis nivea is a rather large, ± white, finely powdery-grainy, ± thimble-shaped coprinoid. The lemon-shaped to ± hexagonal spores have a central germ pore and measure 14–18 × 11–13 µm. Occurs on dung, typically from horse or cow, mostly in dry grasslands.

Coprinopsis pseudonivea ▽ is more grey-brown and has smaller spores; the more untidy-looking *C. cothurnata* ▷▷ has smaller, hexagonal spores and typically a vinaceous-brown stem base.

Widespread and rather common; mainly July–November.

Thomas Læssøe

Coprinopsis pseudonivea is a rather large, clay-pink-tinted coprinoid with a powdery-floccose covering of veil. The basidia are 4-spored. The ± lemon-shaped, ± flattened spores have a truncate and central germ pore and measure 10–13 × 8.5–11 × 6–8 µm. Occurs mostly on cowpats in dry grasslands.

Coprinopsis nivea △ is typically more pure white and has larger spores. *Coprinopsis cothurnata* ▷▷ has ± hexagonal spores and a coloured stem base.

Probably widespread, but rarely recorded; mainly August–November.

Thomas Læssøe

Fred Stevens

Coprinopsis ephemeroides is a small to rather small, slender coprinoid that before opening has a powdery, yellow-brown cap; in older, expanded specimens the colour disappears. It can either have or lack a loose ring. The veil consists of ± spherical cells that lack warts. The spores are ± heart-shaped and measure 6–9 × 5.5–8 × 4.5–5 µm; the germ pore is central. Occurs on dung and compost.

Forms without a ring have been regarded as separate species (*i.e. C. cardiasporus*, *C. cordisporus* and *C. patouillardii*), but following molecular phylogenetic studies they have been lumped. Other species with spherical veil cells that lack warts either have larger fruitbodies, different spore shapes or grow on other substrates.

Widespread and fairly common; mainly May–November.

Thomas Læssøe

Coprinopsis trispora is a coarsely powdery-grainy, whitish, rather small coprinoid that usually has 3-spored basidia. The veil consists of warty, spherical cells connected by thinner hyphae. The spores have a partly loosening, colourless outer wall and measure 7–10 × 4.5–6 µm. Occurs predominantly on horse and cow dung.

Similar species have 2- or 4-spored basidia and mostly occur in other habitats; some have distinct fetid smells.

Rare or overlooked, apparently absent from the boreal zone; May–October.

Thomas Læssøe

Coprinopsis laanii is a rather small, grey-powdered, odourless coprinoid with 4-spored basidia and spherical, warty veil cells. The spores measure 10–11.5 × 5–6 µm and have a distinct, wrinkled, loosening, colourless outer wall. Occurs mostly on stumps from both coniferous and deciduous trees.

Similar species do not occur on stumps and have a less distinct outer spore wall.

Widespread, occasional, possibly more common in Atlantic areas, not recorded in the hemiboreal–boreal zones of Scandinavia; June–November.

Thomas Læssøe

Coprinopsis cothurnata is a rather large, powdery-grainy coprinoid that mostly occurs in rooting clusters. The stem base is usually tinted red-brown. The veil consists of smooth or granulate, spherical cells mixed with branching threads. The somewhat hexagonal spores measure 9.5–15.5 × 6.5–8.5 × 7–7.5 µm. Occurs on middens and compost.

Coprinopsis nivea ⊲⊲ and *C. pseudonivea* ⊲⊲ have wider spores and a somewhat different habit.

Rare, but possibly somewhat overlooked, absent from the boreal zone; mainly June–October.

Thomas Læssøe

Thomas Læssøe

Coprinopsis martinii is a medium-sized, powdery-greyish, somewhat rooting coprinoid with verrucose, spherical veil cells and 4-spored basidia. The spores have a moderately loosening and smooth, colourless outer wall and measure 13–17 × 7.5–8.5 µm; the ends are broadly rounded. Occurs amongst *Carex, etc.* in calcareous fens.

Coprinopsis martinii should probably be split into more than one species. *Coprinopsis semitalis* ⋉ looks very similar but has pointed spore bases. Other similar species have a strong smell and some are deeply rooting or have 2-spored basidia.

Widespread but rare; April–October.

Thomas Læssøe

Coprinopsis tuberosa is a small to medium-sized, powdery-greyish, deeply rooting coprinoid with large, black sclerotia. The odour is rather weak, somewhat like raw potato. The veil consists of spherical, verrucose cells. The basidia are 4-spored. The rounded spores measure 8–11 × 5–6.5 µm and have a more or less distinctly loosening outer wall. Occurs on middens or soil mixed with dung.

Coprinopsis radicans ⋉ and *C. narcotica* ⋉ have a very strong fetid smell and do not form sclerotia.

Widespread, rather rare, but probably also somewhat overlooked; May–October.

The coprinoid genus ***Coprinellus*** includes predominantly yellow-brown to red-brown coprinoids with smooth or hairy caps. The caps of some species have a grainy-threadlike to powdery veil, while others are completely naked except for protruding pileocystidia.

Coprinellus domesticus is a rather large coprinoid with obvious red-brown to whitish veil scales on the cap. The stem is white, widened at the base and often positioned on a vivid orange-yellow, thread-like mycelium, a so-called ozonium. The spores are somewhat bean-shaped and measure 6–9 × 3.5–5 µm. Occurs on wood or litter, mostly deciduous, typically on piles of brush wood or around larger stumps.

Coprinellus radians ⋉ and *C. xanthothrix* ⋉ also have an ozonium but the former has larger spores (8.5–11.5 × 5.5–6.5 µm) and the latter has ± ellipsoid spores that on average are wider than 5 µm. *Coprinellus micaceus* ▷▷ has flattened spores, smaller veil scales and normally no ozonium.

Widespread and rather common; April–November.

Thomas Læssøe

Erik Arnfred Thomsen/Thomas Læssøe

Jens Maarbjerg

Coprinellus micaceus is a yellow-brown, clustered, rather large coprinoid with small, pale, glistening, veil granules on the cap, and small hyaline hairs on the pale stem. The veil consists of ± spherical elements connected by small threads. Lacks pileocystidia. The spores are smooth and distinctly flattened, appearing narrow-ellipsoid from the side and egg-shaped to rhombic from the front; they measure 6.8–10.2 × 5–6.7 × 4.3–5.5 µm. The germ pore is rather large and central.

Unless checked microscopically, may be confused with species in the *C. domesticus* group ◁◁, but those species are usually positioned on a yellow, tufted mycelium (an ozonium). They also typically have larger, browner scales on the cap and are less hairy on the stem. *Coprinellus silvaticus* ▷ is similar but has fine hairs on the cap (pileocystidia) and rough spores. *Coprinellus saccharinus* ⋉ has a smooth stem.

Widespread and very common; mostly April–November.

Coprinellus flocculosus is a rather large, yellow-brown coprinoid, found in troops and small clusters, that has prominent, grainy-flaky veil patches which consist of chains of short, inflated cells. The stem is white and completely smooth. The basidia are 4-spored. The spores are slightly flattened and measure 11–15 × 7–9.5 × 6.5–7.5 µm, and the germ pore is excentric. Typically occurs on wood chips and similar substrates.

The large spores with excentric germ pores in combination with the veil type should make identification straightforward.

Widespread, occasional, becoming scarcer towards the north; April–November.

Erik Arnfred thomsen

Coprinellus silvaticus is a rather large, clustered coprinoid with minutely hairy pileocystidia (use a hand lens) and a finely grainy cap surface. The stem is whitish and also minutely hairy. The veil consists of ± spherical to pear-shaped cells. The basidia are 4-spored. The spores are amygdaliform to ovoid, finely verrucose and measure 10.2–15.0 × 7.2–10 µm. Occurs on and around hardwood stumps.

May resemble a slender version of *C. micaceus* ◁, but that species lacks pileocystidia and has smooth spores.

Widespread, occasional, absent from the boreal zone; April–November.

Jens Maarbjerg/Thomas Læssøe

Tobias Bøllingtoft

Coprinellus angulatus is a ± orange-brown, plicate coprinoid with small, projecting hairs in the cap cuticle (hand lens required) and no veil. The cap remains egg-shaped for a long time. The spores are flattened and, seen in front view, characteristically pentagonal with a wide germ pore; they measure 7.7–10.5 × 6–7.8 × 5–5.5 µm. Occurs on burnt ground and other burnt substrates.

The combination of cap colour, spore shape and habitat should make this species straightforward to identify.

Widespread, occasional to rare; April–November.

Coprinellus impatiens is a rather large, but very thin-fleshed, plicate, quickly applanate coprinoid with small, projecting hairs in the cap cuticle (hand lens required) and no veil. The pileocystidia are hyaline with pointed necks. The spores measure 10.6–11.3 × 6.2–7.6 µm and the germ pore is excentric. Found mainly on leaf-litter on rich soil in deciduous forests, but also in grass.

Coprinellus heterothrix ⋉ is very similar but has a veil and wider spores.

Widespread and common in the nemoral zones; mainly August–November.

Jens Maarbjerg

Coprinellus marculentus is a rather small, rather dark coprinoid with capitate hairs in the cap cuticle (hand lens required) and a veil of spherical cells that quickly disappears. The spores are flattened and hexagonal, and measure 9.7–14.3 × 6.3–8.7 × 6.2–7.3 µm. Occurs on rotten straw, middens and in similar places.

The combination of hexagonal spores and capitate pileocystidia should make identification straightforward.

A rather rare species with only scattered records; May–October.

Coprinellus congregatus is a trooping, medium-sized, yellow-brown to brown-grey coprinoid with small projecting hairs in the cap (hand lens required). The pileocystidia are rather short and pointed. It lacks veil and the hyphae are clampless. The spores measure 9.8–14.2 × 5.6–7.5 µm. Found *e.g.* on decaying wood chips, straw and dung.

Coprinellus ephemerus ⋉ is very similar but has clamps and somewhat larger spores (11.6–15.8 × 6.1–7.9 µm).

Widespread and probably rather common, becoming scarcer towards the north; March–November.

Erik Arnfred Thomsen

Coprinellus disseminatus is a rather small, usually highly trooping, yellowish-grey coprinoid that has both a veil of spherical cells and fine projecting hairs on the cap (hand lens required). The gills are adnexed to adnate. The hairs (pileocystidia) have long, wide, cylindrical necks that lack a widening apex. The spores are rather small, have a projecting, wide, central germ pore, and measure 6.6–9.7 × 4.1–5.8 µm. Occurs on and around rotten stumps and dead trees, mostly on mull soils in shady places, *e.g.* in gardens and parks.

Psathyrella pygmaea (page 574) may look rather similar and occurs in the same habitats, but does not have a round-celled veil or fine projecting hairs.

Widespread and common, although nearly absent from the boreal zone; mainly April–November.

Coprinellus bisporus is a rather small, yellow-brown to pale grey-brown, hairy coprinoid. It does not have veil. The basidia are 2-spored; there are no pleurocystidia or clamps. The pileocystidia have a ± cylindrical top. The spores measure 9.7–13.7 × 6.1–8.4 µm and the germ pore is excentric. Occurs on manure and dung.

Coprinellus bisporiger ⋉ is very similar but occurs on twigs and has pleurocystidia. There are also some similar 4-spored species of *Coprinellus*.

Widespread, occasional, but probably overlooked; mainly May–October.

Thomas Kehlet

Coprinellus heptemerus is a very small, yellow-brown, powdery and hairy coprinoid. The hairs (pileocystidia) are long and pointed, and the veil consists of thick-walled, spherical cells, mostly bearing a spine-like appendage. The spores measure 12.7–16.6 × 7.2–8.5 µm and have an excentric germ pore. Occurs on dung, especially from deer.

Coprinellus pusillulus ⋉ is very similar but has smaller spores.

Widespread and probably common; mainly May–October.

Parasola is a genus of small, highly plicate coprinoids that lack veils but have distinctly free gills, very thin cap flesh and a brown disc at the cap centre. The gills shrivel and curl at maturity, but do not dissolve. Many species have spores that are widest when seen in front view and these dimensions are given below, rather than the narrower side view. The psathyrelloid *Parasola conopilus* (page 579) is a deviating member of the genus; it has the same dark brown setae as *P. auricoma* ▽ but a non-plicate cap.

Parasola plicatilis is a typical *Parasola* with free gills and brownish colours when young, and is acorn-shaped; when mature it is greyish and plicate with a brownish disc. The spores measure 10–14.5 × 7–10.5 µm (av. range 11.1–12.8 × 8.1–9.5 µm) (overall less heart-shaped than in the majority of species). Occurs mostly in lawns and similar grassy places.

The typically earlier-fruiting, usually somewhat more vividly coloured *P. schroeteri* ▷ has larger spores (av. range 11.5–13.6 × 10.2–11.1 µm). The paler *P. lactea* ⋉ has shorter, very angular, heart-shaped spores; *P. kuehneri* ⋉ and *P. crataegi* ⋉ have even smaller spores. There are several other very similar species.

Widespread and relatively common; mainly July–October.

Parasola auricoma is a reddish-brown *Parasola* that retains a brown disc when mature. A good hand lens may reveal brown, thick-walled hairs in the cap cuticle, but a microscope is a better option. The spores are ellipsoid, measure 10–14.5 × 6–8 µm, and have a germ pore. Occurs on ± disturbed mull soils, mostly amongst bark and wood chips.

Parasola conopilus (page 579) also has brown hairs in the cap cuticle, but is non-plicate. Other parasolas lack such hairs and typically have ± heart-shaped spores.

Widespread and rather common, but becoming scarcer towards the north; April–October.

Parasola schroeteri is a typical *Parasola* with free gills and a plicate, smooth cap and, when young, rather red-brown cap colours. The spores measure 10–15.5 × 9–13 µm (av. range 11.5–13.6 × 10.2–11.1 µm). Occurs mostly in open, dry places, occasionally on dung, but mostly on soil.

Parasola auricoma ◁ has similar colours but occurs in more woody habitats and has brown hairs in the cap cuticle. *Parasola plicatilis* ◁ is not as brown and the spores are smaller. Other species have either larger or smaller spores.

Widespread and probably common over much of temperate Europe; April–October.

Jan Vesterholt

Parasola lilatincta is a plicate *Parasola* with a slightly lilac-tinged cap. The spores are indistinctly heart-shaped and measure 9.5–13.5 × 9–11 µm. When the cap is expanded, easily confused with other plicate parasolas, but the presence of oil globules in the cap cuticle cells and elsewhere (best observed in living specimens) is diagnostic. It belongs to a species complex and is possibly generally paler than shown here; it is not yet clear how specimens with distinct lilac stem bases should be classified.

Widespread and uncommon, but distribution not fully understood; summer–autumn.

Thomas Kehlet

Parasola misera is a tiny, pale red-brown to brick-red *Parasola* that, when young, has a rather hairy stem. The finely striate-plicate, almost egg-shaped, but later expanded, umbrella-like cap lacks hairs and has no veil. The spores are ± heart-shaped and measure 7–10.5 × 6.5–10 µm. Occurs on rather fresh dung, *e.g.* from cow and horse.

There are no other very small parasolas or coprinoids with a smooth cap that occur on dung. It also differs from other plicate parasolas by the absence of pleurocystidia.

Widespread and common; March–January.

Psathyrelloids

A group of very fragile agarics, most with black to dark brown, rarely more grey-brown, spore-deposits, mostly dry caps with dull brownish colours, and adnexed, adnate to shortly decurrent gills. A few species have prominent stem rings but most have veil remnants on the cap or, rarely, no veil at all. It includes a number of genera formerly treated in *Psathyrella*: *Cystoagaricus*, *Homophron*, *Kaufmannia*, *Lacrymaria* and *Typhrasa,* as well as the ± classic genera *Panaeolus*, *Psilocybe, Deconica* and part of *Parasola*.

Microscopically, most species have spores with germ pores. In rare cases the spores are ornamented (*Lacrymaria*, *Panaeolus p.p.*). Shape and contents of the cystidia (cheilo- and pleurocystidia) are important for species identification.

Ecologically, vary from lignicolous over soil to dung-inhabiting, and one species is a parasite on *Pluteus*.

OTHER SIMILAR FUNGI:
– coprinoids have ± dissolving gills (page 520).
– other black-spored agarics, *e.g.* *Stropharia*, *Leratiomyces* and *Hypholoma*), have mostly more vivid colours and often slimy caps (page 580).
– *Inocybe* has grey-brown spore-deposits (page 652).
– *Agrocybe* has somewhat paler, grey-brown spore-deposits and mostly sticky cap surfaces (page 625).

FURTHER READING: 86, 154, 155, 156, 171, 182, 225, 370.

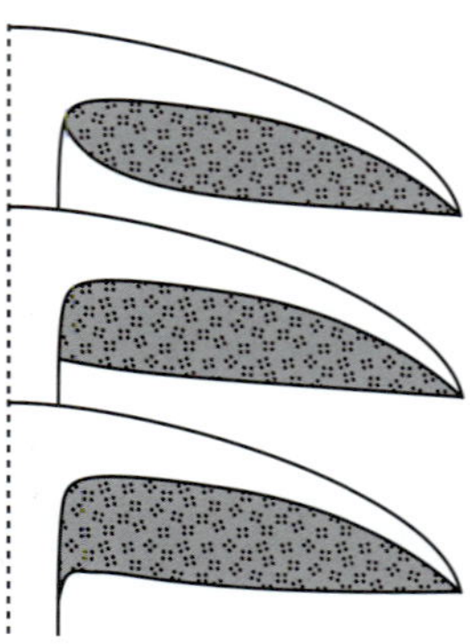

adnexed, adnate and decurrent gills

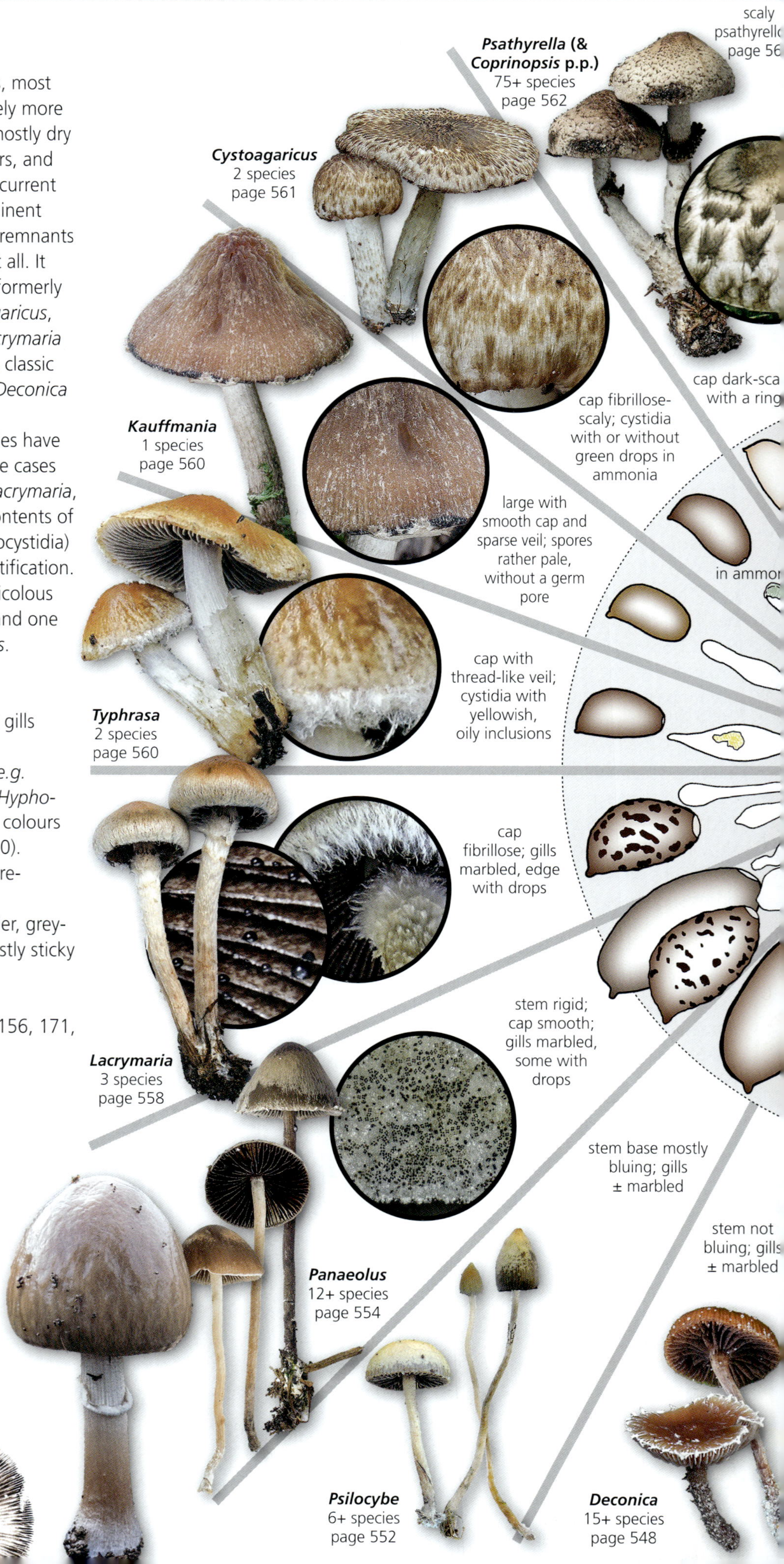

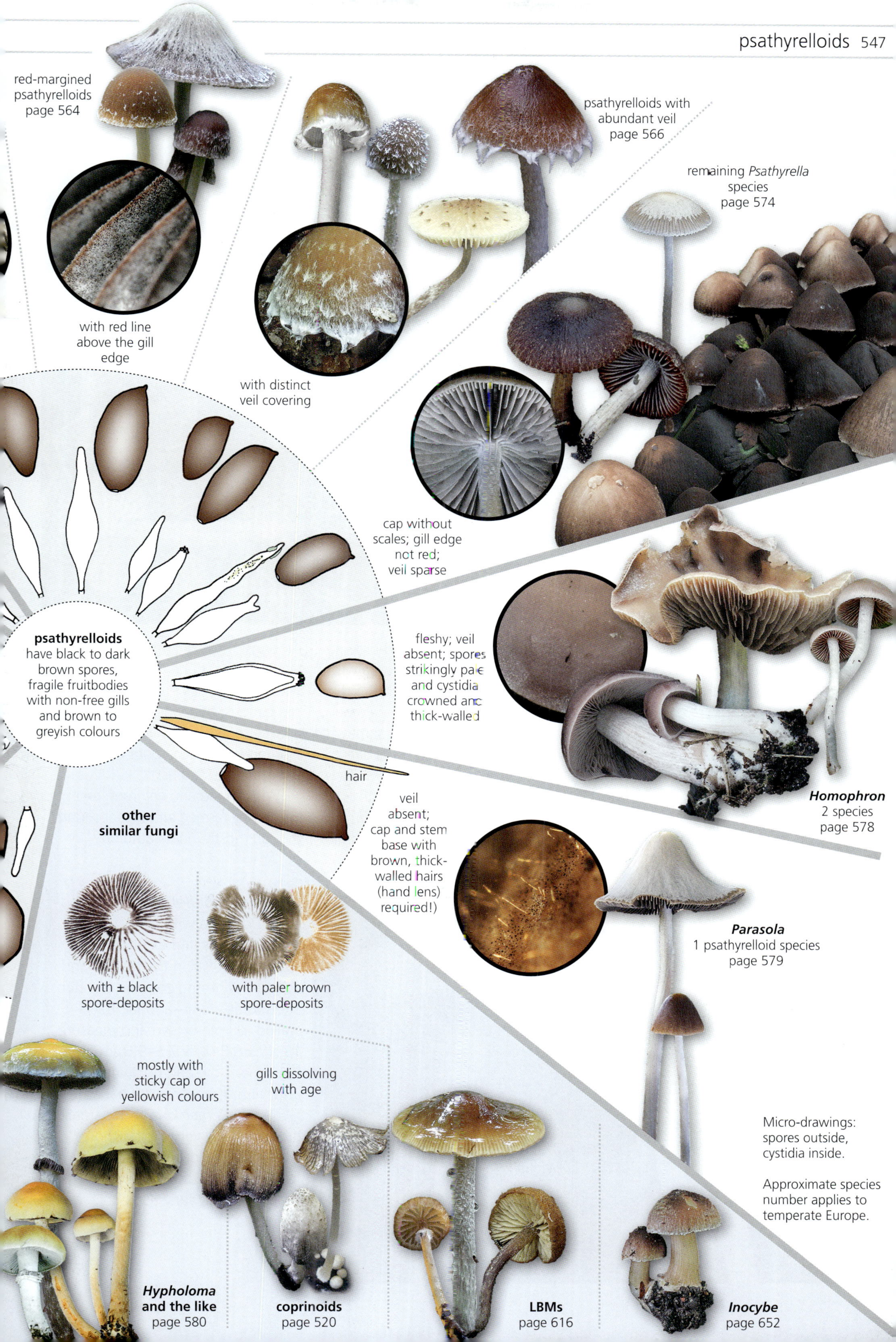

red-margined psathyrelloids page 564
with red line above the gill edge
with distinct veil covering
psathyrelloids with abundant veil page 566
remaining *Psathyrella* species page 574
cap without scales; gill edge not red; veil sparse
fleshy; veil absent; spores strikingly pale and cystidia crowned and thick-walled
Homophron 2 species page 578
veil absent; cap and stem base with brown, thick-walled hairs (hand lens required!)
hair
Parasola 1 psathyrelloid species page 579
psathyrelloids have black to dark brown spores, fragile fruitbodies with non-free gills and brown to greyish colours
other similar fungi
with ± black spore-deposits
with paler brown spore-deposits
mostly with sticky cap or yellowish colours
gills dissolving with age
Hypholoma and the like page 580
coprinoids page 520
LBMs page 616
Inocybe page 652
Micro-drawings: spores outside, cystidia inside.
Approximate species number applies to temperate Europe.

Deconica includes relatively small, often viscid agarics that mostly have adnate, marbled gills. The spore-deposits may vary from pale brown to black. See also the pleurotoid *Deconica* (page 95).

Deconica montana is a red-brown, striate, hygrophanous *Deconica* with verv few remnants of veil on the cap; the stem is dark with a few pale veil remnants. The cap cuticle is somewhat greasy, but not peelable. Gills dark, adnate to somewhat decurrent. The spores are flattened, ± ovoid, thick-walled with a distinct germ pore and measure 7–9 × 5–6 × 4.5–6 µm. Amongst mosses, *Empetrum*, *etc.* in heathlands or dune-like habitats.

Similar species may have peelable cap cuticles, thin-walled spores or non-flattened spores. *Deconica montana* var. *macrospora* ⋉ has 9–12 µm long spores.

Widespread and common except in intensively farmed areas; year-round, though mostly in late autumn.

David Boertmann

Deconica moelleri is a rather large *Deconica* with yellow-brown colours and extensive pale remnants of veil in the margin and as an evident ring zone on the stem. The spores are hexagonal, thick-walled, purple-brown and measure 13–17 × 8.5–11 × 7.5–9.5 µm. Occurs on cow and horse dung in unfertilized, open grassy habitats.

Deconica merdaria ▷ has smaller but otherwise very similar spores.

Probably widespread but frequency and distribution not well known; mainly June–October.

Deconica merdaria is a rather large, slimy-greasy, non-striate, apricot-orange to brownish-yellow *Deconica* with a ± semiglobose cap; young specimens have pale veil triangles at the margin. The spores are hexagonal, thick-walled, purple-brown and measure 11–13 × 7.5–9 × 7–8 µm. Occurs on herbivore dung, mostly in open habitats.

Deconica moelleri ◁ has similar, but larger spores. *Deconica coprophila* ▽ has very similar spores but less veil and normally paler yellow cap colours.

Widespread, occasional; mainly April–October.

Jan Vesterholt

Deconica coprophila is a slimy-greasy *Deconica* with a ± semiglobose cap; the veil disappears quickly. The gills are adnate, and the cheilocystidia are short and acutely lageniform. The spores are thick-walled, brown, hexagonal seen in front view, and measure 10–14 × 6–9 × 5.5–8.5 µm. Occurs on cowpats and other dung, mostly in dry, unfertilized grasslands.

Deconica subcoprophila ▽ has ellipsoid spores. *Deconica merdaria* △ has somewhat shorter spores and distinct veil on the stem. *Deconica moelleri* ◁ has longer spores and more obvious veil on the stem.

Probably widespread, occasional; May–November.

Thomas Kehlet/Jens H. Petersen

Deconica subcoprophila is a rather small, red-brown, striate and hygrophanous *Deconica* with a ± semiglobose cap that has sparse remnants of veil. The gills are adnate, rather distant, and, at maturity, purple-brown. The spores are ellipsoid, measure 15–19.5 × 8–10 µm and have a distinct germ pore. Occurs on dung in unfertilized grasslands, e.g. in dunes.

Deconia coprophila △ has smaller, more angular spores. Other dung-inhabiting species are either yellower or have obvious remnants of veil.

Distribution and frequency poorly known due to confusion with similar species; June–October.

Deconica inquilina is a yellow-brown, sparsely veiled *Deconica* with a peelable, sticky cap cuticle and adnate gills. The spores are relatively pale and thin-walled and measure 7–9(–10) × 4.5–6(–7) × 4–5(–6) µm; the germ pore is distinct and wide. Occurs on the rotten stems of herbs, including grass culms, mostly in open habitats, rarely on wood.

Deconica crobula ▷ has distinct remnants of veil and smaller, mostly more distinctly rhombic spores. Other similar species have non-peelable cap cuticles and typically thicker-walled spores.

Widespread and rather common; mainly July–November.

Jan Vesterholt

Deconica micropora is a small, yellow-brown, very hygrophanous, quickly drying *Deconica* with adnate gills. The veil is evanescent and the cap cuticle is non-peelable. The brown, rather thin-walled, somewhat flattened spores have a small germ pore and measure 5.5–7.5(–8) × 4.5–6 × 4–4.5 µm. Occurs on soil and litter in ± unfertilized grasslands.

Deconica subviscida ⋉ and *D. montana* ◁◁ have wider germ pores.

Rarely reported, but probably widespread; mainly June–September.

Deconica crobula is a red-brown *Deconica* with rather abundant white veil on cap and stem and a slimy, peelable cap cuticle. The cheilocystidia are fairly short and acutely lageniform. The spores are brown, somewhat rhomboid, rather thin-walled and measure 5.5–8 × 4–6 × 3.5–5.5 µm; the germ pore is rather evident and central. Occurs on plant litter, including twigs, *Rubus*-stems and wood chips.

Deconica inquilina ◁ has less veil and somewhat larger spores. *Deconica phyllogena* ▽ has distinctly thick-walled spores. Other species of *Deconica* also typically have less veil and there are small differences in germ pore and spore size.

Widespread and common; mainly June–October

Deconica phyllogena is a rather robust, ± umbonate, rather dark *Deconica* with abundant pale veil, mostly at the cap margin. The cuticle is non-peelable. The adnate to ± emarginate gills acquire a greyish-violet tinge with age. The spores are small, rhombic, thick-walled and measure 6–7.5 × 5–6.5 × 4–4.5 µm. Occurs on woody debris, including wood chips at roadsides, and in various woodland types.

Deconica crobula △ has thinner-walled spores and a peelable cap cuticle.

Widespread but rather rare, perhaps also somewhat overlooked; mainly June–September.

Psilocybe is a genus of small to medium-sized psathyrelloids with dark chocolate-brown to black spore-deposits and marbled gills. Most species stain greenish-blue at the stem base and contain the hallucinogenic compound psilocybin.

Psilocybe semilanceata is a ± olive-grey, striate, quickly drying, greasy *Psilocybe* that typically has a tall cap with a small, pointed papilla. The gills become very dark at maturity. The stem is stiff, pale and typically develops bluish spots at the base. Blue spots may also appear elsewhere on the fruitbody. The spores are ellipsoid, thick-walled, measure 10.5–15 × 6.5–8.5 µm, and have an evident germ pore. Occurs on soil in grazed fields that have only had limited fertilizer applications or in deer-grazed woodland glades, *e.g.* in dense *Deschampsia caespitosa* stands.

May be confused with small species of *Panaeolus* ▷▷ but these are mostly dry and lack a papilla on the cap. *Psilocybe strictipes* ⋉ – if it is genuinely different – has a more expanding cap and the spores are apparently somewhat smaller (10–13 × 5.5–7 µm).

Rather common and widespread; July–November.

Psilocybe semilanceata is a powerful hallucinogen, and its possession and trade is illegal in most countries.

Erik Rald

Psilocybe cyanescens is a yellow-brown, slimy-greasy, quickly applanate, somewhat striate *Psilocybe* with blue-black-staining surfaces. The spores are very dark, thick-walled, measure 9.5–14 × 6.5–8 µm, and have an evident germ pore. Occurs in troops on wood chips or compost.

There are no other typical wood chip agarics with black spores and dark bluing flesh.

Widespread, occasional to very rare, absent from the hemiboreal–boreal zones in Scandinavia; mainly September–November.

Psilocybe liniformans is a ± olive-brown, greasy, semiglobose *Psilocybe* with a peelable, slimy gill edge and a ± evident blue stain at the stem base. The spores are thick-walled, measure 13–14.5 × 7.5–10 µm, and have an obvious germ pore. Occurs mainly on horse dung in dry, unfertilized places, such as dunes and heathlands.

Dung-inhabiting species of *Deconica* ◁◁ are rather similar but do not have a peelable, slimy gill edge or blue stain.

Very rare or overlooked, recorded in the Netherlands, Denmark and England; mainly August–October.

Thomas Læssøe

Psilocybe turficola is a very tall, red-brown, ± striate *Psilocybe*. The stem is finely floccose-girdled, but lacks an evident ring zone. There are no obvious blue stains. The cheilocystidia are ± lageniform, sometimes with small heads. The spores are thick-walled, measure 10–13.5 × 6.5–8 × 6–7 µm, and have an evident germ pore. Occurs on peaty soils in marshes and swampy conifer plantations.

Phaeogalera stagnina (page 646) has a rather evident ring zone and larger spores (12–16 × 7–10 µm).

Widespread but rare, apparently absent from the boreal zone; mainly August–October.

Thomas Læssøe

Thomas Læssøe

Panaeolus is a genus of stiff-stemmed psathyrelloids with dark chocolate-brown to black, smooth or ornamented spores. The gills are marbled due to uneven spore maturation. Many species are coprophilous or prefer dung-influenced soil.

Panaeolus semiovatus is a large, slimy to greasy *Panaeolus* with an egg-shaped cap that barely expands with age. The stiff stem has a distinct, white ring. The spore-deposit is black. The spores are smooth, measure 18–24 × 10.5–13 × 9.5–11.5 µm, and have a ± excentric germ pore. Occurs on cow and horse dung, mainly in coastal grasslands, but may also occur on middens.

The only *Panaeolus* with a ring. *Panaeolus semiovatus* var. *phalaenarum* ⋉ has velar teeth at the cap margin and no ring.

Widespread and common; mainly April–November.

Panaeolus foenisecii is a rather slender, brown, very hygrophanous *Panaeolus*. It lacks veil, has brown gills, and the spore-deposit is dark brown, not black. The cheilocystidia are lageniform and somewhat capitate, widened at the tips. The spores are very distinctly verrucose, ± lemon-shaped, measure (11.5–)14–17(–22) × 7.5–11 µm, and have a somewhat truncate, large germ pore. Occurs on soil in grassland, including lawns *etc.*; can tolerate limited chemical fertilizer use.

This is the only *Panaeolus* with a brown spore-deposit, and is sometimes treated as a separate genus, *Panaeolina*. *Panaeolus fimicola* ▽ occurs in similar places but has smooth, much darker spores. *Panaeolus olivaceus* ⋉ has much darker, slightly ornamented, spores. Species of *Psathyrella* ▷▷ are veiled and have smooth spores that, typically, are smaller.

Widespread and very common; May–November.

Panaeolus fimicola is a rather small, typically rather dark, ± short-stemmed, normally very early fruiting, naked *Panaeolus* with a black spore-deposit. The cystidia may have yellow contents and the cheilocystidia are swollen at the tips. The smooth spores have a ± distinct, excentric germ pore and measure 11–15 × 7–9 × 6–8 µm. Occurs in unfertilized grasslands, including old lawns.

Macroscopically, this species can be confused with the dark-brown-spored, later-fruiting *P. foenisecii* △, but in a microscope that species can easily be recognized by its larger, rough spores.

Widespread and rather common; mostly April–May.

Jens Maarbjerg/Thomas Læssøe

Thomas Kehlet

Panaeolus subfirmus is a sturdy, naked *Panaeolus* with a rather pale, stiff, thick stem. The spores are large (15–20 × 11–12.5 × 8.5–10 µm), opaque, flattened, angular and smooth; the germ pore is central. Occurs on dung, typically cow, in unfertilized, dry grasslands.

Panaeolus cinctulus ▷ has smaller, not nearly such flattened spores. *Panaeolus semiovatus* ◁◁ and *P. papilionaceus* ▷ have either ring or remnants of veil at the cap margin.

Widespread, rather rare, but probably also overlooked; June–November.

Jan Vesterholt

Panaeolus guttulatus is a very dark, naked, sometimes olive-tinged *Panaeolus* with pale, yellowish drops on the gill edge. The spores are very small (7–9 × 4–5.5 µm), very slightly flattened, smooth and have a small, central germ pore. Occurs on calcareous soil and wood chips, in open places in woodland or scrub.

Other dark-spored, wood chip agarics do not have the highly marbled gills typical of *Panaeolus*. This species is also easy to identify due to the combination of substrate, habitat, its guttulate gill edges, sometimes olive colours, and small spores.

Rare and local, absent from boreal–hemiboreal zones; August–November.

Panaeolus acuminatus is a rather variable, typically dark *Panaeolus* that lacks remnants of veil. The stem is dark red-brown and very powdery. The spores are dark brown, smooth, almost opaque, large (13–15 × 9–12 × 6.5–9 µm), distinctly flattened and appear somewhat rhombic in front view and narrowly ellipsoid from the side; the germ pore is central. The cystidia lack yellow contents and are narrowly lageniform. Occurs on dung or dung-influenced grassland, rarely in forests.

Panaeolus cinctulus ▷ is usually more robust and the spores are only slightly flattened; the cheilocystidia tend to be capitate.

Widespread and very common; March–December.

Panaeolus cinctulus is a rather fleshy, zonate, hygrophanous, yellow-brown to red-brown, naked *Panaeolus*. The cheilocystidia are mostly ± capitate. The spores are large (11–14 × 7.5–10 × 6–9 µm), only slightly flattened, smooth, ± rhombic seen in front view and lack drops; the germ pore is central to slightly excentric. Occurs on cow and horse dung, including middens.

Panaeolus acuminatus ◁ is more slender and has distinctly flattened spores. *Panaeolus subfirmus* ◁ is very similar but has distinctly flattened, larger spores.

Widespread, occasional; May–November.

Panaeolus papilionaceus is usually a rather pale, large, tall, dry, smooth or wrinkled, only slightly hygrophanous *Panaeolus* with pale, dentate veil remnants at the cap margin. The opaque, smooth spores are distinctly flattened and angular in front view, and measure 15–18 × 10–12 × 8–9.5 µm. Occurs on dung in grassland, including fertilized fields.

The only other *Panaeolus* species with veil is *P. semiovatus* ◁◁ but it has a ring rather than marginal teeth; it also has a slimy-greasy cap surface and even larger spores. *Panaeolus papilionaceus* var. *parvisporus* ⋉ has smaller spores (13–16 × 8–10 × 6–8 µm), a more evanescent veil and shorter cystidia (25–35 µm long); in *P. papilionaceus* var. *papilionaceus* the cystidia are 30–60 µm long.

Widespread and very common; March–January.

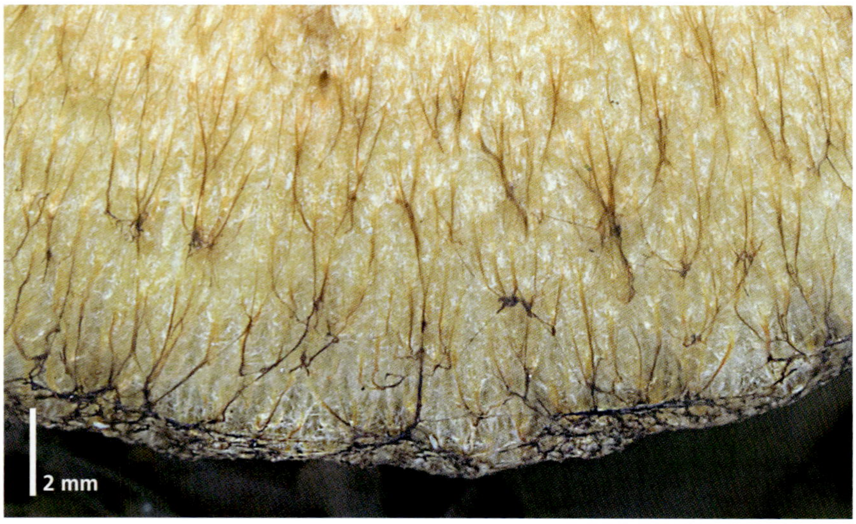

Lacrymaria is recognized by the fibrous-hairy cap with a bearded margin, marbled gills that have drops at the margin, and black, verrucose spores. Occurs on disturbed soil.

Lacrymaria lacrymabunda is a very hairy-floccose, yellow-brown to grey-brown psathyrelloid with very marbled gills with a white, guttulate edge. The stem has an evident ring zone that is mostly black from deposited spores and yellow-brown scaly hairy below. Lacks a distinctive smell or taste. The cheilocystidia have a round to pear-shaped head. The spores are dark brown, coarsely verrucose, measure 8–11 × 5.5–6.5 µm and have a ± truncate, large germ pore. Occurs on disturbed soil at roadsides, in gardens, parks, *etc.*

Lacrymaria pyrotricha ▷ has bright orange hairs on the cap and stem. The smaller *L. glareosa* ⋉ may have somewhat paler fruitbodies and the cheilocystidia have a pointed top.

Common to very common and widespread; mainly July–December.

Lacrymaria pyrotricha is a very radially felty psathyrelloid with ± yellow-orange to orange-red fibres on the cap and on the stem below the ring zone. The gill edge has pale drops that become black from deposited spores. The cheilocystidia are capitate. The spores are dark brown, verrucose, large (8–11 × 5.5–6.5 µm) and have a truncate germ pore. Occurs on ± disturbed soil in parks and forests.

Lacrymaria lacrymabunda ◁ is very similar but the fibres on the cap and stem are more yellow-brown.

Occasional, absent from the boreal zone; June–November.

Enzo Musumeci /Thomas Læssøe

Enzo Musumeci

Typhrasa gossypina is a relatively anonymous, ± dark brown, orange-brown to grey-brown, ± odourless psathyrelloid with rather abundant, pale veil, especially at the cap margin. The cap soon becomes wrinkled. The hyphae have clamps. The pleurocystidia and cheilocystidia have a large, internal oil body and long neck. The spores measure 7–9 × 5–6 × 4.5–5.5 µm. Grows either singly or in small clusters on woody debris that is in contact with soil in ± mull-rich, deciduous and coniferous forests.

Typhrasa nanispora ⋉ has much smaller spores (5–6 × 3–4 µm). *Psathyrella fibrillosa* ⋉ does not have long-necked cystidia with a large oil body.

Distribution and frequency not well known, apparently rather rare; July–November.

Lucie Zíbarová

Kauffmania larga is a large, marginally striate, dark red-brown, hygrophanous psathyrelloid that dries to a much paler colour. The veil at the cap margin is pale, sparse and thread-like. The gills are crowded. Pleurocystidia mostly broadly lageniform. The spores are rather pale, irregularly formed but mostly ellipsoid, measure 7.5–10 × 4.5–6.5 µm and lack a germ pore. Found mostly on wood fragments in mull-rich woodlands.

Best told by its robust, dark fruitbodies and lack of a germ pore.

Presumably rare but distrbution and frequency not well known; summer–autumn.

Jean-Marc Moingeon

Cystoagaricus silvestris is a dark, scaly hairy, non-striate, solitary to almost clustered, large, brownish or olive-brown-tinted psathyrelloid. The stem has brown, thread-like remnants of veil along its length, but lacks a ring. The hyphae have clamps. The pleurocystidia are large (35–60 × 9–15 µm), have a rounded top and the surface deposits stain green in ammonia. The spores are slightly angular and measure 7–10 × 4.5–5.5 µm (average length 4–8.6 µm). Found in mull-rich deciduous forests, on deciduous wood but also directly on soil.

Psathyrella maculata ▷▷ has much smaller spores and the cystidia have a thin extension. *Psathyrella cotonea* ▷▷ has a yellow stem base and grows ± fasciculate. *Psathyrella caput-medusae* ▷▷ has a ring. *Coprinopsis melanthina* ▷▷ has no pleurocystidia. *Inocybe erinaceomorpha* (page 653) has thick-walled cystidia with crystals on the top, and a distinct, cider-like smell.

Very rare and local, apparently absent from the boreal zone; June–November.

Lucie Zíbarová

Rasmus Ejrnæs/Torbjørn Borgen

Psathyrella is a large genus of agarics with dark brown to almost black spore-deposits and narrowly adnate gills that never dissolve at maturity. They have a thread-like universal veil and are mostly extremely fragile.

Recent molecular studies have shown that a few 'psathyrelloids' belong to the coprinoids (page 520) (*Coprinopsis* and *Parasola*) but they are treated here as psathyrelloids.

Psathyrella caput-medusae is a very distinctive, large and dark scaly *Psathyrella* with an unusual sweet, perfumed smell. The scales and stem ring become darker with age. The cystidia have a ± capitate top. The spores are large (8.5–11(–12.5) × 4.5–5.5(–6.5) µm), ± fusiform and lack a germ pore. Found mostly on conifer wood, but is also known from deciduous wood.

May recall scaly species of *Agaricus* (page 500) but can be indentified by its narrowly adnate gills, cystidia and substrate. *Psathyrella cotonea* ▽, which mostly grows on *Fagus*, has a less evident ring and completely different spore and cystidia shapes.

Widespread, but rare to very rare; June–November.

Thomas Læssøe/Jens H. Petersen

Psathyrella cotonea is a large, ± clustered, at first pale-scaled but later brown *Psathyrella* with brown scale girdles on the stem, which is ± yellowish at the base. The cystidia are ± conical. The spores are ± bean-shaped in side view and measure 6–9 × 3–4.5 µm. Mostly on *Fagus* stumps but also found on *e.g. Abies*.

Psathyrella caput-medusae △ smells sweetish and has larger spores; *P. maculata* ▷ is darker and has smaller spores. *Cystoagaricus silvestris* ◁◁ has large cystidia with drops that stain green in ammonia. *Coprinopsis melanthina* ▷ lacks pleurocystidia.

Widespread in the nemoral and into the hemiboreal zones, occasional; mainly September–November.

Thomas Kehlet

Psathyrella maculata is a large, initially pale, but very soon dark, scaly, ± clustered *Psathyrella* with pale stem scales that also soon become dark. The gills are rather crowded and brown. With smell of cacao. The clavate cystidia typically have a narrow extension. The spores are very small (4.5–6 × 3–3.5 µm). Found on all kinds of wood in mull-rich forests.

Psathyrella cotonea ◁ is not as dark and has larger spores. This is also the case with *P. caput-medusae* ◁ and *Cystoagaricus silvestris* ◁◁. *Coprinopsis melanthina* ▽ has no pleurocystidia and much larger spores.

Widespread in the nemoral zones, but rather rare; August–October.

Coprinopsis melanthina is a rather large psathyrelloid member of the *Coprinaceae* with non-deliquescent gills and dark fibres and scales. The stem is brown-scaly and has a ring zone rather low down. Lacks pleurocystidia and has broadly lageniform cheilocystidia. The ± egg-shaped spores are relatively pale, have no germ pore and measure 9–11.5 × 5.5–7 µm. Occurs on soil and around stumps, mostly in damp deciduous forests.

Other dark-scaled psathyrelloids in the genera *Psathyrella* and *Cysto-agaricus* ◁◁ all have pleurocystidia.

Local and rare in the nemoral zones; summer–autumn.

Jean-Marc Moingeon

Leif Örstadius/Thomas Læssøe

Psathyrella corrugis is an elegant, mostly trooping, at first pale red-brown, but soon paler *Psathyrella* with rather limited and rapidly disappearing veil. The gills have a red, usually somewhat broken line just above the proper edge. The stem may be rooting for up to 25 mm. The pleurocystidia are very variable and ± fusiform to acutely lageniform. The spores measure 10–15 × 5.5–7 µm. Typically found on mull-rich soil mixed with remnants of wood, wood chips, *etc.*

Other rooting species with a red line above the gill edge may look very similar. The pinkish-drying *P. microrrhiza* ▷ is darker when expanded and the red line is unbroken. *Psathyrella orbicularis* ⋉ also has an unbroken red line and does not dry pinkish. Other similar species have smaller spores or smaller fruitbodies. *Psathyrella pseudogracilis* ▷ is macroscopically almost identical and can only be separated with certainty by its broadly lageniform cystidia.

Widespread and common; June–December.

Thomas Læssøe

Psathyrella prona is a small, dark, striate *Psathyrella* with rather distant gills that, typically, are red just above the edge. It dries ± pinkish. The veil is sparse and evanescent. The stem is non-rooting and often darker towards the base. Pleurocystidia small, tapering; basidia 2-spored. Spores depressed near the apiculus or a papilla on top, 12.5–17(–20) × 6.5–8 µm. On ± disturbed, rich soils, along roads, in gardens, *etc.*

Psathyrella potteri ⋉ has a paler, less striate cap and does not dry pinkish, it is typically 4-spored; *P. orbitarum* ⋉ has smaller spores; and *P. tenera* ⋉ has more veil and 4-spored basidia.

Widespread and common, at least in the nemoral–hemiboreal zones; May–November.

Psathyrella pseudogracilis is a pale, rooting, large-spored *Psathyrella*. The gills have a red line just above the edge. Very young fruitbodies are red-brown. The pleurocystidia are broadly lageniform with rounded ends. The spores measure 11–16 × 6–7.5 µm. Occurs mostly on disturbed mull soils mixed with woody debris in deciduous forests.

The macroscopically very similar *P. corrugis* ◁ also has pale fruitbodies but the cystidia are more pointed. Other similar psathyrelloids are darker or lack the red line on the gills.

Widespread and rather common; mainly July–November.

Psathyrella microrrhiza is, at maturity, also a rather dark, slender *Psathyrella* with ± obvious remnants of veil and a rooting stem. The cap dries pinkish. The gills have an unbroken red line above the edge. The pleurocystidia are rather variable with both pointed and rounded to capitate tops. The spores measure 10–14 × 5.5–7 µm. Typically occurs on disturbed rich soils intermixed with woody debris.

Psathyrella corrugis ◁ is paler and typically has less prominent veil. The red line on the gills is mostly broken. *Psathyrella orbicularis* ⋉ does not dry pinkish. There are several similar species that either have smaller spores, smaller fruitbodies or lack a red line on the gills.

Widespread and common; mainly September–January.

Thomas Læssøe

Psathyrella sphagnicola is a rather small *Psathyrella* with insignificant amounts of white remnants of veil at the cap margin but with a large projecting ring on the stem. The pleurocystidia are broadly lageniform, 40–60 µm long and have somewhat yellowish walls in ammonia. The spores measure 8–10 × 4.5–5.5 µm (av. range 8.7–9.3 × 4.5–5.1 µm). Occurs directly in *Sphagnum* in bogs and mires.

The very similar *P. vesterholtii* ▽ has slightly larger spores. *Psathyrella fibrillosa* ⋉ occurs in drier habitats and very rarely has a well-defined ring; the cystidia are typically more pointed and have partly thickened walls.

Rare and mostly found in the north, very rare in the nemoral zone; mainly August–September.

Psathyrella vesterholtii is a rather slender, yellow-brown *Psathyrella* with a very prominent and projecting ring. The pleurocystidia are somewhat thick-walled with rounded ends and become yellow in ammonia. The spores measure 9–11 × 5–6 µm (av. range 9.8–10.4 × 5.2–5.7 µm). Found mostly on soil and litter in mossy boreal *Picea*-forests.

Psathyrella sphagnicola △ occurs directly in *Sphagnum* and has somewhat smaller spores.

Recorded from Sweden and the Faroe Islands, probably a rare northern species; late summer–autumn.

Thomas Læssøe

Psathyrella candolleana is a rather large, very thin and fragile *Psathyrella*. When young it is golden-brown and has abundant white veil, but quickly becomes very pale. The stem has thread-like remnants of veil and, rarely, a ring. The gills are very crowded. Lacks pleurocystidia. The spores have a distinct germ pore and measure 6.5–10 × 4–5.5 µm (av. range 7–8.3 × 4.1–5 µm). Occurs on and around rotten, usually deciduous, wood, mostly in disturbed places, including urban settings.

Psathyrella leucotephra ▽ also lacks pleurocystidia but usually has a ring, and the spores lack a germ pore. *Coprinopsis marcescibilis* ▷▷ has far larger spores.

Very common and widespread; May–November.

Psathyrella leucotephra is a large, clustered *Psathyrella* that quickly becomes very pale and has crowded, dark gills. The membranaceous veil is seen as a ring and as flakes at the cap margin. Lacks pleurocystidia. The spores have no germ pore and measure 8–10 × 5–6.5 µm (av. range 8.5–8.8 × 5.5–5.8 µm). Typically occurs at the base of living deciduous trees on mull soils.

Psathyrella candolleana △ normally lacks a ring and the spores are smaller and have a germ pore.

Rare, but rather widespread; June–October.

Wikipedia/Marino Zugna

Thomas Læssøe

Psathyrella spintrigeroides is a dark red-brown, rather spectacular *Psathyrella* with very prominent, triangular veil scales at the cap margin and a white, floccose stem. The pleurocystidia are unusally numerous, very pointed and somewhat thick-walled; the cheilocystidia are ± similar. The spores measure 7–8.5 × 4–5 µm and have a ± indistinct germ pore. Occurs on the trunks of deciduous and coniferous trees in old-growth habitats.

Psathyrella rostellata ⋉ has larger spores and bent or forked pleurocystidia. *Psathyrella pseudocasca* ⋉ has similar remnants of veil but completely different, narrowly fusiform, thin-walled pleurocystidia and some spores are finely ornamented.

Rare, and possibly mostly found in eastern parts of the nemoral zone; mainly September–November.

Psathyrella typhae is a small, slender *Psathyrella* with small, scale-like remnants of veil on the pale yellow-brown, quickly flattening cap. The stem has a ring zone. Pleurocystidia are absent. The spores are large (9–12.5 × 5.5–8 µm) and lack an evident germ pore. Occurs on various herbs, but particularly *Typha*, in marshes and around ponds.

Psathyrella thujina ⋉ may also occur in wet habitats but has pleurocystidia.

Probably rather common, at least in the nemoral zones but rarely recorded in most countries (although frequent in Denmark); May–November.

Psathyrella impexa is a rather small to medium-sized, yellow-brown *Psathyrella* with abundant, white, thread-like veil covering the cap and stem, which may have a ring zone. Dries ± pinkish. The tapering pleurocystidia are thin-walled. The spores measure 8–10.5 × 4.5–5.5 µm. Occurs mostly amongst small, dead branches on rich soils in deciduous woodlands.

Psathyrella fibrillosa ⋉ is very similar. It occurs typically in more acidic habitats and the cystidia have a tendency to be thick-walled and ± yellow-tinged in ammonia.

Widespread and rather common; August–November.

Thomas Læssøe

Coprinopsis marcescibilis is a rather pale, very *Psathyrella*-like, non-deliquescent *Coprinopsis* with rather abundant white veil on the ± bell-shaped cap and on the stem. Lacks pleurocystidia. The spores are somewhat irregularly formed, dark red-brown, measure 10.5–14 × 6–7.5 µm and have an evident germ pore. Occurs on ± disturbed rich, mostly rather damp soil, *e.g.* on wood chips mixed with soil.

Similar species of *Psathyrella* have pleurocystidia or smaller spores.

Widespread, occasional; April–November.

Thomas Læssøe

Coprinopsis canoceps is an elegant, very pale and very felty-hairy *Psathyrella*-like, non-deliquescent *Coprinopsis* with a pointed umbo. It has clamps and large veil cells (30–180 × 4–24 µm). Lacks pleurocystidia but has lageniform to conical cheilocystidia. The spores measure 8–11 × 4.5–6 µm. Occurs on ± calcareous soils and twigs in damp forests and scrub.

Psathyrella effibulata ⋉ lacks clamps and has smaller veil cells. *Psathyrella fibrillosa* ⋉ has ± thick-walled pleurocystidia. *Psathyrella squamosa* ⋉ has smaller spores.

Widespread but uncommon; May–October.

Erik Arnfred Thomsen

Jean-Marc Moingeon

Psathyrella suavissima is a medium-sized, rather colourful, brownish-orange *Psathyrella* with very threadlike veil covering the cap, crowded gills and a strong, sweetish smell somewhat like the flowers of *Prunus padus* (or *Hebeloma sacchariolens*, page 684). The veil cells are very long and wide, 30–450 × 5–50 µm. The spores measure 6.5–9 × 4–5 µm and have a distinct germ pore. Occurs on rotten wood or wood chips on mull or clay soils.

Psathyrella pertinax ▷▷ does not have such a strong smell, occurs mostly on conifer wood and is more red-brown. *Psathyrella fragrans* ⋉ can have a similar fragrance but resembles *P. piluliformis* ▷▷ and occurs mostly with conifers.

Widespread, but very rare; mainly June–September.

Thomas Læssøe

Psathyrella hirta is a small, dark red-brown *Psathyrella* with very prominent, projecting white veil scales on cap and stem. It has 14–26 entire gills that lack a red edge. The pleurocystidia are tapering. The spores measure 10–13 × 5.5–7 µm. Occurs on old cowpats and other kinds of dung in unfertilized grassland.

Several similar species also occur on dung, including: *P. scatophila* ⋉, which has narrower spores and very pointed pleurocystidia; *P. merdicola* ⋉ with smaller spores; *P. romagnesii* ⋉, which has larger spores; and *P. purpureobadia* ⋉, which lacks clamps and has more rounded pleurocystidia.

Widespread, occasional, apparently absent from the boreal zone; mainly July–November.

Jens Maarbjerg

Psathyrella bipellis is a rather small to medium-sized, ± purple-brown to wine-red-tinged, hygrophanous *Psathyrella* with white remnants of veil on the outer third of the cap. The gills have a ± reddish tinge and the stem has fine floccules and may be purple-tinted. The odour can be either peppermint-like or more unpleasant (urine-like). The pleurocystidia are large and variable in shape. The spores measure 11–15.5 × 6.5–8.5 (–9.5) µm and have a central to excentric germ pore. Occurs on ± disturbed, mostly calcareous soil in deciduous forests.

Distinctive and normally easy to identify. *Psathyrella vinosofulva* ⋉ lacks clamps and has narrower spores.

Widespread but becoming scarcer towards the north and apparently absent from the boreal zone; almost year-round, but mainly September–December.

Psathyrella pennata is a rather slender, dark brown *Psathyrella* with rather abundant, ± persistent, pale veil at the somewhat striate cap margin. The stem may have a ring zone and remnants of veil. The pleurocystidia are lageniform to conical, tapering and lack internal drops. The spores measure 6.5–8.5 × 3.5–4.5 µm. Occurs on burnt soil, mostly in forests.

No other similar psathyrelloids occur on burnt soil, but there are several similar species in other habitats.

Widespread but rare; May–December.

Leif Örstacius

Psathyrella piluliformis is typically a rather large, ± clustered, red-brown *Psathyrella* with a thread-like veil at the margin. The stem is rather fragile and whitish. The pleurocystidia are simple. The spores are rather pale and very small (4.5–6.5 × 3–4 µm). Found mostly on the trunks and stumps of deciduous trees in forests and parks, rarely on conifer wood.

Occasionally occurs in confusing dwarf forms. *Psathyrella fragrans* ⋉ is closely related but typically occurs on conifers and generally has a sweetish smell. *Psathyrella laevissima* ⋉ may have a sweetish odour but has beaked pleurocystidia.

Widespread and common, becoming scarcer towards the north; mainly September–January.

Jan Vesterholt

Psathyrella pertinax is a medium-sized, dark red-brown *Psathyrella*, usually with a somewhat wrinkled, thick cap cuticle. It has rather abundant, white veil. The stem has white to brownish remnants of veil along its full length. The crowded gills have a distinct white edge. Usually has a sweetish, marzipan-like smell. The pleurocystidia are variable. The spores measure 6–8.5 × 3.5–5 µm. Occurs on woody debris on ± acidic soil, mostly on conifers, but also on *Betula*.

Psathyrella mucrocystis ⋉ has beaked pleurocystidia; *P. flexispora* ⋉ has ± fusiform spores; and *P. pseudocorrugis* ⋉ usually has a reddish line on the gills and no smell. See also *P. suavissima* ◁◁.

Widespread, rather rare or perhaps overlooked; mainly June–October.

Psathyrella sphaerocystis is a slender, pale cinnamon-buff, almost non-striate *Psathyrella* with a floccose-mealy cap surface. The stem is pale and floccose. The veil cells are spherical and in chains. The spores measure 7.5–9 × 5–5.5 × 4–5 µm. Occurs on horse and cow dung in open habitats.

One of very few psathyrelloids to have spherical veil cells, and the only of these to grow on dung.

Rare, with a few scattered records from a number of countries; summer–autumn.

Thomas Læssøe

Psathyrella globosivelata is a very distinctive, very pale, ± pinkish-tinted, almost mealy *Psathyrella* with rather pale and rather distant gills. The veil consists of spherical cells. The pleurocystidia are variable. The spores measure 7–8.5 × 3.5–4.5 µm and lack an evident germ pore. It has a very unusual ecology – the fruitbodies arise from *Pluteus* primordia on rotten deciduous wood.

Psathyrella tenuicola ⋉ is more downy and has narrowly lageniform to conical pleurocystidia. *Psathyrella alblofloccosa* ⋉ has larger spores and does not grow on fungi. *Psathyrella lyckebodensis* ⋉ has larger spores and is terrestrial.

Very rare and southern; summer–autumn.

Pedro Zapico

Psathyrella lutensis is a rather anonymous red-brown, quickly drying *Psathyrella* with rather limited, pale veil that initialy may reach half way to the centre of the cap. The stem is slightly rooting. Odourless. The pleurocystidia have rounded ends. Drops on the cheilocystidia stain green in ammonia. The spores measure 8–11 × 4.5–6 µm. Occurs on damp, clay soils, *e.g.* in wheel tracks in woodland, often near brush wood.

Psathyrella supernula ▷ has the same green stain in ammonia, but normally has a nauseous smell and a distinctly rooting stem.

Widespread and rather common; mainly August–December.

Psathyrella clivensis is a rather small to medium-sized, red-brown, hygrophanous *Psathyrella* with limited, pale veil at the margin of young specimens. The stem is pale. The pleurocystidia have rounded ends. Cheilocystidia are numerous and clavate. The ± irregularly shaped spores are reddish-yellow in water, have a very indistinct germ pore and measure 8–11 × 5–6.5 µm. Typically occurs in grass in open, calcareous grasslands.

Psathyrella panaeoloides ⋉ has spores with conspicuous germ pores; *Panaeolus foenisecii* ◁◁ has no veil, marbled, dark gills and verrucose spores.

Widespread, occasional; May–December.

Psathyrella pygmaea is a small-spored *Psathyrella* with small, yellow-brown fruitbodies. The white, thread-like veil is rather evanescent on the translucently striate cap. The ± thick-walled, wide pleurocystidia have crystals on the top. The spores are rather pale and measure 6–7.5 × 3.5–4 µm. Occurs on and around stumps of deciduous trees on damp mull soils in gardens, parks and woodland.

Coprinellus disseminatus (page 542) also has small veil flecks but mixed with pale projecting hairs on the cap (hand lens required!).

Widespread, but absent from the hemiboreal–boreal zones, occasional; May–November.

Psathyrella supernula is a rather small and pale, elegant, somewhat rooting *Psathyrella* with a distinct, very unpleasant nauseous odour like sewage water or bad breath. The veil is very limited. The rather distant, greyish gills occasionally have a reddish line just above the edge. The pleurocystidia are tapering and stain green in ammonia. The spores measure 9–12.5 × 4.5–6 µm. Occurs on rich, mostly disturbed soil, amongst brush wood *etc.*

This species was previously known as *P. narcotica*. *Psathyrella lutensis* ⊲⊲ has a similar green stain but the smell is not unpleasant. There are other ± red-lined psathyrelloids (page 564) but these all lack a conspicuous smell and do not stain green in ammonia.

Rare, but possibly overlooked; mainly September–October.

Lucie Zíbarová

Psathyrella ammophila is a medium-sized, half-buried *Psathyrella* that has rather limited veil on the ± yellow-brown cap. It is very dark- and rather large-spored, and has ± broadly lageniform pleurocystidia. The spores measure 9–14 × 5.5–8.5 µm. It is associated with *Ammophila* in white dunes.

The unusual habitat should make this species relatively straightforward to identify, but there are several species of *Stropharia* (page 590), *Agrocybe* (page 625) and *Panaeolus* ⊲⊲ that also occur in the same environment.

Widespread and common in suitable coastal habitats; May–November.

Psathyrella spadiceogrisea is a rather large, ± red-brown, hygrophanous *Psathyrella* with somewhat limited remnants of veil at the ± striate margin. The crowded gills are brownish-black. The cheilocystidia are dominated by clavate types, while the pleurocystidia are predominantly utriform and rarely pigmented. The spores measure 7–9.5(–10.5) × 4–5.5 µm and are typically rather bean-shaped as seen in side view and with a truncate upper end in front view. Occurs on ± buried wood, normally on mull soils.

This is a difficult species to identify and more than one species is probably involved. The somewhat paler *P. fatua* ⋉ has a conical spore end. *Psathyrella pseudocasca* ⋉ has more veil and narrower spores.

Widespread and common; mostly March–June.

Jan Vesterholt

Psathyrella fusca is a rather elegant, solitary *Psathyrella* with a red-brown cap (quickly drying to a brownish-yellow), limited veil and a very pale, finely floccose stem. The gills are crowded and very grey-black, and it is one of the most dark-spored psathyrelloids. The pleurocystidia are large and broadly lageniform to lageniform. The ± angular spores measure 9–11.5(–13.5) × 5–6.5 µm, with distinct germ pore. Occurs on rich soils, perhaps mainly calcareous, amongst litter and brush wood.

Kauffmania larga ◁◁ has red-brown gills and somewhat smaller spores with an indistinct germ pore.

Widespread, occasional; mainly September–December.

Psathyrella multipedata is an extremely clustered and deeply rooting *Psathyrella* with limited veil on the conical, yellow-brown cap. The cheilocystidia are acutely lageniform to conical and surface drops stain green in ammonia. The spores are small and measure 6.5–9 × 4–5 µm. Occurs on clay-rich, typically open, disturbed ground, *e.g.* in grass verges, churchyards, *etc.*

Psathyrella candolleana ◁◁ is often found in the same environment but is always connected to wood; it has a paler and flatter cap with distinct veil when young.

Widespread, occasional, becoming scarcer towards the north; mainly August–December.

Jens H. Petersen/Thomas Læssøe

The genus ***Homophron*** is characterized by fruitbodies that completely lack any form of veil and, for a psathyrelloid, its strikingly pale brown spores. In addition, the pleurocystidia are thick-walled and crowned by crystals. Both species were, until recently, accepted in *Psathyrella*.

Homophron spadiceum is a large, smooth, naked, pale brown to reddish-brown, clustered psathyrelloid with crowded, grey-brown gills and white, fine hairs at the cap margin in young, fresh material (not veil). It usually has a somewhat pinkish tinge. The stem is very pale and rather firm. The pleurocystidia are tapering, thick-walled, and have a small crystal crown at the top. The spores are rather pale brown and measure 6.5–11 × 4–5.5 µm. Occurs at the base of living deciduous trees on disturbed soil, *e.g.* in churchyards, in avenues, *etc.*, but sometimes also on conifer wood.

Homophron cernuum ▽ has pleurocystidia with a rounded top and, typically, smaller, paler fruitbodies.

Widespread, rather common to occasional; mostly September–December, but may be found most of the year.

Thomas Læssøe

Homophron cernuum is a very pale, somewhat shiny, naked, ± clustered psathyrelloid with crowded, grey-brown gills. The cheilocystidia are rounded, thick-walled, and mostly have crystals at the top. The spores are rather pale brown and measure 6–9 × 3.5–5 µm. Occurs on dead trunks and also at the base of living deciduous trees in parks and deciduous forests.

Homophron spadiceum △ is darker brown to reddish-brown and has tapering cystidia.

Widespread, rather common to fairly rare; August–January.

Parasola conopilea is a large, long-stemmed, shiny, dark brown, very hygrophanous, psathyrelloid, non-plicate *Parasola* with long, brown thick-walled hairs at the stem base and in the cap cuticle. The cap becomes pale buff when drying out. The stem is slightly rooting and rather fragile. The very large spores measure 13.5–17 × 6.5–9 µm. Occurs on disturbed soil, mostly mixed with wood chips. Typically found along roads in forests, but also occurs on wood chips in urban settings. It can grow in very large troops, sometimes mixed with species of *Psathyrella*, *e.g. P. microrrhiza* ◁◁.

Parosola auricoma (page 544) is smaller and has a plicate cap surface. The lack of veil and presence of brown hairs are good characters for *Parosola conopilea* in comparison to other psathyrelloids with the same shape. In 2008 it was included in *Parasola* (see page 544), but it has since been suggested that it may deserve to be placed in a separate genus, with *P. auricoma* (page 544).

Widespread and very common; May–December.

Hypholomatoids

This group includes mostly fairly robust agarics with dark chocolate-brown to black spore-deposits and adnate, emarginate or slightly decurrent gills. The caps are mostly slimy and the colours a rather vivid yellowish, greenish or orange-brown. With the exception of *Hypholoma*, the gills are ± marbled due to uneven maturation.

Microscopically, many of the species have dark spores with a germ pore and many have cystidia with yellow contents (chrysocystidia). However, similar cystidia are also common in the brown-spored genus *Pholiota* (page 608).

All species are decomposers. The majority decay wood or wood chips, but a few are decomposers of herbs or *Sphagnum*.

OTHER SIMILAR FUNGI:
– psathyrelloids also have blackish spores but the fragile fruitbodies are less brightly coloured and the caps are mostly dry (page 546).
– gomphidioids are slimy with black spores and deeply decurrent gills (page 596).
– pholiotoids may be both slimy and yellow but have paler, grey-brown spores (page 600).
– the group of little brown mushrooms (LBMs) is separated by their paler brown spore colours (page 616).

FURTHER READING: 156, 169, 181, 225.

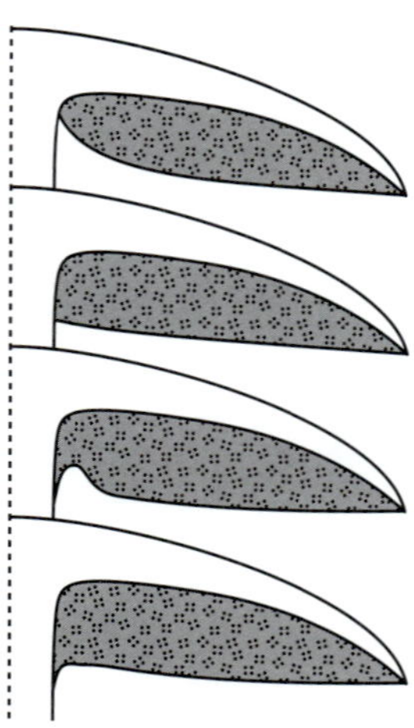

adnexed, adnate, emarginate and shortly decurrent gills

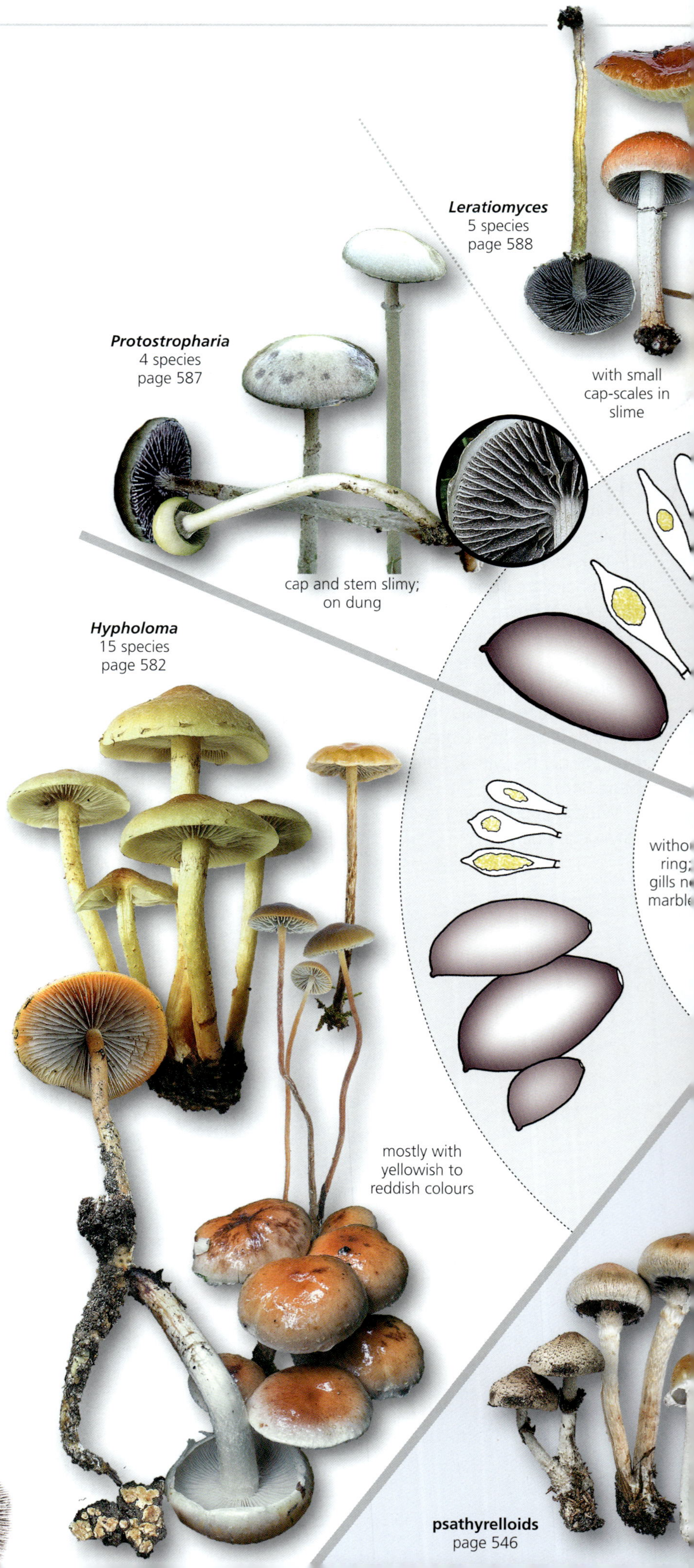

Micro-drawings: spores and cystidia.

Approximate species number applies to temperate Europe.

Stropharia
9 species
page 590

stem mostly woolly-scaly

Hemistropharia
1 species
page 595

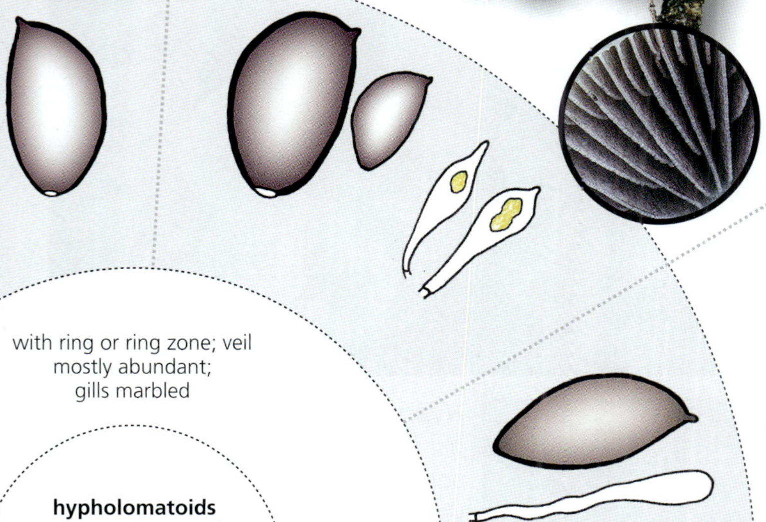

with ring or ring zone; veil mostly abundant; gills marbled

gills with drops; veil prominent, cottony

hypholomatoids are somewhat tough, small to medium-sized, dark-spored agarics with mostly adnate gills and slimy or yellowish caps

other similar fungi

spores ± black | spores ± brown

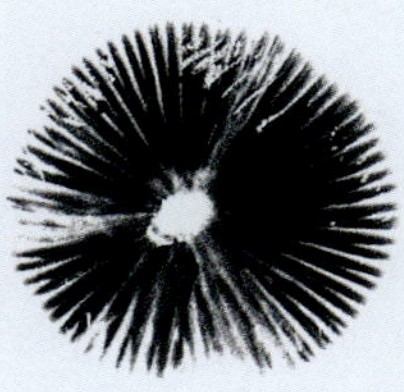

gomphidioids
page 596

pholiotoids
page 600

little brown mushrooms (LBMs)
page 616

Hypholoma is a genus of very dark-spored agarics with ± narrowly adnate gills and typically rather yellowish colours. Some species have a thread-like veil but no ring on the stem. The gills have even spore maturation and thus stay uniform when mature. The spores are thick-walled and have a germ pore, and there are chrysocystidia in the hymenium. Decomposers of wood, *Sphagnum*, *etc*.

† ***Hypholoma fasciculare*** is an almost invariably distinctly clustered, yellow-brown *Hypholoma* with a greenish-yellow tinge on the gills and a strong bitter taste. The cap margin has pale remnants of veil that are mostly stained black by the spores. Has chrysocystidia. The spores measure 6–8 × 4–4.5 µm. Occurs on both deciduous and coniferous wood.

Hypholoma capnoides ▷ has greyish gills and tastes mild. A dwarf form, *H. fasciculare* var. *pusillum* ⋉, can only be separated by its very small size.

Widespread and very common; April–January.

• ***Hypholoma capnoides*** is a rather large, fleshy, mild-tasting, clustered *Hypholoma* with yellow-brown colours and greyish gills that lack green tinges. The cap margin has fine, pale remnants of veil. The stem is mostly smooth and becomes darker towards the base. The clusters are typically slightly rooting into the substrate. Has both cheilo- and pleurochrysocystidia. The spores measure 7–9 × 4–5 µm. Occurs on conifer wood.

Hypholoma fasciculare ◁ is very similar in stature but has greenish-tinged gills and a very bitter taste.

Widespread and common; year-round, but mostly September–November.

Hypholoma marginatum is a typically trooping, rather slender, red-brown to yellow-brown, ± umbonate *Hypholoma* with conspicuous girdles of silvery fibres on the stem and pale remnants of veil in the cap margin. The greyish gills may have an indistinct green tinge. Lacks chrysocystidia on the gill edge. The spores measure 7–9 × 4–5 µm. Occurs on conifer wood, including wood chips.

The combination of dark spores and very conspicuous girdles on the stem make this species easy to identify. *Hypholoma polytrichi* ▷▷ has the same spore dimensions but a very yellow gill edge and less conspicuous girdles on the stem.

Widespread and rather common; August–January.

Hypholoma lateritium is the most fleshy *Hypholoma* and typically has brick-red colours, especially towards the cap centre and grey to bluish-grey gills that lack greenish-yellow tinges. The stem usually has distinct, pale, fibrous girdles, and there are also very pale veil remnants at the cap margin. The taste is slightly bitter. Has chrysocystidia. The spores measure 6–8 × 3–4.5 µm. Occurs in crowded clusters, mostly on stumps from deciduous trees.

Hypholoma fasciculare ⊲⊲ is not brick-red at the cap centre, has greenish-yellow gills and is more bitter-tasting. *Hypholoma capnoides* ⊲⊲ is completely mild-tasting and also less red on the cap; it is restricted to conifer wood.

Widespread and common, becoming scarcer towards the north; most of the year, but mainly September–November.

Thomas Læssøe

Hypholoma polytrichi is a rather slender *Hypholoma* that has a very yellow gill edge due to the presence of numerous chrysocystidia. The relatively small spores measure 7–9 × 4–5 µm. Restricted to acidic turfs, typically in the moss *Polytrichum commune* in damp conifer forests.

Hypholoma elongatum ⊳ and *H. xanthocephalum* ⋉ have similar chrysocystidia on the gill edges but both have larger spores.

Widespread and quite common in the hemiboreal–boreal zones, but rare in the nemoral zones; mainly August–October.

Hypholoma udum is a rather small, but robust, umbonate, ± olive-brown to red-brown *Hypholoma* with first greenish yellow, later sordid brown to brownish-violet gills. Has chrysocystidia and ± bowling pin-shaped cheilocystidia. The ± amygdaliform-fusiform, minutely roughened, purple brown spores have a small germ pore and measure 12–19 × 6–9 µm. Found in *Sphagnum* or on wet peaty soils.

Hypholoma ericaeum ⋉ has somewhat smaller, smooth spores with a large and evident germ pore. *Hypholoma elongatum* ▽ is paler, has a flattened cap and distinctly smaller spores. *Phaeonematoloma myosotis* (page 620) has paler, smooth spores.

Widespread and fairly common; June–November.

Hypholoma elongatum is a slender, pale yellow-brown *Hypholoma* with greyish-yellow gills. The stem is ± reddish-tinged along its entire length. Has chrysocystidia on the gill edge. The ± ellipsoid spores measure 9–12 × 6–7 µm. Found in *Sphagnum* or on wet, peaty soils.

Hypholoma xanthocephalum ⋉ should have amygdaliform spores in side view, but the illustration of *H. elongatum* spores in *Funga Nordica* is also approaching amygdaliform. *Hypholoma laeticolor* ⋉ has narrower spores (average less than 6.5 µm) and the gills are not yellowish (no chrysocystidia on the gill edge). *Hypholoma udum* △ is darker and has much larger spores. See also *H. polytrichi* ◁.

Widespread and common; June–November.

Jens Maarbjerg

Jens Maarbjerg

Hypholoma radicosum is a yellow-brownish, deeply rooting *Hypholoma* with crowded, somewhat serrated, olive-tinged, sordid gills, an odd, perfumed smell and a very bitter and unpleasant taste. Has pleurochrysocystidia. The spores measure 5.5–7 × 3–4 µm. Roots around old conifer stumps, rarely on deciduous wood.

Although distinctive and unlikely to be misidentified, *Hypholoma* may not be the first genus to come to mind.

Widespread, but uncommon; July–November.

Jens Maarbjerg

Hypholoma tuberosum is a ± orange-brown to yellow-brown, somewhat greasy, rather fleshy, umbonate *Hypholoma* with a finely girdled stem that arises from large, folded-gnarled, black sclerotia. Has chrysopleurocystidia. The spores measure 9–11.5 × 5–6.5 µm and have a large, somewhat excentric germ pore. Deeply rooting in thick layers of composting wood chips.

This is the only wood chip-associated agaric with large, folded sclerotia. May recall *H. capnoides* ◁◁ but that species is associated with conifer wood. Occurs on wood chips and could perhaps be confused with the brown-spored *Agrocybe arvalis* (page 627), but that species forms rather small, unfolded sclerotia, and also with *Leratiomyces percevalii* ▷▷, which has larger spores (13–16 × 7–9 µm).

Records in temperate Europe are scattered and probably of alien origin; mainly September–October.

Protostropharia includes pale, slender, slimy, marbled-gilled, coprophilous agarics with chrysocystidia and black, thick-walled spores with a germ pore.

Protostropharia semiglobata is a cream to buff-yellow, slimy-greasy *Protostropharia* with a similar-coloured, sticky, long, stiff stem that is dry above the ring zone. The cap is convex and may have a papilla-like umbo. It has 4-spored basidia, and chrysocystidia on the stem apex. The spores measure 15.5–21 × 9–11.5 µm and have a wide, central germ pore. Occurs on cow and horse dung or on manured soil.

Protostropharia dorsipora ⋉ has a distinctly excentric germ pore and *P. alcis* ▽ has narrower spores and grows on the dung of Eurasian Elk. The shorter-spored *Stropharia albonitens* ▷▷ is dry below the ring.

Widespread and very common; March–December.

Protostropharia alcis is a cream to buff-yellow *Protostropharia* that is sticky to slimy on both cap and stem. The cap has ± evident papilla. Has chrysocystidia. The spores have a central germ pore and measure 12.5–16.5 × 7–8.5 µm. Occurs on *Alces alces* dung in damp forests.

Protostropharia semiglobata △ has wider spores and does not occur on elk dung.

Widespread and rather common within the range of Eurasian Elk – mainly the hemiboreal–boreal zones; summer–autumn.

Leif Stridvall

Jens Maarbjerg

Leratiomyces includes dark-spored agarics, mostly with a convex cap that has small, floccose remnants of veil floating around in the slime, and with marbled gills. Can either have or lack chrysocystidia. The thick-walled spores have a germ pore. Decomposers, mostly on twigs or wood chips.

Leratiomyces ceres is a rather robust, vivid orange *Leratiomyces* with white remnants of veil at the margin when young. The stem, which bears a ± well-defined ring and has white hyphal cords at the base, discolours orange towards the base. The spore-deposit is somewhat purple-tinged. Has chrysocystidia. The thick-walled spores measure 10–14 × (5.5–)6–8 µm and have a large germ pore. Occurs on wood chips in urban settings.

The orange variety of *L. squamosus* ▷ lacks chrysocystidia.

Widespread, occasional, probably introduced to Europe; September–November.

Jens Maarbjerg

Leratiomyces percevalii is a yellow-brown, only somewhat scaly *Leratiomyces* that has a ring zone on its finely fibrillose-floccose stem, but lacks a well-defined ring. Either has or lacks chrysocystidia. The spores are thick-walled, measure 12.5–16 × 7–8.5 µm, and have a large germ pore. Typically occurs on wood chips, mostly in urban environments.

May resemble a species of *Agrocybe* (page 625), but the spores are darker; in contrast to *A. praecox* and *A. rivulosa* (page 625) it has no distinct ring on the stem.

Widespread, occasional; May–December.

Leratiomyces squamosus is an elegant, apricot-orange to orange-red, white-scaled, hygrophanous *Leratiomyces* with pale girdles below a conspicuous stem ring. The young gills have an olive tinge, but later become greyish-black. Lacks chrysocystidia. The spores measure 11–15 × 6.5–9 µm. Occurs on disturbed, mull soils, mostly mixed with remnants of wood, or directly on sticks. Mostly in woodlands.

Reddish forms are called *L. squamosus* var. *thraustus* ⋉. *Leratiomyces ceres* ◁ is more sturdy, the stem ring is not so evident and it has chrysocystidia. The northern *Leratiomyces magnivelaris* ▽ is not hygrophanous and has yellow-brown-tinged scales.

Widespread and rather common; August–December.

Leratiomyces magnivelaris is an elegant, yellow-brown, brown-scaly, mostly papillate, not hygrophanous *Leratiomyces* with a long, stiff stem bearing a distinct, soon black-stained ring (from deposited spores). Lacks chrysocystidia. The spores measure 12–17(–19) × 7–9.5 µm. Typically found along river banks with *Alnus* and *Salix*, on soil or wood.

Leratiomyces squamosus △ is hygrophanous and has paler scales. *Leratiomyces percevalii* ◁ lacks a distinct ring on the stem.

Mainly found in the boreal zone, occasional; late summer–autumn.

Jens H. Petersen/Thomas Læssøe

Species in the genus ***Stropharia*** are generally rather robust with a ± slimy universal veil and a well-defined ring or ring zone. They have dark brown to black spores, marbled gills, chrysocystidia and thick-walled spores with a germ pore. The hyphal cords have star-shaped cells (acanthocytes), which are used to kill nematodes.

Thomas Læssøe

Stropharia coronilla is a rather small, compact, thick-stemmed *Stropharia* with an ochre-yellow, smooth, almost dry cap. The stem is white and has a thick but narrow ring with a furrowed upperside. Chrysocystidia are present in the hymenium. The spores measure 7–9.5 × 4.5–6 µm and have a germ pore. Occurs in open, manured grasslands.

Stropharia albonitens ▽ is paler and more slimy. Species of *Agrocybe* (page 625) lack the thick, narrow ring and have paler, grey-brown spore-deposits.

Widespread and common; May–November.

David Mitchel

Stropharia albonitens is a yellowish-white, slimy, rather slender *Stropharia* with a dry stem that has a ± persistent ring. The gills have a greyish-violet tinge. Has pleurochrysocystidia. The spores measure 7–9 × 4–5 µm and have a germ pore. Found in grasslands and meadows, *e.g.* in connection with the grass *Deschampsia caespitosa*.

Stropharia coronilla △ is drier, more cinnamon-buff, more compact, and has a narrow and very thick ring. *Stropharia melanosperma* ▷ has larger spores.

Widespread, occasional to rare, with most records from the north; September–October.

Stropharia melanosperma is a rather fleshy, very pale, almost white *Stropharia* with a yellow tinge towards the cap centre. The dry stem has a distinct ring, and is somewhat scaly below it. The gills are almost black at maturity. Has cheilo- and pleurochrysocystidia. The spores are large (10–13 × 6.5–9 µm) and have a germ pore. Found in nutrient-rich grasslands but can also occur with wood chips.

Stropharia halophila ⋉, which grows in dunes with *Ammophila*, has no cheilochrysocystidia; *S. albonitens* ◁ has smaller spores.

Apparently widespread, but very rare; late summer–autumn.

Stropharia inuncta is a very slimy, greyish-violet, broadly umbonate *Stropharia* with a ± evident ring or ring zone. Has pleurochrysocystidia, and cheilocystidia that have conspicuous spherical heads. The very dark spores measure 7.5–10 × 4–5 µm and have a germ pore. Occurs mostly in tall, damp grass (*e.g.* in *Dactylis* tussocks) on rich soils and therefore easily overlooked.

May recall the coprophilous *Bolbitius demangei* (page 619), but that species is thinner-fleshed and has much paler spores.

Widespread and probably common, becoming scarcer towards the north; June–November.

Martin Kříž

Thomas Læssøe

Jan Vesterholt

• ***Stropharia hornemannii*** is a very large and fleshy *Stropharia* with rather variable ochre-yellow to wine-reddish cap colours. The cap surface is sticky and the margin has ± distinct velar remnants. The stem is dry and distinctly scaly below a pale, somewhat furrowed ring. The gills are greyish-violet. The spores measure 10–13 × 5–7 µm and have a germ pore. Occurs mostly in damp, mossy conifer forests, and plantations, often in connection with rotten wood; much rarer in boreal deciduous forests.

Stropharia rugosoannulata ▷ is ± smooth below the stem ring. *Stropharia halophila* ⋉ occurs in white dunes and also lacks scales below the ring.

Widespread, especially in the hemiboreal–boreal zones, rare in the nemoral zones; mainly September–November.

Jens Maarbjerg

• ***Stropharia rugosoannulata*** is a large, very fleshy, somewhat sticky, red-brown to purple-brown *Stropharia*. The stem is almost smooth and bears a large, ± furrowed ring. The gills are dark violet-grey. The spores measure 11–15(–17) × 6–8(–9) µm and have a germ pore. Typically occurs on larger piles of wood chips or on various types of compost.

Stropharia hornemannii ◁ is quite similar but has a very floccose-scaly stem surface below the ring.

Widespread, occasional, presumably introduced as it is cultivated commercially for culinary purposes in several European countries; mainly May–October.

Jens Maarbjerg

Stropharia aeruginosa is a very spectacular, turquoise-green, ± scaly, black-spored *Stropharia* with a rather evident stem ring, dark gills with a distinct white edge and, normally, an insignificant smell. With age may become almost cinnamon-buff. Has pleurochrysocystidia. The spores measure 7–9(–10) × 4–5.5(–6) µm and have a germ pore. Occurs on acidic litter, mostly in conifer plantations and forests.

Stropharia cyanea ▷ has paler, browner gills that lack a distinct white edge, and usually lacks an evident ring. *Stropharia pseudocyanea* ▽ smells of freshly ground black pepper and also has a less prominent ring. Aberrant specimens of *S. aeruginosa* that lack an evident ring and are a yellower-brown may be recognized by the absence of cheilochrysocystidia.

Widespread and rather common to common; August–December.

Stropharia pseudocyanea is a ± turquoise, slimy *Stropharia* with rather pale, clay-pink to mouse-grey gills. The odour is distinct, ± like freshly ground black pepper. The gill edge has no cheilochrysocystidia and is, when young, somewhat paler than the sides. The spore-deposit usually lacks black and purple colours. Has pleurochrysocystidia. The spores measure 7–10 × 4–5.5(–6) µm and have a germ pore. Mainly occurs in rather open, herb-rich habitats.

Other turquoise species of *Stropharia* do not smell of freshly ground pepper.

Widespread, occasional; September–December.

Stropharia cyanea is a turquoise to cinnamon-buff *Stropharia* with a slimy cap, a dry, scaly stem that usually lacks a projecting ring, and brownish-grey gills with concolorous edges. Has cheilo- and pleurochrysocystidia, but no normal cheilocystidia. The spores measure 7–10 × 4–5.5(–6) µm and have a germ pore. Occurs on soil and litter, mostly on disturbed rich soils. Also found on wood chips mixed with soil and, rarely, on larger boles.

Stropharia aeruginosa ◁ has darker gills, a white gill edge, normal capitate cheilocystidia and usually a projecting ring. *Stropharia pseudocyanea* ◁ smells strongly of freshly ground pepper.

Widespread and common, becoming scarcer towards the north; September–December.

Thomas Læssøe/Jens H. Petersen

Hemistropharia albocrenulata is a sturdy, fleshy, dark brown, somewhat pholiotoid agaric with whitish veil triangles at the cap margin. The gill edge is white and very serrated and may produce white droplets. The stem is brown and scaly below the ring zone. Lacks chrysocystidia. The spores are very dark, pointed amygdaliform and smooth, measure 11–14 × 5.5–7.5 µm and have a germ pore. Lignicolous, mostly found on *Populus tremula* and other deciduous trees, but also on *Picea*.

Similar species of *Pholiota* (page 608) have spores that are paler and larger, and

Widespread and rare, mostly recorded in the hemiboreal zone from Norway, Sweden and farther east, but also known from the UK; autumn.

Thomas Læssøe

Gomphidioids and *Melanomphalia*

This group includes agarics with dark brown to black spore-deposits and deeply decurrent gills. The two genera of gomphidioids, *Chroogomphus* and *Gomphidius*, have a ± slimy universal veil that covers the entire young fruitbody. They are related to the boletes and have long, somewhat fusiform, typical boletoid spores. The genus *Melanomphalia* has no veils and belongs to the *Agaricales*; it has finely dotted-verrucose spores.

The species of gomphidioids are probably parasites on ectomycorrhizal fungi with conifers (boletes and *Rhizopogon*); the single species in *Melanomphalia* is presumed to be a decomposer.

Other similar fungi:

– *Paxillus* and *Phylloporus* are similar but have brown spore-deposits (page 756).

– *Tapinella* has a brown spore-deposit and a ± excentric or missing stem and grows on wood (page 96).

Further reading: 156, 181, 289.

Chroogomphus rutilus s.l. is a sturdy, fleshy, red-brown, somewhat slimy to dry, almost black-spored agaric with deeply decurrent gills. The stem bears an indistinct ring zone from the ephemeral partial veil, and the flesh has wine-red tinges, but is yellower towards the base. The gills are fimbriate from cystidia. The spores are smooth, ± fusiform, ± dextrinoid and measure 15–22 × 5.5–7 µm. Always found with *Pinus*, but probably parasitic on species of *Suillus* (p. 796) that have *Pinus* as a mycorrhizal partner, *e.g. S. granulatus* (page 800).

Recent studies indicate that there are eight European species in the *Chroogomphus rutilus* complex. In the strict sense, *C. rutilus* should have rather crowded gills (upper picture) and it is the only species with thick-walled cystidia.

The distribution of the eight species is, as yet, unclear. As a whole they are widespread, rather common to occasional; June–November.

Species in the genus **Gomphidius** are characterized by a slimy universal veil that is left as a slimy ring zone. They are probably parasites on *Suillus* mycorrhizae, but the connection is not fully understood.

Gomphidius glutinosus is a reddish-grey *Gomphidius* with canary-yellow stem base and basal stem flesh. The decurrent gills are grey and protected at first by a thick, slimy veil. Later a dark spore-deposit can be seen in the slimy zone just below the gills. The flesh tastes sour. The ± fusiform spores measure 17–20 × 5.5–6.5 µm. Probably a parasite, but if so it is not clear what the host is. It is mainly found near *Picea*.

Gomphidius maculatus ▽ is also greyish, but is more fragile and reddens and blackens after bruising.

Widespread and rather common; June–November.

Thomas Kehlet

Gomphidius maculatus is a rather slender *Gomphidius*, which, when bruised, first reddens, but later blackens. The ± fusiform spores measure 17–23 × 6–8 µm. It is always found with *Larix*, but is probably associated with the *Larix*-mycorrhizal *Suillus grevillei* (page 796).

Distinctive and unlikely to be misidentified.

Widespread, but occasional to rare; June–October.

Gomphidius roseus is a comparatively small, bright red *Gomphidius* that lacks yellow colours in the stem flesh. The gills are grey and decurrent and the stem is white with a somewhat woolly and slimy ring-like zone near the top. The ± fusiform spores measure 17–20 × 5–5.5 µm. Always found in association with the fruitbodies of *Suillus bovinus* (p. 802), which it probably parasitizes.

Distinctive and unlikely to be misidentified.

Widespread, rather common to occasional; July–November.

Melanomphalia nigrescens recalls a species of *Arrhenia* (page 128), but has an almost black spore-deposit. The cap has minute, dark scales, mostly towards the centre. The gills soon become sordid grey from the dark spores, and the stem darkens to black towards the base. Lacks hymenial cystidia. The spores are finely dotted-verrucose, measure (7–)9–12(–15) × 5–7.5(–8) µm and lack a germ pore. Occurs in very calcareous habitats, mostly on disturbed soil, such as in churchyards and in limestone quarries.

Provided the dark spore-deposit is evident, this species is distinctive and unlikely to be misidentified. Young specimens with pale gills can, however, resemble a species of *Arrhenia* (page 128). Despite having dark, ornamented spores the genus *Melanomphalia* is closely related to some white-spored omphalinoids and not to *Crepidotus* (page 90) as was previously assumed.

Very rare, but probably also somewhat overlooked, recorded from *e.g.* Denmark, Germany, Sweden and the UK; June–November.

Thomas Læssøe

Per Schans Christensen

Pholiotoids

This group includes wood-inhabiting agarics with brown (grey-, ochre-, red- to orange-brown) spore-deposits. The principal genera are *Hemipholiota*, *Kuehneromyces*, *Pholiota* and *Gymnopilus*. Many of the species included are rather fleshy, with either slimy or scaly cap surfaces. A number of similar, but mainly terrerestrial, genera are treated under the group 'little brown mushrooms' (LBMs – page 616) and the dividing line between these two groups is not well defined.

All species in the group are decomposers, and almost all degrade wood.

Other similar fungi:

– some 'little brown mushrooms' (LBMs) grow on wood, but at most have only faintly scaly caps (page 616).

– some lignicolous brown-spored agarics are pleurotoid with excentric, lateral or missing stems. See the pleurotoids (page 60).

– lignicolous species with brownish-rose spore-deposits and free gills can be found under the pluteoids (p. 486).

– scaly or slimy species with dark chocolate-brown to black spore-deposits are found under the hypholomatoids (page 580).

Further reading: 112, 156, 181, 225.

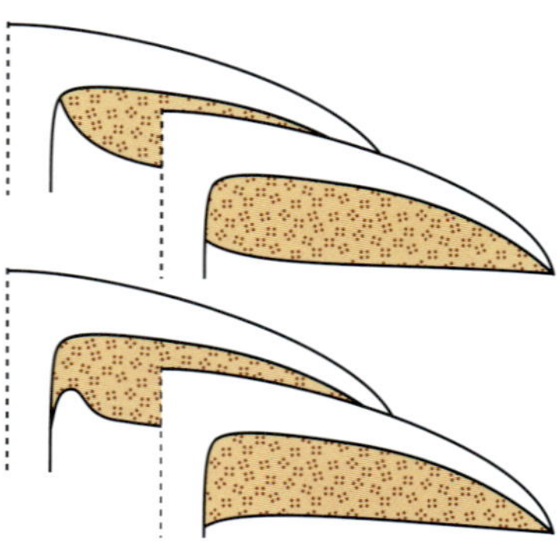

adnexed to adnate, emarginate or short decurrent gills

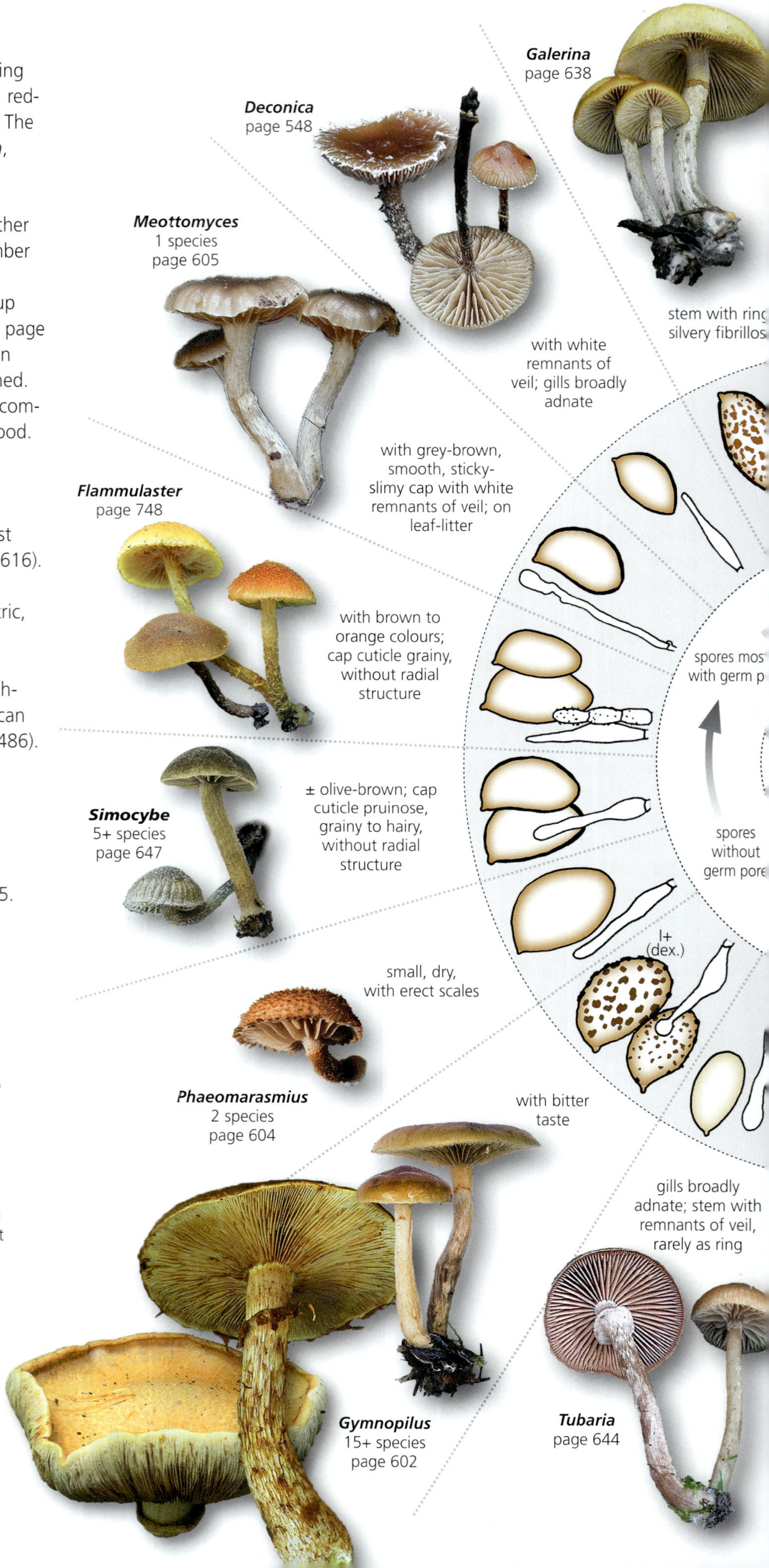

Kuehneromyces
2 species
page 606
stem with ring and
brown scales along
length
Flammula
2 species
page 605
with greasy,
smooth cap;
no pleurocystidia
Micro-drawings:
spores outwards
followed by
cystidia.
Approximate species
number applies to
temperate Europe.
Hemipholiota
2 species
page 607
very fleshy and
large; spore-
deposit grey-
brown; universal
veil thick and
cottony
Pholiota
22+ species
page 608
mostly slimy or
pointed-scaly, ±
yellowish or ± orange
pholiotoids
lignicolous with
rown spores and
easy/slimy or scaly
aps; some have
chrysocystidia
spores with
germ pore and
chrysocystidia
(see more genera with
chrysocystidia under
hypholomatoids, p. 580)
other similar fungi
spore-deposits
± black
spore-deposits
brownish-rose
spore-deposits
± brown
hypholomatoids
page 580
pluteoids
page 486
pleurotoids
page 60
little brown
mushrooms
(LBMs)
page 616

Gymnopilus is a genus of mostly wood-inhabiting, rather fleshy, brown-spored agarics with a bitter taste and orange-brown spore-deposits. The spores are verrucose, dextrinoid and do not have a germ pore. The cystidia are topped with small heads.

Gymnopilus spectabilis is a large, fleshy, mostly minutely scaly, ± clustered *Gymnopilus* with a distinct membranous, ± floccose ring. The taste is very bitter. The spores are ellipsoid to somewhat amygdaliform, measure 8.5–10.5 × 5.5–7 µm and have coarsely ornamented irregular warts. Occurs mainly on deciduous wood, often at the base of living trees, but also on stumps and rarely on conifers.

Phaeolepiota aurea (page 315) is soil-inhabiting and has an even more distinct, flaring ring and a very grainy-powdery surface to the cap and stem. Other species of *Gymnopilus* are less fleshy and most lack a membranous ring. Species of *Pholiota* ▷▷ are not obviously bitter.

Widespread and rather common, becoming scarcer towards the north; July–January.

Gymnopilus penetrans is a medium-sized to large, yellow-brown *Gymnopilus* with ± yellowish, medium-crowded gills and a thin ring zone on the rather pale stem. The taste of the cap cuticle is very bitter. The cap cuticle hyphae are 3–12 µm wide. The roughened spores measure 7–9.5 × 4–5.5 µm. Occurs on wood, including wood chips, mainly from conifers but also from deciduous trees.

The more felty *G. sapineus* ⋉ has cap cuticle hyphae 4–20 µm wide; *G. picreus* ▷ has more contrast between the red-brown cap and more vivid yellow, crowded gills.

Widespread and very common; August–January.

Gymnopilus picreus is a medium-sized, vivid brownish-red *Gymnopilus* with crowded, rather vivid yellow gills and a ± silvery fibrillose stem with an indistinct ring zone. The taste is very bitter. The spores measure 8.5–10.5 × 5.5–6.5 µm. Occurs in damp conifer forests and plantations with much dead wood.

Gymnopilus penetrans ◁ has more distant gills, duller colours and less contrast between gills and cap.

Widespread and common in the hemiboreal–boreal zones, much rarer in the nemoral zones; August–January.

Gymnopilus dilepis is a ± purplish *Gymnopilus* that almost recalls the white-spored *Tricholomopsis rutilans* (page 236), but it has a ring and the purple colours soon fade to orange-brown. The spore-deposit is orange-brown and the spores measure 6–7.5 × 4.8–6 µm. Occurs on large, well-heated piles of wood chips.

This species belongs to a complex of ± purple species, within which further species may turn up in Europe.

Probably accidentially introduced to Europe from SE Asia, mainly recorded in central Europe and in the UK; summer–autumn.

Malcolm Storey

Gymnopilus fulgens is a small, brownish-orange *Gymnopilus.* The stem turns darker towards the base and the ring zone is barely visible. The spores measure 8–11.5 × 5–7 µm. Occurs on wet, peaty soils in open habitats.

Similar species occur on wood in other habitats and have narrower spores.

Widespread but rarely recorded except in the Netherlands – perhaps overlooked elsewhere; mainly July–October.

David Boertmann

Jesus Baz

Phaeomarasmius is a small genus of desiccation-tolerant, small, brown-squamulose, brown-spored agarics that occur predominantly on the bark of living deciduous trees. They have prominent cheilocystidia and the spores are smooth.

Phaeomarasmius rimulincola is a rather pale, felty *Phaeomarasmius* with an almost smooth stem. The spores are large, measuring 11–14 × 7.5–8.5 µm, and lack a germ pore. Occurs on the ± mossy bark of living deciduous trees, *e.g.* together with bark-inhabiting mycenas, such as *Mycena meliigena* (page 211), and liverworts, *etc.*

Other species of *Phaeomarasmius* ▽ have pointed cap scales, and species of *Flammulaster* (page 648) typically have more inflated elements in the cap cuticle and do not occur on living bark.

Widespread, but rarely recorded; may fruit most of the year, except during very cold periods.

Thomas Læssøe

Jan Vesterholt

Phaeomarasmius erinaceus is a rather small, rust-brown, felty-scaly, desiccation-tolerant brown-spored agaric. The cap has pointed, ± upright scales. The cuticle hyphae are not inflated. The spores are smooth, ellipsoid, measure 9.5–12 × 6–8.5 µm and lack a germ pore. Occurs on bark, mostly on attached branches in *Salix* carrs, but also occurs on other hosts in drier woodland types.

Phaeomarasmius rimulincola △ is smaller, has a fibrillose cap surface and grows on bark of larger trees. Species of *Flammulaster* (page 648) are not as tough and not desiccation-tolerant.

Widespread and fairly common to occasional, but rather overlooked; may fruit year-round, but peaks during wet summers.

Thomas Læssøe/David Boertmann

Meottomyces dissimulans is a somewhat anonymous, ± greasy, brown-spored agaric with an evanescent white veil at the margin and a sordid greyish-brown to grey-brown, hygrophanous cap that either has or lacks faint marginal striations. The gills are broad and arched, adnate, greyish to pinkish-buff with a white, ± floccose edge. The stem is ± swollen and, when young, has an indistinct ring zone. The cheilocystidia are irregularly narrowly clavate with gnarled protuberances. The spores are smooth, measure 6.5–9 × 4–6 µm and have an indistinct germ pore. Occurs in damp environements, mostly on leaf-litter from *Populus* or *Fraxinus*.

Recalls a strange-looking *Tubaria* (page 644), although species in that genus are less greasy and the spores lack a germ pore.

Widespread, but occasional to rather rare or possibly overlooked due to its late appearance, absent from the boreal zone; October–March.

Flammula alnicola is a pale greenish-yellow, greasy *Flammula* with a smooth cap that may be more yellow-brown towards the centre; the margin has remnants of veil. The stem is dry, has ± brownish girdles and becomes red-brown towards the base. The taste is typically mild, but may be somewhat bitter. Cheilocystidia are present, but not chrysocystida. The spores are amygdaliform, measure 7–10.5 × 4–5.5 µm and have a germ pore. Typically occurs in clusters on the wood of deciduous trees in damp woodland, mostly on *Alnus* and *Betula*, very rarely on conifers.

Flammula pinicola ⋉ is browner and only known from conifer wood.

Widespread and rather common; July–January.

Thomas Læssøe

Kuehneromyces includes hygrophanous pholiotoids with a ring or a ring zone and small, evanescent veil floccules on the cap surface. Spores smooth. Cystidia narrowly lageniform and lack yellow contents. Lignicolous.

• **_Kuehneromyces mutabilis_** is a hygrophanous, brown-spored agaric with small, buffish veil scales on the ± greasy cap. The stem has an evident flaring ring that is mostly brown on top from deposited spores; there is a dense covering of small, pointed, brown scales below the ring. The odour is aromatic. The cheilocystidia have a long, sinuous neck. The spores are smooth, measure 6–7.5 × 3.5–4.5 µm and have a wide germ pore. Occurs in large clusters and troops on stumps and fallen trunks of deciduous trees.

Kuehneromyces leucolepidotus ⋉ is accepted by some as a separate species based on abundant remnants of veil on the cap and a quickly disappearing ring. *Galerina marginata* (page 638) has a ± smooth, silvery-fibrillose stem, a more farinaceous smell and ornamented spores. *Kuehneromyces lignicola* ▽ has a silvery-fibrillose stem.

Widespread and common; April–January.

Steve Trudell

Kuehneromyces lignicola is a clustered *Kuehneromyces* with a ring zone but no proper ring. Below the ring the stem is fibrillose, not scaly. The spores are smooth, measure 6–7.5 × 3.5–4.5 µm and have a wide germ pore. Occurs almost exclusively on conifer wood, including wood chips.

Galerina marginata (page 638) and other similar galerinas can recall this species, and are closely related, but have ± ornamented spores. Species of *Gymnopilus* ◁◁ have ornamented spores and taste bitter.

A boreal species with stray records from farther south; occurs from late spring to early autumn.

Hemipholiota is a genus of pholiotoids with thick, woolly universal veil and large, dry, woolly-scaly caps. The spores are brown and have a germ pore. The cystidia are lageniform and lack yellow contents. Lignicolous.

Hemipholiota populnea is a very large, very fleshy, brown-spored agaric with abundant white veil on the cap and stem. The cap margin is almost bearded by the veil. The odour is insignificant. Lacks pleurocystidia and chrysocystidia but has lageniform to clavate cheilocystidia. The spores are smooth, measure 7–9.5 × 4.5–6 µm and have a germ pore. Solitary or in small clusters on *Populus*, mostly on the sawn-off ends of thick branches or where branches have fallen off, or on trunks; very rarely on other hosts.

Hemipholiota heteroclita ▽ has a strong, perfumed smell and typically occurs on *Betula* or *Alnus* trunks.

Widespread, occasional, becoming scarcer towards the north; September–January.

Hemipholiota heteroclita is a large, fleshy, pale-scaly, dry agaric with a striking sweetish, aromatic smell. Lacks pleurocystidia, has clavate cheilocystidia. The ± ellipsoid spores measure 7.5–10 × 4.5–6.5 µm. Occurs on living or dead deciduous trees, mostly *Betula* and *Alnus*.

Hemipholiota populnea △ is even larger, has very conspicuous pale scales, lacks a distinct smell and grows on *Populus*.

Widespread and rather common in the Scandinavian hemiboreal–boreal zones but becomes scarcer towards the south and west; August–January.

Jens Maarbjerg

Species in the large genus ***Pholiota*** often have a threadlike and/or slimy universal veil that is left as scales and slime on cap and stem. They mostly have pleurocystidia with yellow contents (chrysocystidia) and the smooth, brown spores usually have a germ pore. The majority of the species are lignicolous and often clustered.

Pholiota squarrosa is a large, distinctly clustered, dry *Pholiota* with coarse, recurved, brown scales, on both the cap and stem. Young gills are yellowish. The odour is indistinct and the taste mild. Has chrysocystidia and lageniform cheilocystidia. The ± bean-shaped, smooth, brown spores measure 6–8 × 3.5–5 µm. Mostly found at the base of living deciduous trees, but also around stumps, rarely on conifer wood.

Pholiota squarrosoides ▷ is somewhat greasy-slimy, smells of bread yeast, is slightly paler, has whitish gills at first, smaller spores and typically occurs on fallen trunks.

Widespread and common; August–December.

Pholiota squarrosoides is a large, very spectacular, slightly greasy-slimy *Pholiota* with pale yellow-brown, recurved scales on the cap and stem, and a very pale cap surface that can be seen clearly between the scales. The young gills are whitish. Smells of bread yeast. Has a few chrysocystidia and clavate cheilocystidia. The ellipsoid to bean-shaped spores measure 4.5–6.5 × 2.5–3.5 µm. Occurs on dead trunks of deciduous trees, mostly *Fagus* and *Betula*, usually in old-growth habitats.

Pholiota squarrosa ◁ is completely dry and not as pale between the scales; it also has pale yellowish gills that acquire a greenish tint when coloured by the dark spores, and somewhat larger spores (6–8 × 3.5–5 µm).

Widespread, but very rare or absent in western parts of temperate Europe; increasing towards the east; absent from the UK and the Netherlands; mainly September–October.

Pholiota flammans is a medium-sized, vivid orange-yellow, dry *Pholiota* with yellow, recurved, pointed scales on the cap and stem (may be sparse on older specimens, which are also a more yellow-brown). The gills are at first bright yellow, later rusty brown. Has chrysocystidia and ± lageniform cheilocystidia. The spores measure 4–5 × 2–3 µm. Occurs on conifer wood, mostly in contact with soil, rarely on wood from deciduous trees.

Distinctive and unlikely to be misidentified.

Widespread and rather common; July–November.

Pholiota adiposa is a large, distinctly clustered, apricot-orange, very slimy *Pholiota* with ± adpressed, scattered, brown scales on the cap and stem. Has chrysocystidia and clavate cheilocystidia. The spores measure 7.5–10 × 5–6 µm. Occurs on dead or living deciduous wood, mostly *Fagus*. The mycelium grows in the centre of the trunks and the fruitbodies mostly appear high up where branches have broken off, or from split, fallen trunks or the sawn-off ends of trunks.

The name of this fungus has been much debated, but is now accepted as *P. adiposa*. *Pholiota jahnii* ▽ also occurs on *Fagus* but mostly appears close to the soil and has pointed, almost black, cap scales and smaller spores.

Widespread and rather common, but becomes scarcer towards the north and east; April–June & August–December.

Pholiota jahnii is a large, distinctly clustered, slimy *Pholiota* with dark, pointed cap scales. Has chrysocystidia and clavate cheilocystidia. The spores measure 5–7 × 3–4 µm. Occurs on ± buried wood of deciduous trees, mostly *Fagus*, typically on clay or mull soils.

Pholiota adiposa △ has larger spores, paler, pointed cap scales and typically occurs higher up without contact with the soil.

Widespread, occasional to rare in the nemoral zones and absent from the boreal zone; September–December.

Jens H. Petersen/Thomas Læssøe

Pholiota gummosa is a rather small, greasy, pale greenish-yellow *Pholiota* with indistinct adpressed, pale brownish scales. The stem is pale and finely scaly below the ring zone. The chrysocystidia are ± cylindrical and have a ± distinct head. The spores measure 5.5–8 × 3.5–4 5 µm. Typically grows in clusters from buried wood in lawns in parks and gardens, along roadsides, *etc.* May also occur directly on wood, mostly deciduous.

Pholiota connissans ▷▷ is generally more slender and lacks the greenish shades. It normally occurs in damp or wet habitats near *Salix*.

Widespread and rather common, becoming scarcer towards the north; May, August–January.

Pholiota lucifera is a medium-sized, ± clustered, yellowish to orange-brown, somewhat slimy *Pholiota* with small, adpressed, brown scales. The taste is bitter. The cheilocystidia are ± clavate but may be constricted. Pleurocystidia (chrysocystidia) absent. The spores measure 7–9 × 4–5.5 µm, and either lack or have a very small germ pore. Occurs on deciduous wood that is usually buried.

Other ± slimy, ± yellow pholiotoids have pleurocystidia, mostly as chrysocystidia.

Widespread but rarely recorded; May, July–October.

Jean-Marc Moingeon

Pholiota astragalina is a medium-sized, vivid brownish-orange, slimy, ± clustered *Pholiota* with a few whitish velar remnants on the cap and almost apricot gills. The stem has fine, pale scale girdles and an indistinct ring zone. Has chrysocystidia and very long- and thin-necked cheilocystidia. The spores measure 5.5–8 × 3.5–4.5 µm. Occurs on conifer wood, mostly in contact with the soil.

Similarly coloured species of *Hypholoma* (page 582) have darker spores, and other pholiotoids do not have the distinct apricot colours on the gills.

Widespread and occasional, apparently rare in the UK; July–November.

Jan Vesterholt

Pholiota mixta is a medium-sized, pale red-brown, slimy to greasy *Pholiota* with a pale cap margin and pale stem scales. The stem stains red-brown towards the base with age. Lacks true chrysocystidia, but has many cystidia with yellow contents and long necks. The spores are ellipsoid, measure 6–8.5 × 3.5–4.5 µm and have an evident germ pore. Occurs on soil and litter in coniferous and mixed forests, occasionally on wood chips and similar substrates; also in urban settings.

Pholiota spumosa ▷ is more yellow-brown.

Mostly found in the north with more scattered occurences in the nemoral zones; April–May, July–November.

Pholiota lenta is a medium-sized, very slimy *Pholiota* with very indistinct, pale, triangular veil scales that float in the slime close to the cap margin. The gills are at first pale yellow, later a more rusty-brown. The stem is ± dry and has an indistinct ring zone, below which are pale, woolly scales. The cystidia may have yellow contents in the neck, but true chrysocystidia are not present. The spores measure 6–8 × 3–4.5 µm and are ± bean-shaped in side view. Occurs on litter in deciduous and coniferous forests, mostly on rich soil.

Other similar pholiotoids have more vivid colours. *Meottomyces dissimulans* ◁◁ has more sordid olive-brown colours.

Widespread and common; August–December, peaking late in the season.

Thomas Læssøe

Pholiota spumosa is a medium-sized, slimy-greasy, yellow-brown *Pholiota* with a red-brown cap centre and rather indistinct, scattered scales in the slime. The stem is yellow above and red-brown towards the base. Cystidia have long, ± apically widened necks and mostly have yellow contents in the upper parts. Spores measure 6–8.5 × 3.5–4.5 µm. On conifer wood, mostly buried or as wood chips; also in urban settings.

Pholiota mixta ◁ is more red-brown and lacks the yellow tones.

Widespread and fairly common in the north, but much rarer in the nemoral zones; mainly September–November.

David Boertmann

Thomas Læssøe

Pholiota conissans is a small, slender, apricot-orange *Pholiota* with a greasy cap surface. The gills are pale yellow, and the stem somewhat floccose below the ± indistinct ring zone. Has chrysocystidia and very long-necked cheilocystidia. The spores measure 6–8 × 3–4.5 µm. Occurs on ± buried *Salix* wood and apparently on soil in *Salix* carrs and meadows.

Pholiota pityrodes ⋉ occurs on *Phragmites* and is even more slender. *Pholiota gummosa* ◁◁ normally has a greenish tinge on the cap and is somewhat more fleshy.

Widespread and rather common in nemoral zones, but apparently rare in the north; mainly August–November.

Pholiota highlandensis is a rather small, ± yellow-brown to red-brown, greasy *Pholiota* with rather indistinct velar scales; when young, the cap margin is white from remnants of veil. The long-necked cystidia may have yellow contents in the neck – not proper chrysocystidia. The spores are ellipsoid (not bean-shaped) and measure 6–8 × 4–5 µm. Occurs on burnt ground, mostly in forests.

Pholiota gummosa ◁◁ typically has colder colours. If the habitat is not known, it may be diffucult to separate *P. highlandensis* from *P. spumosa* ◁◁ and *P. mixta* ◁◁.

Widespread, occasional; almost all year, but mostly September–October.

Pholiota scamba is a very small, aberrant, pale, yellowish, solitary or trooping *Pholiota* with rather indistinct pale velar scales on the greasy cap. The gills are at first pale yellow, and the thin stem is white-flocculose above the indistinct ring zone. The lageniform cystidia may have yellow contents in the neck, but proper chrysocystidia are absent. The spores measure 7–10 × 4.5–6 µm. Occurs on very decomposed conifer wood, mostly with distinct hyphal cords in the wood.

Young fruitbodies may recall pale species of *Inocybe* (page 652).

Widespread, locally common to occasional, becoming scarcer towards the south; June–November.

Pholiota henningsii is a small, somewhat anonymous, pale yellow-brown to red-brown *Pholiota* with small, rust-coloured, adpressed scales and pale remnants of veil in the cap margin. The gills are very pale and the stem is fibrillose below an indistinct ring zone. Has lageniform to fusiform cystidia without yellow contents. The spores measure 8–10 × 4.5–5.5 µm. Occurs in mires, mostly deep in *Sphagnum*.

May recall a pale *Hypholoma* (page 582), but the spores are not nearly as dark.

Widespread, but rarely recorded, apparently absent from the UK and Norway; mainly September–October.

Thomas Læssøe

Pholiota tuberculosa is a small, completely dry, yellow-brown to rust-brown *Pholiota* with pale yellow, ± brown-spotted gills and a finely scaly stem that is red-brown towards the base. Lacks pleurocystidia but has ± capitate cheilocystidia. The somewhat bean-shaped spores measure 7–9 × 4–5.5 µm and usually lack a germ pore. Occurs on *Quercus* branches, mostly attached to living trees, but also on wood from other deciduous trees.

Several characters indicate that this species does not belong in the genus *Pholiota*. The very similar *Flammulaster limulatus* (page 648) is somewhat more rust-coloured, less distinctly scaly and the spores are more bean-shaped and have a small germ pore.

Widespread, occasional to locally common and apparently increasing; June–November.

Thomas Læssøe

Little brown mushrooms

This group includes mostly small, predominantly brownish agarics, *e.g. Tubaria*, *Naucoria*, *Galerina*, *Conocybe*, *Pholiotina*, *Simocybe*, *Agrocybe* and *Flammulaster*. The label LBM that is used for this group of fungi was coined many years ago. The more vividly coloured *Bolbitius* species are also included here, and also the more fleshy *Agrocybe*, *Cyclocybe* and *Phaeocollybia*. The group overlaps somewhat with the pholiotids (page 600).

The majority of the species are decomposers but *Descolea* and *Naucoria* are ectomycorrhizal and *Phaeocollybia* is possibly a root parasite.

OTHER SIMILAR FUNGI:
– psathyrelloids and hypholomatoids have darker, purple-brown to black spores (pages 546 & 580).
– pholiotoids are mostly wood-inhabiting, larger and have slimy or scaly caps (page 600).
– *Inocybe* has mostly coarsely fibrillose cap surfaces and spermatic or strong fruity odours (page 652).
– *Hebeloma* has mostly earthy, fruity or flower-like odours (page 680).
– *Cortinarius* has mostly larger fruitbodies with red-brown spore-deposits and threadlike partial veils (page 690).

FURTHER READING: 102, 103, 113, 113, 121, 122, 123, 156, 181, 182, 201, 227, 296, 331, 341.

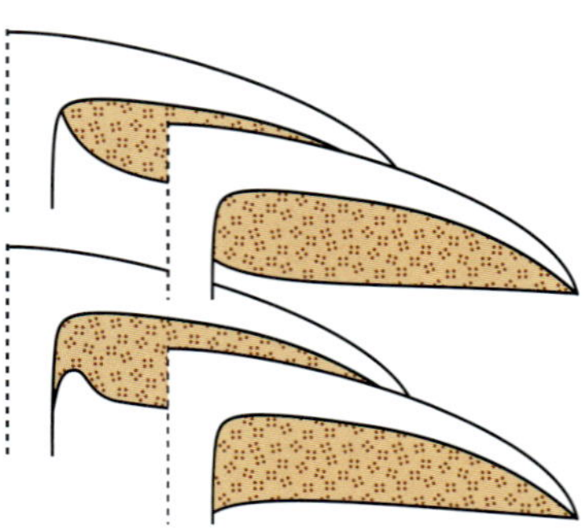

free, adnexed to adnate, emarginate or short decurrent gills

Micro-drawings: spores outwards followed by cystidia. Dex. = dextrinoid.

Approximate species number applies to temperate Europe.

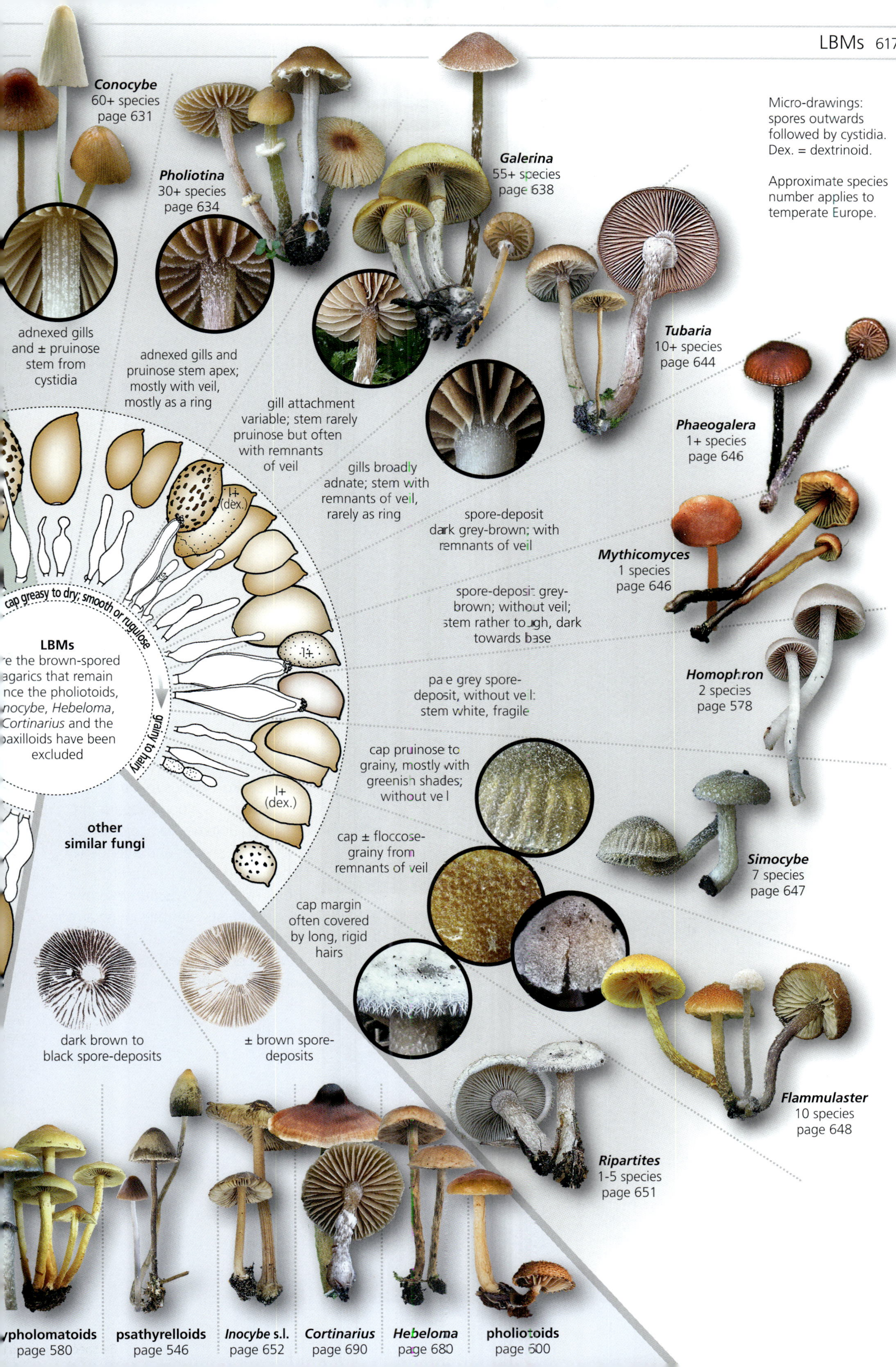

Jens Maarbjerg

Bolbitius is a genus of pale, brown-spored agarics with slimy caps and rather ephemeral, almost coprinoid, thin-fleshed, mostly vividly coloured fruitbodies with narrow, free to adnexed gills. The spores are smooth and have a prominent germ pore. Decomposers on middens, manured soil, rotten vegetation and decaying wood.

Bolbitius titubans is a pure yellow, but soon paler *Bolbitius* that occurs in variable sizes depending on the substrate. The cap is initially egg-shaped and very slimy, later expanded and radially furrowed with a more intensely coloured centre. The stem is mostly paler and finely mealy-floccose. The fruitbodies shrivel or rot very quickly. The spore-deposit is yellow-brown. The thick-walled spores measure 9.5–15 × 5.5–9 µm and have a germ pore. Has a rather wide ecology and may occur on dung, rotten grass, rotten wood or wood chips, in both open habitats and in woodlands.

Partly sterile forms may look similar to yellow species of the pinkish-spored genus *Pluteus* (page 494) or species of the white-spored genus *Leucocoprinus* (page 336), but the cap slime is a good separating character. *Bolbitius variicolor* ▷ has olive-yellow shades and a tendency to be veined at the centre of the cap and *B. callistus* ⋉ has purple-brown colours at first, later becoming a more yellowish-brown.

Very common and widespread; April–November.

Bolbitius reticulatus is a slender to rather compact, somewhat pluteoid, grey, ± veined and slimy *Bolbitius*. *Bolbitius reticulatus* f. *aleuriatus* ⋉ is a gracile form that lacks veins on the cap surface. The spores measure 7–12 × 4–5.5 µm and have a germ pore. Occurs on very decomposed wood from deciduous trees, often partly hidden in cavities in large fallen trunks.

Species of *Pluteus* (page 491) look rather similar but are not slimy on the cap and have brownish-rose spore-deposits and much smaller spores that lack a germ pore.

Widespread, occasional, absent from the boreal zone; May–January.

Bolbitius demangei is a lead-grey, rather large *Bolbitius*. The spores measure 8.5–14 × 6–8.5 µm and have a germ pore. Occurs on middens, perhaps mostly on horse dung.

Bolbitius excoriatus ⋉, which has been reported from Sweden, is smaller and the cap cuticle splits into radial fibres to reveal the white flesh below. *Bolbitius coprophilus* ▷▷ may lose its pink colours and then strongly recalls *B. demangei*, but has larger spores.

Very rare, but perhaps expanding its range; mainly June–September.

Jette Anitha Hansen

Bolbitius variicolor is a sturdy, olive-yellow, mostly veined *Bolbitius*. The stem is often yellow along its full length. The spores measure 10.5–14 (–15) × 6.5–8.5(–9) × 6–7.5 µm and have a germ pore. Occurs mostly on wood chips, straw mixed with dung, *etc*.

This species is sometimes treated as a variety of *B. titubans* ◁ under the name *B. titubans* var. *olivaceus*. *Bolbitius titubans* may occur in large forms but it lacks olivaceuous tinges on the cap and is not veined at the cap centre.

Widespread, but rare, although perhaps spreading; May–September.

Frede Scheye

Bolbitius coprophilus is a rather large and ± flamingo-pink *Bolbitius*. The stem is paler than the cap and mostly lacks red tinges. The spores have a distinct, ± excentric germ pore and measure 11.5–16 × 8–11 × 7–9 µm. Occurs on middens or large piles of wood chips.

Older, faded fruitbodies may be confused with *B. demangei* ⊲⊲ but that species has somewhat smaller spores.

Widespread in the nemoral zones with a few records from farther north, but perhaps expanding its range; March–October.

Phaeonematoloma myosotis is a rather elegant, olive-brown, slimy, umbonate, not hygrophanous agaric with a slimy stem that is increasingly darker towards the base. Lacks purple tinges on the gills. Has many chrysocystidia, mainly in the hymenium. The spore-deposit is ± greyish-brown. The smooth, ovoid to amygdaliform spores measure 14.5–17 × 7.5–9.5 µm and have a rather small germ pore. Occurs on peaty damp soils or directly in *Sphagnum*.

Hypholoma udum (page 585) has similar large spores but they are purple-black and finely ornamented, its cap is only greasy, and there is a ± purple tinge on the gills.

Widespread, fairly common to occasional; May–October.

Thomas Læssøe/Jens H. Petersen

Descolea antarctica is a somewhat *Agrocybe/Cyclocybe*-like, brown-spored agaric with a red-brown, smooth, slimy cap with white velar floccules, and a hanging, striate ring on the fibrous stem. The smell is insignificant. Has ± clavate cheilocystidia and no pleurocystidia. The lemon-shaped spores are dextrinoid, verrucose, measure 11–13.5 × 6–7.5 µm and lack a germ pore. Mycorrhizal with *Nothofagus*.

Agrocybe ▷▷, *Cyclocybe* ▷▷ and *Pholiotina* ▷▷ are generally ± dry and lack white velar remnants swimming in slime, have smooth spores that are a different shape, and are all decomposers.

Very rare and introduced from Patagonia with *Nothofagus* seedlings that were destined for a forestry experiment on the Faroe Islands but were overwintered in an arboretum north of Copenhagen, Denmark; it was subsequently found fruiting under some of the trees left behind (top image), and also in a plantation near Torshavn (lower image); October–November.

Phaeocollybia is a genus of rooting, acute, naked agarics with brown, ornamented spores. The genus is often considered to be mycorrhizal, but it is more likely that the species concerned have a parasitic connection with tree roots.

Phaeocollybia christinae is a brownish-orange to orange-brown, very pointed-conical *Phaeocollybia* with a very long and very deeply rooting, yellow-brown to orange-brown stem. It has a marzipan-like smell. The cheilocystidia are ± capitate. The amygdaliform to lemon-shaped, rough spores measure 8.5–12(–14) × 4.5–6 µm. Found mainly in mossy conifer forests, probably parasitic on roots.

The two other orange-brown species of *Phaeocollybia*, *P. jennyae* ⋉ and *P. cidaris* ⋉, have smaller spores and more farinaceous or radish-like odours.

Widespread in the hemiboreal–boreal zone, very rare to absent from the nemoral zones, absent from the UK; mainly September–October.

Phaeocollybia arduennensis is a small, very dark, ± purple-brown *Phaeocollybia* with a red-brown to purple-brown, deeply rooting stem. The odour is earth-like and the taste mild to somewhat bitter. The cheilocystidia are distinctly cylindrical-capitate and have clamps. The spores are small, rough, broadly ellipsoid to amygdaliform and measure 5–6(–7) × 3.5–4.5 µm. Occurs in *Picea* and Abies forests and plantations on mull soils, rarely in deciduous forests, probably parasitic.

The very dark colours, small size and presence of clamps should make the species easily recognizable.

Widespread, rare or perhaps somewhat overlooked, ± absent from the boreal zone; August–October.

Phaeocollybia festiva is a medium-sized, green-tinged, later more red-brown *Phaeocollybia* with a greasy cap. The deeply rooting stem may have greenish to bluish tinges. It has a radish-like odour. The cheilocystidia are ± clavate. The amygdaliform, roughened spores measure 7–8 × 4.5–5 µm. Occurs on rich soils, with both coniferous and deciduous trees, and is probably parasitic.

Phaeocollybia lugubris ⋉ is a similar, but more robust, more slimy *Phaeocollybia* with lecythiform cystidia (shaped like bowling pins).

Mostly recorded in the hemiboreal zone in Scandinavia with stray records from farther south; mainly May–September.

The genus ***Cyclocybe*** includes medium-sized brown-spored agarics with well-defined, persistent rings. The spores either lack or have a very small germ pore.

Cyclocybe erebia is a brown to dark brown, greasy, farinaceous agaric with a large, projecting stem ring and fine, pale girdles below. Has club-shaped, lageniform to urticiform cheilo- and pleurocystidia. The basidia are mostly 2-spored. The spores measure (9–)11–5(–17) × (5–)6–7.5(–8) µm and lack a germ pore. Occurs on rich, disturbed soils, mostly amongst *Urtica* in parks and forests.

Agrocybe firma ▷▷ is recognized by its ringless stem, dry cap and occurence on wood. May also recall some psathyrelloids (page 546), but the combination of a farinaceous smell and absence of a germ pore should make identification straightforward.

Widespread, fairly common to occasional, becoming scarcer towards the north; mainly June–November.

Jan Vesterholt

• ***Cyclocybe cylindracea*** is a large, rather pale, aromatic fragrant agaric with a large stem ring and minute scales on the stem below. There are occasionally remnants of veil at the cap margin. The spores measure 8–12(–15) × 5–6.5(–8) µm and have a very small germ pore (up to 0.5 µm wide). Found mostly on wood and wood chips of *Populus* and *Salix*.

Agrocybe rivulosa ▷ and *A. praecox* ▷ have wider germ pores and a farinaceous smell and taste.

Widespread and fairly common to rare in the nemoral zones, just reaching Denmark, Norway and Sweden, possibly expanding in range; spring-autumn.

The genus ***Agrocybe*** includes small to large agarics that lack radial structure in the cap surface and may have a ring. The spores usually have a germ pore.

Agrocybe praecox is a medium-sized to large, fleshy, hygrophanous, farinaceous *Agrocybe* with a distinct stem ring. The stem base normally has prominent white hyphal cords. Has rather few lageniform cheilo- and pleurocystidia. The spores measure 8–10(–12) × 5–7(–8) µm and have walls up to 1 µm thick and a wide germ pore. Typically occurs on ± disturbed soil or decaying wood chips.

Agrocybe elatella ⋉ is generally more gracile and has a more flaring stem ring; its spore walls are 1.5 µm thick, and it has more cystidia on the gill edge. *Agrocybe rivulosa* ▽ has a veined cap surface and somewhat longer spores (10–14 × 6.5–8 µm). *Agrocybe putaminum* ▷▷ has no ring on its felty stem. *Agrocybe praecox* forms a complex of about four biological species that cannot be separated morphologically.

Widespread and common; April–October, mainly early in the season.

Agrocybe rivulosa is a large and fleshy *Agrocybe* with pale colours, a characteristic wrinkled/veined cap surface and a prominent stem ring. The spores measure 10–14 × (6–)6.5–8(–8.5) µm and have a wide germ pore. Occurs on thick piles of composting wood chips.

Agrocybe praecox △ has somewhat smaller spores and lacks the veined cap surface pattern. *Leratiomyces percevalii* (page 588) has darker gills and blackish-brown spores. *Bolbitius demangei* ◁◁ and *A. putaminum* ▷▷ lack a ring on the stem.

Widespread and expanding in range, not recorded in the hemiboreal–boreal zones; April–November.

Thomas Læssøe

Agrocybe putaminum is a fleshy, farinaceous *Agrocybe* with a pubescent cap and stem surface. The cystidia are lageniform to broadly lageniform and do not have extensions. Spores 9.5–13(–15) × 5.5–7(–8) µm; with a germ pore. Typically occurs on wood chips on rich soil in gardens and parks.

The velvety cap surface in combination with the rather fleshy fruitbodies should make this species easy to recognize. Other species of *Agrocybe* and *Leratiomyces percevalii* (page 588) that occur on wood chips have distinct rings or ring zones on the stem.

Widespread in the nemoral zones, occasional to very rare; mainly May–September.

Tobias Bøllingtoft

Agrocybe dura is a very pale coloured *Agrocybe* with a fugacious partial veil that often is attached to the cap margin. The cap surface soon cracks irregularly. The smell is faint. The spores are thick-walled with a big germ pore and measure 10–14(–18) × 6–8(–9) µm. In calcareous grasslands, including along road verges.

Other fleshy *Agrocybe* species have a stem ring and typically a distinct farinaceous smell.

Widespread, occasional and becoming rare in the boreal zone, mainly in the summer period.

Agrocybe pediades is a rather slender, farinaceous, buff, ringless *Agrocybe* with a floccose stem that lacks a root-like extension. The gills are rather distant and wide, and pale grey-brown. The cheilocystidia typically have a small, round head; no pleurocystidia. Spores measure 10–16(–20) × 7–11(–15) µm and have a wide germ pore. Found in open grasslands, including road verges and lawns, rarely on old cowpats.

Agrocybe ochracea ⋉, a poorly known species, has pleurocystidia. *Agrocybe pediades* var. *fimicola* ⋉ has a veil and occurs on old dung. *Hypholoma subericaeum* ⋉ has darker and much smaller spores.

Widespread and common; April–October, mostly early in the season.

Agrocybe pusiola is a very small, slender, farinaceous *Agrocybe*. The spores are small, 7–10(–11.5) × 4.5–6(–6.5) µm, and lack a germ pore. Occurs on soil in open, dry habitats, perhaps mostly calcareous.

Agrocybe vervacti ⋉ is also small-spored but is somewhat larger and not farinaceous. *Agrocybe pediades* ◁ has far larger spores and the fruitbodies are more fleshy. Species of *Tubaria* ▷▷ normally have velar remnants at the cap margin.

Widespread but infrequently recorded, perhaps overlooked, absent from the boreal zone; June–November.

Thomas Læssøe

Agrocybe firma is a dark, ringless, farinaceous, hygrophanous *Agrocybe* that lacks veil. The stem is covered by fine floccules. The spores are small, measuring (5.5–)6–8(–8.5) × 4–5 (–5.5) µm, and either lack or have only an insignificant germ pore. Occurs on the wood of deciduous trees in damp woodland with plenty of dead wood.

Cyclocybe erebia ◁◁ has a prominent ring and much larger spores. Psathyrelloids (page 546) normally have a threadlike veil, either no or a different smell, and darker spores.

Widespread, but rarely recorded; (August–) September–October.

Thomas Læssøe/Jens H. Petersen

Agrocybe arvalis is a rather slender, farinaceous *Agrocybe* with a smooth cap and a finely felty-floccose stem that has root-like extensions to a dark, round sclerotium. The pleurocystidia have finger-like extensions on top; the caulo-, pileo- and cheilocystidia are ± lageniform. Spores measure 8–10.5(–12) × 4.5–6 µm and have a small germ pore. Occurs on soil and wood chips in a variety of woodlands, mostly on rich soils.

No other *Agrocybe* has such unusual pleurocystidia; otherwise best detected by its rooting, floccose stem. The rooting *Hypholoma tuberosum* (page 586) has darker spores.

Widespread in the nemoral zones, occasional; July–October.

Ole Martin

Thomas Læssøe

Naucoria is a genus of small, brownish agarics with tapering or clavate cheilocystidia and rust-brown, ± verrucose spores. They are ectomycorrhizal and most species favour damp or wet soils, *e.g.* in *Alnus* and *Salix* carrs.

Naucoria escharioides is a pinkish-buff to buff, almost non-striate *Naucoria* with sparse remnants of veil. The gills are ± pinkish-buff. The odour is indistinct. The cheilocystidia taper to a point. The amygdaliform, rough spores measure 9–12 × 5–6.5 µm and lack a germ pore. Occurs on damp or wet soils, mycorrhizal with *Alnus*.

This species is also known as *Alnicola escharioides* and *N. melinoides*. *Naucoria striatula* ⋉ is striate to the cap centre and *N. luteolofibrillosa* ▷ has abundant threadlike veil on cap and stem.

Widespread and very common; July–November.

Jacob Heilmann-Clausen

Naucoria scolecina is a pale orange-brown, somewhat striate *Naucoria* with orange-brown-tinged gills and a similarly coloured stem. The odour is indistinct. The cheilocystidia are narrow and tapering. The somewhat fusiform, rough spores measure 10–14.5 × 5–6.5 µm and lack a germ pore. Mycorrhizal with *Alnus* on wet soil.

Naucoria sphagneti ⋉ and *N. subconspersa* ⋉ have similar gill colours but are mostly non-striate.

Widespread and common; May–November.

Naucoria luteolofibrillosa is a ± pinkish-buff *Naucoria* with abundant threadlike universal veil on both cap and stem. The cheilocystidia are narrow and tapering. The rough spores measure 8–12 × 5–7 µm and lack a germ pore. Mycorrhizal with *Alnus* on damp soil.

The rather pale gills and abundant veil should make this species fairly easy to recognize. *Tubaria conspersa* ▷▷ may look similar but has smooth spores and rounded cystidia.

Widespread, uncommon; August–November.

Jacob Heilmann-Clausen

Naucoria fellea is a dark, red-brown, finely rugose to minutely scaly *Naucoria* with red-brown gills. The stem is similarly coloured and floccose. The taste is very bitter. The cheilocystidia are narrow and tapering. The rough spores measure 11.5–13.5(–15) × 5.5–7 µm and lack a germ pore. Mycorrhizal with *Salix* and *Betula* on sandy soils.

Naucoria amarescens ▽ also tastes bitter but has smaller spores, as does the mild-tasting *N. geraniolens* ⋉. *Naucoria tantilla* ⋉ is mild-tasting and apparently occur in alpine habitats.

Widespread but very rarely recorded; mainly August-September.

Thomas Læssøe

Naucoria amarescens is a ± brownish-orange, pallescent, bitter-tasting *Naucoria*. The stem darkens towards the base. The odour may recall that of *Pelargonium* leaves. The cheilocystidia are tapering. The rough spores measure 8.5–10.5 × 4.5–5.5 µm and lack a germ pore. Occurs on burnt ground and on calcareous, not very damp soil in *Salix* thickets.

Naucoria geraniolens ⋉ has a similar smell but a mild taste. *Naucoria fellea* △ has larger spores.

Widespread, rather rare, but probably somewhat overlooked; April–June and September–October.

Naucoria celluloderma is a ± clay-buff to pale brownish-orange, distinctly striate *Naucoria*. It has clamps and 2-(4-)spored basidia. The cheilocystidia are ± clavate to lageniform but not pointed. The rough spores measure 13–14.5 × 7–7.5 µm and lack a germ pore. Mycorrhizal with *Alnus*, mostly on rather dry ground.

Recalls the species with attenuated cystidia on the previous page spread but has less pointed cystidia. Other similar species have 4-spored basidia, *e.g. N. submelinoides* ⋉, or have darker fruitbodies.

Widespread, occasional, easily overlooked; July–November.

Thomas Læssøe

Naucoria salicis is an orange-brown *Naucoria* with a paler cap margin and a rather pale, somewhat fibrillose stem. Basidia 2–4-spored; cheilocystidia ± clavate, 6–10(–16) µm wide at the top. The rough spores measure 14–19 × 7–9 µm (from 2-spored basidia) or 13–14 × 6–7 µm (from 4-spored basidia) and lack a germ pore. Occurs on damp soils, mycorrhizal with *Salix*.

The very similar *N. bohemica* ▽ has only 2-spored basidia and the spores are smaller. *Naucoria badiolateritia* ⋉ has cystidia 7–24 µm wide at the top.

Widespread and fairly common, less frequent in the boreal zone; May–November.

Jens H. Petersen/Thomas Kehlet

Naucoria bohemica is a ± orange-brown to somewhat olive, marginally striate, smooth *Naucoria* with a pale, floccose stem. Has 2-spored basidia and ± clavate cheilocystidia. The rough spores measure 11–13 × 7–8.5 µm and lack a germ pore. Occurs on rather dry soils with *Salix*, *Betula* and *Corylus*.

Naucoria salicis △ and *N. badiolateritia* ⋉ typically occur in wetter habitats with *Salix* and have both 2- and 4-spored basidia and larger spores.

Widespread, occasional; April–November, mainly in the autumn.

Molecular data indicate that the three species on this page should be placed in their own genus.

Conocybe is a large genus of mostly ± conical, ± brown agarics with rust-brown spores. The cap cuticle does not have a radial structure and the stem is mostly striate from rows of hair-like or flask-shaped cystidia. All have lecythiform cheilocystidia. The spores are smooth to warty (rarely) and usually have a germ pore. All species are decomposers and found mostly in grassland, gardens or along roads through scrub or forests.

Conocybe subpubescens is a rather large, golden-orange, minutely hairy *Conocybe* with a mix of hair-like and lecythiform cystidia on the stem, which is pale but usually darker towards a bulbous base. Has 4-spored basidia and cheilocystidia with 2.5–5 µm wide heads. The spores measure 9–16 × 5–9.5 µm (av. length less than 14 µm) and the germ pore is 1.5 µm wide. Occurs on soil, remnants of wood, *etc.*, mostly in scrub.

Conocybe pulchella ⋉ is more thimble-shaped, more gracile and has larger spores. *Conocybe pubescens* ⋉ is very similar, but occurs mostly on dung or on manured soil and has larger spores; *C. subpallida* ▷▷ is typically rather paler, has only hair-like cystidia on the stem and somewhat shorter spores.

Widespread and common; April–November.

Conocybe rickeniana is a brightly coloured, orange-brown, striate, slender *Conocybe* with lecythiform cystidia on the stem, which is ± two-coloured or ± brownish along its entire length and has a ± distinct bulb. The basidia are 4-spored. The spores are thin-walled, measure 7–11.5 × 3.5–5.5 µm and have germ pore 1 µm wide. Occurs on soil in forests and in more open places.

Conocybe echinata ⋉ is less vividly coloured and has a rather dark cap centre; *C. juniana* var. *subsejuncta* ▷▷ has somewhat larger spores that stain darker in KOH.

Distribution and frequency not fully understood; August–November.

Jan Vesterholt/Jens H. Petersen

Conocybe dumetorum is a slender, thin-fleshed, translucently striate *Conocybe* with a distinctly two-coloured stem (white at the top, reddish below). It has lecythiform cystidia on the stem and gill edges and very small (4.5–7 × 3–4 µm), ± ellipsoid, warty spores that have an indistinct germ pore. Occurs on mull soils along tracks in forests, *etc*.

Conocybe dumetorum var. *phaeoleiospora* ⋉ has almost smooth spores; *C. dumetorum* var. *laricina* ⋉ has a truncate germ pore; *C. microspora* ⋉ has smooth, ± bean-shaped spores.

Widespread in the nemoral zones with a few more northern records, possibly overlooked; July–November.

Conocybe juniana is a rather dull, ± grey-brown, at most slightly striate, rather tall-capped *Conocybe* with lecythiform caulocystidia and cheilocystidia that measure 17–33 × 8–20 µm with 4.3–10.5 µm wide heads. No crystals develop with ammonia. Spores 8–13.5 × 4.5–7.5 µm, thick-walled; germ pore 1–1.5 µm wide. On soil, mostly in open habitats.

Conocybe macrocephala ⋉ has thinner-walled spores that are a somewhat different shape and mostly form needle-shaped crystals in ammonia. There are several other similar species.

Frequency and distribution poorly understood; April–November.

Conocybe subpallida is a medium-sized, almost non-striate, ± unevenly pigmented, rather pale *Conocybe* with a smooth to somewhat wrinkled cap surface. The pale stem has lecythiform cystidia and also hair-like cystidia in the upper part; the cap surface has lecythiform cystidia; the cheilocystidia have 3.5–5 µm wide heads. Basidia 4-spored. The spores meaure 9–13.5 × 5–5 µm and are pale in KOH; germ pore 1 µm wide. Occurs on soil in deciduous forests.

Conocybe subalpina ⋉ and *C. tenera* ⋉ have darker colours; *C. macrocephala* ⋉ has smaller spores.

Widespread, but rare or overlooked; mainly August–September.

Conocybe apala is a large, very tall, umbonate, strongly hygrophanous, somewhat greasy and shiny, yellow-brown, pallescent to almost white *Conocybe*. The whitish, hairy stem lacks lecythiform cystidia but has a large bulb. The basidia are 4-spored. The spores measure 9–14.5 × 6–8 µm and have a wide germ pore. Occurs on ± calcareous soils in scrub and forests.

Typical, fresh specimens are distinctive and unlikely to be misidentified. *Conocybe albipes* ▽ has sterile cells around the basidia.

Widespread, but fairly rare; mainly July–October.

Thomas Kehlet

Conocybe inocybeoides is a large, almost fleshy, red-brown, but quickly pale cinnamon-buff, finely pruinose *Conocybe* with a conical and umbonate cap. The pale stem has a large, marginate bulb and the surface only has hairs (no lecythiform cystidia). The basidia are 2-spored. The spores measure 10.5–20 × 6.5–11 µm and have a wide germ pore. Occurs on clay soils in parks, *etc*.

The more elegant *C. apala* △ is even more pallescent and has 4-spored basidia.

Widespread, but rare or overlooked; mainly June–October.

Thomas Læssøe

Conocybe albipes is a ± thimble-shaped, milky-white to dark cream *Conocybe*. The stem is white and somewhat bulbous at the base. The 4-spored basidia are surrounded by sterile cells (physaloids). The spores measure 9–16 × 6.5–9 µm. Occurs typically in open, fertilized grasslands, including lawns.

Conocybe pseudocrispa ⋉ has 2-spored basidia and is somewhat less thimble-shaped but has similar physaloids. *Conocybe apala* △ is more richly coloured when fresh and has a larger bulb at the stem base.

Widespread and common; June–October, mainly in the summer after rains.

Thomas Læssøe

Pholiotina are *Conocybe*-like agarics with rust-brown spores. The cap cuticle lacks radial structure and the stem rarely has lecythiform cystidia. Most species are veiled, often with a ring. Decomposers on rich, ± disturbed soil or on wood.

†? ***Pholiotina arrhenii*** is an archetypical *Pholiotina* with a striate, rather low-positioned, ± projecting stem ring and an orange-brown, ± umbonate cap. The stem is silvery-fibrillose. The basidia are 4-spored. Cheilocystidia are rather narrow, lageniform and have a sinuous neck. The spores are somewhat citriform to amygdaliform, measure 6–9 × 3.5–5 µm and have a germ pore. Occurs on rich, ± disturbed soil, in forests and parks.

Pholiotina rugosa ▷ has a darker stem and somewhat larger spores, which darken in KOH. *Pholiotina hadrocystis* ⋉ is very similar, but has highly variable cheilocystidia that are typically wider than in *P. arrhenii*.

Widespread and rather common; June–November.

Erik Rald

†? ***Pholiotina aporos*** is a sturdy, rather dark, usually somewhat wrinkled, early-occurring *Pholiotina* with an evident, flaring, rather loose ring. Cheilocystidia cylindrical to narrowly clavate. Spores measure (7–)8–10 × 4.5–5.5 µm and lack a germ pore. On mull soils, often calcareous, at roadsides in forests and parks.

Other species of *Pholiotina* have a ± evident germ pore. Species of *Cyclocybe* ◁◁ and *Agrocybe* ◁◁ are more fleshy or lack a ring, and have germ pores and much wider cystidia. *Tubaria furfuracea* ▷▷ does not have a membranous ring, and has capitate cystidia.

Widespread in nemoral–hemiboreal zones, rather common; March–July.

††*Pholiotina rugosa* is a typical *Pholiotina* with a ring on the stem and a smooth to wrinkled, ± yellow- to orange-brown cap surface. The stem darkens towards the base. The cheilocystidia are pointed and the basidia are mostly 4-spored. The spores are ± fusiform and rather pigmented, measure 8–12 × 4–6 µm, and have a rather wide germ pore. Occurs on disturbed rich soil in gardens, parks and forests.

The wrinkled cap surface is diagnostic. *Pholiotina arrhenii* ◁ has a paler stem and somewhat smaller, paler spores. *Pholiotina vexans* ⋉ is very similar but has somewhat larger and darker spores. *Pholiotina hadrocystis* ⋉ has more variable cystidia and smaller, paler spores. All these species typically have less wrinkled cap surfaces.

Widespread and fairly common in the nemoral zones; June–November.

Malcolm Storey

†?*Pholiotina teneroides* is a rather small and slender *Pholiotina* with a stem ring, 2-spored basidia and rather variable cystidia ranging from broadly lageniform to ± clavate or almost cylindrical. The narrowly ellipsoid spores measure 9–14 × 4.5–7.5 µm and have a germ pore. Occurs on ± disturbed mull soils in gardens, parks and along roadsides in forests.

This is the only *Pholiotina* with a stem ring and 2-spored basidia.

Widespread but rather rare, probably absent from the boreal zone; June–December.

†?***Pholiotina velata*** is a rather pale, yellow-brown, ± marginally striate *Pholiotina* with a veiled cap margin and no stem ring. The cheilocystidia are ± lageniform and have a ± sinuous neck. The spores are small, measuring 6–9 × 3.5–5.5 µm, and have a germ pore. Occurs on rich, mostly disturbed soils in gardens, parks and along forest roads, rarely on decomposed wood.

Pholiotina vestita ⋉ has more permanent and larger remnants of veil, more regular cheilocystidia, and spores that lack a germ pore. *Pholiotina exannulata* ⋉ has very variable cystidia and darker colours, and *P. nemoralis* ⋉ has larger spores.

Widespread, but mainly in the nemoral zones, occasional; June–December.

Thomas Kehlet

Pholiotina sulcata is an aberrant, rather small, naked, deeply sulcate-plicate, yellow-brown *Pholiotina* with a somewhat pruinose cap surface. The cheilocystidia are narrowly lageniform and the basidia are 4-spored. The spores measure 6.5–10.5 × 4–5.5 µm and have a germ pore. Occurs on soil in grasslands, including in parks.

Dried-out specimens of *Parasola* (page 544) and *Conocybe* ◁◁ may recall *P. sulcata* and identification can only be confirmed with the aid of a microscope.

Widespread but rare to very rare, probably also overlooked; May – October.

Pholiotina coprophila is a coprophilous, naked, conocyboid *Pholiotina* with a somewhat greasy, rather pale, ± semiglobose cap and a pale, non-rooting stem. The cheilocystidia are lageniform and do not have a widened head. The basidia are 4-spored. The spores are thick-walled, measure 9.5–15 × 5.5–8 µm, and have a germ pore. Occurs on old cowpats, *etc.* in dry grassland.

Species of *Conocybe* ◁◁ have lecythiform cystidia on the gill edge. *Agrocybe pediades* ◁◁ also has small heads on the cystidia but rarely occurs on dung.

Widespread, but rare to very or possibly overlooked; July–November.

Thomas Læssøe

Pholiotina striipes is an aberrant, large, fleshy, guttulate, solitary or clustered, naked *Pholiotina*. The cheilo- and pileocystidia are narrowly lageniform. The basidia are 4-spored. The spores measure 6–9 × 3.5–5 µm and have a distinct germ pore. Occurs on newly established lawns, in grassy roadsides, *etc.*

Pholiotina pygmaeoaffinis ⋉ is very similar, and best told by its larger spores.

Widespread in the nemoral-hemiboreal zones, occasional to rare; mainly September–November.

Pholiotina aeruginosa is an almost steel-blue, naked *Pholiotina* with a more intensely blue stem base. The cheilocystidia are ± cylindrical. The spores are rather pale, measure 8.5–11.5 × 4.5–7 µm, and have a germ pore. Occurs In damp deciduous forests, and also in connection with burnt ground.

Distinctive and unlikely to be misidentified. *Pholiotina cyanopus* ⋉ has a blue-green stem base but the cap is brownish.

A southern, rare species, *e.g.* recorded in the Netherlands and a few times in the Nordic countries; late summer–autumn.

Enrique Rubio Domínguez

Galerina mostly produce rather small, brownish fruitbodies with fairly distant gills and rust-brown, smooth to verrucose spores. The stem may have a ring but is rarely striped with cystidia, as seen in *Conocybe* ◁◁ and partly in *Pholiotina* ◁◁. Often moss-associated but also on wood.

†† ***Galerina marginata*** is a large *Galerina* with a hygrophanous cap up to 40 mm wide. The stem has a narrow, membranous ring, below which it is silvery-fibrillose The smell and taste are rancid-farinaceous. The pleurocystidia are ± lageniform. The rough to verrucose spores have a ± loosening outer wall and measure 7.5–10.5(–13) × 4.5–6.5(–7) µm. Occurs mostly on wood, including wood chips, but also found directly on turf in open habitats.

Galerina walleyniana ⋉ has smaller spores (shorter than 7 µm); *G. pseudo-mycenopsis* ⋉ has larger spores (wider than 5.5 µm); G. *salicicola* ⋉ is not farinaceous and has shorter, bladder-shaped cystidia; and *G. uncialis* ⋉ is very similar, but has no pleurocystidia.

Widespread and common; August–December.

Galerina sideroides is a relatively large *Galerina* with an up to 30(–40) mm wide, sticky, orange-brown and ± translucently striate cap. The stem is ringless, 2–6 mm wide and silvery-fibrillose, discolouring darker towards the base. The cheilocystidia are ± lecythiform and have small heads. The spores measure 6–8 × 4–5 µm. Occurs on wood, wood chips and litter from conifers.

The much rarer *G. camerina* ⋉ also has small spores and lecythiform cystidia with small heads but has a beige to cinnamon-buff cap. *Galerina marginata* ◁ and similar species have a stem ring, pleurocystidia and larger, verrucose spores. The 2-spored *G. badipes* ⋉ has verrucose spores and narrowly lageniform cystidia and pleurocystidia.

Widespread and rather common to occasional; September–January.

Galerina jaapii is a very small, slender *Galerina* with a white ring on the stem. The basidia are 2-spored. The cheilocystidia are somewhat bowling pin-shaped. The large, rough spores measure 10–14(–19) × 5–7(–7.5) µm and have a germ pore. Occurs with mosses (but not *Sphagnum*) in wet habitats.

Galerina marginata ◁ and other galerinas with a ring-like zone on the stem are larger or occur in different habitats.

Widespread, occasional, rare in the boreal zone; June–October.

Thomas Læssøe

Jan Vesterholt

Galerina calyptrata is a small *Galerina* with a bell-shaped, orange-brown to red-brown cap up to 15 mm wide. The stem is pale with indistinct velar fibres. The odour is farinaceous. Has 4-spored basidia and clamps, but no pleurocystidia. The spores have a distinctly loosening outer wall (*i.e.* it is calyptrate) and measure 10–12.5 × 6–6.5 µm. Predominantly moss-associated, mostly with *Dicranum*, more rarely with *Sphagnum*, in acidic, damp forests or in more open habitats.

Sometimes treated as a variety of *G. cerina* ⋉ although, when young, that species has a yellow-brown to clay-buff cap, a fainter smell and a yellow-brown stem. *Galerina hypnorum* ▽ has narrower spores (< 5.5 µm), fewer of which have a less obviously loosening outer wall, and occurs mostly on moss-covered rotten trunks.

Widespread, occasional; May–November.

Martin Vestergaard

Galerina hypnorum is a small, very typical *Galerina* with a bell-shaped, yellow-brownish cap and a farinaceous smell. The spores have a rather indistinctly loosening outer wall, are very finely ornamented and measure (7.5–)8–10(–10.5) × (4.5–)5–5.5(–6) µm. Predominantly associated with mosses, and occurs mostly on rotten trunks in rather damp woodland habitats.

Galerina cerina ⋉ and *G. calyptrata* △ have wider spores, which have a more obviously loosening outer wall, and a larger number of gills; they also occur amongst mosses but mostly on rather poor soils.

Widespread and common; April–November.

Galerina pumila is a medium-sized *Galerina* with a warm yellow-brown, translucently striate cap that has a yellow margin, and a yellow-brown stem with scattered white veil fibres. The cheilocystidia are large, lageniform and usually have characteristic snakehead-shaped tops and/or repeated constrictions. The spores are smooth and measure (8.5–)10–13.5(–15) × (4.5–)5–7(–8) µm. Occurs amongst mosses in dry to moist, acidic habitats, including unfertilized grasslands.

Galerina allospora ⋉ has a band-like thickening on the spore, near the apex; *G. lacustris* ⋉ has a darker red-brown cap, is found mostly on litter in wet *Salix* thickets, and has cystidia with a simple head-like thickening.

Widespread and common; June–December.

This species may be parasitized by *Squamanita contortipes* (page 228).

Thomas Læssøe

Galerina vittiformis is a small to rather small, translucently striate, ± orange-brown *Galerina* with cystidia along the entire length of its stem (hand lens required!) and pleurocystidia. Pileocystidia can be present in the cap margin. The spores are verrucose and measure 8–10.5 (–11) × 5.5–7 µm (from 4-spored basidia) or 10–13 × 5.5–7 µm (from 2-spored basidia). Occurs on and with various mosses, mostly in rather rich sites, but also in wet habitats with upwelling water.

Galerina atkinsoniana ⋉ and *G. perplexa* ⋉ have pileocystidia over the entire cap surface (hand lens or microscope required!). *Galerina caulocystidiata* ⋉ has remnants of veil on the stem and lacks pleurocystidia.

Widespread and common; June–January.

Thomas Læssøe

Galerina triscopa is a ± small, acutely umbonate, translucently striate, relatively dark *Galerina*. Stem darkens towards the base and has small veil floccules. Lacks pleurocystidia. Spores small, verrucose, ± amygdaliform and measure 6–9 × 3.5–5(–5.5) µm. Typically on very rotten conifer wood in damp forests, but may also grow on deciduous wood and on humose soil.

Galerina ampullaceocystis ⋉ has larger, ± smooth spores and 2-spored basidia. Species of *Phaeocollybia* ◁◁ are larger and deeply rooting. Species of *Deconica* (page 548) have darker, smooth spores with a germ pore.

Widespread, occasional, but rare in western parts; June–December.

Thomas Kehlet

Galerina mniophila is a medium-sized *Galerina* with an olive-yellow to pale brown, pallescent, matt cap. The stem is smooth and has scanty veil fibres. The smell and taste are indistinct. The cheilocystidia, up to 70 µm long, are lageniform and lack heads. Clamps are present. Spores indistinctly rough, (8.5–)9–11(–11.7) × (4.9–)5.5–6.5(–6.8) µm. Found in mosses in open grassland.

Galerina cephalotricha ⋉ has a farinaceous taste and smell; *G. pseudomniophila* ⋉ has a brownish-ochraceous to cinnamon-brown cap and shorter cystidia (< 45 µm); *G. clavata* ⋉ is a purer yellow, and lacks clamps.

Widespread and common; mostly September–November.

Thomas Læssøe

Galerina graminea is a rather small *Galerina* with an applanate-umbonate, buff cap up to 10 mm wide, and a pale, fibrous stem. The cheilo-, caulo- and pileocystidia are lageniform and have a widened head. Pleurocystidia and clamps absent. Spores pale, thin-walled, almost smooth and measure 7–9.5(–10) × 4–5(–5.5) µm. Found amongst mosses and grasses in low, open, mostly calcareous vegetation.

Galerina clavata ⋉ and *G. subclavata* ⋉ also lack clamps, but are less applanate and have distinctly rough and thick-walled spores.

Widespread, rather common to ± absent from the boreal zone; mostly September–November.

Galerina hybrida is a medium-sized *Galerina* with a vivid orange-brown cap with a low umbo, and a brown stem with fine white veil fibrils when young. The cystidia are lecythiform and have a small, delimited head. Spores narrowly amygdaliform with a low ornament, (7.5–)8.5–11(–11.5) × 5–6(–6.5) µm. Occurs in mires, deep in *Sphagnum*.

Frequently reported as *G. tibiicystis*. *Galerina mairei* ⋉ has a more cinnamon-buff to yellow-brown cap, a pale stem without fibres, and amygdaliform to ovoid spores (wider than 6 µm) that have a conical apex and a distinct ornament. *Galerina paludosa* ▽ and *G. sphagnorum* ⋉ are farinaceous and have cystidia that lack a head, and almost smooth spores.

Widespread, common to occasional; May–September.

Thomas Læssøe

Galerina paludosa is a relatively large, long-stemmed, ± felty *Galerina* with obvious remnants of veil at the cap margin. The stem has a ring zone and pale veil girdles. The taste and smell are farinaceous. The ± lageniform cheilocystidia are capitate; lacks pleurocystidia. The amygdaliform spores are almost smooth and measure (7.5–)9–12.5 × 5–7(–8) µm. Occurs deep in *Sphagnum* in various types of bogs and mires.

Galerina sphagnorum ⋉ has, at most, indistinct remnants of veil on the stem, and fruits mostly late in the season. *Galerina hybrida* △ and *G. mairei* ⋉ have lecythiform cystidia with small, delimited heads, distinctly roughened spores, clearer cap colours and lack a farinaceous taste. Small species of *Cortinarius* subgenus *Telamonia* (page 732) have no or less differentiated cheilocystidia, and normally have darker, distinctly ornamented spores.

Widespread, rather common; May–November.

Tubaria is a genus of mostly small agarics with broadly adnate gills, brown, ± smooth spores that lack a germ pore, and prominent, thin-walled cheilocystidia. The cap and stem are mostly floccose from veil or have a ring. They are decomposers on soil, plant debris or wood.

Tubaria confragosa is a large, ± clustered, red-brown *Tubaria* with a large, flaring, membranous stem ring and broadly adnate gills. Cheilocystidia narrowly lageniform to cylindrical. Spores smooth, 6–8.5 × 5.5–6.5 µm, and lack a germ pore. On deciduous and coniferous wood or on wood chips, mostly in rather damp habitats.

Galerina marginata ◁◁ is more yellow-brown and has ornamented spores. Species of *Pholiotina, Agrocybe* and *Cyclocybe* ◁◁ with stem rings have adnexed gills, and most have spores with a germ pore.

Mainly hemiboreal–boreal, rarer in the nemoral zones; August–December.

Thomas Læssøe

Tubaria conspersa is a relatively pale yellow-brown to pale red-brown, mostly very floccose-felty, not particularly striate, medium-sized *Tubaria* with adnate gills. The cystidia are lageniform. The spores are ± amygdaliform, measure 7–9.5 × 4.5–7 µm, and lack a germ pore. Occurs on woody debris in woodlands, rarely in more open habitats.

Tubaria furfuracea ▷ is less felty, has ellipsoid spores and the cystidia have a swollen head. Species of *Flammulaster* ▷▷ normally have spores with a germ pore and constricted cap cuticle hyphae that have distinct incrusted pigment.

Widespread, rather common; May–January.

Tubaria furfuracea is a red-brown to pale yellow-brown, at most somewhat striate *Tubaria* with a zone of pale remnants of veil at the cap margin. The gills are pale brown, rather distant and broadly adnate. The cheilocystidia have a ± well-defined capitate top. The spores are ellipsoid, 6.5–10 × 4.5–6 µm, and lack a germ pore. Occurs on wood chips, disturbed soil with remnants of wood, stubble fields, *etc.*

Tubaria conspersa ◁ has a more felty cap, ± amygdaliform spores and lageniform cheilocystidia. Species of *Deconica* (page 548) have darker, thicker-walled spores with a germ pore.

Widespread, very common; all year, mainly during mild winter periods.

Tubaria dispersa is a small, rather smooth, pale-stemmed *Tubaria* with adnate to slightly decurrent, yellowish gills. The spores measure 5–7 × 3–4 µm and have very fine warts. Occurs on ± clay-rich or mull soils amongst litter from *Crataegus*.

Tubaria conspersa ◁ is also pale but has a more felty cap and duller brownish gills. Similar species of *Pholiotina* ◁◁ have adnexed gills.

Widespread, rather common in the nemoral zones, absent from the boreal zone; April–November.

Tubaria minutalis is a small *Tubaria* that lacks a stem ring but has fine remnants of veil on the cap and stem. The gills are adnate. The cheilocystidia are clavate to utriform. The spores are ovoid to amygdaliform and measure, 4.5–6.5 × 3–4.5 µm (highly variable). Occurs on rich, ± disturbed soils in open habitats.

At least two species seem to be included under this name. Species of *Conocybe* ◁◁ and *Pholiotina* ◁◁ have adnexed gills and, typically, other types of cystidia and cap cuticle. Species of *Flammulaster* ▷▷ have a different spore shape; *Deconica* ◁◁ have darker spores with a germ pore.

Apparently widespread, but frequency unknown; July–November.

Thomas Læssøe

Jukka Vauras

Mythicomyces corneipes is a somewhat greasy, umbonate, orange-brown, rather small, naked agaric with a grey-brown spore-deposit and yellowish, adnexed gills. The stiff stem becomes much darker towards the base. Has thick-walled pleuro- and cheilocystidia topped with crystals. The almost smooth spores measure 6–8.5(–10) × 3.5–6 (–6.5) µm and lack a germ pore. Occurs on litter and mosses in damp, *Picea*-dominated old-growth forests.

May recall a *Hypholoma* (page 582) or a *Phaeocollybia* ⊲⊲ but can be identified immediately by its unusual cystidia. Molecular phylogenetic studies seem to indicate a relationship with the genus *Psathyrella* (page 546).

Restricted to the boreal (–hemiboreal) zone in Scandinavia, very rare; summer–autumn.

Jean-Marc Moingeon

Phaeogalera stagnina is a dark red-brown, somewhat *Galerina*-like agaric with white, triangular veil remnants at the striate cap margin. The stem is rather stiff and dark and has white floccules and a ring zone. The cheilocystidia are lageniform and have long necks that may be widened towards the top. Spores thick-walled, pale brown, smooth, measure 12–16 × 7–10 µm, and have a germ pore. Spore-deposit is grey-brown. Amongst mosses in bogs and mires.

Psilocybe turficola (page 553) is usually larger, has smaller, purplish-brown spores, and the cystidia lack long necks. Species of *Galerina* ⊲⊲ have yellower-brown spores.

Widespread, rare in the nemoral zones, more freqent in the north; May–October.

Simocybe species are rather small to small, mostly ± olive-brown agarics that occur on wood or soil. The cap and stem have a pruinose surface due to projecting, mostly capitate and often guttulate cystidia. The spores are pale brown and have no or only an indistinct germ pore.

Simocybe centunculus is a rather variable, ± olive, olive-brown to brownish, slender *Simocybe* with a central stem and a cap cuticle that lacks obvious heads on the cystidia. The basidia are 4-spored. The spores are ± kidney- to bean-shaped and measure 6–8.5 × 4–6 µm. Occurs on very decayed wood, herbaceous debris and soil both in forests and in more open habitats.

The often more fleshy *S. sumptuosa* ▽ has obviously capitate cheilo- and pileocystidia. Although the main variety of *S. centunculus* occurs on wood, other varieties are found on soil or on the remains of monocots. *Simocybe geraniolens* ⋉, described from France, should smell strongly like *Pelargonium* leaves. It is at present unclear how many species should be recognized in the *S. centunculus* complex.

Widespread, occasional, becoming scarcer towards the north; May–November.

Simocybe sumptuosa is mostly a rather sturdy, soon applanate, usually rather dark, ± olive *Simocybe* with obviously capitate elements in the cap cuticle. The gills are ± olive-tinged, ± emarginate, and have a floccose edge due to protruding, capitate cheilocystidia. The bean-shaped, smooth spores measure 7–10(–11) × 4–5.5 um. Occurs on very rotten wood from deciduous trees in damp, mostly old-growth woodlands.

Simocybe centunculus △ is more slender and lacks obviously capitate elements in the cap cuticle.

Widespread in the nemoral zones, rather rare farther north and absent from the boreal zone; June–October.

Jan Vesterholt

Simocybe haustellaris is a small, olive to sordid olive-brown, finely floccose *Simocybe* with a short, ± excentric stem. The cap cuticle has erect cystidia with a small, rounded head and no inflated elements.The basidia are 2-spored. The spores are smooth, somewhat almond- to kidney-shaped, pale brown, measure 7–10 × 4.5–6.5 µm and lack an evident germ pore. Occurs on deciduous wood, rarely on the stems of herbs.

Other simocybes are typically 4-spored and do not have excentric stems. *Deconica horizontalis* (page 95) is browner, has darker spore-deposits, no pileocystidia and the spores have a germ pore.

Widespread, occasional, nearly absent from the boreal zone; May–December.

Thomas Læssøe

Flammulaster is a genus of rather small to very small agarics with either pale brown or rust-brown, smooth spores. The cap surface is ± scaly or powdery from remnants of veil; the gill attachment is variable, but mostly broadly adnate. The hyphae are strongly incrusted. Decomposers on *e.g.* wood, wood chips and *Fagus* cupules (husks), rarely on mull soils.

Flammulaster limulatus is an orange-brown, felty-scaly, brown-spored, lignicolous *Flammulaster* with scales that consist of cylindrical elements with strongly incrusted pigment. Cheilocystidia clavate. The spores measure 7–9 × 4–4.5 µm, are ± bean-shaped and have an almost invisible germ pore. Occurs on very decomposed trunks of deciduous trees, mostly in places with plenty of dead wood; also on heaps of sawdust.

The very similar *Pholiota tuberculosa* (page 615) is normally more yellowish,somewhat more scaly, and the spores are less bean-shaped and lack a germ pore. The cheilocystidia have obvious heads. *Flammulaster muricatus* ▷ has spherical elements in the cap scales.

Local and rare, apparently more common in the hemiboreal zone; June–November.

Flammulaster muricatus is a golden- to rust-brown, brown-spored *Flammulaster* with distinct triangular veil-patches at the cap margin and pointed scales covering the entire surface. The scales consist of chains of ± spherical elements with strongly incrusted pigment. The spores measure 7–8 × 4–5 µm and are ± amygdaliform with a small germ pore. Occurs on the rotten wood of deciduous trees (*e.g. Fagus*), usually on larger trunks, and mostly in old-growth habitats.

Flammulaster limulatus ◁ has oblong elements in the cap scales.

Widespread, but rare to very rare; mainly July–October.

Thomas Læssøe/Jens H. Petersen

Flammulaster novasilvensis is a vividly honey-coloured, finely felty-scaly, dark-stemmed and brown-spored *Flammulaster*. The veil elements have cylindrical cells with incrusted pigment. The spores are distinctly amygdaliform, measure 7–8.5 × 4–5 µm and have a germ pore. Occurs on the wood of deciduous trees, but also on conifer wood chips.

Recalls *Galerina marginata* ◁◁ but is obviously felty-scaly on the cap. *Flammulaster limulatus* ◁ has more bean-shaped spores that have only a very small germ pore.

Local and rare; July–September.

Flammulaster ferrugineus is a medium-sized, red-brown, ± marginally striate, mealy grainy, brown-spored *Flammulaster*. Has loose chains of oblong, incrusted cells in the cap cuticle; the cheilocystidia typically have a distinct, delimited head. The spores measure 6–8 × 4–5 µm, are conical towards the apiculus, and lack an evident germ pore. On rich soil, often with woody fragments intermixed.

Flammulaster granulosus ▷▷ has a non-striate margin and larger spores. *Tubaria conspersa* ◁◁ does not have loose chains of oblong cells in the cap cuticle and has different cheilocystidia.

Widespread, occasional to rare, absent from the boreal zone; July–December.

Thomas Læssøe

Flammulaster granulosus is a yellow-brown, very grainy-mealy, not marginally striate, brown-spored *Flammulaster*. The cheilocystidia are narrowly lageniform; the cap cuticle dissolves into globose to pear-shaped, loose cells with incrusted pigment. The spores measure 8–10 × 4.5–5.5 µm. Occurs on rich soils in forests and scrub, often around stumps.

Flammulaster ferrugineus ◁◁ is more red-brown, marginally striate, and the cap cuticle does not dissolve into ± spherical elements.

Widespread, occasional; mostly August–November.

Flammulaster carpophilus is a rather pale, small and slender, ± mealy *Flammulaster* with a very pale brown spore-deposit. The cheilocystidia are lageniform and lack capitate heads (apex less than 3.5 µm wide). The spores are ± amygdaliform and measure 7–10 × 4.5–6 µm. Found amongst litter fragments in woodlands and in more open habitats, including dunes.

Flammulaster subincarnatus ▽ has more capitate cheilocystidia that are wider than 4 µm at the apex. *Flammulaster rhombosporus* ⋉ has spores with an apical constriction.

Widespread, probably common; mainly October–November.

Flammulaster subincarnatus is a small, ± pinkish-tinged, pale brown-spored *Flammulaster* that tends to have capitate cystidia wider than 4 µm at the apex. The spores measure 7–10 × 4.5–5.5 µm, are ovoid, fusiform to rhombic, very thin-walled and collapsing. Found on litter fragments, mostly deep in dense, damp vegetation, *e.g. Salix repens* stands.

Other similar flammulasters either have narrower cystidia or different spores. Belongs to a difficult and not yet fully resolved complex of closely related species that also includes *F. carpophilus* △.

Widspread, probably very common; May–December.

Jens Maarbjerg

Ripartites tricholoma is a rather small agaric with pinkish-buff-tinged gills and a ± distinct white 'beard' at the cap margin. The stem is tinged brown and ± floccose. The brownish, finely echinulate spores typically measure 4–6(–6.5) × 3–4.5 µm. Occurs on rich soil, mostly in conifer plantations amongst mosses, also on burnt ground, along roadsides, *etc.*

Belongs to a poorly understood species complex, where up to six species or varieties could be recognized. *Ripartites albocarneus* ⋉ is the name used for small-spored specimens with a smooth, shiny cap centre and rose tinges; *R. serotinus* ⋉ and *R. krieglsteineri* ⋉ are smooth forms that lack rose colours; *R. odorus* ⋉ is described as having a nauseous-sweet smell (as in *Hebeloma sacchariolens*, page 684). There is very considerable variation in spore size and fruitbody size from one collection to the next. A strange form, or undescribed species, has more decurrent gills and the stem is powdery from rose conidia.

Widespread, occasional; mostly September–November.

Lucie Zíbarová

Inocybe, Inosperma, Mallocybe and *Pseudosperma*

The genera include ectomycorrhizal agarics with fibrillose to scaly caps, grey-brown spore-deposits and mostly conspicuous cystidia. Many species have spermatic or, more rarely, flower- or fruit-like, strong odours.

The species may have a thread-like universal veil and a similar partial veil and cystidia above the veil zone. Species without partial veil are typically downy from cystidia along the entire lengh of the stem. Young fruitbodies with untouched stems are important when assessing these characters. The genus can be divided into subgenera based on cystidia characters.

Many species are poisonous and contain, among other compounds, muscarin – none of the species should be consumed.

OTHER SIMILAR FUNGI:
– *Cortinarius* has more rust-brown spore-deposits (page 690).
– LBMs have different odours or more rust-brown spores (page 616)

FURTHER READING: 156, 173, 184, 315.

†?***Inocybe corydalina*** is a large, pale *Inocybe* with greenish, appressed fibrous scales mostly towards the the umbonate centre. The smell is sweetish, perfume-like, nauseating. The flesh is ± reddening. The stem is pale, but darkens ± towards the base, sometimes greenish. The pleurocystidia are thick-walled (to 2 µm) with crystals on top. The spores are amygdaliform and measure 7.5–9.5 × 5–6 µm. Occurs on ± calcareous soils or clay, mostly with *Fagus* and other deciduous trees, rarely with conifers.

Other large inocybes with sweetish odours, *e.g. I. fraudans* ▷▷ and *Inosperma bongardii* ▷▷, are not green at the cap centre, redden less obviously, or have distinct brown scales. *Inocybe erinaceomorpha* ▽, sometimes treated as a variety of *I. corydalina*, differs in having a dark brown cap centre and mostly dark brown fibrous, adpressed scales.

Widespread, occasional, absent from the boreal zone; mainly July–October.

†?***Inocybe erinaceomorpha*** is a rather fleshy, cinnamon-brown, scaly-fibrous *Inocybe* with a nauseatingly strong smell of cider. The stem is ± cylindrical, at first white, later tinged in the cap colour. The flesh can redden slightly. The pleurocystidia are ± clavate with up to 2 µm wide walls. Spores, smooth, amygdaliform and measure 7.5–9.5 × 5–6 µm. Occurs on clay soils with *Fagus* and *Quercus*.

Inocybe corydalina △ has a grey-green cap centre. Other somewhat similar species have flesh that reddens strongly or thin-walled pleurocystidia.

Widespread, occasional to rather rare, absent from the boreal zone; August–October.

Thomas Læssøe

Inocybe fraudans is a rather large, umbonate, ± reddening, nauseatingly cider-smelling, yellow-brown *Inocybe* with a white veil coat on the young cap. The stem usually has a wine-red tinge. The pleurocystidia have wall up to 2 µm thick and crystals on the top. The amygdaliform, smooth spores measure 8–12 × 5.5–7.5 µm. Occurs on clay soils, often on slopes with phlegmacias (page 698-721), *etc.*; mycorrhizal with deciduous trees or conifers.

Inocybe incarnata ⋉ is even larger, up to 10 cm wide, while *I. fraudans* is normally less than 5.5 cm wide. Other similar species with a ± similar smell have ± green tinges or conspicuous cap scales.

Widespread, occasional, rare in the boreal zone; July–November.

Jan Vesterholt

Inosperma cookei is a medium-sized *Inocybe* with yellowish colours and a smell of honey. The cap is radially fibrillose and has partial veil. The gills have a prominent, pale edge. The cheilocystidia are clavate, thin-walled. The spores are bean-shaped, smooth, and measure 7–9 × 4–5 µm. Found in forests on mull soils, with both coniferous and deciduous trees.

Similar to species in the *Pseudosperma rimosa* group ▷▷ but differs in having a honey-like smell. *Inocybe quietiodor* ⋉ also looks similar, but smells of stink-bugs or *Lactarius quietus* (page 432).

Widespread and rather common, becoming scarcer towards the north; mainly May–October.

Inosperma bongardii is a large, brownish, fibrillose-scaly *Inocybe* with a distinct, sweetish, nauceous soap-like smell, a scaly cap and reddening flesh. The stem is brown-fibrous and darkens along its length. The cheilocystidia are clavate, thin-walled and lack crystals. The spores are ellipsoid, smooth, and measure 11–16 × 6.5–9 µm. Occurs with deciduous trees, rarely *Picea*, in rich deciduous forests, mostly on calcareous soils, often in the company of phlegmacias (page 698).

Inocybe fraudans ◁ is less scaly. *Inocybe erinaceomorpha* ◁◁ either reddens less or not at all. Both have thick-walled cystidia. *Inocybe grata* ⋉ has thin-walled cystidia, a pale stem and a ± fishy smell.

Widespread, locally common to uncommon; June–October.

†?*Inocybe hirtella* is a small, ochre-yellow, naked *Inocybe* with a distinct smell of marzipan or bitter almond, and sometimes a ± pinkish stem apex. The stem is downy along its entire length from cystidia. The pleurocystidia are thick-walled, broad in the middle, and have crystals on top. The spores are amygdaliform and measure 9–12 × 4–7 µm. Occurs on ± calcareous soils with deciduous trees.

This is the only small, yellowish *Inocybe* with such a distinctive smell.

Widespread, occasional, mostly absent from the boreal zone; mainly July–October.

Tobias Bællingtoft

Inosperma adaequatum is a large, vinaceous, discolouring *Inocybe* with a radially fibrillose and splitting, brown, dark red-brown to purple-brown cap with a dark centre. The stem discolours wine-red from the base upwards. The smell is rather strong, earthy-sweetish or like beetroot. The cystidia are thin-walled, clavate. The smooth spores are ellipsoid to ± amygdaliform and measure 9–13 × 5.5–7 µm. Occurs on ± calcareous, often disturbed soils with deciduous trees, mostly in parks.

May recall species of *Tricholoma* (page 258), but these have white spore-deposits. Most reddening *inocybes* are smaller. *Inocybe erubescens* ▷ has a paler, cream to clay-buff cap, a sweet smell, and is normally found earlier in the year. *Inosperma bongardii* ◁◁ has a perfumed smell.

Widespread, occasional, absent from the boreal zone; June–October.

☠? ***Inocybe godeyi*** is a rather small to medium-sized, white, but very soon red-stained, naked, umbonate *Inocybe* with an innately radially fibrillose cap surface. The stem is pale, downy from cystidia, and has a ± marginate, pale bulb. The smell is somewhat sour. The pleurocystidia are thick-walled (up to 1.5 µm), topped with crystals. Spores amygdaliform, measure 9–12 × 6–7 µm. Occurs on ± calcareous soils, with e.g. *Fagus*, *Quercus*, *Tilia* and *Corylus*, more rarely with conifers.

Inocybe whitei ▷ and *Inosperma erubescens* ▷ do not have a marginate bulb.

Widespread in the nemoral–hemiboreal zones, occasional; July–November.

Thomas Læssøe

††*Inosperma erubescens* is a large, pale, reddening, veiled *Inocybe* with a radially-fibrillose and splitting, cream to pale brownish cap. The smell is unpleasant, somewhat ± sweetish-soap-like. The flesh is reddening. The cystidia are thin-walled and clavate. The ± ellipsoid, smooth spores measure 10–14 × 5.5–7.5 µm. Occurs on ± calcareous soils, often disturbed, mostly in parks, churchyards and open deciduous forests.

Most reddening *inocybes* are smaller. *Inosperma adaequatum* ◁ has a darker cap, a rather strong, earthy-sweetish or beetroot-like smell and is mostly found later in the year.

Widespread, occasional in the nemoral zones, becoming scarcer towards the north; May–September, mainly during the early part of this period.

†*Inocybe whitei* is a white to somewhat flesh-pink, reddening, innately fibrillose, veiled *Inocybe* with a spermatic smell. The stem is pale and at most clavate, widened at the base. The pleurocystidia have walls up to 2.5 µm thick and are topped with crystals. The spores are ovoid to ellipsoid and measure 8–9.5 × 4.5–5.5 µm. Occurs on mull soils with coniferous and deciduous trees.

Inocybe godeyi ◁ has a ± marginate bulb. *Inosperma erubescens* △ is larger and mostly found in late spring. *Inocybe sindonia* ▷▷ does not redden.

Widespread and rather common to occasional; June–December.

Benny Tom Olsen

† ***Inocybe haemacta*** is a dull-coloured, veiled *Inocybe* with greenish-blue tinges and a strong smell of horse stable. The flesh is reddening. The pleurocystidia have walls up to 2 µm thick and are topped with crystals. The spores are smooth, almond-shaped, and measure 8–12 × 5–7 µm. Occurs on ± calcareous soils in open forests and parks with deciduous trees.

Inocybe corydalina ◁◁ is similar, but has a completely different, ± perfumed, cider-like smell.

Widespread in the nemoral zones, rare; mainly July–October.

Thomas Læssøe

Inocybe calamistrata is a distinctive, dark, fibrillose-scaly *Inocybe* with a partial veil, a greenish to bluish discolouring stem base and a faint to strong fish-like smell. The gills are dark with white, fimbriate edges. The cystidia are ± clavate, thin-walled. The smooth, ± amygdaliform spores measure 10–13.5 × 5–6(–6.5) µm. Typically occurs in rather damp habitats in marshes and scrub.

Other inocybes with similar odours do not have greenish-blue colours at the stem base and are mostly paler and less scaly.

Widespread and quite common in the hemiboreal–boreal zones, much rarer in the nemoral; June–October.

Felix Hampe

† ***Inocybe aeruginascens*** is a fibrillose, yellow-brown, veiled *Inocybe* with a pale stem that slowly stains green from the base. The stem has cystidia towards the top and veil-threads farther down. The smell is soap-like. The pleurocystidia are ± fusiform and rather thin-walled. The smooth, amygdaliform spores measure 8–10 × 4–5.5 µm. Occurs on sandy soils with deciduous trees, *e.g. Salix* and *Populus*.

Inosperma calamistratum △ is much darker, scaly and smells of fish.

Very local and rare, apparently not known from the boreal zone; summer–autumn.

Jan Vesterholt

Jens Maarbjerg

† ***Inocybe mixtilis*** is a rather small, dull apricot-orange, rather finely fibrillose, naked *Inocybe*. The stem has cystidia along its entire length and the base has a white, marginate bulb. The odour is ± spermatic. The pleurocystidia have walls up to 3 µm thick and crystals at the top. Spores distinctly nodulose, 7–10 × 5–7 µm. Occurs on ± clay-rich soil in gardens, parks and forests, with both coniferous and deciduous trees.

Inocybe margaritispora ▷▷ has a more scaly cap; other similar species have larger spores.

Widespread and rather common; June–November.

†? ***Inocybe praetervisa*** is a yellow-brown, innately radially fibrillose, naked *Inocybe* with a pale, marginate stem bulb. The flesh is unchanging and smells ± spermatic. The pleurocystidia have walls up to 2.5 µm thick and are topped with crystals. The spores measure 9–12 × 7–10 µm and have prominent, rather pointed nodules. Occurs on calcareous and more acidic soils with both coniferous and deciduous trees.

Inocybe mixtilis △ has shorter spores.

Widespread and fairly common; mainly July–October.

Jan Vesterholt

Jens Maarbjerg/Jan Vesterholt

† ***Inocybe asterospora*** is a medium-sized *Inocybe* with a radially fibrillose, cracking, dark red-brown, umbonate cap and a paler, red-brown, marginate bulbous stem with cystidia along entire length. Lacks partial veil. The spores are distinctly nodulose, almost star-shaped, and measure 9–12 × 7–10 µm. Occurs with various deciduous trees on ± clay-rich, often disturbed soils in parks or at road verges in woodlands.

Inocybe napipes ▷ is superficially similar, but is a more pure brown, has partial veil and occurs on acidic soils. *Inocybe pseudoasterospora* ⋉ has similar spores, but no marginate bulb.

Widespread and quite common, ± absent in boreal zone; June–November.

Thomas Læssøe

† ***Inocybe paludinella*** is a small, umbonate, very pale, ivory-white, finely fibrillose *Inocybe*. The gills are often tinged yellow. The stem has cystidia along its entire length and a small basal bulb. The pleurocystidia have walls up to 5 µm thick. The spores are nodulose and measure 8–10 × 5–6.5 µm. Occurs in damp or wet marshes and carrs with *Salix* and *Alnus*.

Inocybe geophylla ▷▷ is less yellow, has, at most, a clavate stem base, and smooth spores. *Inocybe umbratica* ⋉ occurs with *Picea*.

Widespread, apparently rare but probably somewhat overlooked; July–October.

Jean-Marc Moingeon

† ***Inocybe margaritispora*** is a medium-sized *Inocybe* with a marginate bulb and brownish-yellow colours. The cap surface has small brown scales on a yellowish background. The stem is somewhat tinged yellow. The pleurocystidia have walls up to 3 µm thick. The spores are nodulose and measure 8–10 × 6–8 µm. Occurs on ± calcareous soils, mostly in deciduous forests.

Inocybe mixtilis ◁◁ does not have small brown scales on the cap.

Widespread, but uncommon in the nemoral zones, with scattered occurences in the hemiboreal zone; mainly August–November.

†? ***Inocybe acuta*** is a small, dark brown *Inocybe* with a pointed-umbonate cap (up to 20 mm wide). Has partial veil. The pleurocystidia are ± lageniform with walls up to 2 µm thick. The spores are distinctly nodulose and measure 8.5–11.5 × 5–7.5 µm. Occurs on damp, acidic soils with *Salix*.

Recalls a small, dark version of *Inocybe napipes* ▽, which has more distinctly umbonate spores, pleurocystidia with thinner walls, and typically occurs in forest.

Widespread, occasional but probably somewhat overlooked; mainly July–October.

Tom Smidth

†† ***Inocybe napipes*** is a medium-sized, brownish, mostly somewhat shaggy-margined *Inocybe* with a radially fibrillose cap that has a distinct umbo. Young fruitbodies have a distinct partial veil from cap margin to stem (see inset circle). The stem stains brownish and has a basal, onion-shaped bulb. Pleurocystidia walls are up to 1 µm thick. The spores are very distinctly nodulose, 8–11 × 6–8 µm. Mostly in forest on ± acidic soils.

Inocybe acuta △ is smaller, with a darker cap, paler stem, smaller spore nodules and occurs with *Salix*. *Inocybe assimilata* ⋉ has smaller and less nodulose spores and often pale remnants of veil on the cap centre. *Inocybe asterospora* ◁ is naked.

Widespread and common; June–November.

Jan Vesterholt/Jens Maarbjerg

Jan Vesterholt

†? ***Inocybe muricellata*** is a rather small, dull brown, somewhat scaly, naked *Inocybe*. The stem has cystidia along its entire length and is somewhat widened at the base. The odour is spermatic. The cystida are very long-necked with walls up to 3 µm thick. The spores are amygdaliform, smooth, and measure 8–12 × 5–6.5 µm. Occurs on mull soils, mostly with deciduous trees.

Similar species do not have the very long cystidial necks.

Widespread, rather rare, but perhaps overlooked; June–October.

Thomas Læssøe

†? ***Inocybe petiginosa*** is a small, ± zonate, pale felty-hairy, umbonate, naked *Inocybe* with, initially, pale mustard-yellow gills, and a reddish-brown stem that lacks a bulb but has cystidia along its entire length. The pleurocystidia are thick-walled and rounded at the base. The spores are finely nodulose and measure 6–8 × 4.5–5.5 µm. Occurs on mor soils in clay-rich deciduous forests, mostly with *Fagus* or *Quercus*.

A rather unique species. *Inocybe nematoloma* ⋉ is more orange-brown and lacks fine hairs.

Widespread and rather common, ± absent from the boreal zone; mainly July–November.

†?***Inocybe pelargonium*** is a medium-sized, *Pelargonium*-smelling, naked *Inocybe* with ± yellow-brown colours and a smooth cap centre and a fibrillose to minutely scaly cap margin. The stem has cystidia along its entire length. The pleurocystidia are lageniform with walls up to 2 µm thick. The amygdaliform spores measure 6.5–10 × 4–6 µm. Occurs on calcareous soils, with both coniferous and deciduous trees.

If the smell is not detected confusion with a number of similar species is possible.

Widespread, rather rare to rare; mainly July–October.

Jan Vesterholt

† ***Inocybe calospora*** is a small, dark brown, naked *Inocybe* with small, upright scales around the cap centre. The stem is 1–3 mm wide and has cystidia along its entire length. The pleurocystidia are clavate and only moderately thick-walled. The nearly spherical spores have conspicuous spines and measure 11–14 × 9.5–12.5 µm (including the spines). Occurs mostly in damp woodland on dark soils.

The spherical, spinulose spores separate this species from all other inocybes in temperate Europe.

Widespread, but rare or perhaps somewhat overlooked; mainly July–October.

Thomas Læssøe

†?***Inocybe jacobi*** is a small, yellow-brown, rather scaly, naked *Inocybe* that may have an olive-brown or buff tinge towards the cap margin. The stem has cystidia along its entire length. The gills are pale grey-brown. The pleurocystidia are thick-walled with a rounded base, 40–60 × 12–16 µm. The spores are nodulose and measure 7.5–10 × 5–7 µm. Occurs on sandy soils with conifers or *Betula*.

Inocybe castanea ⋉ has less distinct nodules on the spores and larger pleurocystidia (50–80 × 12–25 µm).

Widespread, but mainly to the north and east, apparently absent, or nearly so, from the UK; mainly June–October.

Lucie Zíbarová

Jan Vesterholt

†? **Inocybe tenebrosa** is a medium-sized, grey-brown, ± olive tinged, ± umbonate, fibrillose to scaly, naked *Inocybe* with a grey-brown, blackening stem that has cystidia along its entire length. The ± clavate pleurocystidia have walls up to 2.5 µm thick. The spores are amygdaliform and measure 7–9.5 × 4.5–5.5 µm. Occurs with deciduous trees on calcareous soils.

Inocybe vulpinella ▽ has larger spores and no olive tinges. The somewhat paler *I. melanopus* ▷▷ has a partial veil and no cystidia on lower part of the stem.

Widespread in the nemoral zones, uncommon, apparently absent from the boreal zone; September–October.

Jan Vesterholt

†? **Inocybe splendens** is a rather robust-fleshy, yellow-brown to grey-brown *Inocybe* with a ± persistent pale cover of veil on the cap and a long, inrolled margin. Lacks partial veil. The stem is pale, has cystidia along its entire length, is somewhat bulbous and may stain brown from the base. The pleurocystidia are rather long-necked, fusiform, and have walls up to 2 µm thick. The amygdaliform spores measure 9–12 × 5.5–6.5 µm. Occurs on ± calcareous soils in deciduous forests.

Inocybe splendens var. *phaeoleuca* ⋉ is more slender and lacks evident veil on the cap.

Widespread, but apparently uncommon; June–November.

Ditte Bandini

†? **Inocybe vulpinella** is a medium-sized, dark brown, fibrillose-scaly *Inocybe*; when young has a pale veil cover on the cap, but lacks partial veil. The stem stains brown from the base and has cystidia along its entire length. The pleurocystidia are large, ± lageniform to fusiform, and have walls up to 5 µm thick. The spores are very large, smooth, ellipsoid-ovoid, and measure 12–18 × 7–9 µm. In dunes with *Salix repens* or in other habitats with sandy soils with *Salix* or *Populus*.

Inocybe tenebrosa △ has a ± olive tinge and smaller spores.

Widespread, occasional, becoming scarcer towards the north; mainly May–October.

†? ***Inocybe serotina*** is a rather large *Inocybe* with a thick stem with cystidia in the upper half. The yellow-brown cap has a white veil coat. Possibly lacks partial veil. Smell spermatic or unpleasantly sweetish, like *Amanita phalloides* (page 359). Pleurocystidia clavate to inflated with walls up to 3 µm thick. Spores ± amygdaliform, 12–16 × 6.5–8.5 µm. Occurs in sand with both coniferous and deciduous trees, including *Salix repens* in dunes.

Inocybe vulpinella ◁ stains ± dark brown from the stem base upwards, is ± odourless, and has thicker pleurocystidia walls (up to 5 µm).

Scattered, but locally common; mainly June–October.

† ***Inocybe sindonia*** is a pale, rather yellow-brown-tinged, fibrillose to somewhat scaly, umbonate *Inocybe* with a partial veil and with cystidia on the upper half of the ± pinkish stem. The ± yellow, very slender pleurocystidia are up to 90 µm long and have walls up to 3 µm thick. The amygdaliform spores measure 7.5–10 × 4.5–5.5 µm. Occurs with both coniferous and deciduous trees on damp, ± clay-rich soils.

Other pale inocybes either redden, have nodulose spores, or have cystidia along the entire length of the stem.

Widespread and common; June–November.

Ditte Bancini

Jan Vesterholt

† ***Inocybe flocculosa*** is a medium-sized, fibrillose to minutely scaly, rather pale brown *Inocybe* with a partial veil. The pale stem stains brownish with age and has cystidia above the veil. The gills are initially very pale, later pale grey-brown. The pleurocystidia are lageniform to fusiform and have walls up to 2.5 µm thick. The spores are amygdaliform and measure 8–11 × 4.5–6 µm. Not especially demanding, tolerating a range of soil types, and found with both coniferous and deciduous trees.

This is not a very distinct species, and is easily confused with other species based on macroscopical features alone.

Widespread and common; June–November.

Jan Vesterholt

† ***Inocybe lacera*** is a rather floccose-scaly, grey-brown *Inocybe* with a partial veil. The stem is fibrillose and increasingly dark towards the base. The pleurocystidia have walls up to 2 µm thick. The rather narrow, irregular fusiform spores are characteristic, and measure 10–15 × 4–6 µm. Occurs on sandy, poor soils with both coniferous and deciduous trees.

This species can be divided into a number of varieties: *I. lacera* var. *maritima* ⋉ is hygrophanous; *I. lacera* var. *helobia* ⋉ occurs in wet habitats and has very thick-walled pleurocystidia and *I. lacera* var. *rhacodes* ⋉ has incrusted pigment on the cystidia.

Widespread and common; May–November.

† ***Inocybe lanuginosa*** is a dark red-brown, scaly, rather small to medium-sized *Inocybe* with a somewhat fibrillose, red-brown stem that lacks cystidia, and with a partial veil. The pleurocystidia are club- to pear-shaped and measure 30–50 × 13–25 µm. The distinctly nodulose spores measure 8–11 × 6–8 µm. Occurs on poor soils, with coniferous and deciduous trees.

Amongst the nodulose-spored species the very similar *I. stellatospora* ⋉ is best told by its cylindrical to lageniform pleurocystidia (which measure 50–80 × 15–25(–30) µm). *Inocybe leptophylla* ⋉ lacks pleurocystidia. *Inocybe subcarpta* ▽ is more woolly rather than scaly.

Widespread and common; June–October.

Jan Vesterholt

†? ***Inocybe subcarpta*** is a medium-sized to rather fleshy, dark brown *Inocybe* with a woolly-scaly cap surface and a partial veil. The stem is grey-brown and longitudinally fibrose. The pleurocystidia are thin-walled with no, or only a few, crystals on top. The spores have a few nodules and measure 8–11 × 5–7 µm. Occurs on acidic, poor soils in conifer forests, mostly in young plantations.

Several similar species are more scaly: *I. leptophylla* ⋉ lacks pleurocystidia; *I. teraturgus* ⋉ occurs in *Sphagnum*; and *I. lanuginosa* △ and *I. stellatospora* ⋉ have more nodules on the spores.

Widespread and fairly common; June–November.

Jan Vesterholt

Jan Vesterholt

†?*Inocybe curvipes* is a small to medium-sized *Inocybe* with a radially fibrillose to minutely scaly cap with a dark umbo, a pale margin and a grey-brown discolouring stem. Has a partial veil. The stem lacks a marginate bulb, but has white veil fibres. Pleurocystidia broadly lageniform with a ± pointed apex. Spores slightly nodulose, measure 8.5–12 × 5.5–7 µm. Occurs on clay-rich soils in parks and deciduous forests.

Inocybe lacera ◁◁ is usually darker, and has a monochrome cap and cylindrical spores. *Inocybe subcarpta* ◁◁ occurs in conifer plantations and is darker.

Widespread and rather common; June–October.

Jan Vesterholt

†?*Inocybe soluta* is a grey-brown, fibrillose to somewhat scaly, conical to more applanate *Inocybe* with a partial veil. The cylindrical stem is dark and has cystidia on the upper part only. The pleurocystidia are almost fusiform and have walls up to 1 µm thick. The spores are nodulose, almost square, and measure 6–8 × 5–6.5 µm. Occurs on acidic, sandy soils with conifers and *Betula*.

Inocybe curvipes △ and *I. rennyi* ⋉ have larger spores.

Widespread and rather common; June–October.

Thomas Læssøe

†?*Inocybe auricoma* is a small, rather fleshy, yellowish *Inocybe* with a radially fibrillose cap and partial veil. The white, later yellowish stem has cystidia near the top. The ± fusiform pleurocystidia have walls up to 3 µm thick. The spores are amygdaliform and measure 8–9.5 × 4.5–5.5 µm. Occurs with deciduous trees on mull or clay soils in parks and forests.

Pseudosperma flavellum ▷▷ has ellipsoid spores and thin-walled cystidia; *Inosperma cookei* ◁◁ smells of honey and has thin-walled cystidia. *Inocybe appendiculata* ⋉ has triangular veil remnants at the cap margin.

Widespread, but rather rare; mainly June–September.

Jean-Marc Moingeon

☥? **_Inocybe queletii_** is a large, fleshy *Inocybe* with a pale pinkish-buff cap that has a whitish velar coating. It has partial veil, initially whitish gills and a whitish stem with cystidia at the top. The pleurocystidia have walls up to 3 µm thick. The spores are amygdaliform and measure 8.5–13 × 5.5–7 µm. Occurs in coniferous forests on calcareous soils, mostly with *Abies*.

Inocybe sambucina ▽ is similar but has smaller spores and occurs on sandy soils with *Pinus*.

Southern and rare; late summer–autumn.

☥? **_Inocybe sambucina_** is a large, fleshy, ivory-white *Inocybe* with a ± silk-like cap surface and partial veil at the margin. The stem has cystidia on the upper part only, and the smell is ± spermatic. The pleurocystidia have walls up to 2.5 µm thick. The spores are almost cylindrical and measure 7–10 × 3.5–5 µm. Occurs with *Pinus* on sandy soils.

Inocybe queletii △ has larger spores, is somewhat more pinkish-buff and occurs on richer soils, *e.g.* with *Abies*.

Widespread, but rather rare, becoming scarcer towards the north; July–November.

David Boertmann

† ***Inocybe geophylla*** is a rather small, umbonate *Inocybe* with partial veil and a radiating silky cap surface. The cap is white, but, with age, may become slightly sordid yellow towards the centre. The odour is spermatic. The gills are pale brownish-grey. The pleurocystidia have walls up to 2.5 µm thick. The ± ellipsoid, smooth spores measure 7.5–10.5 × 4.5–6.5 µm. Occurs on ± disturbed, mull soils along roads in both deciduous and coniferous forests, and in gardens and parks.

Inocybe umbratica ⋉ and *I. paludinella* ▷▷ both have nodulose spores. *Inocybe whitei* ◁◁ is reddening.

Widespread and very common; July–November.

Jan Vesterholt

†? ***Inocybe pusio*** is a small, rather dark brown *Inocybe* with ± distinct violet colours on the stem and only a slightly fibrillose cap surface with veil remnants at the margin. The stem is downy from cystidia on the upper part only. Pleurocystidia lageniform, ± hyaline, with walls up to 2 µm thick. The ± amygdaliform spores measure 8–11 × 4.5–6 µm. Occurs on rich soils in parks and deciduous forests.

Inocybe cincinnata ▷▷ also has violet stem colours but has at most cystidia at the extreme stem apex, and its cheilocystidia are ± coloured brown.

Widespread, occasional, but becoming scarcer towards the north; mainly June–October.

† *Inocybe lilacina* is a strikingly lilac, umbonate and rather small *Inocybe* with a silky cap surface and partial veil. The cap centre may be tinged brownish or brownish-yellow. The stem is also lilac and has cystidia on the upper part only. The smell is distinctly spermatic. The pleurocystidia have walls up to 2.5 µm thick. The smooth, ± ellipsoid spores measure 7.5–10.5 × 4.5–6.5 µm. Occurs on disturbed, mull soils at roadsides in both deciduous and coniferous forests.

In older literature often treated as a variety of *I. geophylla* ◁. Other inocybes with lilac or violet colours do not have a silky, smooth cap surface.

Common and widespread; June–November.

†? *Inocybe griseolilacina* is a smallish *Inocybe* with a grey-brown to clay-buff, minutely scaly cap and no scales on the violet-tinged stem. Has partial veil. The pleurocystidia walls are approx. 2 µm thick. Spores amygdaliform, measure 8–11 × 5–6 µm. Occurs on ± calcareous soils in deciduous forests and scrub, mostly with *Fagus* and *Corylus*.

Inocybe amethystina ⋉ has violet tinges only on stem apex and a darker red-brown cap. *Inocybe huijsmanii* ▷▷ has a smoother, pale clay-buff cap with an almost white margin. *Inocybe cincinnata* ▷▷ has scales on the stem.

Widespread and rather common in the nemoral zones, much rarer in northern parts; June–November.

Jan Vesterholt

† *Inocybe cincinnata* is a small *Inocybe* with partial veil, a brown, scaly cap and a brownish, more or less distinctly scaly stem with a violet apex. Usually has brown gill edges and thus mostly brown-coloured cheilocystidia. The pleurocystidia stain yellowish in KOH and have walls up to 2 µm thick. The spores are amygdaliform, smooth, and measure 8–11 × 5–6 µm. Occurs mostly on rich soils in deciduous forests, and in parks and scrub on calcareous soils.

The variety *I. cincinnata* var. *major* has stems more than 3 mm wide and distinct brown gill edges. *Inocybe huijsmanii* ▷, *I. amethystina* ⋉ and *I. griseolilacina* ◁◁ have stems that lack scales; *I. griseolilacina* also has a paler, grey-brown cap and violet tinges along the stem.

Widespread and rather common; June–November.

†? *Inocybe nitidiuscula* is a small, slender, ± umbonate, red-brown, fibrillose *Inocybe* with white partial veil at the cap margin and a pinkish-tinged stem. The gills are initially whitish. The pleurocystidia can be somewhat yellowish in KOH, fusiform to lageniform, and have walls up to 2.5 µm thick. The smooth, amygdaliform spores measure 9–12 × 5–6.5 µm. Occurs on calcareous soils with conifers, but also with deciduous trees.

Inocybe tarda ⋉ is very similar, but has distinctly yellow pleurocystidia in KOH; it occurs with *Pinus* and deciduous trees.

Widespread and fairly common; July–November.

Christian Lange

† ***Inocybe hystrix*** is a rather large *Inocybe* with recurved, pointed scales on both cap and stem. It has remnants of partial veil at the cap margin. The pleurocystidia are lageniform with walls up to 2 µm thick. The amygdaliform spores measure 8–12 × 5–7 µm. Occurs on ± calcareous soils with various deciduous trees, often in the company of other rare ectomycorrhizal agarics.

Other very scaly inocybes are typically smaller or, when young, have a pale cover from universal veil on the cap.

Widespread but rare, almost absent from the boreal zone; mainly July–October.

†? ***Inocybe huijsmanii*** is a small, very pale brownish, umbonate *Inocybe* The stem is whitish with faint violet colours at the top, and has cystidia only at the apex. The cap is almost smooth with traces of veil at the margin. The pleurocystidia are lageniform and have walls approx. 1 µm thick. The spores are smooth, amygdaliform, and measure 8.5–11 × 5.5–6 µm. Occurs on rich soils in deciduous forests.

Sometimes lacks violet tinges, and may then be confused with a number of other species. *Inocybe griseolilacina* ◁◁ has violet shades on the entire stem and has a darker grey-brown, ± scaly cap; *I. cincinnata* ◁ has scales on the stem; *I. amethystina* ⋉ has a darker red-brown cap.

Nemoral, rare or overlooked; mainly July–September.

Jan Veste holt

Jan Vesterholt

† *Inocybe fuscidula* is a medium-sized, dull grey-brownish, fibrillose, but not scaly *Inocybe* with partial veil. The gills and stem stay whitish for a long time. The ± lageniform pleurocystidia have walls up 2 µm thick. The spores are smooth, ± amygdaliform, and measure 8–10.5 × 5–6 µm. Occurs on rich soils in deciduous and coniferous forests.

A 2-spored variety also occurs, *I. fuscidula* var. *bisporigera* ⋉. *Inocybe pseudodestricta* ⋉ is darker and more shiny brown at the cap centre.

Widespread and common; May–November.

Jan Vesterholt

†? *Inocybe glabripes* is a rather small, indistinctly umbonate, dark cinnamon-buff *Inocybe* with partial veil. The pale, ± pinkish, fibrous stem has cystidia only at the very apex. The gills are yellowish-brown. The odour is spermatic. The pleurocystidia are lageniform with walls up to 2 µm thick. The smooth, ± amygdaliform spores measure 6.5–8 × 4–5 µm. Occurs in deciduous and mixed forests, on rich soils.

The small spores are distinctive. *Inocybe microspora* is apparently a synonym.

Widespread, rare or perhaps overlooked, becoming scarcer towards the north; mainly June–October.

Jan Vesterholt

†? ***Inocybe melanopus*** is a medium-sized, rather fleshy, ± cinnamon-buff, felty-scaly and cracking *Inocybe* with partial veil. The stem has cystidia on the upper part and soon stains dark grey-brown from the base upwards. The odour is not striking. Has clavate, ± capitate, rather thin-walled pleurocystidia. The spores are smooth, amygdaliform and measure 8–10 × 4.5–6 µm. Occurs on calcareous soils, mostly in conifer plantations.

The name is sometimes given as *I. melanopoda*. The darkening stem is a good field character. *Inocybe splendens* ◁◁ and *I. tenebrosa* ◁◁ also darken on the stem but lack partial veil.

Widespread, but rather local and rare; mainly June–October.

†? ***Inocybe aurantiifolia*** is a medium-sized, yellow-brown to grey-brown *Inocybe* with obvious remnants of veil at the cap margin and on the stem. The gills are strikingly yellow-orange. The pleurocystidia are ± lageniform, have ± yellow, rather thin walls, and are topped with crystals. The smooth, amygdaliform spores measure (8.5–) 9–10.5 (–11) × (5–) 5.2–5.8 (–6.1) µm. Mycorrhizal with both deciduous trees and conifers.

Pseudosperma rimosum ▷▷ and *I. flocculosa* var. *crocifolia* ⋉ have pale yellow gills (never deeply yellow-orange).

Very local and rarely recorded; autumn.

Thomas Læssøe

† ***Pseudosperma rimosum*** is a rather large, ± acutely umbonate, distinctly radially fibrillose, cinnamon-buff to grey-brown *Inocybe*, mostly with a darker centre. The pale flesh can be seen clearly between the fibres. The gills are tinged greenish-yellow and have a very distinct, white, floccose edge (from protruding cheilocystidia). The stem is pale, lacks a bulb, and has cystidia on the upper part. The odour is spermatic. Has ± clavate, thin-walled cheilocystidia, 10–22 µm wide. Pleurocystidia absent. The ± ellipsoid, smooth spores measure 9–15 × 6–8 µm. Occurs on rich, mostly disturbed soils in parks and along roads in deciduous forests.

Belongs to a difficult species complex. Closely related species lack the greenish-yellow tinges on the gills. The golden-orange-gilled *Inocybe aurantiifolia* ◁◁ has pleurocystidia topped with crystals. *Inocybe flocculosa* var. *crocifolia* ⋉ has lageniform to fusiform pleurocystidia and narrower spores.

Widespread and common; June–November.

† ***Inosperma maculatum*** is a ± dark grey-brown, very radially fibrillose *Inocybe* with whitish remnants of pale veil at the cap centre. The gills are pale grey-brown with white, fimbriate edges. The cheilocystidia are thin-walled, clavate, and measure 30–60 × 10–24 µm. Lacks pleurocystidia. The spores are smooth, ± bean-shaped, and measure 8–12 × 5–6.5 µm. Occurs on rich, disturbed soils at roadsides in deciduous and coniferous forests, and in parks, *etc.*

Part of a species complex that is difficult to understand and requires further study.

Widespread and common; mainly July–October.

Jens Maarbjerg

†? ***Pseudosperma flavellum*** is a rather large, pale yellow-brown, conical-umbonate, non-scaly *Inocybe* that lacks veil patches at the cap centre, but has partial veil. The gills are greyish-yellow. The cheilocystidia are almost cylindrical, thin-walled, and measure 30–65 × 8–16 µm. The ± bean-shaped spores measure 9–13 × 5–6 µm. Occurs with deciduous trees, particularly *Salix*, mostly in ± damp sites.

Pseudosperma rimosa ◁ has more distinct greenish-yellow tinges on the gills, somewhat larger spores and wider cystidia.

Widespread, occasional; May–October.

Jan Vesterholt

The species on this page belong in genus *Mallocybe*. They have short, mostly septate cheilocystidia, no pleurocystidia, do not smell spermatic and generally have abundant partial veil.

† ***Mallocybe agardhii*** is a medium-sized, buff to dull apricot-orange, soon applanate *Mallocybe* with a felty cap and prominent partial veil that forms a ring zone on the ± brown-staining stem, Has thin-walled, ± septate, clavate cystidia on the gill edge. Lacks pleurocystidia. The spores are smooth and measure 8–11 × 5–6.5 µm. Occurs in marshes and dunes with *Salix*; ± calcicolous.

Mallocybe dulcamara ▷ has less prominent partial veil and no distinct ring zone. *Mallocybe terrigena* ▷ has projecting brown scales on the stem.

Widespread and rather common; June–November.

Jan Vesterholt

Mallocybe terrigena is a red-brown, medium-sized *Mallocybe* with projecting brown scales on the stem below a distinct ring or ring zone of partial veil. The gills are initially mustard-yellow. The clavate cheilocystidia are thin-walled. The ± ellipsoid, smooth spores measure 9–12 × 5.5–7 µm. Occurs with *Pinus* and *Picea* on calcareous soils, but apparently also with deciduous trees.

Other species of *Mallocybe* have less obvious stem scales.

Local, occasional to rare; mainly August–October.

† **Mallocybe dulcamara** is a relatively large, dull apricot-orange *Mallocybe* with a felty cap and a paler stem, which discolours somewhat brownish with age and when handled. Has abundant partial veil. The cheilocystidia are short, clavate, septate and thin-walled. Pleurocystidia absent. The smooth spores measure 8–11 × 5–6.5 µm. Occurs in marshes, dunes and scrub with a range of different trees.

Mallocybe agardhii ◁ has even more abundant partial veil that forms a distinct ring zone on the stem. The similar *Mallocybe fuscomarginata* ⋉ has somewhat thick-walled, brown cystidia on the gill edge and possibly a dark gill edge. There are further similar species in this poorly understood group.

Widespread and common; May–November.

Jan Vesterholt

Hebeloma

This genus includes mycorrhizal agarics with ± grey-brown spore-depopsits, mostly smooth, greasy to slimy caps, emarginate gills and fruity, flower-like, radish-like to earth-like odours. The spores are ± verrucose and may be ± dextrinoid and have a ± loosening outer wall; the gill edge is covered with cystidia which, in some cases, produce a clear liquid that traps the dark spores.

More than 55 species in temperate Europe.

OTHER SIMILAR FUNGI:
– *Cortinarius* may look similar, but has more rust-brown spores (page 690).
– the remaining group of brown spored mushrooms (LBMs) has different odours and mostly more rust-brown spores (page 616).

FURTHER READING: 29, 156, 184.

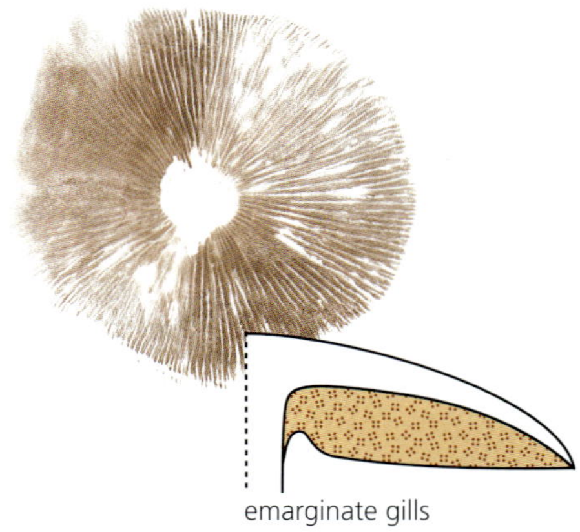

Micro-drawings: spores first followed by cystidia.

Hebeloma radicosum is a very distinctive, large *Hebeloma* with a deeply rooting stem, a striking marzipan- or bitter almond-like smell and a floccose ring. The cap is dull yellowish brown to pinkish buff with scaly remnants of veil. The spores measure 8–10 × 4.5–5.5 µm. Occurs in deciduous forests, mostly with *Fagus*, appearing from latrines or middens made by moles or other small forest-dwelling mammals; a so-called 'ammonia fungus'.

May recall some pholiotoids, particularly *Hemipholiota populnea* (page 607), but the smell, its soil-inhabiting nature with a very deeply rooting stem and ornamented spores are good separating characters.

Widespread and fairly common in the nemoral zone but becoming scarcer towards the north and absent from the boreal zone; July–November.

Hebeloma mesophaeum is a small to medium-sized, dull orange- to cinnamon-brown *Hebeloma* with a thread-like veil at the pale margin. The veil is later seen on the stem as threads and as a ring zone. The stem is pale on the surface, but stains brown when bruised and towards the base with age; not rooting. The odour is radish-like. The cheilocystidia are predominantly lageniform. The ellipsoid, somewhat ornamented, non-dextrinoid, brown spores measure 7–12 × 4.2–7 µm. Occurs in troops in all kinds of habitats in association with a number of partner trees, both deciduous and coniferous, and various bushes.

A genetically very variable species. *Hebeloma dunense* ▷, *H. psammophilum* ⋉ and *H. subtortum* ⋉ are very closely related species and the separating characters are very subtle.

Very common and widespread; mainly August–December.

Squamanita odorata (page 228) is a parasite on this species and may occasionally be seen with the host (lower left corner of top image).

Hebeloma dunense is a dull orange- to cinnamon-brown, rather slender *Hebeloma* with thread-like veil at the cap margin and on the stem, which darkens towards the base. Sand is typically gathered in a lump at the base. Has fewer than 50 entire gills. The cystidia are lageniform. The ellipsoid to ± amygdaliform spores measure 8.6–13.5 × 5.1–7.9 µm. Mycorrhizal with *Salix*, both in dunes and in more wooded habitats.

Hebeloma mesophaeum ◁ has, on average, shorter, ellipsoid spores; *H. psammophilum* ⋉ is more robust and has more than 50 entire gills.

Widespread, mainly coastal, occasional; May, August–November.

Jan Vesterholt

Hebeloma sordescens is a sordid orange-brown *Hebeloma* with thread-like veil at the cap margin and on the stem; it dries to a brownish-black. The cystidia are ± lageniform. The dextrinoid, amygdaliform to lemon-shaped spores measure 9–13.7 × 5–7.2 µm. Occurs both in the open and with deciduous and coniferous trees in forests, mostly at roadsides, but apparently not with *Salix*.

Hebeloma mesophaeum ◁ has ellipsoid, smaller spores; it and other similar species are paler when dry.

Widespread and fairly common; August–October.

Jan Vesterholt

Hebeloma nigellum is a small *Hebeloma* with thread-like veil, a blackish-brown cap centre and a paler margin. The slender stem darkens towards the base. The cheilocystidia are lageniform. The amygdaliform, dextrinoid spores measure 10.1–15.2 × 6.4–8.1 µm. Typically occurs in wet *Salix* carrs.

This species is recognized by the relatively few entire gills and its wide spores, on average wider than 7 µm. *Hebeloma hygrophilum* ⋉ occurs in the same environment but has narrower spores.

Widespread, but apparently somewhat local; June–November.

Jan Vesterholt

Hebeloma sacchariolens is a rather pale, pinkish-buff to cinnamon-brown *Hebeloma* with a very strong nauseous-sweet smell, somewhat like the flowers of *Prunus padus*. Lacks veil and drops on the gill edge. The cheilocystidia are ± clavate and short. The amygdaliform to fusiform, coarsely ornamented, dextrinoid spores have a loosening outer wall and measure 8.6–13.5 × 5.4–7.8 µm; Qav. 1.6.

This and four other species of *Hebeloma* belong to a group that has the same nauseous-sweet smell: *H. odoratissimum* ▽ differs in having a dry, felty, quickly cracking cap and larger spores; *H. fusisporum* ⋉ and *H. ischnostylum* ⋉ have narrower spores with a Qav. above 1.8; and *H. nauseosum* ⋉ has spores that are 7–9 µm wide.

Rather common and widespread, except in the boreal zone; July–November.

Thomas Læssøe

Hebeloma odoratissimum is a dry, downy to felty, smallish, naked *Hebeloma* with a strong nauseous-sweet smell. Cystidia small and clavate. The ± lemon-shaped, dextrinoid spores have a loosening outer wall and measure 10.5–15.6 × 6.7–9.2 µm. Occurs mostly in damp woodlands with *Alnus*, *Salix* and *Populus*.

Also known as *H. tomentosum*. *Hebeloma nauseosum* ⋉ is rather sticky and at most somewhat downy at the margin. Both *H. nauseosum* and *H. odoratissimum* have larger spores than *H. sacchariolens* △ and *H. ischnostylum* ⋉.

Widespread, rather rare, but possibly somewhat overlooked; September–November.

Jan Vesterholt

Hebeloma laterinum is a medium-sized, naked *Hebeloma* with rather pale, dull pinkish-buff to yellowish-brown colours. The stem is floccose along its entire length and strongly brown-staining. It has crowded gills and is ± rooting. The smell is rather variable, from earthy or soap-like to cacao-like, but is sometimes almost absent. The taste is ± bitter. The cheilocystidia are small and somewhat widened above. The amygdaliform, dextrinoid spores have a ± loosening outer wall and measure 7.5–12.2 × 4.7–7.1 µm. Occurs on calcareous soil, with both coniferous and deciduous trees, also with *Helianthemum*, mostly in fairy rings.

Hebeloma danicum ⋉ has fewer entire gills (less than 65).

Widespread, locally quite common; September–November.

Hebeloma cylindrosporum is a ± cinnamon-buff, almost or completely naked *Hebeloma* with a paler cap margin. The stem is mostly distinctly rooting. The odour is variable, often somewhat cacao-like. The spores are cylindrical, dextrinoid, have a ± loosening outer wall and measure 7.4–10.9 × 4–5.7 µm. Typically mycorrhizal with *Pinus* on nutrient-poor, sandy soil.

Told mainly by the cylindrical spores and the ± rooting stem.

Widespread, but with scattered records, possibly somewhat overlooked; June–November.

Thomas Kehlet

***Hebeloma crustuliniforme* s.l.** is a large, greasy-sticky, pale, naked *Hebeloma* with drops on the gill edge and a strong radish-like smell. It has more than 60 entire gills. The stem is floccose along its entire length and not rooting. The cheilocystidia are long, sinuous and widened at the top, on av. 6.5–7.9 µm wide at the apex. The amygdaliform, slightly dextrinoid, coarsely ornamented spores measure 9.6–13.5 × 5.5–7.2 µm. Occurs with both deciduous and coniferous trees, mostly on rather disturbed soils in open habitats along roads, *etc.*

This species belongs to a very difficult complex. Some similar species are mostly somewhat paler with fewer than entire 60 gills. These include *e.g. H. aanenii* ⋉, *H. geminatum* ⋉, *H. eburneum* ⋉ and *H. alpinum* ⋉. *Hebeloma velutipes* ▷ is also similar, but has clavate, not sinuous cystidia and more dextrinoid spores. *Hebeloma helodes* ▷ is smaller with 30–60 entire gills.

Widespread and common – at least as a broadly defined super-species; mostly September–November.

Jan Vesterholt

Hebeloma velutipes is a pale, relatively slender, but rather large, sticky, naked *Hebeloma* with drops on the gill edges and a radish-like smell. The cheilocystidia are long, narrowly clavate, on average 6.6–8.9 µm wide at the apex. The amygdaliform, finely ornamented, dextrinoid spores measure 8–14.5 × 5.1–8. µm. Has a very wide ecology, but is normally mycorrhizal with deciduous trees.

The most likely confusion species are *H. leucosarx* ⋉ and *H. incarnatulum* ⋉: the former has a very umbonate, darker cap, but is microscopically identical; the latter has very narrow cheilocystidia. *Hebeloma crustuliniforme* s.l. ◁ is more robust, has more sinuous cystidia and less dextrinoid spores.

Widespread and common; April, July–November.

Hebeloma helodes is a small, pale, ± umbonate, almost dry, naked *Hebeloma* with a floccose stem. It has 30–60 entire gills and a radish-like smell. The cheilocystidia are clavate, with a delimited, wide (on average 8.3–11.4 µm), thick-walled head. The pointed-amygdaliform, mostly non-dextrinoid spores measure 8.2–12.4 × 4.3–6.5 µm. Mycorrhizal with various deciduous trees, mostly in damp woodlands.

The wide, thick-walled heads to the cystidia is the best character for seperating this from similar species. *Hebeloma pusillum* ▷▷ has a darker cap centre, occurs only with *Salix* and has fewer than 30 entire gills.

Widespread, occasional; August–November.

Jan Vesterholt

Hebeloma pusillum is a small, naked, ± umbonate *Hebeloma* with a distinctly two-coloured cap that has a dull orange-brown centre. The stem is pale and floccose. Has fewer than 30 entire gills; the edges are floccose and have drops. The odour is radish-like. The cheilocystidia are clavate, widened above (on average 8–10 µm wide), and thin-walled. The amygdaliform to fusiform, slightly dextrinoid spores measure 9.8–14.9 × 5–7.3 µm. Mycorrhizal with *Salix* on damp soils.

Hebeloma helodes ⊲⊲ has cystidia heads that are more delimited and thicker-walled, and a larger number of entire gills. *Hebeloma salicicola* ⋉ is somewhat more robust and has distinctly dextrinoid spores.

Widespread and rather common; July–November.

Hebeloma theobrominum is a fairly robust, rather monochrome, red-brown *Hebeloma* with a somewhat greyish, pruinose cap surface. The odour is radish- to cacao-like and the taste bitter. The cheilocystidia are small and almost bowling pin shaped. The finely ornamented, very dextrinoid spores measure 7.8–10.9 × 4.1–6.3 µm, Qav. 1.8–1.95. Mycorrhizal with various trees, *e.g. Quercus,* but also with *Helianthemum*, on ± calcareous soils.

Hebeloma vesterholtii ⋉ has a paler cap and a higher spore Q (av. 1.64–1.79). *Hebeloma griseopruinatum* ⋉ is also mycorrhizal with *Helianthemum*, but has wider spores (on average more than 5.5 µm).

Widespread and rather common, rare in the boreal zone; June–November.

Jan Vesterholt

Hebeloma sinapizans is a large, ± cinnamon-buff, naked *Hebeloma* with a paler cap margin. The gills are crowded and lack drops on the edge. The robust, bulbous stem is floccose-scaly along its entire length and is hollow with a downwards-pointing tooth in the top part (seen in section in lower image). Smells strongly radish-like. The cheilocystidia are rather short, lageniform. The coarsely ornamented, very dextrinoid spores have a loosening outer wall and measure 8.9–14.2 × 5.3–8.7 µm. Occurs on clay soils with deciduous trees, mostly *Fagus*, but also with conifers and in open areas with *Helianthemum*.

Separated from other similar species by the very crowded gills (>80 entire). *Hebeloma quercetorum* ⨯ usually has drops on the gill edge, fewer entire gills and is less fleshy.

Widespread and rather common; July–November.

Cortinarius

Cortinarius is a genus of ectomycorrhizal agarics with ± emarginate gills, rusty-brown, verrucose, mostly dextrinoid spores and well-defined veils. The partial veil is mostly thread-like, while the universal veil leaves a woolly, thread-like or slimy covering on the lower stem and on the cap. Only very rarely cystidia (of a simple type).

The genus is enormous and difficult to deal with. New species are constantly being described and new synonymies published. Safe identification is in many cases not possible without sequencing. Important macroscopical characters include the type of veil (slimy or thread-like), the gill and flesh colour when young, the overall shape, the smell, and chemical colour reactions with KOH.

OTHER SIMILAR FUNGI:

- *Inocybe* has fibrillose to coarsely fibrillose caps and grey-brown spores. The spores are smooth, nodulose or star-shaped, and all species have cystidia (page 652).
- *Hebeloma* likewise has grey-brown spores and smells mostly earth- or radish-like. The spores are verrucose and all species have prominent cystidia on the gill edge (page 680).
- little brown mushrooms (LBMs) are typically decomposers, and generally have cystidia (page 616).
- *Leucocortinarius bulbiger* recalls a white-spored phlegmacia (page 257).

FURTHER READING: 46, 69, 124, 125, 126, 127, 156, 184, 305, 323.

More than 500 species in temperate Europe.

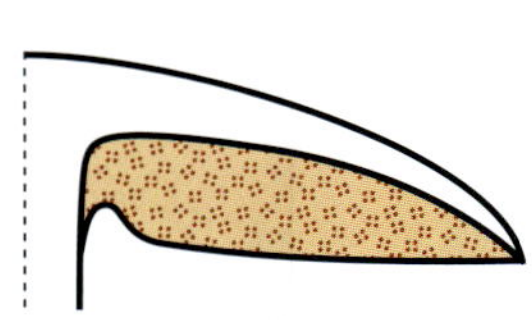

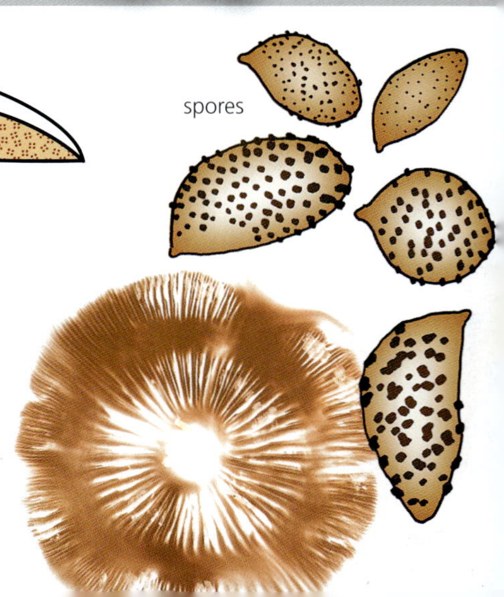

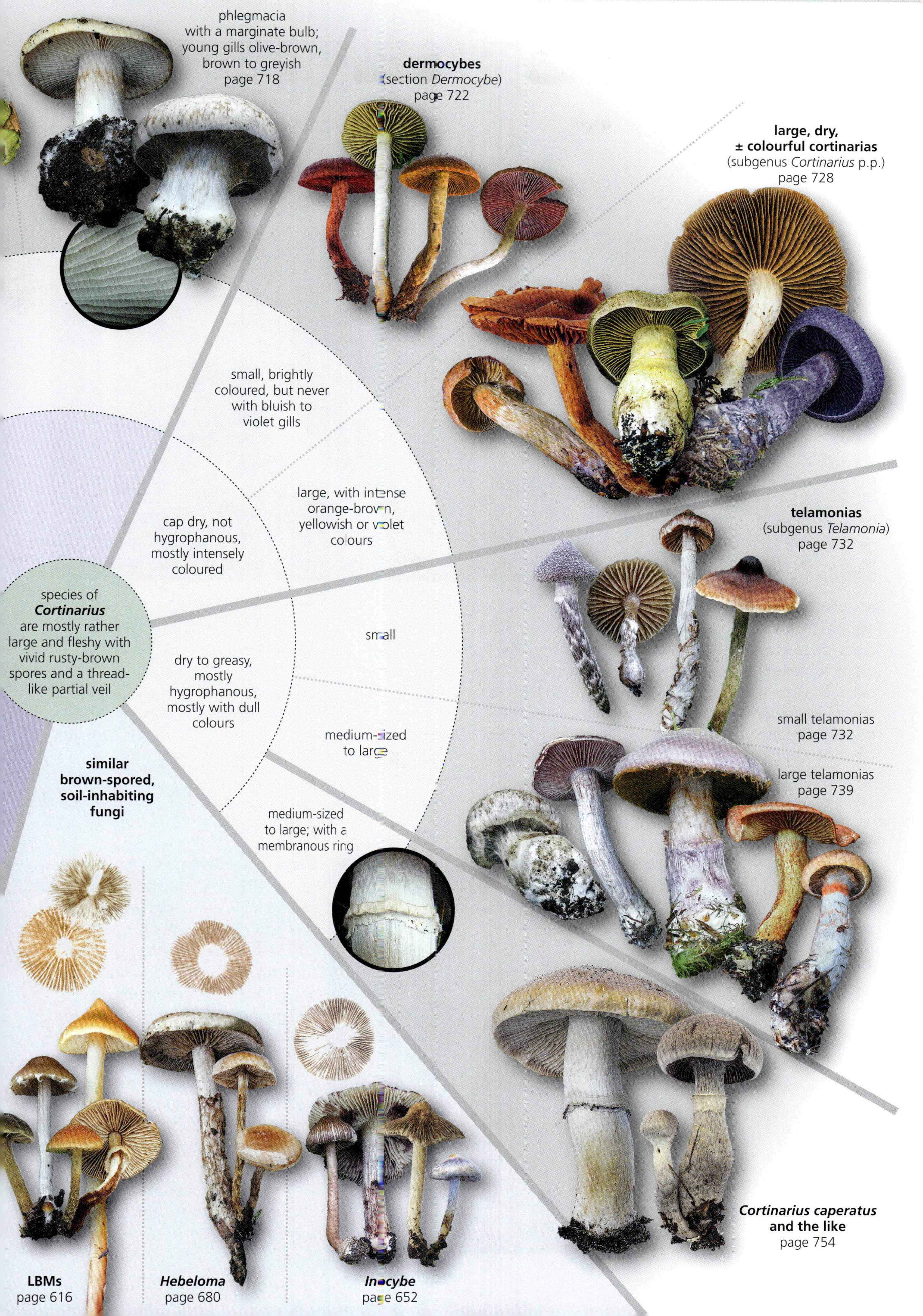
phlegmacia
with a marginate bulb;
young gills olive-brown,
brown to greyish
page 718
dermocybes
(section *Dermocybe*)
page 722
large, dry,
± colourful cortinarias
(subgenus *Cortinarius* p.p.)
page 728
small, brightly
coloured, but never
with bluish to
violet gills
large, with intense
orange-brown,
yellowish or violet
colours
cap dry, not
hygrophanous,
mostly intensely
coloured
telamonias
(subgenus *Telamonia*)
page 732
species of
Cortinarius
are mostly rather
large and fleshy with
vivid rusty-brown
spores and a thread-
like partial veil
small
dry to greasy,
mostly
hygrophanous,
mostly with dull
colours
small telamonias
page 732
medium-sized
to large
large telamonias
page 739
similar
brown-spored,
soil-inhabiting
fungi
medium-sized
to large; with a
membranous ring
Cortinarius caperatus
and the like
page 754
LBMs
page 616
Hebeloma
page 680
Inocybe
page 652

Jens Maarbjerg

Myxacia (*Cortinarius* subgenus *Myxacium*) are surrounded by a slimy universal veil, which is left as slime on the cap and on most of the stem.

Cortinarius vibratilis is a relatively slender and small, yellow to brownish-yellow myxacium with bitter-tasting cap cuticle and flesh. The stem is very pale. The ellipsoid spores measure 7–7.5(–8) × 5–5.5 µm; barely dextrinoid. Mycorrhizal with *Picea* in rather rich *Picea* forests and plantations. (However, specimens with similar characters occur with deciduous trees.)

Cortinarius causticus ▷ has more red-brown cap colours and is finely rugose around the cap centre. *Cortinarius pluvius* ⋉ is smaller, more grey-brown, and pale at the cap margin. *Cortinarius vibratilis* constitutes a species complex that includes several morphologically barely discernible species.

Widespread and rather common, becoming scarcer towards the south; late summer–autumn.

Jan Vesterholt

Cortinarius barbatus is a small to medium-sized myxacium with persistent, whitish colours and a very bitter taste (both cap cuticle and flesh) and mostly with a fusiform stem. The spores are ± amygdaliform, finely ornamented, and measure 7.5–8(–8.5) × 4.5–5 µm; barely dextrinoid. Mostly in clusters with *Fagus*, *Quercus* and *Carpinus*.

Cortinarius emollitoides ⋉ has a less bitter-tasting cap cuticle, mild or somewhat bitter flesh, a less sticky cap and quickly acquires cinnamon-buff cap colours.

Widespread, occasional, becoming scarcer towards the north; August–December.

Cortinarius causticus is a small, bitter-tasting myxacium with an apricot-orange to brownish-orange cap that is finely rugose around the centre. The small, ellipsoid, rather coarsely ornamented spores measure 6.5–7.5 × 4.3–4.8 µm; barely dextrinoid. Occurs on poor soils with *Pinus*.

Differs from other warm-coloured bitter-tasting myxacia in conifer forests by the rugose cap and small spores. The name has been variously interpreted.

Widespread, common in the hemiboreal–boreal zones but rare in the nemoral zones; mainly July–October.

Jan Vesterholt

Cortinarius croceocaeruleus is a rather small but striking myxacium with vivid blue colours on the cap and bitter-tasting cap cuticle and flesh. Starting in spots, the cap becomes increasingly yellow with age. The stem is ± fusiform and pale. The ± ovoid-ellipsoid spores measure 7.5–8.5 × 4.5–5(–5.5) µm; somewhat dextrinoid. Occurs mainly with *Fagus* on mineral-rich to calcareous soils.

Cortinarius salor ▷▷ has a mild taste. *Inocybe lilacina* (page 671) is smaller, mostly dry, has a grey-brown spore-deposit and smooth spores.

Widespread in the nemoral zones but rather rare, rare in the hemiboreal zone; August–November.

Federico Calledda

Cortinarius salor is a rather large and fleshy myxacium with a striking bluish-violet cap and a paler stem with the same shade of blue. The taste is mild. The subglobose, strongly ornamented spores measure 7.5–10 × 6–8.5 µm. Occurs in calcareous forests with *Fagus*, *Quercus*, *Tilia* and *Corylus*, but also in herb-rich *Picea* forests.

Cortinarius emunctus ⋉ has no contrast between the cap and stem, and the slimy veil is more persistent. *Cortinarius betulinus* ▿ is more slender and less colourful. *Cortinarius croceocaeruleus* ◁◁ is very similar but has a bitter-tasting cap cuticle.

Rather rare in the hemiboreal zone, almost absent from the nemoral zones; August–November.

Emilio Pini

Cortinarius betulinus is a small, slender myxacium with delicate pale blue colours fading to yellowish on the cap and stem. The taste is mild. The subglobose to broadly ellipsoid, fairly coarsely ornamented spores measure 8–10 × 6.5–8.5 µm. Occurs with *Betula*, mostly in damp conifer forests on poor soils.

Other bluish myxacia are either larger or bitter-tasting.

Common in the hemiboreal–boreal zone but very rare in the nemoral zones; mainly August–October.

***Cortinarius trivialis* s.l.** is a medium-sized myxacium with a brownish cap, usually with a low, broad umbo, and mostly with bluish gills when young. The stem is cylindrical and has a thick slime layer that breaks into girdles; it discolours brownish from the base. The amygdaliform to citriform, coarsely ornamented spores measure 10.5–12.5(–13) × 6–7.5 µm. Mycorrhizal with *Salix*, *Populus*, *Quercus*, *Betula* and rarely *Fagus* on both rich and poor soils.

Constitutes a species complex, where the taxa are ± partner-specific. It was originally described with *Salix* from Denmark. The fruitbodies depicted here were growing with *Populus*. Similar species from deciduous forests have a honey-like smell at the stem base.

Widespread, rather common to occasional; June–December.

Cortinarius delibutus is a small to medium-sized myxacium with a yellow cap, yellowish slime on the stem and greyish to violet-grey gills. The taste is mild. The subglobose to broadly ellipsoid, coarsely ornamented spores measure 7–9 × 6–8 µm. Occurs with *Betula Fagus* and *Picea* on poor soils.

This is the only mild-tasting, yellow myxacium. The phlegmacia *C. cliduchus* ▷▷ has yellow girdles on a dry stem, a grainy, cracking cap surface and smells of maize.

Rather common and widespread; July–November.

Jens Maarbjerg

Cortinarius elatior is a medium-sized to large myxacium with a characteristic conical, brown cap, which is usually radially wrinkled at the margin. The gills are pale brown and mainly ± interveined. The stem is whitish with stripes at the top and usually has a quickly fading, violet slime. After bruising, it has a smell of honey at the stem base. The lemon-shaped spores have a mixture of small and large warts, and measure 11–14 × 6.5–9 µm. Occurs with *Fagus*, *Quercus* and also with conifers.

Possibly constitutes a species complex. Several names are available for specimens with small fruitbodies. *Cortinarius mucosus* ▷ lacks the smell of honey. *Cortinarius stillatitius* ▽ lacks obvious wrinkles on the cap, interveined gills and stripes on the stem apex.

Widespread and common; July–December.

Katriina Bendiksen

Cortinarius stillatitius is a medium-sized myxacium with a brown cap, a pale stem with pale violet slime, and a strong smell of honey at the stem base. The densely verrucose, lemon-shaped spores measure 13–16 × 7–9 µm. Occurs with both coniferous and deciduous trees.

Cortinarius collinitus ▷ lacks the smell of honey. *Cortinarius mucifluus* ⋉ is paler and has a white stem. *Cortinarius elatior* △ has an obviously, radially, wrinkled cap, stripes on the stem apex and interveined gills.

Widespread and common in the north, but less frequent farther south; August–November.

Cortinarius collinitus is a medium-sized to large myxacium with a reddish-brown cap and violet colours to the slime on the stem; the stem base is odourless. The amygdaliform, coarsely ornamented spores measure 13–16 × 7.5–9.5 µm. Occurs with *Picea* on poor, often sandy soils.

Cortinarius mucosus ▽ occurs with *Pinus* and has a white stem. Other similar species have a honey-like smell at the stem base.

Widespread and very common to rather common; mainly July–October.

Cortinarius mucosus is a medium-sized, but relatively fleshy myxacium with a chestnut-brown cap and a white, mostly rather short stem that is odourless at the base. The narrow spores measure 11.5–14 × 5.5–7(–7.5) µm. Occurs on poor, sandy soils with *Cladonia* (*Cladina*) lichens and is mycorrhizal with *Pinus*.

Cortinarius collinitus △ has a violet-tinged stem and occurs with *Picea*. Told from other similar species by the combination of a lack of a honey-like smell, cap colour, stem colour, habitat (with *Pinus*) and narrow spores.

Widespread, most frequent in northern continental parts of temperate Europe, occasional in more Atlantic areas; August–November.

Morten Christensen

Phlegmacia (*Cortinarius* subgenus *Phlegmacium*) have slimy universal veil that is left as slime on cap and at stem base (*e.g.* at the margin of a marginate bulb). Stem dry, typically with abundant partial, thread-like veil.

•• ***Cortinarius praestans*** is a very fleshy phlegmacium with a clavate stem and orange-brown cap colours. The cap margin has abundant white veil, which is also seen as flakes towards the cap centre; the surface is ± wrinkled. Young gills are ± grey. The stem has white or somewhat violet veil girdles. The flesh is pale and ± tinged blue-grey in the stem apex. No reaction in KOH. The spores are amygdaliform and measure 14–17 × 8–10 µm. Occurs on calcareous soils with *Corylus*, *Quercus*, *Tilia* and *Carpinus*.

The massive size and large spores make this species easy to recognize.

Widespread and occasional in the hemiboreal zone, very rare in the nemoral zones; mainly August–October.

Thomas Stjernegaard Jeppesen

Cortinarius cumatilis is a large, ± violet phlegmacia with greyish gills. The stem has a non-marginate bulbous base and ± obvious, violet velar girdles. Surfaces and flesh do not have a KOH reaction. The amygdaliform spores measure 10–12 × 5–6 µm. Mycorrhizal with *Picea*, rarely *Tilia,* on calcareous soils.

Cortinarius serarius ⋉ is not as fleshy and has a darker cap centre.

Widespread and mainly found in the hemiboreal zone, nearly absent from the nemoral zones; mainly August–October.

Thomas Stjernegaard Jeppesen

Cortinarius balteatus is a large, fleshy and firm, thick-stemmed, bulbless phlegmacia with a bluish, almost dry to felty cap, which later turns brownish from the centre. The gills are pale. The flesh stains yellow with KOH. The amygdaliform to citriform spores measure 9.5–11.5 × 5–6.5 µm. Occurs in dry conifer forests and plantations.

Cortinarius balteatocumatilis ⋉ and *C. chromataphilus* ⋉ are similar but occur with deciduous trees on clay-rich to calcareous soils. Other similar species have violet gills when young.

Widespread in hemiboreal Scandinavia, but rare elsewhere and largely absent from the nemoral zones; mainly August–October.

Cortinarius cliduchus is a medium-sized phlegmacia with a brownish-yellow cap, the surface of which breaks into small flakes. The whitish, clavate stem has yellowish veil girdles. The gills are pale. The odour recalls raw maize cobs. The spores are ± amygdaliform and measure 9.5–10.5 × 5.5–6.5 µm. Occurs with deciduous trees, typically *Fagus*, on mobile, clay-rich soils.

Also known as *C. olidus*. *Cortinarius anomalus* ▷▷ is odourless and has almost spherical spores. *Cortinarius delibutus* ▷▷ has a yellowish slime and also lacks a smell. *Cortinarius rhizophorus* ⋉ has only a yellowish slimy girdle and lacks a smell.

Occasional to rather rare and local, absent from the boreal zone; August–November.

Jens H. Petersen/Jan Vesterholt

Thomas Stjernegaard Jeppesen

Cortinarius argutus is a medium-sized to large, whitish to pale cinnamon-buff phlegmacium with a somewhat sticky cap, whitish gills and a ± rooting, non-bulbous stem. The surfaces and white flesh discolour reddish and blackish. It dries ± blackish, especially in the flesh. The amygdaliform spores measure 9.5–11.5 × 5–6.5 µm. Mycorrhizal with *Populus tremula*.

Cortinarius rosargutus ⋉ is a similar species associated with conifers. Species around *C. barbatus* ⊲⊲ may look similar, but are bitter-tasting, have sticky stems and only rarely occur with *Populus*.

Rather common in the hemiboreal zone extending into the boreal zone, but rare in the nemoral zones; mainly August–October.

Thomas Stjernegaard Jeppesen

Cortinarius leucophanes is a relatively small phlegmacium with a distinctly sticky, whitish to cream cap. The gills are pale, mostly with a rose tinge. The stem is clavate. Lacks a distinctive smell or taste. No reaction in KOH. The small, pale spores measure 5.5–7.5 × 3.5–4.5 µm. Found mostly in dry *Pinus* forests on poor soils, but also with *Picea* on ± calcareous soils.

Cortinarius lustratus ⋉ smells and tastes farinaceous. Other similar species have larger spores.

Occasional in the hemiboreal zone, very rare in the nemoral zones; mainly August–October.

Cortinarius vulpinus is a medium-sized to large phlegmacium with a brownish-orange cap, which usually has abundant white remnants of universal veil at the margin. The gills are pale greyish to violet-grey. The clavate, rooting stem has prominent white velar girdles. Smells earthy-musty – especially during drying and when old. The large, amygdalform spores measure 11–13 × 6.5–7.5 µm. Occurs with *Fagus* on clay-rich and/or calcareous soils.

Cortinarius obsoletus ⋉ has less warm cap colours, weaker girdles, an indistinct smell and ellipsoid spores that measure 11.5–13.5 × 7–8.5 µm. *Cortinarius pseudovulpinus* ⋉ has yellowing flesh that smells of old cheese, and occurs with *Carpinus* and *Corylus*.

Rather rare to locally fairly common, but only in the nemoral zones; August–November.

Thomas Stjernegaard Jeppesen

***Cortinarius infractus* s.l.** is a medium-sized, bitter-tasting phlegmacium with brownish to olive-greenish colours, a radially fibrillose cap and strikingly dark, ± olive-brown gills. The stem is clavate and mostly paler than the cap. The flesh is pale but may be grey or violet-tinged at the stem apex. No reaction in KOH. Usually odourless. The spores measure 7–8.5 × 5.5–6.5 µm. Found mostly with deciduous trees on calcareous or clay-rich, mostly disturbed soils; rarely with conifers.

Constitutes a large species complex, where most of the taxa remain undescribed. *Cortinarius subtortus* ▷ is also bitter-tasting, has dark gills and ± globose spores, but occurs exclusively on poor soils and has a characteristic cedar wood-like smell. *Cortinarius purpurascens* ▷▷ may look similar, but is mild-tasting.

Widespread, rather common to occasional, extending into the boreal zone, but more common in the south; July–November.

Cortinarius sphagnophilus is a medium-sized phlegmacium with a rounded bulb. The slimy-sticky cap is ± orange-brown with hygrophanous spots and radial streaks. The gills are grey-brown to bluish-grey when young. The veil on the bulb and the basal mycelium is ± greenish-yellow. No reaction in KOH. The odour is honey-like. The broadly ellipsoid spores measure 9–10.5 × 6–7 µm. Occurs typically with *Picea* in damp forests and plantations.

Cortinarius scaurus ⋉ has distinctly green colours at the cap margin and on the gills. *Cortinarius herpeticus* ⋉ is larger and has a paler cap.

Widespread and common in the hemiboreal–boreal zones, but rare farther south; July–October.

Cortinarius subtortus is a medium-sized to large phlegmacium with an olive-yellow, somewhat sticky cap and olive-brown gills. The paler, clavate to cylindrical stem lacks a bulb. The taste is bitter and it has a characteristic cedar wood-like smell. and. No reaction in KOH. The almost spherical spores measure 7.5–9 × 5.5–6.5 µm. Occurs with *Fagus* and *Picea* on mor soils.

The colour of the immature gills, smell, taste and spore shape make identification straightforward. *Cortinarius infractus* ◁ occurs on richers soils and is usually odourless.

Widespread, rather common in the north to rather rare in the nemoral zones; August–November.

Jens Maarbjerg

Tobias Frøslev

Cortinarius nanceiensis is a yellow-brown phlegmacium with pale sulphur-yellow gills and a grainy-sqamulose cap centre. The stem base is, at most, somewhat bulbous. The flesh stains reddish with KOH. The odour is rather like that of unripe bananas. The amygdaliform spores measure 10–12 × 6–7 µm. Occurs with deciduous trees on calcareous soils, farther south also with *Abies*.

Cortinarius cliduchus ◁◁ has whitish to greyish gills.

Very rare and always at sites with many other rare species, absent from the boreal zone; mainly August–October.

Thomas Stjernegaard Jeppesen

Cortinarius percomis is a medium-sized phlegmacium with yellowish-brown to apricot-orange cap colours tinged greenish. The gills are pale sulphur-yellow, and the pale stem lacks a bulb. Has a very distinct smell of marjoram (*Origanum*) or apples. No reaction in KOH. The ± amygdaliform spores measure 11–12.5 × 6–7 µm. Occurs with *Picea* in calcareous forests.

The similar *C. majoranae* ⋉ occurs with *Tilia* and *Corylus* and perhaps also *Quercus*.

Widespread and common in Scandinavia, very rare or absent from the nemoral zones; mainly August–October.

Cortinarius varius is a medium-sized phlegmacium with yellow-brown cap colours and a smooth cap surface that lacks distinct fibres. The young gills are violet. The whitish, non-bulbous stem does not discolour when bruised and may have ± distinctly buff veil girdles. The stem flesh is white or yellowish and does not have a distinctive smell. No reaction in KOH. The amygdaliform spores measure 9.5–11.5 × 5.5–6.5 µm. Occurs with *Picea* on rich, ± calcareous soils.

Cortinarius spadicellus ⋉ has a violet-tinged stem when young, the stem stains brown when bruised, and the cap is ± radially fibrillose.

Widespread and fairly common in mainly hemiboreal Scandinavia, but absent or nearly so from the nemoral zones; mainly August–October.

Thomas Sternegaard Jeppesen

Cortinarius triumphans is a large phlegmacium with a yellow-brown cap and, usually, pale remnants of universal veil at the cap margin. The pale gills have a violet tinge when young. The clavate, long stem has brownish-yellow, scaly girdles of universal veil. The flesh is pale and does not have a distinctive smell. No reaction in KOH. The amygdaliform spores measure 10.5–12.5 × 6–7 µm. Mycorrhizal with *Betula* on dry to ± damp, rich soils.

Cortinarius variiformis ⋉ occurs with *Quercus* and has significantly less remnants of yellow veil on the stem, mostly only as a thin band, and more distinctly violet gills.

Widespread and rather common in the hemiboreal–boreal zones, but much rarer in the nemoral zones; August–November.

Margaretha Liebmann/Jens H. Petersen

Cortinarius largus is a medium-sized to large phlegmacium with relatively pale, greyish-violet to brownish cap colours and ± violet gills. The pale, cylindrical to clavate stem is rather hard. The flesh has violet colours, especially in the stem base. Stains yellow, later vinaceous-brown with KOH (inset image). The odour is faint and the taste mild. The amygdaliform to somewhat citriform spores measure 9.5–11.5 × 5–7 µm. Occurs mostly with *Fagus* on ± calcareous or clay-rich soils.

Cortinarius variicolor ▽ has a strong earthy-musty smell, darker cap colours and a radially fibrillose cap. *Cortinariius balteatus* ◁◁ and *C. balteatocumatilis* ⋉ have whitish gills.

Widespread, fairly common to occasional, absent from the boreal zone; August–October.

Thomas Stjernegaard Jeppesen

Cortinarius variicolor is a medium-sized to large phlegmacium with greyish-violet to brownish cap colours and radial fibres. The gills are ± violet and the stem non-bulbous, cylindrical to clavate. The flesh has violet colours. Stains yellow to vinaceous-brown with KOH. The smell is strong, unpleasant earthy-musty, or like banana peel. The ± amygdaliform spores measure 10–11.5 × 5.5–6.5 µm. Occurs with deciduous trees and conifers on ± calcareous soils.

The paler *C. largus* △ is less fibrillose, harder stemmed, has a faint smell and grows with deciduous trees.

Rather rare and confined to southern parts of the hemiboreal zone and the nemoral zones; mainly August–October.

Laurent Francini

Cortinarius cyanites is a sturdy phlegmacium with a non-marginate bulb and an almost dry, grey-brown cap. The gills are dark blue and the flesh is bluish, discolouring reddish when cut. The ellipsoid to amygdaliform spores measure 8.5–10.5 × 5.5–6.5 µm. Occurs with deciduous trees and conifers on rich, ± calcareous soils.

Cortinarius subpurpurascens ▽, *C. purpurascens* ▷▷ and *C. subporphyropus* ⋉ discolour violet on the stem and cap, but do not redden in the flesh. *Cortinarius violaceus* ▷▷ is violet on both cap and stem, is completely dry and does not discolour. Most other bluish phlegmacia also lack the reddening reaction in the flesh.

Occasional from the southern part of the boreal zone to the hemiboreal zone, very rare in the nemoral zones; mainly August–October.

Cortinarius subpurpurascens is a medium-sized to large phlegmacium with an ochraceous to red-brown, radially fibrillose cap, which may have hygrophanous spots or streaks and violet colours at the margin. The stem and gills have purple-brown to violet colours and discolour violet when bruised. The stem lacks a basal bulb. The spores measure 8.5–10 × 5–6.5 µm. Occurs on clay-rich, ± calcareous soils with *Quercus*, *Corylus*, *Tilia* or *Fagus*.

Cortinarius purpurascens ▷▷ occurs mostly with conifers, is more violet and has a basal bulb. *Cortinarius subporphyropus* ⋉ is much smaller and has rather distant gills.

Occasional, absent from the boreal zone; August–November.

Jens H. Petersen/Jens Maarbjerg

Jens Maarbjerg

Cortinarius purpurascens is medium-sized to large phlegmacium with a violet-brown, radially fibrillose cap, sometimes with hygrophanous spots. The stem and gills are ± violet and discolour violet when bruised. The stem has a basal bulb with violet remnants of veil on the margin. Spores measure 8–9 × 5–6 µm. Mostly with conifers on ± calcareous or clay-rich soils.

Cortinarius subpurpurascens ◁◁ occurs in deciduous forests, is less violet and has a clavate stem. *Lepista nuda* (page 251) is naked, with pinkish-brown, less ornamented spores.

Widespread, occasional to rather rare, becoming scarcer towards the north; July–October.

Thomas Stjernegaard Jeppesen

Cortinarius suaveolens is a medium-sized phlegmacium with a characteristic smell of *Prunus padus* flowers (or as *Hebeloma sacchariolens*, page 684), a mixture of yellow and violet colours and pale flesh. Does not react with KOH. The spores measure 11.5–13 × 6.5–7.5 µm. Occurs with *Fagus* and *Quercus* on clay-rich or calcareous, ± mobile soils.

Other similar-smelling phlegmacia are either greenish-yellow or lack any yellow or violet colours.

Very rare and ± confined to the nemoral zones; August–November.

Thomas Læssøe/Thomas Stjernegaard Jeppesen

Cortinarius sodagnitus is a small to medium-sized, violet-blue phlegmacium that later fades brownish; with a smooth cap and distinct dark spots at the margin. It stains deep pink with KOH (inset image). The gills are greyish to violet. Lacks a distinct smell. Spores amygdaliform, 9–10 × 5–6 µm. Occurs with *Fagus* on mobile, ± calcareous clays.

Cortinarius eucaeruleus ▷ and *C. caerulescens* ▷▷ are larger, radially fibrillose and lack pink KOH reactions; *C. nymphicolor* ⋉ is paler and usually has flakes of veil on the cap; *C. suaveolens* △ smells of *Prunus padus* flowers.

Rare, only nemoral–hemiboreal zones; August–November.

Cortinarius eucaeruleus is an impressive, massive phlegmacium with a purple-blue cap that has innate darker blue fibres and bluish gills. The stem has a distinct, marginate bulb. The odour is ± metallic. Does not have any conspicuous reactions with KOH. The ellipsoid spores measure 9.5–12 × 6–7 µm. Occurs mostly with deciduous trees on mobile, ± calcareous clays, usually where *Corylus* and/or *Quercus* are present.

Cortinarius caerulescens ⊳⊳ has amygdaliform spores, is a paler blue and has obvious flakes of universal veil on the cap. *Cortinarius sodagnitus* ◁ stains pink with KOH on the cap and bulb, is smaller and has smaller spores (< 9 µm). *Cortinarius moenne-loccozii* ▽ has an almost membranous universal veil, which forms a volva around the bulb, and amygdaliform spores. The name *C. terpsichores* ⊳⊳ is sometimes applied to this species.

Very rare and confined to the nemoral–hemiboreal zones; mainly September–November.

Cortinarius moenne-loccozii is a sturdy, purple-blue, striate-speckled phlegmacium with a marginate bulb and a white volva from a very thick universal veil. The cap has velar flakes. The volva may be violet on the inside. Flesh whitish. Smell malty or indistinct. Spores amygdaliform and measure 9.5–10.5 × 5.5–6.5 µm. Occurs on mobile, clay-rich soils with *Corylus* and perhaps *Quercus* and *Fagus*.

The combination of conspicuous white, volvate, universal veil and blue colours is diagnostic. *Cortinarius eucaeruleus* △ and *C. sodagnitus* ◁ have similar colours, but the veil is less prominent.

Very rare and southern; August–November.

Cortinarius caerulescens is a large phlegmacium with a radially fibrillose, blue cap that turns yellowish with age, and usually has pale flakes of universal veil. The stem, gills and flesh are pale blue. No pink reactions with KOH. Spores amygdaliform and measure 9.5–12 × 6–7 µm. Occurs with *Fagus* on mineral-rich or calcareous soils, mostly on slopes where the soil is mobile.

Cortinarius eucaeruleus ⊲⊲ and *C. moenne-loccozii* ⊲⊲ are deep purple-blue: the former has ellipsoid spores and only sparse universal veil; the latter has amygdaliform spores but more universal veil. *Cortinarius luhmannii* ⊳⊳ and *C. anserinus* ⊳⊳ have brownish caps. *Cortinarius foetens* and *C. coerulescentium* ⊳⊳ have paler caps that almost lack blue tinges. *Cortinarius terpsichores* ▽ grows mostly with conifers.

Widespread in the nemoral–hemiboreal zones, but absent from the boreal zone, rather rare to locally common; July–November.

Thomas Stjernegaard Jeppesen

Cortinarius terpsichores is a large phlegmacium with pale blue colours at the cap margin and, soon, a buff-tinged cap centre. The surfaces stain yellow when bruised. The flesh is pale and does not have pink KOH reactions. Has a metallic odour. Spores ellipsoid and measure 8.5–10 × 5.5–6.5 µm. Occurs with *Pinus* on calcareous soils in northern parts, but in central Europe also with deciduous trees.

Cortinarius eucaeruleus ⊲⊲ is deeper violet and has larger spores; *C. lepistoides* ⋉ has amygdaliform to citriform spores and blue gill edges; *C. caerulescens* △ usually has flakes of universal veil on the cap.

Very rare and scattered; absent from the boreal zone; August–November.

Cortinarius vesterholtii is a rather small phlegmacium with an indistinctly radially fibrillose, yellow-brown to rather dark red-brown cap. The gills are violet. The stem is mostly deep lilac-violet and has whitish hyphal cords. Does not have conspicuous KOH reactions. Spores amygdaliform to citriform; measure 10–11.5 × 5.5–6.5 µm. Mostly or exclusively with *Fagus* on clay-rich and/or calcareous soils in forests with many other rare mycorrhizal fungi.

Cortinarius calochrous ▷▷ has lemon-yellow hyphal cords; *C. piceae* ⋉ occurs with *Picea*. Other similar species are paler.

Very rare; absent from the boreal zone; mainly September–October.

Thomas Stjernegaard Jeppesen

Cortinarius magicus is a rather large to large phlegmacium with a tapering, marginate bulb that is up to twice as wide as the stem. Has greenish to golden-brown colours and abundant olive-brown to brownish universal veil on the cap and bulb margin. The gills are pale blue. The bluish stem discolours yellow-brown to orange-brown. Does not have conspicuous KOH reactions. The ellipsoid spores measure 7–8 × 4.5–5.5 µm. Occurs with deciduous trees on mobile, ± calcareous soils.

Cortinarius glaucopus ⋉ has a narrower, non-tapering bulb and much less universal veil. *Cortinarius pansa* ⋉ occurs in conifer forests on poor soils.

Widespread in the nemoral–hemiboreal zones but very rare; mainly August–October.

Jens H. Petersen/ Thomas Stjernegaard Jeppesen

Cortinarius anserinus is a large phlegmacium with a smooth, only somewhat radially fibrillose, pale brownish to brown cap, although mostly with a somewhat sordid yellowish/greenish tinge. The gills, flesh, and also the stem are bluish, and the bulb has a somewhat rounded margin. The cap cuticle tastes bitter and the smell may recall plums or celeriac, or later be earthy-musty. Does not have conspicuous KOH reactions. The spores are citriform with distinct warts/spines and measure 10–12 × 6–7 µm. Almost always occurs with *Fagus* on ± mobile, ± calcareous or clay-rich mull soils.

Easily recognized by the bitter-tasting cap cuticle. *Cortinarius luhmannii* ▽ is smaller and has a fibrillose cap with flakes of universal veil. Other bitter-tasting phlegmacia have different colours and spores.

Widespread and one of the most common phlegmacia in the nemoral zones, rare in the hemiboreal zone and absent from the boreal zone; August–November.

Thomas Stjernegaard Jeppesen

Cortinarius luhmannii is a rather large phlegmacium with a marginate, olive-grey bulb and an at first greyish, later olive-grey to yellow-brown cap that has pale cinnamon-buff velar flakes. The gills are violet-grey. The stem is at first greyish-blue but quickly discolours red-brown. Lacks a distinct smell; the cap cuticle tastes bitter when young. The spores are amygdaliform, measure 9–10 × 5–6 µm. Occurs with *Fagus* on clay-rich, ± calcareous soils.

Cortinarius riederi ⋉ and *C. argent-eolilacinus* ⋉ have rounded bulbs and larger, ellipsoid spores; *C. arcifolius* ⋉ has somewhat larger spores and blue remnants of veil on the bulb margin.

Rather rare and mainly nemoral; August– November.

Thomas Stjernegaard Jeppesen

Cortinarius calochrous is a small to medium-sized phlegmacium with a smooth, vivid yellow cap that lacks obvious fibres. The gills are lilac. The stem is pale with a wide, marginate bulb and lacks conspicuous violet tinges. The mycelium is lemon-yellow. The cap surface becomes dark brown with KOH but there is no reaction on the bulb surface. The ± amygdaliform spores measure 9–10 × 5–6 µm. Occurs with *Fagus* on mineral-rich, ± mobile soils.

Belongs to a very difficult complex with a number of extremely similar species. *Cortinarius cisticola* ⋉ has larger spores and reacts somewhat pink with KOH on the surface of the bulb and wine-red on the bulb margin. *Cortinarius catharinae* ⋉ reacts pink on the cap and bulb surface. Other similar species have paler cap colours.

Widespread, occasional to rare, although the distribution is not clear due to confusion with similar species; August–November.

Thomas Stjernegaard Jeppesen/Jan Vesterholt

††*Cortinarius splendens* is a small to medium-sized phlegmacium entirely saturated with yellow colours, including in the flesh. The cap centre becomes somewhat darker with age. The bulb is marginate and the basal mycelium is very obviously yellow. The cap surface reacts olive and the flesh somewhat reddish with KOH. The odour is insignificant. The coarsely ornamented, amygdaliform spores measure 9.5–11.5 × 5.5–6 µm. Occurs typically with *Fagus* on mobile calcareous or clay-rich soils, rarely with other deciduous trees.

Cortinarius citrinus ▽ has greenish shades, is a less saturated yellow on the stem apex and in the flesh, and does not have distinctly yellow hyphal cords. *Cortinarius xanthochlorus* ⋉ has greenish colours, almost colourless cap flesh and much larger spores (>11 µm long); *C. flavovirens* ▷▷ smells farinaceous. Other yellow phlegmacia react red to purple on the cap with KOH.

Widespread, but rare; September–November.

Cortinarius citrinus is a small to medium-sized phlegmacium with a distinct marginate bulb, greenish-yellow colours throughout and a faint smell of fresh pepper. Basal mycelium pale. Hardly stains with KOH. Spores amygdaliform to ± citriform, coarsely ornamented, 8.5–10 × 5–5.5 µm. Occurs with *Fagus* on mineral-rich soils.

Cortinarius xanthochlorus ⋉ has ± colourless cap flesh and larger spores; *C. flavovirens* ▷▷ smells farinaceous; *C. splendens* △ has deep yellow colours and a yellow mycelium. Other yellow phlegmacia react red to purple on the cap with KOH.

Nemoral, occasional; mainly August–October.

Jens H. Petersen/Margaretha Liebmann

Cortinarius bergeronii is a large phlegmacium with a smooth, yellow cap that, with age, becomes yellow-brown from the centre. The gills are yellow. The stem is greenish-yellow, often with a violet tinge on the upper part, and has a marginate bulb. Has violet shades in the central stem flesh and sometimes also in the cap. Cap surface stains purple-red, later blackish with KOH (inset image). The ± amygdaliform spores measure 10–11.5 × 6.5–7.5 µm. Occurs with *Fagus* on mineral-rich, usually on mobile soils.

Cortinarius elegantissimus ⋉ may also have violet shades in the flesh but has larger spores (>12 µm long). Other yellow phlegmacia lack such shades.

Rare and mainly nemoral; August–November.

Cortinarius alcalinophilus is a rather large phlegmacium with a yellow cap that lacks greenish tinges and has small, brown scales/flakes on the surface. Has a pale, marginate bulb and yellow bulb flesh. All surfaces stain red with KOH. The ± amygdaliform spores measure 10–11.5 × 5.5–6.5 µm. Occurs mostly with *Fagus* on mobile, mineral-rich soils.

Cortinarius olearioides ⋉ becomes orange-yellow in damaged parts and has citriform spores. *Cortinarius elegantissimus* ⋉ and *C. bergeronii* △ do not have yellow bulb flesh all the way through. *Cortinarius splendens* ◁ stains olive on the cap with KOH; *C. citrinus* ◁ hardly stains at all.

Nemoral–hemiboreal, rather rare; August–November.

Thomas Stjernegaard Jeppesen

†? **Cortinarius atrovirens** is a large, fleshy, ± olive-brown phlegmacium with ± pale sulphur-yellow to greenish-yellow gills. The stem is greenish-yellow and has a marginate bulb. The cap surface stains black with KOH. The smell recalls black pepper. The amygdaliform spores measure 9.5–11 × 5.5–6.5 µm. Typically mycorrhizal with *Pinus* or *Abies* on calcareous soils, but is also known to form mycorrhiza with *Dryas*.

Cortinarius ionochlorus ⋉ occurs with *Fagus* and *Quercus* and has greyish to violet gills. *Cortinarus flavovirens* ▽ is paler and smells farinaceous.

Primarily a southern species but known from a few scattered sites in Northern Europe and from The Burren in Ireland (with *Dryas*); mainly August–November.

Thomas Stjernegaard Jeppesen

Cortinarius flavovirens is a medium-sized phlegmacium with a greenish-yellow, smooth cap. The gills are ± pale yellow and the flesh whitish or somewhat greenish-yellow. The stem has a marginate bulb. Smells and tastes distinctly farinaceous. Does not react with KOH. The amygdaliform spores measure 9–10.5 × 5–6 µm. Occurs with *Corylus*, *Quercus* or *Fagus* on ± mobile, mineral-rich soils.

Other greenish phlegmacia do not have a farinaceous taste and smell.

Nemoral–hemiboreal, rather rare; mainly August–October.

Cortinarius cupreorufus is a medium-sized phlegmacium with at first greyish, later orange-brown cap colours. The gills are straw-yellow to ± olive-green. The stem has greenish colours and a marginate bulb. Reacts purple-black on the cap in KOH. The amygdaliform to citriform spores measure 10–11.5 × 6.5–7.5 µm. Occurs with conifers on calcareous soils.

Cortinarius rufo-olivaceus ▽, *C. xanthophyllus* ⋉ and *C. prasinus* ⋉ occur with deciduous trees.

Mainly hemiboreal, occasional to rare; August–October.

This species is included on the IUCN international Red List.

Thomas Stjernegaard Jeppesen

Cortinarius rufo-olivaceus is a medium-sized to large, variable phlegmacium with grey-rose to coral, later wine-red cap colours and sordid straw-yellow to grey-olive gills. Reacts olive-green or purple-red to black with KOH. The citriform spores measure (10.5–)11–12.5 × 6.5–7.5 µm. Mostly with *Fagus* on mobile, mineral-rich soils.

When young, may look like *C. sodagnitus* ◁◁ but that species is smaller, has smaller spores, more violet gills and stains pink with KOH. Other similar species have more greenish or yellowish cap colours.

Nemoral–hemiboreal, occasional to rare; August–November.

Thomas Læssøe

Cortinarius odorifer is a medium-sized olivaceous brown phlegmacium with tinges of olive-yellow on cap and gills. The odour is markedly aniseed-like. The cap surface stains dark red with KOH. The ± amygdaliform spores measure 10.5–12 × 6.5–7.5 µm. Occurs with conifers on calcareous soils.

Other similar olive-yellow species do not smell of aniseed.

Mainly in more continental parts of Scandinavia, ± absent from the nemoral lowlands; mainly August–October.

Thomas Stjernegaard Jeppesen

Federico Calledda

Cortinarius talus is a medium-sized phlegmacium with a brownish-yellow, radially fibrillose cap that has a thin velar covering at the cap centre and, usually, hygrophanous streaks. The gills are pale. The stem is whitish and has a marginate bulb. Has an evident honey-like smell, especially from the stem base after bruising. The spores are ellipsoid and relatively small, measuring 7.5–9.5 × 4.5–5.5 µm. Occurs with *Fagus*, *Quercus* and *Betula* on rather poor to ± rich soils.

Similar species lack the velar coating or honey-like smell, or have larger or different-shaped spores. *Cortinarius xantho-ochraceus* ▷ has no coating on the cap, a sharper-edged bulb and a malty to yeast-like smell. *Cortinarius multiformis* ▷ occurs with conifers and is darker brown on the cap.

Widespread, fairly common to rather rare; August–November.

Steen A. Elborne/Thomas Stjernegaard Jeppesen

Cortinarius langeorum is a medium-sized, ± cinnamon-buff phlegmacium with greyish-white, ± serrated gills. The bulb is ± yellowish. Reacts red on the cap to KOH (inset image). Lacks a distinct smell. The citriform spores measure 11–12.5 × 6.5–8 µm. Occurs with *Fagus* on mineral-rich, mobile soils.

Cortinarius caesiocortinatus ⋉ has rounded spores. Other similar species have different reactions to KOH.

Nemoral and rare; mainly September–October.

Thomas Stjernegaard Jeppesen

Cortinarius xantho-ochraceus is a medium-sized, yellow to cinnamon-buff phlegmacium that lacks obvious fibres or grains on the cap, but usually has whitish flakes of universal veil. The gills are greyish-white and the bulb sharp-edged. The odour is malty to yeast-like. The ellipsoid to amygdaliform spores measure 8–9 × 5–6 µm. Occurs in older, open deciduous forests on calcareous or clay-rich, ± mobile soils, where it is mycorrhizal with *Fagus*.

Cortinarius saporatus ▷▷ looks similar and has the same smell, but the spores are larger. *Cortinarius osmophorus* ▷▷ smells of *Prunus padus* flowers *Cortinarius talus* ◁ and *C. multiformis* ▽ smell of honey, while *C. langeorum* ◁ is usually odourless.

Nemoral, rather rare; August–November.

Cortinarius multiformis is a medium-sized phlegmacium with a cinnamon-buff to yellow-brown, radially fibrillose cap and pale gills. The whitish stem has a rather indistinct, rounded or somewhat marginate bulb. Has a faint, honey-like smell. Does not react with KOH. Spores ellipsoid to amygdaliform, weakly ornamented, measure 8–9.5 × 5–5.5 µm. Occurs with *Picea*.

Belongs to a very difficult complex. *Cortinarius talus* ◁ is more yellowish and occurs with deciduous trees. Other very similar species may have somewhat bluish gills. *Cortinarius inexspectatus* ⋉ occurs in the Nordic countries and fruits earlier (May–June).

Widespread and rather common; mainly July–October.

Jan Vesterholt

Thomas Stjernegaard Jeppesen

Cortinarius osmophorus is a medium-sized to large phlegmacium with a cream to pale yellow-brown cap. The gills are greyish-white and the stem whitish with a wide, marginate, soft-fleshed bulb. The bulb has white universal veil at the margin. The odour is heady, like *Prunus padus* flowers (or as *Hebeloma sacchariolens*, page 684). Does not react with KOH. The amygdaliform to citriform spores measure 8.5–10 × 5–6 µm. Occurs on calcareous mobile soils with *Fagus*, *Quercus* or *Tilia*.

The two other phlegmacia with the same odour – *C. odoratus* ⋉ and *C. suaveolens* ⊲⊲ – have more intense colours.

Mainly nemoral, rare; mostly September–October.

Thomas Stjernegaard Jeppesen

Cortinarius saporatus is a large phlegmacium with a smooth, brown cap, usually with remnants of universal veil as white spots or flakes. The gills are greyish-white and the stem and flesh whitish. Mostly with a brownish (slimy) zone of universal veil near the bulb edge. Has a malt-like smell. Does not react with KOH. Spores amygdaliform, measure 10–12 × 6–7 µm. With deciduous trees on calcareous or clay-rich, mobile soils.

Cortinarius xantho-ochraceus ⊲⊲ has the same smell but smaller spores. Other similar species have radially fibrillose caps, are smaller or have intense colours on gills or flesh.

Nemoral–hemiboreal zones, rare; mainly September–October.

Cortinarius foetens is a large, pale phlegmacium with a radially fibrillose, greyish to cinnamon-buff cap. Usually has pale flakes of universal veil on the cap or on the bulb margin. The gills, flesh and stem apex may have faint violet tinges. The odour is at first malt-like, later unpleasant earthy-musty. The spores are amygdaliform, and measure 9.5–10.5 × 5.5–6.5 µm. Occurs with *Fagus* on mineral-rich, mobile soils.

Cortinarius caerulescens ◁◁ has obvious blue colours on the cap; *C. coerulescentium* ▽ has ellipsoid spores and usually lacks remnants of veil on the cap; *C. luhmannii* ◁◁ has distinctly blue-tinged gills and is smaller.

Nemoral, rather rare; August–November.

Thomas Stjernegaard Jeppesen

Cortinarius coerulescentium is a large, pale phlegmacium with a radially fibrillose, greyish cap with bluish to pink tinges. Gills grey, sometimes with faint violet tinges; stem and/or flesh usually with bluish to violet tinges. Cap rarely has flakes of universal veil. The odour is somewhat metallic. Spores ellipsoid, measure 9–11 × 5.5–6.5 µm. Occurs with *Quercus*, *Corylus* and *Fagus* on mineral-rich, mobile soils.

Shape as *C. eucaeruleus* ◁◁ but lacks distinct blue colours. *Cortinarius foetens* △ has flakes of pale universal veil on the cap and amygdaliform spores. Other pale phlegmacia lack faint bluish shades, are more intensely coloured or are bitter-tasting.

Nemoral–hemiboreal zones, rare; mainly September–October.

Thomas Stjernegaard Jeppesen

Dermocybes (subgenus *Cortinarius*, section *Dermocybe*) are slender and have dry, non-hygrophanous, vividly coloured caps. They contain variable amounts of anthraquinones and are ± poisonous. Good for dyeing wool.

†? ***Cortinarius semisanguineus*** is a small to medium-sized dermocybe with a yellow-brown cap. The gills are intense blood-red. The yellow-brown stem may have orange-yellow veil remnants and a carmine basal mycelium. The ovoid to amygdaliform spores measure (5–)5.5–7.5(–9) × 3.5–5(–6) µm. Occurs mostly with conifers on poor, sandy soils but occasionally with *Betula* or *Fagus*.

Cortinarius phoeniceus ▽ has a more fleshy stem and blood-red stem girdles; *C. sanguineus* ▷ is blood-red throughout; *C. fervidus* ⋉ has orange-red gills and usually a characteristic broad umbo.

Widespread and common; July–December.

Torbjørn Borgen/Thomas Læssøe

†? ***Cortinarius phoeniceus*** is an eyecatching dermocybe with an orange-brown cap, blood-red gills and red girdles of universal veil on the relatively thick stem. The ± amygdaliform spores measure (5.5–)6–8(–9.5) × 3–4.5(–5) µm. Occurs mostly with *Picea* and *Pinus*, mainly on sandy, poor soils.

Cortinarius semisanguineus △ lacks red girdles on the stem and is mostly more slender.

Widespread, occasional, becoming scarcer towards the north; June–November.

†? **Cortinarius puniceus** is a dermocybe with blood-red colours throughout, usually with a violet tinge, and a rose basal mycelium. The veil is ± cinnamon-buff. The cuticle has incrusted pigment. The spores measure 7.6–8.6 × 4.6–5.1 µm, Qav. >1.63. Occurs in deciduous forests, mostly with *Fagus*.

Cortinarius cruentus and *C. rubrosanguineus* are synonyms. *Cortinarius sanguineus* ▽ normally occurs in conifer forests, has wider spores (Qav. <1.63) and lacks incrusted pigment in the cap cuticle; *C. cinnabarinus* ▷▷ is ± hygrophanous.

Widespread but rather rare; August–November.

Christophe Lécuru

†? **Cortinarius sanguineus** is a distinctive, ± uniform, blood-red, relatively slender dermocybe with a finely scaly cap surface. The cap cuticle hyphae lack incrusted pigment. The spores measure (6–)6.5–8.5(–10) × (4–)4.5–5.5 (–6) µm, Qav. <1.63. Occurs with *Picea*, and perhaps rarely *Fagus*, on mor soils.

Cortinarius puniceus △ occurs with *Fagus* and can only be recognized with certainty by spore shape and the presence of incrusted pigment in the cap cuticle. *Cortinarius cinnabarinus* ▷▷ is more orange-red, and is a hygrophanous telamonia from deciduous forests.

Widespread, occasional to rather rare; August–January.

Jan Vesterholt

☨? ***Cortinarius malicorius*** is a medium-sized dermocybe with a dark red-brown cap, usually with obvious orange remnants of universal veil at the margin, which remains somewhat inrolled for a long time. The gills are orange, and the fruitbody is typically somewhat more compact in stature than other dermocybes. The flesh is ± dark olive-green. The spores are small, measuring (5–)5.5–7(–7.5) × 3.5–4.5(–5) µm. Occurs mostly with *Picea* and *Abies*, but also with other coniferous and deciduous trees.

Other dermocybes do not have the combination of olive-green flesh, an orange cap margin and orange gills; the small spores are also characteristic. *Cortinarius cinnamomeus* ▷ has some green in the cap flesh, but larger spores.

Widespread and common; July–January.

Thomas Kehlet

†? *Cortinarius cinnamomeus* is a dermocybe with orange-yellow gills and a red-brown cap. The stem base lacks distinctive colours and the stem flesh usually does not have green colours, although the flesh above the gills is ± olive-green. The spores measure (5.5–)6–8.5(–9.5) × (3.5–)4–5(–6) µm. Occurs mostly on acidic soils with *Picea*, *Pinus* and *Betula*, rarely with other partner trees.

Cortinarius sommerfeltii ⨯ has a darker brown, more thin-fleshed cap, usually with evident zonation. *Cortinarius malicorius* ◁ has dark olive-green stem and cap flesh and an orange cap margin. *Cortinarius pratensis* ▽ occurs in open grassland and has larger spores.

Widespread and common; June–January.

†? *Cortinarius pratensis* is a small to medium-sized dermocybe with a dull, red-brown to orange-brown cap and orange-yellow gills. The flesh is sordid yellow-brown and, at most, only slightly green-tinged. The rather large spores measure (6.5–)7–10(–10.5) × (4–)4.5–5.5(–6) µm (there are always a few spores longer than 9.5 µm). Found typically in grasslands, *e.g.* inner dunes, mostly without any of the classic mycorrhizal partners (but *Carex* always seems to be present).

Similar dermocybes do not occur in the same habitat and have smaller spores.

Rare and mostly found in nemoral dunes; mainly September–November.

Thomas Læssøe

Erik Arnfred Thomsen/Jens H. Petersen

†?Cortinarius uliginosus is a distinctive dermocybe with either a vivid red-brown or yellow cap (*C. uliginosus* f. *uliginosus* and *C. uliginosus* f. *lutea*, respectively). The gills and stem are yellow. In f. *uliginosus* the stem is ± covered in orange-red universal veil, whereas f. *lutea* is paler, with a more brownish to rose veil. The spores measure (7–)8–11.5(–12) × (4.5–)5–6.5(–7) µm. Occurs with *Salix* on damp soil; the two forms are often found together.

Cortinarius croceoconus ⋉ occurs with *Picea*, and has reddish universal veil and a distinctly umbonate cap; it is typically somewhat smaller and less vividly coloured.

Widespread, occasional to locally common; July–November.

†?Cortinarius bataillei is a medium-sized dermocybe with orange-yellow gills and a distinct orange colour on the lower part of the stem, including the flesh. The spores measure (6.5–)7.5–9.5(–10.5) × (4–)4.5–5.5(–6) µm. On damp soil with conifers or *Betula*.

Cortinarius croceus ▷ has, at most, a slight orange tinge to the very base of the stem.

Fairly widespread, occasional, becoming scarcer towards the south; mainly August–October.

†?*Cortinarius croceus* is a dermocybe with yellow gills, a yellow-brownish cap and yellow to brownish universal veil. The flesh is yellowish and lacks green tinges. The spores measure (5.5–)6.5–9(–11) × (3.5–)4–5.5(–6) µm. Occurs with deciduous and coniferous trees on poor, often sandy soils.

Cortinarius croceoconus ⋉ has a more acute umbo and darker cap colours. *Cortinarius bataillei* ◁ has obvious orange colours on the lower part of the stem.

Widespread and common; July–November.

Guillaume Eyssartier

†?*Cortinarius chrysolithus* is a dermocybe with dull, greenish-grey-brown tinges and ± yellowish gills (var. *huronensis*) or olive-coloured gills (var. *chrysolithus* – depicted). It has a characteristic red-brown to carmine-red reaction with KOH on the gills, and some basidia contain a grainy, yellowish pigment. The spores measure (7–)7.5–11(–14) × (4–)4.5–6(–7.5 µm. Occurs in mires and boggy woodland with *Betula* and *Pinus*, mostly deep in *Sphagnum*.

Cortinarius tubarius ⋉ occurs in the same habitats but is more yellowish, has a dark, yellow- to red-brown KOH reaction and the basidia lack yellow contents.

Widespread, but threatened by drainage; July–October.

Thomas Kehlet

Thomas Læssøe

The remaining species of **subgenus *Cortinarius*** (this and the next page spread) are rather large and dry with violet or intensely yellow-brown to red-brown colours. Many have fleshy, clavate stems. Some species are among the most poisonous of all agarics.

Cortinarius violaceus is a medium-sized to large, very distinctive *Cortinarius* that is saturated throughout with dark violet colours. The cap is broadly umbonate, dry and finely scaly. Has a characteristic smell of cedar wood. The spores are amygdaliform and measure 11.5–13.5 × 7.5–8(–8.5) µm. Occurs on acidic soil with *Populus tremula* and *Fagus*, but also with other deciduous trees.

Cortinarius hercynicus ⋉ is a very closely related and almost identical species associated with *Picea*; it is recognized by its ovoid, more coarsely ornamented spores that measure 11–13.5 × (7.5–)8–9 µm.

Widespread, rather common to occasional; mainly July–October.

Cortinarius humicola is a medium-sized *Cortinarius* with saturated brownish-yellow colours and distinct, erect scales on stem and cap. The gills are pale and yellowish. Odour is faint, like cedar wood. The spores measure 8.5–9.5 × (5–)5.5–6 µm. Occurs with *Fagus* on clay-rich or calcareous soils.

Once recognized as a *Cortinarius*, distinctive and unlikely to be misidentified. The superficially similar but lignicolous *Pholiota squarrosa* (page 608) has smooth spores, and *Leucopholiota decorosa* (page 230) has a white spore-deposit. *Inocybe hystrix* (page 673) has grey-brown spores.

Mainly nemoral and rare; mainly August–October.

Cortinarius venetus is a medium-sized, entirely olivaceous-buff *Cortinarius* with a dry, finely felty-scaly cap. The stem is cylindrical. The odour is faintly of radish. The almost spherical spores measure 6.5–7.5 × 5.5–6 µm. Occurs on clay-rich or calcareous soils, with *Fagus* in the nemoral zones or with *Picea* and sometimes *Pinus* farther north.

Cortinarius cotoneus ▽ has a more fleshy, mostly clavate stem, evident radish-like smell and larger spores. *Cortinarius olivaceofuscus* ⋉ is smaller, has a smooth cap and narrower spores.

Widespread, but occasional to rare; August–October.

Cortinarius cotoneus is a large, entirely olivaceous-buff to olivaceous-yellow *Cortinarius* with a dry, finely felty/woolly cap and a relatively thick stem. Smells of radish. The spores measure 8.5–9.5(–10) × (7–)7.5–8 µm. Occurs with various deciduous trees on ± calcareous soils in species-rich deciduous forests.

Both *C. olivaceofuscus* ⋉ and *C. venetus* △ may occur in the same habitat but have smaller and more slender fruitbodies and smaller spores.

Nemoral–hemiboreal zones, rather rare to rare; August–November.

Jens H. Petersen/Michael Krikorev

††***Cortinarius rubellus*** is a medium-sized *Cortinarius* with warm, red-brown colours and red-brown to orange-brown gills. The cap is relatively acutely umbonate and fibrillose to minutely scaly. The universal veil is seen as distinct zigzag patterns on the ± tapering stem. Smells fainty of radish. The broadly ellipsoid spores measure 9–10.5(–11) × 7–8(–8.5) µm. Occurs on poor, acidic, mostly mossy soil with *Picea*, and on mor soils with *Fagus*, but also in *Betula-Pinus* forests.

Cortinarius orellanus ▽ has a convex or broader umbonate cap and less obvious girdles on the stem; it only occurs in deciduous forests. Other red-brown, dry species, *e.g.* *C. limonius* ▷, have yellow-brown gills.

Widespread, common to occasional – has increased in some areas; July–November.

Michael Krikorev

††***Cortinarius orellanus*** is a medium-sized *Cortinarius* with orange-brown colours and red-brown to orange-brown gills. The cap is convex and fibrillose to minutely squamulose, and the stem lacks obvious veil girdles. The amygdaliform spores measure 9–10(–10.5) × 6–6.5 µm. Occurs on mor soils in older *Fagus* forests, but also with *Corylus*, *Quercus* and *Tilia*.

Cortinarius rubellus △ has a more acutely umbonate cap and distinct veil girdles on the stem; it occurs with both deciduous and coniferous trees. Other red-brown, dry species, *e.g.* *C. limonius* ▷, have yellow-brown gills.

Nemoral–hemiboreal zones, rare and local; August–November.

†?**Cortinarius limonius** is a medium-sized *Cortinarius* with saturated yellow to orange and brownish colours and yellow-brown gills. The cap is smooth and somewhat hygrophanous. The universal veil is ± yellow-brown and forms indistinct girldes on the ± tapering stem. Odour indistinct. The ± subglobose spores measure 7.5–8.5 × 6–6.5 µm. Occurs with *Picea* and *Pinus* on poor to somewhat richer soils, and with *Fagus* on mor soils.

Cortinarius rubellus ◁ has an umbonate, finely scaly cap and darker, red-brown to orange-brown gills, and obvious veil girdles on the stem.

Widespread and common in the north, rarer southwards; July–January.

†?**Cortinarius callisteus** is a medium-sized, yellow-brown *Cortinarius* with pale yellow gills and a distinct smell of ozone (warm metal, steam engine or extinguished candle). The ± subglobose spores measure 7–7.5(–7.7) × 5.5–6.5 µm. Occurs with conifers on mineral-rich to calcareous soils.

Cortinarius limonius △ and other similar species have ± different odours. *Cortinarius citrinofulvescens* ⋉ is a more vivid yellow and typically has a ± apple-like smell.

Widespread, fairly common in the hemiboreal–boreal zones, but rare in the nemoral zones; August–October.

Jean-Marc Moingeon

†?**Cortinarius rubicundulus** is a medium-sized, but sturdy, smooth to felty, yellow-brown to apricot-orange *Cortinarius* with very pale, narrow gills and a pale stem. The flesh and the surface initially stains distinctly yellow after bruising and later reddens. Has clusters of cystidia on the gill edge. The ellipsoid to amygdaliform spores measure 6.5–9.5 × 3.5–5 µm. Occurs with *Fagus* and *Quercus* on mor soils, or with *Picea* on somewhat richer soils.

Cortinarius crassus ⋉ does not stain yellow after bruising.

Mainly found in the hemiboreal zone, locally quite common; mainly August–October.

Leif Stridvall

Telamonias (*Cortinarius* subgenus *Telamonia*) are small to large, dry species with rather dull, greyish or brownish colours. They may have violet gills, violet or orange flesh or orange veils. Some species have hygrophanous caps. This subgenus is very species-rich and complicated.

Cortinarius flexipes is a small telamonia with pale scales on a brown, conical cap and variable amounts of greyish-violet colours on the cap, stem and gills. Usually has a distinct *Pelargonium*-like smell. The spores measure 8–9(–9.5) × 5–6 µm, Qav. < 1.5. Often in damp forests, with both deciduous and coniferous trees.

Cortinarius flexipes var. *flexipes* has a violet mycelium and gills; *C. flexipes* var. *flabellus* lacks violet colours; and *C. flexipes* var. *inolens* is odourless. *Cortinarius diasemospermus* ▷ also smells of *Pelargonium* but has smaller fruitbodies and narrower spores (Qav. >1.5). *Cortinarius hemitrichus* ▽ is odourless, grows with *Betula* and has narrower spores.

Widespread and very common; mainly August–November.

Cortinarius hemitrichus is a small to medium-sized telamonia with a dark grey to dark brown cap that has fine, white scales, and pale brown gills. It is odourless. The spores measure 7–8.5 × 4–5 µm, Qav. < 1.5. Occurs with *Betula* on poor soils.

Cortinarius gossypinus ▷ is smaller, paler and browner and occurs with *Salix*. Species/varieties around *C. flexipes* △ may look similar but mostly have a distinct *Pelargonium*-like smell and broader spores.

Widespread and common; July–November.

Cortinarius gossypinus is a rather small telamonia with pale brown cap and gill colours and abundant white universal veil. The cap typically has a conical umbo. The odour is indistinct. The ± ellipsoid to slightly amygdaliform spores measure (8–)8.5–9.5 × 5–5.5 µm. Occurs with *Salix* in damb scrub and marshes.

Cortinarius hemitrichus ◁ typically occurs in drier habitats with *Betula*.

Probably widespread and common; mainly August–October.

Cortinarius diasemospermus is a small, brown, almost mycenoid telamonia with a faint smell of *Pelargonium* that completely lacks violet colours. The spores measure 8–9(–9.5) × 4.5–5.5(–6) µm, Qav. < 1.7 (var. *diasemospermus*). Occurs with various deciduous trees, mostly on rich soils.

There are several similar-smelling species: *C. diasemospermus* var. *leptospermus* ⋉ has narrower spores, Qav. > 1.7; *C. flexipes* ◁ is larger with Qav. <1.5; and *C. violilamellatus* ⋉ has a minutely scaly cap and occurs with *Pinus*.

Widespread and probably very common; August–November.

André Bidaud

Cortinarius parvannulatus is a small, slender, mostly long-stemmed telamonia that is brown throughout. The cap is hygrophanous and thin-fleshed. The stem has a white, cuff-like ring. Smells strongly of cedar wood. The spores measure (7–)8–9 × 4.5–5.5 µm. Occurs mainly with deciduous trees, mostly *Salix*, on rather damp soil.

There are numerous similar-looking, small, brown telamonias, but these should lack the combination of smell and stem ring; it is possible that more than one species is involved.

Widespread and probably quite common; June–November.

Jan Vesterholt

Jan Vesterholt

Cortinarius obtusus is a small telamonia with a marginally striate, warm yellow-brown to orange-brown and strongly hygrophanous cap. The gills are ± yellow-brown and usually have a pale edge. It smells of iodoform during drying. The amygdaliform spores measure (7.5–)8–9 × 4.5–5 µm. This species is treated here in a broad sense, as it includes a group of species that grow both with conifers and in deciduous forests.

Cortinarius acutus ▷ is smaller, has a pointed umbo, and the gill edge has balloon-shaped cystidia. *Cortinarius albovariegatus* ⋉ is very similar but has wider spores (> 5 µm).

Widespread and very common; June–January.

Thomas Læssøe

Jan Vesterholt

Cortinarius saniosus s.l. is a rather small telamonia with saturated yellow-brown colours and a ± golden universal veil that is seen as woolly girdles on the stem. The ellipsoid spores measure 8.5–10 × 5–6.5 µm, Qav. < 1.8. Occurs with various deciduous trees, including *Salix* and *Populus*.

Cortinarius aureovelatus ⋉ occurs with conifers and has narrower spores, less than 5 µm wide. *Cortinarius helvelloides* ▷ has distant gills, minute scales on the cap and occurs with *Alnus*.

Widespread and fairly common; May–October.

Cortinarius acutus is a small *Galerina*-like (see page 638) telamonia with a conical cap and a conical umbo. The cap is marginally striate, warm yellow-brown and strongly hygrophanous. Smells of iodoform during drying. The gill is edge pale and has balloon-shaped cystidia. Spores 8–9(–9.5) × 4.5–5.5 µm. Typically with *Picea* in damp, mossy plantations and forests.

Several similar species occur – also in deciduous forests – some of which lack formal names. *Cortinarius obtusus* s.l. ◁ is larger, usually lacks a conical umbo and does not have balloon-shaped cystidia at the gill edge.

Widespread and common; June–November.

Cortinarius casimiri is a rather small, slender telamonia with a dark, hygrophanous cap that is finely floccose when young and usually somewhat radially wrinkled when old. The stem apex generally has a pale violet tinge. The spores are ellipsoid and large, measuring 10–11.5 × 6–7 µm. Occurs with both deciduous and coniferous trees on various soils.

Most similar species have smaller spores.

Widespread and ± common; August–November.

Tobias Frøslev

Cortinarius helvelloides is a small, dull brown telamonia with yellowish veil fibres on the stem and cap, and distant gills. It typically has faint, dark violet tinges on the gills and stem apex when young. Occurs with *Alnus* on ± damp soil.

Several small telamonias, *e.g.* *Cortinarius lilacinopusillus* ▷▷, occur with *Alnus*, but these are darker or have distinctly violet colours. *Cortinarius saniosus* ◁ and other small telamonias with yellowish remnants of veil do not grow with *Alnus* and have more crowded gills.

Widespread and rather common; mainly August–October.

Thomas Læssøe/Jens H. Petersen

Jan Vesterholt

Cortinarius lilacinopusillus is a very small telamonia with a pale violet to lilac cap and rather thick and distant lilac gills. Smell indistinct. Gill flesh pale and lacks incrusted pigment. Spores ellipsoid, 9–10(–10.5) × 5–6.5 µm. Occurs with *Alnus* on damp soil

A number of similar species occur with *Alnus*: *C. americanus* ⋉ has a darker, with age blackish cap, incrusted pigment in the gill flesh and smaller spores; *C. helvelloides* ⪡ is larger, more brownish and has thicker, more distant gills; *C. alnetorum* ▽ is also larger, normally lacks violet colours and occurs mostly on drier soils.

Widespread, occasional to locally common; June–November.

Thomas Stjernegaard Jeppesen

Cortinarius alnetorum is a small, dark, umbonate telamonia with distinct greyish-white veil on the stem and young cap. It may have faint violet tinges on the gills and at the stem apex. The stem flesh is almost black. The gill edge is ± sterile. The spores measure 9–10.5 × 5–5.5 µm. Occurs with *Alnus* on rather dry to somewhat damp soils.

Both *C. lilacinopusillus* △ and *C. americanus* ⋉ occur with *Alnus*, but are smaller and have obvious violet colours; they tend to grow on damper soils.

Widespread, occasional; August–October.

Cortinarius atrocoeruleus is a small telamonia with an almost black, hygrophanous cap that has obvious greyish-white remnants of veil, a pale stem that darkens, and a smell recalling cedar wood. The ± ellipsoid spores measure 7–8.5 × 5–5.5 µm. Occurs mostly in young stands of deciduous trees.

Also treated as a variety of *C. decipiens* ⋉, which has less well-developed universal veil, larger spores (longer than 8.5 µm) and a faint smell. There are numerous other similar small, dark telamonias.

Widespread and probably quite common; mainly August–October.

Cortinarius umbrinolens is a small, ± blackish telamonia with a hygrophanous cap that is somewhat paler at the margin. The stem is silky-fibrillose and darkens from the base upwards. Has a distinct strong, earth-like smell recalling raw beetroot. The ± amygdaliform spores measure 8.5–9.5 × (4.5–)5–5.5 µm. Mycorrhizal with *Betula* on ± acidic soils.

There are many superficially similar species, but none have the same combination of smell and partner association. *Cortinarius carbunculus* ⋉ has the same strong smell, but occurs with conifers and has somewhat larger spores (>5.5 µm wide).

Widespread, rather common; June–November.

Jan Vesterholt

Cortinarius vernus is a small telamonia with a very dark cap, grey-brown, somewhat distant gills and a pale stem that becomes red-brown from the base upwards. The rounded, strongly ornamented spores measure 7.5–8.5 × 5–6 µm. Occurs with deciduous trees on dry soils.

There are various other small dark telamonias: *C. erubescens* ⋉ has a reddening veil, narrower spores (4–5 µm) and paler gills; *C. praestigiosus* ⋉ also has veil girdles and reddens with age but is smaller; *C. depressus* ⋉ occurs with conifers and has narrower spores and paler gills.

Widespread, common to occasional; March–November.

Jan Vesterholt

Thomas Læssøe

Cortinarius anthracinus is a rather small, orange-veiled telamonia with a purple-brown to blackish, hygrophanous cap that has a brownish-orange margin. The gills are yellow-brown and there are orange-red fibres (remnants of universal veil) on the stem. The ellipsoid to slightly amygdaliform spores measure 8–10 × 5–6 µm. Occurs on rich soils with many different types of deciduous trees, including *Corylus*, *Fagus* and *Quercus*; probably also able to associate with *Picea*.

Cortinarius danicus ▽ has blood-red universal veil, a red-brown cap and smaller spores. Other telamonias with red colours on the stem are larger or have paler colours.

Widespread but rare, absent from the boreal zone; mainly September–November.

Thomas Læssøe

Cortinarius danicus is a rather small telamonia with blood-red remnants of universal veil on the stem and a dark, red-brown cap. The ellipsoid to slightly amygdaliform spores measure (6.5-) 7.5–8.5 × 5–5.5 µm. Occurs with *Fagus* and *Quercus* on clay-rich to calcareous soils.

Cortinarius anthracinus △ is darker on the cap and has orange universal veil. *Cortinarus bulliardii* ▷ has larger fruitbodies.

Nemoral, rare, but locally quite common; mainly August–October.

Jan Vesterholt

Cortinarius spilomeus is a small to medium-sized telamonia with silvery grey colours, a hygrophanous cap and characteristic reddish universal veil, which is seen as girdles on the stem. The almost spherical spores measure (6.5–)7–8 × (5–)5.5–6.5 µm. Occurs with *Picea*, mostly in young stands.

The combination of almost spherical spores and reddish universal veil is diagnostic. Most other red-veiled species occur with deciduous trees.

Widespread and common in the north, becoming scarcer towards the south; mainly August–October.

Cortinarius bulliardii is a medium-sized, fleshy, pale brownish telamonia with distinctly reddish colours on the lower half of the stem. The stem is clavate. The ± amygdaliform, strongly ornamented spores measure (8.5–)9–10 × 5.5–6 µm. Occurs on clay-rich, ± calcareous soils, mostly with *Fagus* and *Quercus*.

There are only a few similar species in the same habitat: *Cortinarius danicus* ◁ and *C. anthracinus* ◁ are both smaller and somewhat darker.

Nemoral–hemiboreal zones, rare; mainly August–October.

Cortinarius bolaris is a medium-sized, rather fleshy telamonia with yellowish and reddish colours. The cap and stem are covered with scarlet scales. The pale flesh stains yellow after bruising. The veil quickly becomes red. The subglobose, strongly to moderately ornamented spores measure 6.5–7.5 × 5–5.5 µm. Occurs with *Quercus*, *Fagus* and *Betula* on dry, poor soils.

No other telamonia has a yellowish base colour with scarlet scales.

Widespread and rather common, very rare to absent from the boreal zone; mainly July–October.

Cortinarius armillatus is a large telamonia with a dry, hygrophanous cap and a dry stem with obvious orange-red to scarlet girdles of universal veil. The somewhat amygdaliform spores measure 10–12 × 6–7 µm. Occurs with *Betula* on poor soils, mostly in rather damp plantations and forests.

Cortinarius paragaudis ⋉ and *C. pinigaudis* ⋉ are typically smaller, have shorter spores (<8.5 µm long), less distinct girdles on the stem, and occur typically with conifers. *Cortinarius roseoarmillatus* × has smaller spores. *Cortinarius hinnuleoarmillatus* ⋉ has more widely spaced gills that that are often tinged violet.

Widespread, common to occasional; July–November.

Cortinarius cinnabarinus is a medium sized, hygrophanous telamonia with orange-red colours throughout. The densely warty, amygdaliform, dextrinoid spores measure 7.5–8.5 x 4.8–5.3 µm. Occurs with *Fagus*, *Tilia*, *Quercus* and *Corylus* on mull soils.

Red dermocybes, notably *C. puniceus* ◁◁, that can occur in the same habitats are not hygrophanous and have somewhat darker red colours.

Widespread and fairly common in the nemoral–hemiboreal zones; mainly August–November.

Jacob Heilmann-Clausen

Jens Maarbjerg

Cortinarius traganus is a large, fleshy telamonia with violet surface colours, warm yellowish to brownish-orange colours in the flesh and a characteristic, pleasant, fruity smell. The ± amygdaliform spores measure 8–9(–9.5) × 5–6 µm. Occurs mostly with *Picea* on poor soils in forests and plantations, but may also be associated with other conifers.

Cortinarius camphoratus ▷ does not have orange tinges in the flesh and has an unpleasant, pungent smell. *Cortinarius venustus* ▽ has a similar fruity smell, but is more gracile and has a ± brownish cap.

Widespread, common in the hemiboreal–boreal zones but becomes scarcer towards the south; mainly August–October.

Leif Stridvall

Cortinarius venustus is a small to medium-sized telamonia with a brown cap, cylindrical, violet stem with abundant remnants of veil and a characteristic, sweetish, fruity smell. The flesh is ± brownish-orange, sometimes more bluish in the upper part of the stem. The ellipsoid spores measure (9–)9.5–10.5 × 5.5–6.5 µm. Mycorrhizal with *Betula* and *Picea* on poor soils.

Cortinarius traganus △ is larger, has a clavate stem, violet colours on the cap and brownish-orange flesh. *Cortinarius camphoratus* ▷ is violet on all parts and has a pungent smell.

Widespread, occasional in the hemiboreal–boreal zone, very rare farther south; August–October.

Cortinarius camphoratus is a large, fleshy telamonia with violet colours on all parts, and a silky cap surface; the colour fades with age. The distinctive pungent odour recalls burnt horn or hair. The flesh is at first bluish-violet but later becomes more brownish-orange. The spores measure (9–)9.5–10.5 × (5.5–)6–6.5 µm. Mycorrhizal with *Picea* (and possibly *Abies*) on relatively poor soils in forests and plantations.

Cortinarius traganus ◁ and *C. venustus* ◁ have orange colours in the flesh and smell fruity. *Cortinarius alboviolaceus* ▽ and *C. malachius* ▷▷ have insignificant odours.

Widespread and common in the hemiboreal–boreal zones, rare in the nemoral zones; mainly August–October.

Cortinarius alboviolaceus is a rather large, ± odourless telamonia with a fine, silky-fibrous, non-hygrophanous, pale blue cap. The stem is tinged pale blue and the flesh is also bluish. The gills have a serrated, pale edge. The spores measure 8.5–10 × (5.5–) 6–6.5 µm. Occurs mostly with *Fagus* and *Betula* on different soil types.

Cortinarius traganus ◁, *C. venustus* ◁ and *C. camphoratus* △ have characteristic odours; *C. malachius* ▷▷ has minute scales; the smaller *C. geophilus* ⋉ is more pale lilac-tinged, is somewhat sticky and bitter-tasting, and has smaller spores (< 8.5 µm long).

Widespread and rather common; August–November.

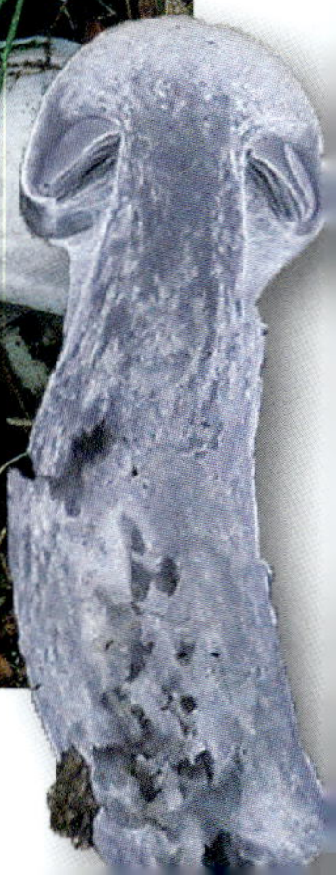

Jens H. Petersen/Jens Maarbjerg

Cortinarius pholideus is a distinctive rather small to medium-sized telamonia with brown scales on the cap and stem. When young, usually has a violet tinge to the stem flesh and on the stem apex. The spores are broadly ellipsoid and measure 7–8 × 5–6 µm. Occurs with *Betula* on poor soils.

An easily recognizable species. *Inocybe hystrix* (page 673) has a more grey-brown spore-deposit, larger, smooth spores and lageniform, thick-walled cystidia.

Widespread and fairly common in the north, but has shown a distinct decline in the nemoral zones; August–November.

Michael Krikorev

Cortinarius fuscoperonatus is a rather compact telamonia with brown veil and a grey-brown, umbonate, dry, ± hygrophanous, striate cap with a margin that remains inrolled for a long time. The stem has brown girdles of universal veil and can be bluish at the apex. The flesh is marbled brown and the basal mycelium is white. The ± ellipsoid spores are strongly verrucose and measure 10.5–12 × 6.5–7.5 µm. Occurs with *Picea* and *Pinus* in mossy, calcareous forests.

Cortinarius pholideus △ has distinct scales on the cap.

Widespread but rare, absent from the nemoral zones; July–October.

Cortinarius raphanoides is a medium-sized, olive-yellow-brown telamonia that sometimes has faint violet tinges on the stem apex and in the flesh. The gills are medium- to rather distant. The odour is radish-like. The broadly ellipsoid spores measure 7–8 × 5–5.5(–6) µm. Typically occurs with *Betula* on poor soils.

Cortinarius valgus ⋉ has more distinct violet colours, fainter olive tinges, and lacks the radishy smell; it also has somewhat larger spores (>8 µm long).

Widespread, occasional to rather common; August–October

Jan Vesterholt

☨? ***Cortinarius gentilis*** is a small to medium-sized, intensely yellow-brown to red-brown telamonia with a long, slender, mostly rooting stem that has red brown flesh. When young, the cap has a yellow margin and the stem has golden veil girdles. The spores are ± subglobose and measure 7.5–9 × 6–7 µm. Occurs with conifers on poor soils.

When old, the yellow/golden remnants of universal veil may be lost, and it may then be confused with numerous other telamonias.

Widespread and common in the hemiboreal–boreal zones but rare to absent from the nemoral zones; August–November.

Jan Vesterholt

Jan Vesterholt

Cortinarius caninus is a medium-sized to large, fleshy telamonia with a dry, matt, non-hygrophanous greyish-brown cap. The stem has a brownish girdle of universal veil. The young gills may have a bluish tinge, as has the flesh in the upper part of the stem. The odour is weak. The spores are almost spherical and measure (7.5–)8–9 × 6–7 µm. Mycorrhizal with *Picea* on relatively rich soils in forests and plantations.

There are several similar species of telamonia, but most have a slightly sticky cap surface. *Cortinarius anomalus* ▷ and *C. lepidopus* ▽ both have dry caps, but differ by having yellowish rather than brownish veil girdles on the stem.

Widespread and common in the north, but becomes scarcer towards the south; August–January.

Jan Vesterholt

Cortinarius lepidopus is a medium-sized, relatively slender, but fleshy, brownish telamonia with a dry, non-hygrophanous cap and yellow-brown girdles of universal veil on the stem, which may have bluish flesh in the top. The coarsely ornamented spores are almost spherical and measure 7–8 (–8.5) × 5.5–6.5(–7) µm. Occurs with deciduous trees, mostly on poor soils.

Cortinarius anomalus ▷ is more fleshy (stem typically >8 mm wide), paler, and occurs on richer soils. *Cortinarius caninus* △ has brown remnants of veil on the stem. Other similar species have slightly sticky caps.

Widespread and probably rather common; mainly August–October.

Jan Vesterholt

Cortinarius anomalus is a medium-sized, relatively fleshy, greyish telamonia with yellow and bluish tinges, a dry, non-hygrophanous, somewhat shiny cap, and yellow-brown girdles of universal veil on the stem, which may have bluish flesh in the top. The spores are almost spherical and measure 7.5–8.5 × 6–7 µm. Occurs with deciduous trees, *e.g. Fagus*, mostly on rather rich soils.

There are several similar species that either have different spores, sticky caps, no yellow veil girdles, and/or occur in a different habitat *Cortinarius lepidopus* ◁ is more brownish and slender (stem rarely > 8 mm wide) and occurs on poorer soils. *Cortinarius caninus* ◁ has brown remnants of veil on the stem.

Common and widespread in the nemoral zones but becomes scarcer to the north; August–November.

Cortinarius xanthocephalus is a medium-sized to large telamonia with a yellow-brown, sticky cap and yellow-brown veil remnants on the stem. The flesh is whitish, rarely bluish in the stem top. The spores are almost spherical and measure 8–9(–9.5) × 6.5–7.5(–8) µm. Occurs with deciduous trees on rich soils.

Similar species in the *Cortinarius anomlaus* △ group have either dry caps or less yellowish colours on the cap and universal veil. *Cortinarius tabularis* ⋉ also becomes yellowish with age, but has less universal veil and is smaller and less fleshy.

Widespread and probably fairly common; August–October.

Thomas Læssøe

Jens Maarbjerg

Cortinarius laniger is a medium-sized to rather large telamonia with a brownish-orange, non-hygrophanous cap, which, for a long period of time is woolly to minutely scaly from abundant pale universal veil. The gills are brick-red. The stem has obvious white girdles and a well-defined ring zone, and the flesh is pale brown. The ellipsoid, strongly dextrinoid spores measure 10–11.5(–12) × 6–6.5(–7) µm. Occurs with *Picea*, mostly on rich soils in damp forests and plantations.

Cortinarius bivelus ▽ occurs with *Betula*, has a hygrophanous cap, less evident veil on the stem and less colourful gills. *Cortinarius alborufescens* ⋉ also occurs with *Betula* but has narrower spores and pale yellow stem girdles.

Widespread and fairly common in the hemiboreal–boreal zones, rare in the nemoral zones; August–October.

Jens Maarbjerg

Cortinarius bivelus is a medium-sized to rather large telamonia with an orange-brown, somewhat hygrophanous cap that usually becomes dark-dotted with age. The gills are rust-red. The stem has obvious white remnants of universal veil and the flesh is pale brown. The ellipsoid spores measure 8.5–9.5 × 5.5–6 µm. Occurs with *Betula* on poor soils.

There are several superficially similar species. *Cortinarius subbalaustinus* ⋉ has less obvious remnants of universal veil on the stem, is distinctly hygrophanous and has narrower spores. *Cortinarius laniger* △ grows with conifers.

Widespread and common in the north, becoming scarcer in the south; June–October.

Cortinarius armeniacus is a medium-sized telamonia with a distinct apricot-orange to orange-brown, ± umbonate, somewhat sticky, hygrophanous cap. The gills are rather pale. The stem is white, finely silky-fibrillose and clavate. The flesh is speckled. The odour is indistinct. The spores are ellipsoid and measure 8–9 × 5–5.5 µm. Occurs with *Pinus* and *Picea* in sandy forests and plantations.

Cortinarius subbalaustinus ⋉ can have similarly bright cap colours but occurs with *Betula*; *C. carneinatus* ⋉ has a blue tinge.

Common in the north, but rare in the nemoral zones; August–October.

Jan Vesterholt

***Cortinarius duracinus* s.l.** is a medium-sized telamonia with a pale brown, hygrophanous cap. The gills are medium-spaced and the ± rooting stem is hard, white, and almost lacks remnants of veil. The flesh is white to pale brown. The odour is indistinct. The spores measure 7.5–10 × 4.5–6 µm. Occurs with coniferous and deciduous trees.

The name is applied here in a broad sense. *Cortinarius fragrantior* ⋉ and *C. acetosus* ⋉ smell of cedar wood and iodoform, respectively; *C. dolabratus* ⋉ has larger spores.

Fairly common in the north, but rare in the nemoral zones; August–November.

Jan Vesterholt

***Cortinarius hinnuleus* s.l.** is a medium-sized, hygrophanous telamonia with yellowish-brown colours. The gills are strikingly distant. Has obvious remnants of universal veil on the stem, mostly as a ring-zone; the flesh is pale yellow-brown, darker towards the base. Smell pungent, earthy. Spores ± ovoid, 7–8.5 × 5–6.5 µm. Occurs with various deciduous trees on rich soils.

Constitutes a large and complicated species complex; the name is applied here in a broad sense. *Cortinarius safranopes* ⋉ lacks the earthy smell. Other brownish telamonias do not have the combination of earthy smell, distant gills and plentiful universal veil.

Widespread and common, less so in the north; July–November.

Jan Vesterholt

Cortinarius biformis is a medium-sized telamonia with a very hygrophanous cap, which, initially, is dark with almost greyish-violet tinges, but later more yellow-brown to cinnamon-brown, usually with frost-like remnants of the pale universal veil at the margin. The gills are medium-spaced and may, when young, have a violet tinge. The ± clavate stem has ± obvious girdles of white universal veil and the flesh is somewhat greyish and usually violet in the stem apex. The ± ellipsoid to slightly amygdaliform spores measure (7–)7.5–8 × 4.5–5 µm. Occurs with *Pinus* and *Picea* in damp, mossy plantations and forests.

There are a number of superficially similar species. *Cortinarius melitosarx* ⋉ is consistently yellowish and occurs with *Pinus* on sandy soils.

Widespread and common in the north to rather rare in parts of the nemoral zones; July–December.

Jan Vesterholt

Cortinarius brunneus is a medium-sized telamonia with dark brown colours throughout and a very hygrophanous cap. The gills are medium-spaced and dark. Pale remnants of universal veil on the stem darken with age. The flesh is dark red-brown with ± violet tinges. Stains blackish when dry. The broadly ellipsoid spores measure 8–9.5 × 6–7 µm. Occurs typically with *Picea* on poor soils.

Cortinarius glandicolor ⋉ is paler and less fleshy and occurs typically with *Pinus*. Other similar species do not become blackish when dry.

Widespread and common in the north, but rare in the nemoral zones; August–November.

Cortinarius lucorum is a large telamonia with a strongly hygrophanous cap and a mixture of greyish-violet and dark brown colours throughout. The cap margin generally has fibrous remnants of universal veil. The flesh is usually ± violet. Gills rather distant. The spores are ± ellipsoid and measure 9.5–11 × 5.5–6.5 µm. Occurs with *Populus*, mostly in troops and fairy rings.

Cortinarius saturninus ▽ is smaller, occurs with *Salix* (and *Corylus*), mainly in clusters and rings; it has smaller spores (<9.5 µm long).

Fairly common in the hemiboreal zone, rarer to the north and south; mainly August–October.

Cortinarius saturninus is a medium-sized telamonia with a hygrophanous cap and a mixture of greyish-violet and dark brown colours throughout. The cap margin has fibrous remnants of universal veil. The flesh is ± violet, mainly in the stem apex. The gills are medium-spaced. The spores are narrowly ellipsoid and measure 8–9(–9.5) × 4.5–5 µm. Occurs with *Salix*, mostly in clusters, often in ± urban or disturbed habitats.

Cortinarius lucorum △ is larger, has more distant gills, occurs with *Populus* and has larger spores (>9.5 µm long).

Widespread, rather common to occasional; August–November.

Cortinarius evernius is a medium-sized telamonia with a dark brown, hygrophanous cap. The gills are medium-spaced to distant. The stem is slender, tapering and has white remnants of universal veil. The flesh is violet, at least in young specimens. The ± amygdaliform spores measure 9.5–11 × 5.5–6.5 µm. Occurs with *Picea* on damp, poor soils.

Cortinarius tortuosus ⋉ differs in having a faint smell of cedar wood; *C. imbutus* ⋉ has a strong smell of damp cellar; *C. cagei* ⋉ grows in deciduous forests.

Widespread and common in the north, becoming scarcer towards the south; mainly August–October.

Jan Vesterholt

Cortinarius stipitemirus s.l. is a medium-sized to large, fleshy, sturdy telamonia with prominent, whitish universal veil on young fruitbodies; the innately fibrillose, pale brown cap surface later turns darker brown to grey-brown. The flesh and stem discolour ± blackish. Smell weak og flower-like. The spores measure (8.5–)9–10 × 6–6.5 µm. Occurs on rich soils, mostly with phlegmacia in deciduous forests, often *Fagus*.

The names *C. aprinus, C. sociatus* and *C. suillonigrescens* have also been used for this species. Similar species have less developed veil and are not whitish when young. This species belongs to a poorly understood complex. *Cortinarius turgidus* ▷ has whitish flesh and a persistent pale cap, and does not discolour dark on the stem.

Widespread in the nemoral–hemiboreal zone, occasional to locally common; August–November.

Jan Vesterholt

Cortinarius quarciticus is a robust, medium-sized to large telamonia with a silky-fibrillose, grey-brown to brown cap that has violet tinges and hygrophanous striations. The gills are crowded. The stem is clavate, usually somewhat bulbous and silky-fibrillose, and has violet tinges at the apex. The ± ellipsoid spores measure 7.5–8.5 × 4.8–5.3 µm. Occurs with *Pinus* on sandy soils.

Cortinarius malachius ▷ has a minutely scaly cap surface; *C. suberi* ⋉ has distinct, innate, radial fibres and lacks violet tinges.

Widespread and fairly common in the north, but rare to ± absent from the nemoral zones; mainly July–October.

Cortinarius phaeosmus is a small to medium-sized telamonia with white veil and a distinctive smell of parsley or nutmeg. The cap is brown, hygrophanous and pale-rimmed. The stem is paler, ± clavate. The flesh is brownish. The ± ellipsoid spores measure 8–9(–9.5) × 4.8–5.3 µm. Occurs with *Fagus* on rich soils.

Species around *C. stipitemirus* ◁ may look similar but have larger spores and different smells. *Cortinarius turgidus* ▽ and *C. urbicus* ⋉ are paler.

Mostly nemoral–southern hemiboreal, occasional to rare; August–November.

Cortinarius turgidus is a large, fleshy, pale telamonia with a whitish, cream to pale brown, somewhat hygrophanous, silky-fibrillose cap. The gills are pale. The stem is whitish and clavate, usually tapering into the soil below the swollen part, and has sparse remnants of universal veil and whitish flesh. The ± amygdaliform spores measure (9–)9.5–10.5 × (5.5–)6–6.5 µm. Occurs with *Fagus etc.* on rich soils.

Cortinarius stipitemirus ◁ has darker flesh; *C. urbicus* ⋉ has narrower spores.

Widespread and fairly common to occasional in the nemoral zones, August–November.

Thomas Læssøe

Cortinarius malachius is a medium-sized to large telamonia with a greyish cap that has small, fine, radiating scales from remnants of veil. When young, usually has a violet tinge on the cap and on the clavate, pale stem. The stem flesh is also ± violet in the apex. The spores measure (9.5–)10–11 × 6–7 µm. Occurs in rich *Picea* forests.

Cortinarius quarciticus ◁ has a smooth cap and a clavate stem with a bulbous base. *Cortinarius traganus* ◁◁ and *C. camphoratus* ◁◁ have obvious odours and darker flesh.

Widespread and fairly common in the hemiboreal zone, rarer southwards; August–October.

Jan Vesterholt/Torbjørn Borgen

Cortinarius torvus is a rather large, fleshy telamonia with a brown, innately radially fibrillose, at most somewhat hygrophanous cap. The gills are distant and there are abundant remnants of universal veil on the stem, mostly as a projecting ring zone. The flesh is pale with violet tinges in the stem apex, and there are varying amounts of violet on the surface of the stem apex and on the gills. Has a distinctive, fruity odour. The ovoid-ellipsoid spores measure 10–11.5 × 6–7(–7.5) µm. Occurs with various deciduous trees, particularly *Fagus*, mostly on somewhat better soils.

Cortinarius caperatus ▷ has an even more distinct ring, a scurfy-felty cap surface, more crowded gills and is typically even larger. *Cortinarius vulpinus* ◁◁ may also have a ± well-developed ring, but it has a slimy cap and dry stem (and belongs with the phlegmacia).

Widespread and common in the hemiboreal–boreal zones, rare or absent from the boreal zone; June–November.

•• ***Cortinarius caperatus*** is an unusual, rather large to large, fleshy, yellow-brownish *Cortinarius* with a membranous stem ring. The ± radially wrinkled cap has a cracking, floccose veil cover that may have a somewhat bluish shade. The gills are rather crowded, narrow and ± pale-edged. The spore-deposit is cinnamon-buff. The ± lemon-shaped, rather pale brown, fairly coarsely ornamented, rather dextrinoid spores measure 12–13.5 × 7–8(–8.5) µm. Occurs on mor soils, with both deciduous and coniferous trees, mostly *Fagus* and *Pinus*.

This species was formerly placed in the genus *Rozites*. No other *Cortinarius* species have such a well-defined, membranous stem ring. *Cortinarius torvus* ◁ has an upturned, less prominent stem ring and lacks the felty cap surface. *Cortinarius vulpinus* ◁◁ has a more floccose ring and is slimy on the cap surface.

Widespread and very common in the boreal zone, becoming scarcer towards the south; July–November.

Paxillus and the like

This group includes agarics with a ± brown spore-deposit, deeply decurrent gills and a ± central stem. Both of the genera included, *Paxillus* and *Phylloporus*, are ectomycorrhizal and belong phylogenetically to the *Boletales*.

Other similar fungi:
– gomphidioids have dark brown to black spore-deposits (page 596).
– the brown-spored genus *Tapinella* has ± excentric, lateral or missing stems (pleurotoids page 96).
– *Lepista* & *Paralepista* are more smooth-capped and have whitish to brownish-rose spore-deposits and verrucose spores (page 112).

Further reading: 97, 133, 333.

Paxillus is a genus of ectomycorrhizal, brown-spored agarics with fairly crowded, decurrent gills that can be loosened from the cap flesh. The colour of a fresh spore-deposit is important for identification (although this can change over time). Repeated consumption of paxilloids may cause a life-threatening autoimmune reaction.

††***Paxillus involutus*** is a relatively slender, rather pale *Paxillus* with a downy, inrolled, wrinkled margin. The stem is ± smooth; lacks conspicuous reddish spots or yellow bands. The gills are pale straw-yellow to yellowish-brown and stain brown when bruised; they can be loosened from the pale cap flesh. The spore-deposit is yellow-brown to grey-olive. The smooth spores measure 7–11 × 5–6.5 µm. Mostly mycorrhizal with *Picea* and *Betula*, and usually on poor acidic soils.

Paxillus filamentosus ▷▷ is more yellow-fleshed and occurs with *Alnus*. Other species are shorter and thicker-stemmed.

Widespread and very common; June–November.

Ebbe Kristensen

††*Paxillus filamentosus* is a relatively slender, very strongly ribbed *Paxillus* with a distinct yellow tinge to the flesh. The cap surface often cracks into a net-like pattern. The spores measure 5.5–8.5 × 4–5 µm. Occurs on damp soils, mycorrhizal with *Alnus*.

Paxillus involutus ◁◁ occurs in drier habitats and not with *Alnus*. It has somewhat larger spores and less yellow tinges in the flesh. *Alnus*-associated *Paxillus* species are sometimes treated as more than one species, and the name *P. rubicundulus* is often used instead of *P. filamentosus*.

Widespread and rather common; May–October.

††*Paxillus obscurisporus* is a very fleshy, short- and thick-stemmed *Paxillus* that typically has wine-red spots (bottom inset) and blotches at the stem base, and often a yellowish belt at the stem apex. The cap margin is felty but scarcely ribbed; stays deeply inrolled for a long time. The fresh spore-deposit is red-brown (top inset). Spores measure 7–10.5 × 5–6.5 µm. Typically on rich soil in parks, *etc.*, where it is mycorrhizal with *e.g. Tilia*.

Paxillus ammoniavirescens ▷ is similar in stature but has a yellow-brown to grey-olive spore-deposit when fresh. *Paxillus involutus* ◁◁ is more slender, less thick-fleshed, lacks red spots or yellow bands on the stem and has a yellow-brown to grey-olive spore-deposit; it occurs in more acidic habitats, mainly with *Betula* or *Picea*. *Paxillus filamentosus* △ has distinctive ribs in the cap margin and distinct yellow tinges to the flesh.

Widespread, occasional, probably absent from the boreal zone; May–November.

††**Paxillus ammoniavirescens** is a very fleshy, short-stemmed *Paxillus* with ± brownish-vinaceous cap colours and a ± yellow-brown to olive-brown spore-deposit. The flesh stains ± green in ammonia. The spores measure 7–11 × 5–7 µm. Occurs on rich soil in open forests, parks and churchyards, mostly with *Tilia*, *Carpinus* and *Populus*, but also with *Salix repens* in dunes.

Paxillus validus is a synonym. Other species of *Paxillus* may also stain with ammonia. *Paxillus obscurisporus* ◁ has a red-brown spore-deposit and a less vinaceous cap; *P. involutus* ◁◁ is a more slender species from acidic soils.

Widespread and rather common, at least in the nemoral zones; May–November.

Phylloporus pelletieri is a rather small, convex, fleshy bolete-like agaric with brownish cap colours and yellow, thick and distant, somewhat decurrent gills with numerous connecting ribs. The stem is ± tapering and tinged red-brown. The flesh is rather pale and does not stain blue. The taste is mild. The brownish spores measure 11–13 × 4–5 µm. Occurs on mor soils in older deciduous forests, but also known from coniferous forests.

Distinctive and unlikely to be misidentified. It is the only lamellate true bolete (*Boletaceae*) in Europe but there are many closely related species in E and SE Asia and N America. It is relatively closely related to boletes in the genus *Xerocomus* (page 784).

Widespread in the nemoral zone, but rare; July–October.

Thomas Læssøe

Boletes

The boletes is a group of fungi that form rather soft-fleshed fruitbodies that have tubes rather than gills below the cap and a sturdy central stem. The tubes are generally not joined firmly to each other, or to the cap flesh, from which the entire tube layer can easily be loosened. The spores form inside the tubes and are mostly brown. Many boletes change colour when bruised or sectioned. These colour changes, the colour of the pores (the visible part of the tubes) and the cap surface structure are very important characters.

The boletes constitute a monophyletic entity and are classified in the order *Boletales*. The order also includes some false truffles (*e.g. Rhizopogon*), some corticioids (*e.g. Serpula*) and some species with gills (*e.g. Paxillus* and the gomphidioids).

Almost all boletes are ectomycorrhizal with deciduous trees or conifers. The genus and sometimes the species of the host is important for identification, especially in the genera *Leccinum* and *Suillus*. A few species are parasitic: *Pseudoboletus* parasitize *Scleroderma*; *Buchwaldoboletus* is probably parasitic on *Phaeolus;* and *Chalciporus* perhaps parasitizes *Amanita muscaria*.

There are many choice edibles among the boletes, but also a few bitter, inedible species (*e.g. Tylopilus felleus*, *Caloboletus radicans* and *C. calopus*) and most species of *Rubroboletus* and *Imperator* are moderately poisonous.

Other similar fungi:
– polypores with a stem (*e.g. Albatrellus* and *Boletellus*) may look very similar. They have tubes that are firmly joined – both with each other and the flesh. They also often have decurrent tubes and tough flesh – see page 826.

Further reading: 71, 150, 156, 170, 194, 205, 228.

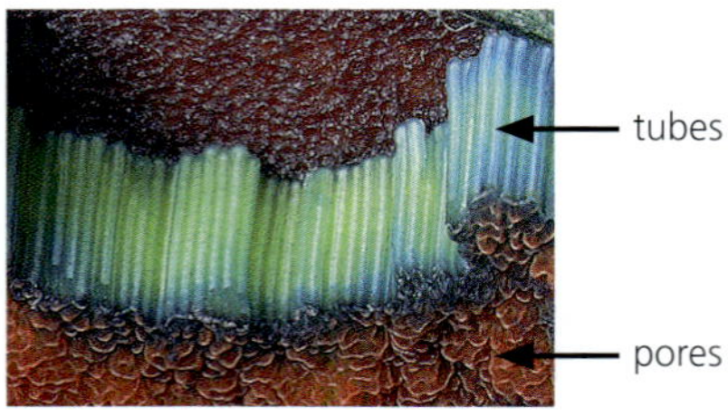

Micro-drawings: spores.

Approximate species number applies to temperate Europe.

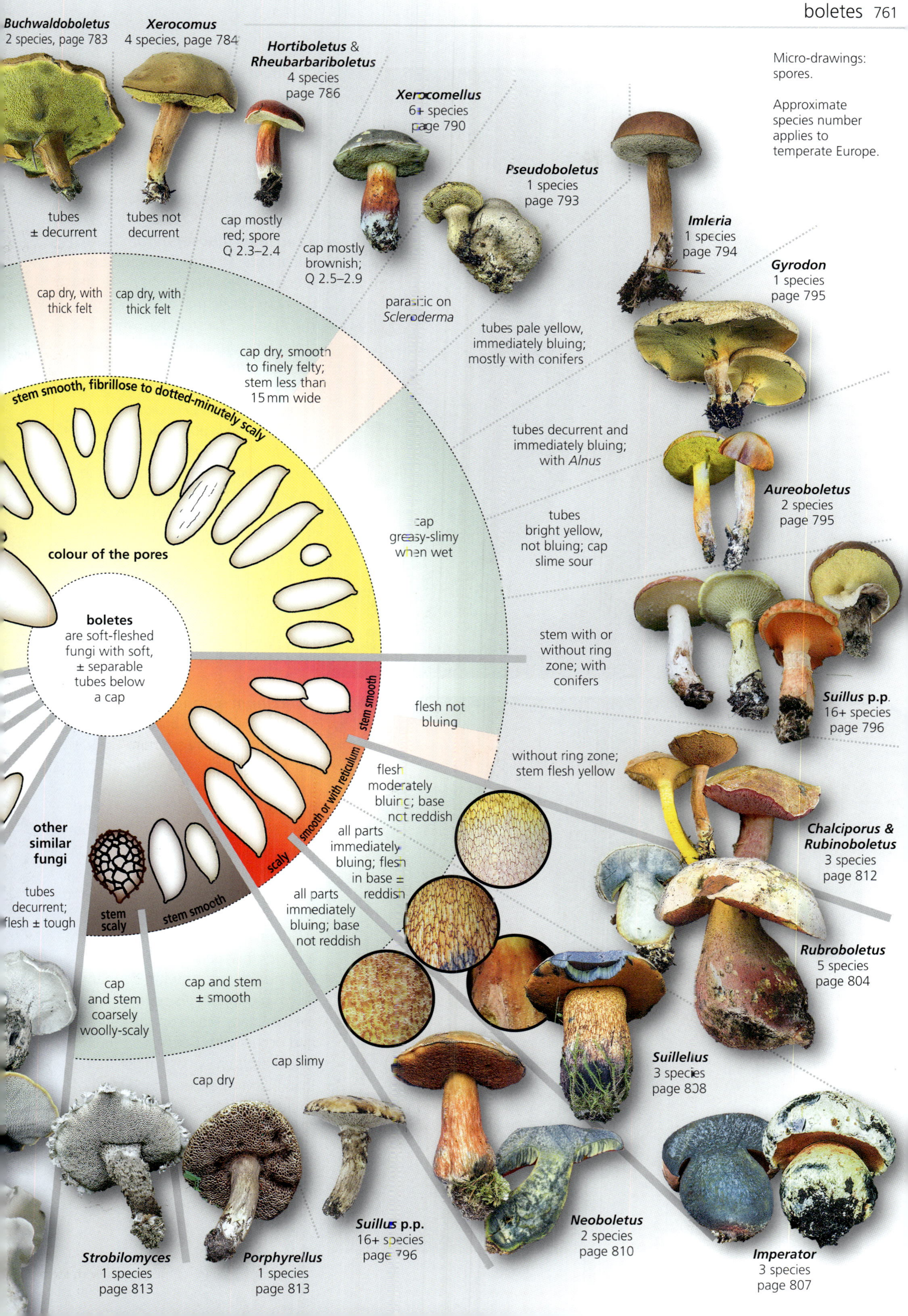

Leccinum and ***Leccinellum*** are genera of dry boletes with strikingly long, scaly stems. The flesh stains black, pinkish or turquoise.

Orange species (this page spread) may cause stomach upsets but are tolerated by most if well cooked.

• ***Leccinum albostipitatum*** is a rather large to large, brownish-orange *Leccinum* with blackening flesh, and with cap cuticle that overhangs the cap margin. The stem has whitish scales when young, which only later become ± brown. The spores measure 11–17 × 4–5 µm, Qav. 3.1. Occurs with *Populus*, including *Populus tremula*.

The name *L. aurantiacum* has been used for this species, but also for *L. aurantiacum* ▽. The latter may also grow with *Populus*, but has rust-red scales on the stem, even in young fruitbodies. It is important to note that the scales of *L. albostipitatum* stain brown, so it is essential to obtain young, fresh specimens.

Widespread, occasional to rather rare; July–November.

• ***Leccinum aurantiacum*** is a large, firm-fleshed, brownish-orange to orange-brown *Leccinum* with blackening flesh, and with cap cuticle that overhangs the cap margin. It has rust-brown scales on the stem, even in young fruitbodies. The spores measure 12.5–18.5 × 3.5–5 µm, Qav. range 3.2–3.6. Occurs with various deciduous trees, mostly *Populus* and *Quercus*.

The name *L. quercinum* has been used for this species. *Leccinum vulpinum* ▷ occurs with *Pinus* and *Picea*. *Leccinum versipelle* ▷ has greyish-black scales on the stem and occurs with *Betula*. *Leccinum albostipitatum* △ has whitish, rather than brownish scales on the stem of young fruitbodies.

Widespread, occasional; June–November.

• ***Leccinum versipelle*** is a large, firm-fleshed, brownish-orange to yellow-brown *Leccinum* with blackening flesh, and with cap cuticle that overhangs the cap margin. It may also occur in very pale forms, from almost white to pale yellowish-brown or vinaceous-brown. The stem has conspicuous greyish-black scales. The spores measure 12.5–16 × 4–5 µm, Qav. range 3–3.3. Occurs with *Betula* in rather dry habitats.

Leccinum versipelle is the only brownish-orange *Leccinum* that occurs with *Betula*. The similar *L. albostipitatum* ◁, *L. aurantiacum* ◁ and *L. vulpinum* ▽ occur with other hosts and have whitish or brownish stem scales. The pale forms of *L. versipelle* have been called *L. percandidum*, *L. roseotinctum*, *L. cerinum* and *L. callitrichum*.

Widespread, very common to occasional; June–October.

• ***Leccinum vulpinum*** is a medium-sized to large, brownish-red to orange-brown *Leccinum* with blackening flesh, and with cap cuticle that overhangs the cap margin. The stem is very firm-fleshed and has brownish-red scales, even in young fruitbodies. The spores measure 12.5–15 × 3–4 µm, Qav. range 3.2–3.4. Occurs with *Picea* and *Pinus* on ± sandy, poor soils.

The other ± brownish-orange species of *Leccinum* occur with deciduous trees, and have either blackish-grey, rust-brown or whitish scales on the stem and usually more vivid cap colours. Populations of *L. vulpinum* occurring with *Picea* may be referred to a separate species, *L. piceinum* ⋉.

Widespread, fairly common in the north, but rather rare in southern parts of temperate Europe; June–October.

Jens Maarbjerg/Jan Vesterholt/Jens H. Petersen

The brown-capped ***Leccinum*** species are edible (but soft-fleshed) and are less prone to causing stomach upsets than their orange-capped relatives. Do not eat any *Leccinum* uncooked.

• ***Leccinum scabrum*** is a medium-sized *Leccinum* with a ± dark clay-buff cap and flesh that either does not change colour or reddens only slightly; it never discolours bluish or greenish. The long, somewhat clavate stem has grey-brown to greyish-black scales that become coarser towards the base. Elements of the cap cuticle do not float freely in preparations. The fusiform spores measure 14–20 × 5–6 µm, Qav. range 3–3.3. Occurs with *Betula*, mostly in rather dry habitats.

Specimens with a blackish-brown cap, a grey stem surface and very dark scales are sometimes treated as a separate species, *L. melaneum*. *Leccinum rotundifoliae* ⋉ is smaller, has very pale scales on the stem, and occurs with *Betula nana* in alpine and subarctic climates. Other brown-capped species of *Leccinum* have bluish or greenish spots in the flesh and on the stem surface, mostly at the base.

Widespread and common; May–October.

• ***Leccinum cyaneobasileucum*** is a medium-sized, pale grey-brown *Leccinum* with rather soft flesh that usually has bluish or greenish spots. It has whitish to pale grey-brown scales on the stem. The cap cuticle elements float freely in preparations. The spores measure 14–19.5 × 4–6.5 µm, Qav. 3.1–3.5. Occurs with *Betula*, mostly in damp habitats.

Leccinum niveum ▷▷ has shorter spores, Qav. below 3; *L. schistophilum* ▽ is very similar, but also has short spores. Both lack the free-floating cap cuticle elements. *Leccinum scabrum* ◁ never has bluish or greenish colours in the flesh or on the stem.

Widespread and common; June–November.

• ***Leccinum schistophilum*** is a pale grey-brown *Leccinum* with rather pale, grey-brown stem scales. The flesh stains bluish or greenish at the stem base. The cap cuticle cells do not float float freely in preparations. The spores measure 13.5–17 × 5–6.5 µm, Qav. range 2.8–2.9. Mycorrhizal with *Betula* on ± dry to damp soils.

Leccinum cyaneobasileucum △ is very similar, but has higher spore Q values (range 3.1–3.5) and the cap cuticle elements float ± freely in preparations. *Leccinum scabrum* ◁ never has bluish or greenish colours on the stem or in the flesh.

Widespread and probably rather common; mainly August–October.

Jens H. Petersen/Jan Vesterholt

• ***Leccinum variicolor*** is a medium-sized, dark grey-brown *Leccinum* with dark and pale blotches on the cap, and with bluish- or greenish-spots on the stem and in the flesh. The stem has coarse, grey-brown to greyish-black scales. The spores measure 13.5–17.5 × 5–6.5, Qav. range 2.6–2.8. Occurs with *Betula* on ± dry to damp, peaty soils.

Leccinum cyaneobasileucum ◁◁ and *L. schistophilum* ◁◁ also have bluish- or greenish-spotted stem and flesh, but their caps lack blotches and are a more monochrome brown.

Widespread, rather common to occasional; June–November.

• ***Leccinum niveum*** is a rather small, whitish to pale grey-brown *Leccinum*, usually with pale greenish shades on the cap. Always has bluish- or greenish-staining flesh and whitish stem scales. Cap cuticle elements do not float freely in preparations. The spores measure 17.5–20 × 5.5–6.5 μm, Qav. range 2.6–2.8; cells in stem scales up to 13.5 μm wide. Occurs with *Betula* in marshes and mires, mostly in *Sphagnum*.

Leccinum cyaneobasileucum ◁◁ has loosening cap cuticle elements and a spore Q above 3; *L. schistophilum* ◁◁ is darker coloured and has cells in the stem scales up to 20 μm wide.

Widespread, common to rather rare; June–October.

• ***Leccinum duriusculum*** is a large, brown, rarely somewhat orange-tinged *Leccinum* with reddening, later blackening flesh, and with cap cuticle that overhangs the cap margin uniformly. The spores measure 14.5–16 × 4.5–6 µm, Qav. range 2.3–2.7. Occurs with *Populus*, often on clay soils.

Other leccinums that occur with *Populus* typically have more orange tinges and the cap cuticle overhangs the cap margin unevenly. *Leccinellum pseudoscabrum* ▽ blackens but the cap cuticle does not overhang the margin and it occurs with *Carpinus* and *Corylus*.

Widespread, uncommon, ± absent from the boreal zone; May–October.

• ***Leccinellum pseudoscabrum*** is a rather small to medium-sized, yellow-brown to brown *Leccinellum* with blackening flesh and whitish to pale sordid brown pores. The cap cuticle is very cracking but does not overhang the cap margin. The somewhat rooting stem is brownish and has rather small scales. The spores measure 15–19 × 5–6 µm, Qav. range 2.7–2.9. Occurs with *Carpinus* and *Corylus*, typically on clay-rich soils.

Also known as *Leccinum/Leccinellum carpini*. *Leccinum duriusculum* △ also has a brown cap and blackening flesh, but the cap cuticle distinctly overhangs the cap margin and it occurs with *Populus*. *Leccinellum crocipodium* ▷▷ has blackening flesh but the pores are yellow and it occurs with *Quercus*.

Widespread and fairly common to rather rare, becoming scarcer towards the north; July– October.

Guillaume Eyssartier

Tylopilus felleus is a rather small to rather large, cinnamon-buff, finely felty, dry bolete with at first white, later pale rose, rather fine pores. The stem has a coarse, dark reticulum. The taste is very bitter. The flesh does not stain after sectioning or bruising. The spores are smooth, ± fusiform and measure 11–15 × 4–5 µm. The spore deposit is brownish-rose. Ectomycorrhizal with both deciduous and coniferous trees, often on poor, sandy soils.

This is the only bolete in temperate Europe that acquires rose pores at maturity. Young specimens with whitish pores resemble young specimens of *Boletus* ▷ but these are mild-tasting, and the reticulum on the stem is usually paler. Species of *Caloboletus* ▷▷ have a similar bitter taste but the pores are yellow.

Widespread, rather common to common, less so in the boreal zone; June–October, with a peak early in the season.

The genus ***Boletus*** is recognized by fleshy fruitbodies, initially whitish, later yellowish pores, and a whitish reticulum on the stem. The flesh is white, mild-tasting and does not stain when bruised. All are choice edibles.

•• ***Boletus edulis*** is a large bolete with a greasy, clay-buff to orange-brown cap with a ± tough, elastic cuticle (upper inset picture). The very fine pores are initially whitish, later yellowish to pale olive. The fleshy, mostly barrel-shaped stem has a whitish reticulum that is most pronounced near the top. The ± fusiform spores measure 14–17 × 4.5–6 µm. Occurs with deciduous and coniferous trees, mostly on poor, acid soils, but it has a wide tolerance. Favours *Fagus*, *Picea* and *Abies* and is very frequent in younger *Abies* plantations. Specimens with a small cap and a very large swollen, soft stem are attacked by *Hypomyces chrysospermus* (page 1543). Such specimens later become yellow from conidia.

Very similar to *B. reticulatus* ▷▷ and *B. aereus* ▷▷ but these species lack elastic cap cuticle. *Boletus pinophilus* ▷▷ may have somewhat elastic cap cuticle but typically has a very rugose-bumpy, more reddish-brown cap surface and a darker stem. *Boletus edulis* may occur in almost white forms. A number of other taxa have been described around *B. edulis*, e.g. *B. betulicola* and *B. pinetorum*, but since this is not supported by molecular evidence, and the microscopical characters are rather weak, they are treated here as synonymous with *B. edulis*.

Widespread and common; June–November.

•• ***Boletus reticulatus*** is a large bolete with a dry, clay-buff, mostly cracking, finely felty cap. The fine pores are initially whitish, later yellowish, and the ± barrel-shaped stem has a whitish to somewhat brownish, rather raised reticulum. The ± fusiform spores measure 13–17 × 4–5.5 µm. Occurs with deciduous trees, mostly in thermophilic forests on clay soils.

The name *B. aestivalis* is a frequently used synonym. Separated from *B. edulis* ◁◁ by its pale, dry, finely felty, rather than greasy, elastic cap cuticle. *Boletus aereus* ▽ has much darker colours; *B. pinophilus* ▷ occurs only with conifers. *Tylopilus felleus* ◁◁ soon acquires a pinkish tinge on the pores, has a darker reticulum and tastes very bitter.

Widespread, rather common, becoming scarcer towards the north and almost absent from the boreal zone; June–October, peaking early in the season.

•• ***Boletus aereus*** is a large bolete with a dry, dark grey-brown, very finely felty cap that usually has small, whitish crystalline granules (inset picture). The stem is ± yellowish-brown and has a whitish reticulum. The spores measure 14–17 × 4.5–6 µm. Occurs with *Fagus* and *Quercus* in thermophilous habitats on clay-rich soils.

Boletus edulis ◁◁ is paler and has a greasy, elastic cap cuticle. *Boletus reticulatus* △ has paler, less contrasting colours. *Boletus pinophilus* ▷ only occurs with conifers.

Mainly a southern species, locally common, absent from the boreal zone; July–October.

•• ***Boletus pinophilus*** is a large to very large bolete with a somewhat greasy to dry, rugose-bumpy, red-brown cap. The fine pores are initially whitish, later yellowish. The brown-tinged stem has an extensive, whitish or brown reticulum. The ± fusiform spores measure 15–19 × 4.5–5.5 µm. Occurs with conifers in sandy forests and plantations.

Boletus edulis ⊲⊲ also occurs with conifers but has a distinctly greasy-tough cap cuticle, a less rugose cap surface and less red-brown colours. *Boletus aereus* ⊲ occurs in deciduous forests in the south and has a finely felty, not rugose, more greyish cap.

Widespread and rather common in the north, becoming scarcer towards the south; July–December.

Caloboletus includes species with pale, brownish-grey caps, yellow, rather strongly bluing pores and bitter tastes. However, not everyone can easily detect the taste, and prolonged chewing with the sample in the back of the mouth may be needed.

Caloboletus calopus is a medium-sized, pale, ± brownish-grey, finely felty *Caloboletus* with yellow, bluing pores, a bitter taste, rather pale bluing flesh and a reticulate stem that is yellow towards the apex and reddish towards the base. The ± fusiform spores measure 12–16 × 4.5–6 µm. Occurs with deciduous and coniferous trees, typically on rather mor soils.

The only other bitter-tasting *Caloboletus* with yellow pores is the somewhat larger *C. radicans* ▷, which has a pale yellow stem with reticulum only at the top, and sometimes a narrow, red belt around the centre.

Widespread, common to occasional, nearly absent from the boreal zone; June–November.

Caloboletus radicans is a very large, heavy, firm-fleshed, pale brownish-grey *Caloboletus* with fine, yellow pores and bitter-tasting flesh. The barrel-shaped, yellow stem has a fine yellow reticulum at the top and sometimes a narrow red zone in the middle. If the taste is not recorded, the strongly bluing flesh and tendency for the cap cuticle to crack are good characters. It may also have a pungent, curry-like smell, mostly from the cap cuticle. The ± fusiform spores measure 11–15 × 4.5–6 µm. Occurs with deciduous trees, mostly *Fagus* and *Quercus*, in thermophilic open forests and parks, mostly on clay-rich soils.

Caloboletus kluzakii ⋉ is ± similar but with a reddish zone below the cap cuticle. When touched, the cap becomes pinkish. Known from the Czech Republic. *Caloboletus calopus* ◁ has an obvious reticulum along the entire length of the stem with vivid orange-red colours on the lower part.

Widespread in the nemoral zones extending into the hemiboreal zone, occasional; July–October.

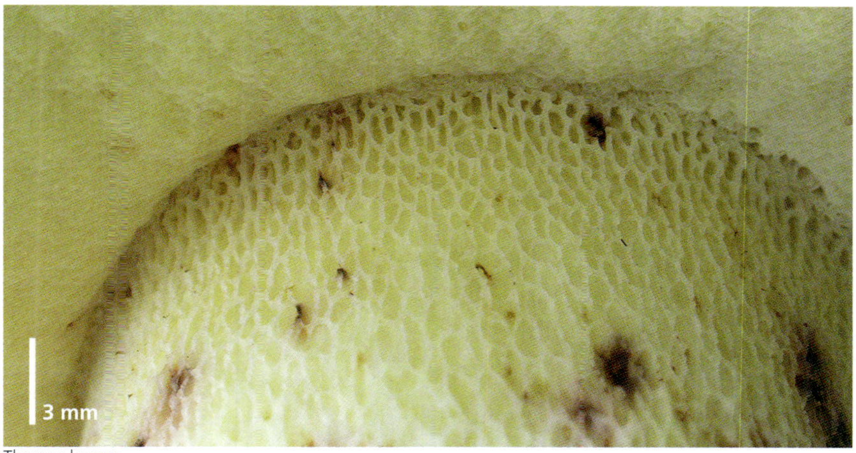

Thomas Læssøe

Butyriboletus includes medium-sized to large, reticulate boletes with vivid yellow pores and mild tastes. They often stain blue on the pores and/ or in the flesh. They are all choice edibles.

●● ***Butyriboletus appendiculatus*** is a medium-sized, cinnamon-brown to clay-buff *Butyriboletus* with yellow, bluing, fine pores. The stem has a fine, yellow reticulum on the upper part. The flesh does not normally turn blue and the taste is mild. Spores ± fusiform, 12–14.5 × 3.5–4.5 µm. With deciduous trees, typically on clay soils, mostly with *Fagus* or *Quercus*.

Caloboletus radicans ◁◁ is paler, bitter-tasting and the flesh turns blue. *Hemileccinum impolitum* ▷▷ lacks a reticulum and smells of iodoform at the stem base. *Butyriboletus fechtneri* ▷▷ and *B. fuscoroseus* ▷▷ have moderately bluing flesh and, typically, a red belt around the stem. The similar *B. subappendiculatus* ⋉ occurs in calcareous conifer forests.

Widespread in the nemoral and hemiboreal zones, but absent from the boreal zone, occasional to locally common; June–October.

José Ignacio Fernández Simón

●● ***Butyriboletus regius*** is a very large, vivid rose to grey-rose *Butyriboletus* with fine, yellow pores. The bright yellow stem has a yellow reticulum. The cap surface becomes speckled-scaly with age. The flesh does not change colour and tastes mild. The spores measure 10.5–16 × 3–5 µm. Occurs with deciduous trees in warm habitats.

Butyriboletus appendiculatus ◁ is a rather similar species with flesh that does not change colour, but it has a clay-buff cap that lacks rose shades. *Butyriboletus fuscoroseus* ▷▷ and *B. fechtneri* ▷▷ are also similar but have moderately bluing flesh and rose tinges on the stem.

Southern, extending as far north as halfway up Germany, rare; late summer–autumn.

Michal Mikšík

Michal Mikšík

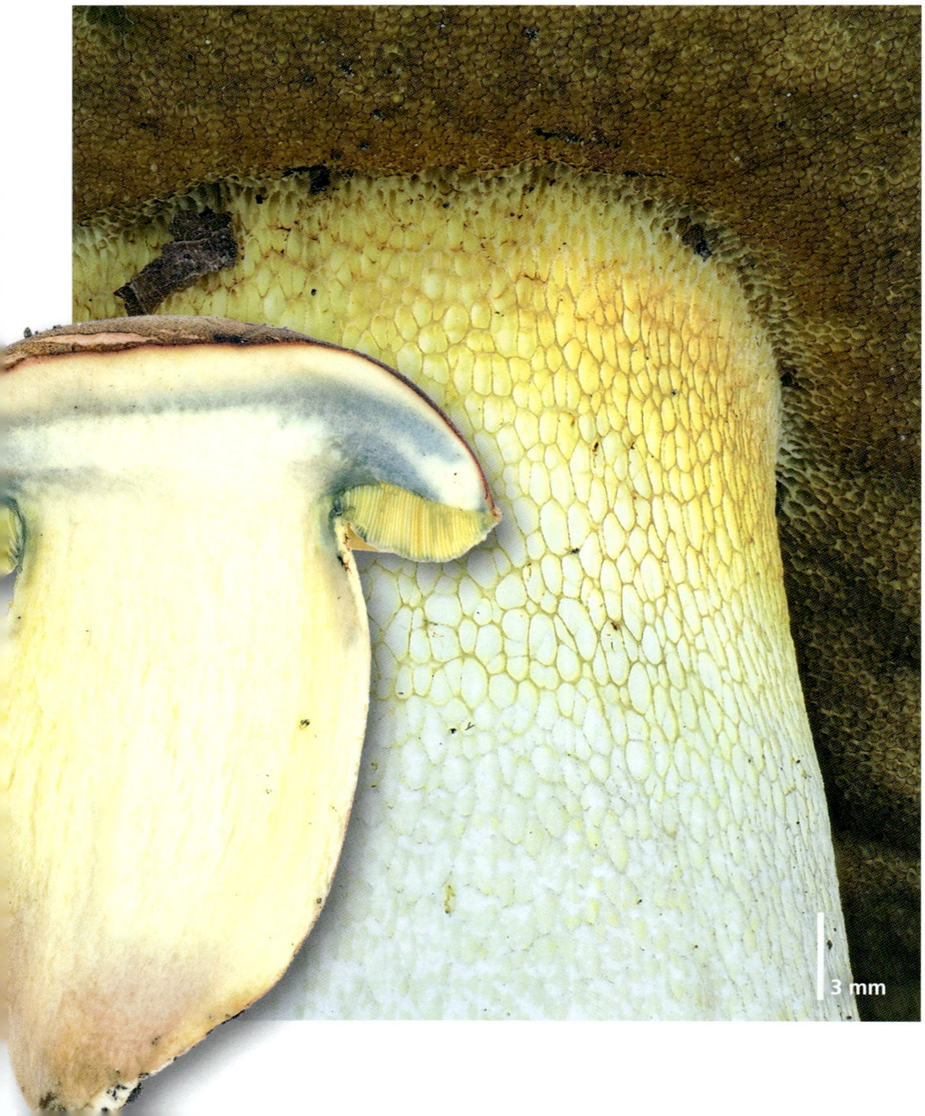

•• ***Butyriboletus fuscoroseus*** is a medium-sized, grey-rose to vinaceous-brown *Butyriboletus* with small, yellow pores. The stem has a yellow reticulum towards the apex and typically a reddish belt lower down. The flesh is mild and moderately bluing. The spores measure 12.5–14 × 4.5–5 µm. Occurs with deciduous trees, mostly *Quercus* and *Fagus*, on clay-rich soils in thermophilic forests.

The most similar species, *B. roseogriseus* ⋉, occurs predominantly with *Abies* in upland areas of central Europe and should completely lack red on the stem. *Butyriboletus fechtneri* ▷ is also very similar but has a pale grey, almost silvery, cap surface. *Butyriboletus appendiculatus* ◁◁ has a ± clay-buff cap, and the flesh does not stain blue; *B. regius* ◁◁ also has flesh that does not change colour. *Caloboletus radicans* ◁◁ looks similar, but is bitter-tasting and has strongly bluing flesh.

Southern and local, extending as far north as southern Denmark; mainly July–October.

Michal Mikšík

●● ***Butyriboletus fechtneri*** is a medium-sized, greyish-silvery *Butyriboletus* with small, yellow pores. The stem has a yellow reticulum and normally a reddish belt below the middle. The mild-tasting flesh is moderately bluing. The spores measure 12–14.5 × 5–5.5 µm. Occurs with deciduous trees in warm places, e.g. with *Fagus* and *Quercus*.

The most similar species, *B. fuscoroseus* ◁, has a more vinaceous-brown, non-silvery cap surface. *Butyriboletus appendiculatus* ◁◁ has a ± clay-buff cap, the flesh does not stain blue and the yellow stem is typically brownish rather than reddish towards the base and lacks a red belt. *Caloboletus radicans* ◁◁ is bitter-tasting, has strongly bluing flesh and produces somewhat larger fruitbodies.

Widespread but rare in the nemoral and hemiboreal zones, absent from the boreal zone; mainly August–September.

Michal Mikšík Michal Mikšík

Jiří Polčák

Imperator has extremely blue staining flesh and surfaces. The colours are often vivid and pores may be either yellow, orange or red, or a combination (see also page 807).

† ***Imperator torosus*** is a very large, bulky, strongly blue-staining bolete with a olive-brown cap and yellow pores. The yellow stem has a yellow reticulum. The spores measure 13–16 × 5–6.5 µm. Occurs with deciduous trees in warm habitats.

Forms of *I. rhodopurpureus* ▽ with yellow tubes are rather similar, but the caps are paler and often have rose tinges; the tubes frequently have an orange tinge over part of the pores. *Cyanoboletus pulverulentus* ▷▷ has the same colours and also stains a similar blue, but is much smaller and lacks a reticulum on the stem. Species of *Butyriboletus* ◁◁ do not stain as much and never on the cap surface.

Southern, extending as far north as southern parts of England, rare; late summer–autumn.

Jan Vesterholt

† ***Imperator rhodopurpureus*** is a medium-sized *Imperator* that may occur in forms with either yellow pores (pictured here), or in forms with orange to orange-red pores (see page 807 for a full account of the orange-pored form).

Imperator torosus △ has a darker, grey brown cap without pinkish colours. *Rubroboletus satanas* ▷▷, *Suillellus luridus* ▷▷ and *Neoboletus erythropus* ▷▷ may also occur in forms with yellow pores, but only rarely.

Michael Löhner

• ***Aureoboletus projectellus*** is a medium-sized, slender bolete with orange-brown to brownish-orange caps, pale yellow pores and a buff stem that has a very raised, coarse reticulum almost to the base. The spores are huge: 27.3–36.4 × 7.8–13.3 µm. Occurs with two-needled *Pinus* species on sandy soils, *e.g.* in coastal dune plantations.

The shape recalls species of *Leccinum* ◁◁ and *Leccinellum* ▷▷ but these have scales rather than a reticulum on the stem and only *Leccinellum crocipodium* ▷▷ has yellow pores. Species of *Xerocomus* ▷▷ may have a coarse reticulum on the upper parts of the stem but the cap surface is felty and the pores are a brighter yellow. Other European *Aureoboletus* ▷▷ are very different.

Mostly confined to the coasts of the Baltic Sea, where, locally, it can be very common; it was apparently introduced from North America to the eastern part of the Baltic region sometime before 2007 and is spreading westwards and northwards; mainly September–October.

Michael Löhner

Thomas Læssøe

Leccinellum can be separated from from *Leccinum* ◁◁ by a combination of a wine-red to black-staining flesh, a tapering-rooting stem and a tendency to develop cracks in the cap cuticle. The tubes can be vivid yellow, or whitish as in *L. pseudoscabrum* ◁◁. Found only with deciduous trees.

• ***Leccinellum crocipodium*** is a medium-sized, yellowish or yellow-brown *Leccinellum* with fine, yellow pores. The flesh stains initialy wine-red, later black. The cap cuticle soon develops a pattern of cracks. The stem scales are at first yellowish, later somewhat brown-tinged. The spores measure 12–18 × 5–7 µm, Qav. range 2.3–2.4. Occurs with *Quercus* on clay-rich soils, mostly in warm habitats.

Could perhaps be confused with *Hemileccinum impolitum* ▷ but that species does not blacken in the flesh and smells of iodoform at the stem base. Also rather similar to *Neoboletus erythropus* var. *pseudosulphureus* ▽ but that stains dark blue. In southern Europe two further yellow-pored leccinellums occur: *L. corsicum* ⋉ and *L. lepidum* ⋉, which grow with *Cistus* and southern species of *Quercus*.

Widespread, but local in the nemoral zones, with scattered occurences in the hemiboreal zone; July–October.

• ***Neoboletus erythropus* f. *pseudosulphureus*** is a medium-sized *Neoboletus* with completely or predominantly bright yellow colours and minute, ± yellow dots on the stem. All parts stain blue immediately. The spores measure 13–17 × 4–6 µm. Found mainly with *Fagus*.

This is a colour form of *N. erythropus* ▷▷, which is typically orange- to red-brown; intermediate forms also occur. *Neoboletus xanthopus* ▷▷ looks like a somewhat yellow *N. erythropus* but has a smoother cap and very fine scales on the stem.

Widespread, but rare; May–October.

Thomas Læssøe

• ***Hemileccinum impolitum*** is a medium-sized bolete with a pale vinaceous-brown cap and relatively fine, yellow pores. The stem has fine, brownish-red dots on the upper part but no reticulum (inserted picture), and the stem base smells strongly of iodoform. No parts of the fruitbody stain. The ± fusiform spores measure 11–16 × 4–5 µm. Occurs on clay soils with deciduous trees, e.g. *Quercus* and *Fagus*, mostly in warm habitats.

The smell of iodoform unique to this species and *Hemileccinum depilatum* ⋉. The latter has a "hammered" cap surface. May look similar to *Caloboletus radicans* ◁◁ but has species a very fine reticulum on the stem (hand lens needed!) and a bitter taste. *Leccinellum crocipodium* ◁ and *Neoboletus erythropus* f. *pseudosulphureus* ◁ have more distinct scales or dots on the stem and the flesh either blackens or turns blue.

Widespread but local, absent from the boreal zone; June–November.

• ***Cyanoboletus pulverulentus*** is a medium-sized bolete with a brown, dry cap and fine, yellow pores. The stem has yellow and red tinges and is smooth to finely dotted but lacks a reticulum. Immediately stains dark blue, including the cap surface. Spores smooth, measure 11–15 × 4–6 µm. Occurs with deciduous trees, both on clay and more sandy soils, mostly with *Fagus*, *Corylus* and *Quercus*.

Neoboletus erythropus f. *pseudosulphureus* ◁ has more dotty stem scales and a ± yellow cap. Other strongly bluing species have a reticulum.

Widespread, occasional, very rare or absent north of the nemoral zones; mainly June–October.

Species of ***Gyroporus*** are recognized by their chambered stems. The pores are initially white but become buff with age. The cap and stem are dry, and the stem does not have dots or a reticulum. Some species have strongly blue-staining flesh.

• ***Gyroporus cyanescens*** is a rather small to medium-sized, dry, finely felty bolete with pinkish-buff to buff colours. The stem is smooth, the same colour as the cap, but often with a narrow, pale zone around the top, and has internal chambers. All parts, and particularly the flesh, stain a characteristic ultramarine-blue immediately. The ellipsoid, smooth, hyaline spores measure 9–11 × 4.5–6 µm. Occurs with deciduous and coniferous trees on poor, mostly sandy soils.

Gyroporus lacteus ⋉ is a paler, almost white species that has recently been raised from varietal status. It stains blue even more markedly and is found in parts of temperate Europe. There may also be further cryptic species of *Gyroporus* with bluing flesh in the region that have yet to be recognized.

Widespread, occasional to rather rare, ± absent from the boreal zone; June–December.

Jens H. Petersen/Jan Vesterholt

• ***Gyroporus castaneus*** is a small to medium-sized bolete with a cinnamon-brown to orange-brown cap and stem, and white, later buff, fine pores. The stem is smooth and has chambered flesh that does not stain. The spores measure 8–11 × 5–6 µm. Occurs with deciduous trees, quite often *Quercus*.

May perhaps be confused with *Imleria badia* ▷▷ or species of *Xerocomellus* ▷▷ but these have massive stems and most stain ± blue. In the Mediterranean, a very similar species, *G. ammophilus* ⋉, occurs, which is considered to be poisonous.

Widespread but local, absent from the boreal zone; July–November.

Buchwaldoboletus lignicola is a rather large, cinnamon-brown to yellow-brown, dry, deeply felty bolete with yellowish-brown to curry-yellow somewhat decurrent tubes with coarse pores that slowly turn blue after bruising. The cheilo- and pleurocystidia have yellow contents. The hyphae have clamps. The somewhat fusiform, smooth spores measure 6.5–8.5 × 3–3.5 µm. Occurs on or at the base of conifer trunks or stumps, normally associated with the brown-rotting polypore *Phaeolus schweinitzii* (page 853), which it probably parasitizes.

Gyrodon ▷▷ and species of *Suillus* ▷▷ may also have decurrent tubes, but their caps are ± greasy or slimy and they are not associated with brown-rotted wood. Species of *Xerocomus* ▷▷ have felty cap surfaces but the tubes are hardly decurrent. *Buchwaldoboletus sphaerocephalus* ⋉ is very similar, but can be recognized by its vivid sulphur-yellow to lemon-yellow colours and the absence of clamps.

Widespread but rare and local, and absent from the boreal zone; mainly July–October.

Xerocomus forms medium-sized, hardly staining fruitbodies with felty cap surfaces, coarse, vivid yellow tubes and ± smooth cap cuticle hyphae.

• ***Xerocomus ferrugineus*** is a rather small to medium-sized bolete with a brownish or apricot-orange, rarely greenish or greyish, mostly rather pale, felty cap. The pores are coarse and yellow and the flesh is pale, rather whitish to somewhat rose, but often has a pale yellow zone above the tubes. The stem may have a ± distinct, coarse reticulum and the base has sulphur-yellow hyphal cords. The spores are fusiform and measure 10.5–13.5 × 4–5 µm, Qav. 2.7. Occurs with deciduous and coniferous trees.

Xerocomus subtomentosus ▷ has yellowish cap flesh and more brownish-rose flesh in the stem base; the hyphal cords are mostly white. *Xerocomus ferrugineus* has the narrowest spores: *X. subtomentosus* has spores Qav. 2.5; and *X. silwoodensis* ▽ and *X. chrysonemus* ▷ have a Qav. of 2.3 and 2.2, respectively.

Widespread and common; June–November.

Alan E. Hills/Jiří Polčák

• ***Xerocomus silwoodensis*** is a medium-sized bolete with a red-brown to orange-brown, dry, felty cap, often mottled at the margin, yellow pores and a pronounced, red-brown reticulum on the stem. The cap flesh is at first whitish, later (especially on drying) yellowish, and the stem flesh is ± brownish-rose at the base. Basal felt and hyphal cords are cream to yellow. The spores measure 9–14.5 × 4–6 µm, Qav. 2.3. Occurs with *Populus*.

Other species of *Xerocomus* are generally paler, have a less pronounced reticulum and grow with other hosts. *X. ferrugineus* △ has narrower spores (Qav. 2.7).

Known from England, Denmark and the Czech Republic; summer–autumn.

• ***Xerocomus subtomentosus*** is a small to medium-sized bolete with a dull apricot-orange, yellow-brown or ± olive, felty, dry cap. The stem may have a ± distinct, coarse reticulum. The yellow pores are rather coarse. The flesh is pale yellow in the cap and stem apex and noticeably brownish-rose in the stem base, which usually has ± white, hyphal cords. The fusiform spores measure 11–13 × 4.5–5.5 µm, Qav. 2.5. Occurs with deciduous trees, *e.g. Fagus*, *Quercus* and *Betula*.

This species can be hard to identify. *Xerocomus ferrugineus* ◁ has paler, whitish flesh that lacks brownish-rose colours in the stem base, and has bright yellow hyphal cords. *Xerocomus silwoodensis* ◁ is red-brown, has a very pronounced reticulum and grows with *Populus*. *Xerocomus chrysonemus* ▽, which also occurs with *Quercus*, has paler yellow flesh, yellow hyphal cords and relatively wider spores (Qav. 2.2).

Widespread and probably quite common, but due to the remarkably high level of confusion with similar species the precise distribution is not known; June–December.

Jan Vesterholt/Jens H. Petersen

• ***Xerocomus chrysonemus*** is a medium-sized bolete with a yellow-brown, curry-yellow to golden, felty cap and yellow, rather coarse pores. The cap flesh is whitish, but the stem flesh is sulphur-yellow at the base, as are the hyphal cords. The stem may have a ± distinct, coarse reticulum. The spores measure 9–14.5 × 4.5–7 µm, Qav. 2.2. Occurs with *Quercus* in parks.

The yellowish colours and the yellow flesh at the stem base are characteristic. The spores are, on average, somewhat wider than other species in *Xerocomus*.

Known from *e.g.* England, Sweden, Germany and the Czech Republic, probably rare; summer–autumn.

Martin Kříž

Hortiboletus, ***Rheubarbariboletus*** and ***Xerocomellus*** are recognized by their small fruitbodies (stem normally less than 1.5 cm wide), cracking or pruinose cap cuticle, pale yellow tubes and cap cuticle hyphae with striate incrustations (but see *Rheubarbariboletus armeniacus* ▷▷).

Hortiboletus species have rather wide spores (Qav. 2.3–2.4), while *Xerocomellus* ▷▷ species have a spore Qav. of 2.5–2.9. In addition, *Hortiboletus* species usually have orange grains in the stem base, a character that apparently never occurs in *Xerocomellus* species.

• ***Hortiboletus bubalinus*** is a rather small bolete with a pruinose to finely felty, pinkish-buff, clay-buff to vinaceous-brown cap. The pores are pale yellow and rather coarse, and the stem is yellowish with ± coral-red fibres and sometimes a few orange grains in the flesh at the base (lower inset picture). The cap flesh reddens in a zone below the cap cuticle and then stains blue towards the tubes; the stem flesh also stains blue. The ± fusiform spores measure 11.5–12 × 4.5–5 µm; Qav. 2.4. Occurs with deciduous trees, often *Betula*, in parks and gardens.

The colour of the cap flesh (in longitudinal section) with the so-called 'sunrise above the sea' effect is the best diagnostic character. The very similar *H. engelii* ▷ has flesh that does not change colour and a somewhat more felty cap surface, Other similar species have mostly more distinctly felty caps, *e.g. Xerocomellus chrysenteron* ▷▷, or much darker colours, *e.g. X. pruinatus* ▷▷. The closely related *H. rubellus* ▷▷ usually has a reddish cap and abundant orange dots or grains in the flesh at the stem base.

Widespread and quite common in the nemoral zones, becoming scarcer towards the north; June–October, peaking early in the season.

• ***Hortiboletus engelii*** is a relatively small bolete with a finely felty, rather pale grey-brown, cinnamon-brown to coral-red cap and pale yellow, somewhat bluing, rather coarse pores. The stem is yellowish with coral-red fibres, and the pale yellow flesh remains ± the same colour. The flesh in the stem base has numerous small, orange grains. The spores measure 12–13.5 × 5–5.5 µm; Qav. 2.4. Occurs with deciduous trees, especially *Quercus*, in parks and other open, disturbed places.

Hortiboletus rubellus ▷▷ also has orange grains at the stem base; they are also sometimes present in *H. bubalinus* ◁ but no other bolete has this character. *Hortiboletus rubellus* very quickly stains blue on the pores (and usually also on other parts) and mostly has a blood-red cap. *Hortiboletus bubalinus* generally has the same colours, but the cap is more distinctly pruinose, and the flesh displays a distinctive bluing pattern. *Hortiboletus engelii* was, for a long time, mistakenly called *Xerocomus armeniacus* and has also gone under the name *X. communis*. The true *armeniacus*, now *Rheubarbariboletus armeniacus* ▷▷, does not have orange grains in the stem base but has red-spotted cuticle hyphae in Congo Red; it is a southern species.

Widespread and at least locally common, but often overlooked, becoming scarcer towards the north; mainly June–September.

Hortiboletus engelii and other species in the *Xerocomus-Xerocomellus*-complex are often attacked by species of *Hypomyces* (page 1543) - the bolete flesh is transformed into a mass of yellow, asexual *Hypomyces* spores.

Thomas Læssøe

Jan Vesterholt

Jens H. Petersen/Thomas Læssøe

• ***Hortiboletus rubellus*** is a small bolete with finely felty, blood-red to brownish-coral-red cap and pale yellow, quickly bluing pores (in less than 10 sec.). The yellowish, ± bluing stem is covered by coral-red dots and fibres. The cap flesh is pale yellow, and the flesh at the stem base has small, orange grains. The flesh stains somewhat blue in the cap and the upper part of the stem. The spores measure 11.5–12.5 × 5–5.5 µm, Qav. 2.3. Occurs on ± disturbed soils with deciduous trees in parks, along road sides, *etc.*; mostly with *Quercus* and *Fagus*.

The orange grains at the base of the stem is a very important character, but are not always easy to locate, and several sections are often necessary to confirm their presence. The very fast blue-staining is also diagnostic. The very closely related *Hortiboletus engelii* ◁◁ also has orange grains in the stem base, but the cap colours are browner and the pores stain blue more slowly and not as strongly.

Xerocomellus ripariellus ▷, *X. intermedius* ⋉ and *Rheubarbariboletus armeniacus* ▷ have similar blood-red caps and mostly distinctly bluing pores but lack orange grains in the stem base. They also have longer spores (Q 2.6–2.9) and occur in damper habitats.

In southern Europe, a similar species can be found, *Xerocomellus redeuilhii* ⋉: this occurs predominantly with *Quercus* and has the same spore Q as *H. rubellus,* although the spores are somewhat smaller. However, the most important difference is that the stem base of *X. redeuilhii* has wine red flesh and lacks orange grains.

Widespread and rather common, becoming scarcer towards the north and ± absent from the boreal zone; June–November, mostly early in the season.

• ***Rheubarbariboletus armeniacus*** is a medium-sized bolete with a brick-red cap, pale yellow, bluing pores, a yellowish stem covered in coral-red dots and fibres, and somewhat blue-staining, pale yellow cap flesh. The basal stem flesh stains dark green with $FeSO_4$. The cuticle hyphae have red spots in Congo Red. The spores measure 11.5–13 × 4.5–5 µm, Qav. 2.6. Occurs with deciduous trees in warm habitats such as parks.

The red spots on the cuticle hyphae are unique. Other red species often stain strongly blue on the stem and do not become green with $FeSO_4$.

Southern and rare; summer–autumn.

Jiří Polčák

• ***Xerocomellus ripariellus*** is a very small to small bolete with a finely felty, blood-red, rarely deep orange-brown, white-cracking cap and pale yellow, quickly bluing pores. The yellowish stem is covered in coral-red dots and fibres. The pale yellow flesh stains blue in both cap and stem and lacks orange grains. The spores have indistinct ridges/stripes and measure 12.5–14 × 4.5–5 µm, Qav. 2.9. Occurs with deciduous trees, *e.g. Salix* and *Alnus*, in damp habitats.

Recalls a small *Hortiboletus rubellus* ◁ but that species has orange grains in the stem base, occurs in parklands and has wider spores with a Qav. of 2.3. *Xerocomellus internedius* ⋉ (also known as *X. fennicus*) is very similar but 5–100% of the spores are truncate and they are somewhat narrower (Qav. 2.6). *Rheubarbariboletus armeniacus* △ also lacks coloured grains in the stem base but is rather larger and the cap cuticle hyphae have red spots in Congo Red.

Widespread, but apparently very rare; July–October.

Jan Vesterholt

Jiří Polčák

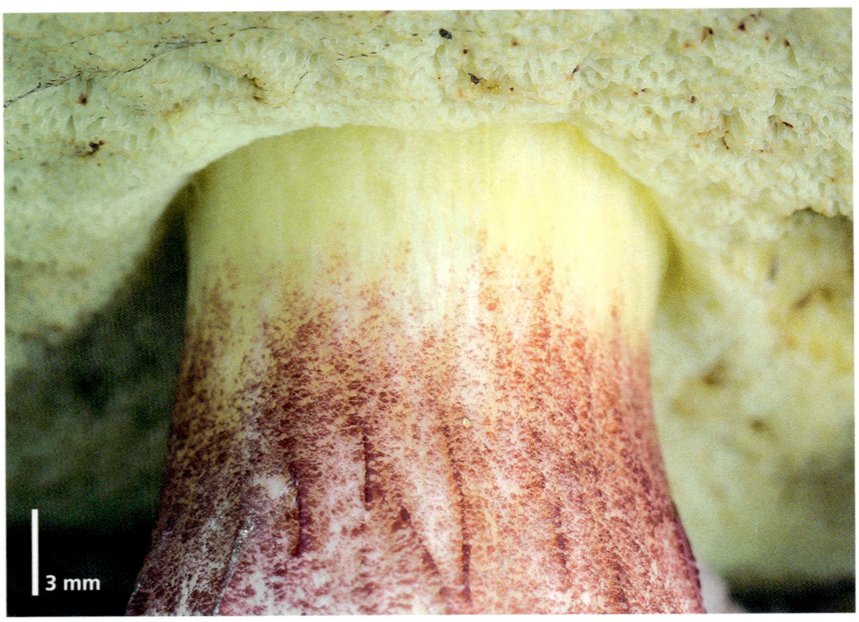

• ***Xerocomellus chrysenteron*** is a small to medium-sized bolete with a felty, ± grey-brown, reddish cracking cap and pale yellow, rather coarse pores. The stem is yellow at the apex but covered by reddish dots and fibres below. The flesh normally hardly stains, but a bluish zone is often present where the stem joins the cap flesh. The smooth spores measure 13–14.5 × 4.5–5.5 µm. Occurs predominantly with *Fagus*, but also in conifer forests.

Xerocomellus cisalpinus ▷ also cracks in a red pattern but stains distinctly blue, particularly on the stem. *Xerocomellus porosporus* ▷ also cracks but lacks the reddish colours and has a very characteristic grey-brown stem with narrow red and yellow belts around the upper part. The most similar species is perhaps *X. pruinatus* ▷▷ but this can be told by a narrow, red zone at the cap margin; the cap also cracks less distinctly and mostly along the margin; in young fruitbodies the surface is pruinose rather than felty. The caps of *Hortiboletus engelii* ◁◁ and *H. rubellus* ◁◁ may also crack, but are mostly redder, and they normally have orange grains in the stem base.

Widespread and common, becoming scarcer towards the north and ± absent from the boreal zone; May–December, peaking early in the season.

• ***Xerocomellus porosporus*** is a small to medium-sized bolete with a felty, ± grey-brown, rarely yellow-brown, whitish-cracking cap and yellow, rather coarse pores. The stem has a yellow apex, sometimes a narrow red zone, and is grey-brown below. The pores and stem stain blue. The spores are truncate and measure 13–15 × 5–6 µm. Occurs with deciduous and coniferous trees, but predominantly with *Fagus*.

The only brown-capped bolete with truncate spores. *Xerocomellus chrysenteron* ◁ and *X. cisalpinus* ▽ have other stem colours and red cracks.

Widespread and rather common, but absent from the boreal zone; June–November.

• ***Xerocomellus cisalpinus*** is a small to medium-sized bolete with a felty, brown, reddish-cracking cap and pale yellow, rather coarse pores. The stem is yellow above and covered by reddish dots and fibres towards the base. All parts stain blue, but particularly the stem flesh. The spores are very finely striate and measure 10.5–16 × 4–5.5 µm. Occurs predominantly with *Fagus* and *Quercus* in forests and parks.

Similar to a rather reddish version of *X. chrysenteron* ◁ but can be told by its prominent blue staining. The ornamented spores are also characteristic (smooth in *X. chrysenteron*) but the stripes can only be observed with a high quality microscope after staining with Melzer's reagent (× 1,000 needed!). *Xerocomellus pruinatus* ▷▷ also has striate spores but does not stain as blue and has a thinner cap felt. *Xerocomellus porosporus* △ may stain strongly, but usually lacks red colours on the cap and stem, and has smooth, truncate spores.

Widespread and rather common, but absent from the boreal zone; May–November.

Jan Vesterholt

The name "*cisalpinus*" implies a distribution south of the Alps, but soon after the name was coined this species was also shown to be widespread in northern Europe.

Thomas Læssøe

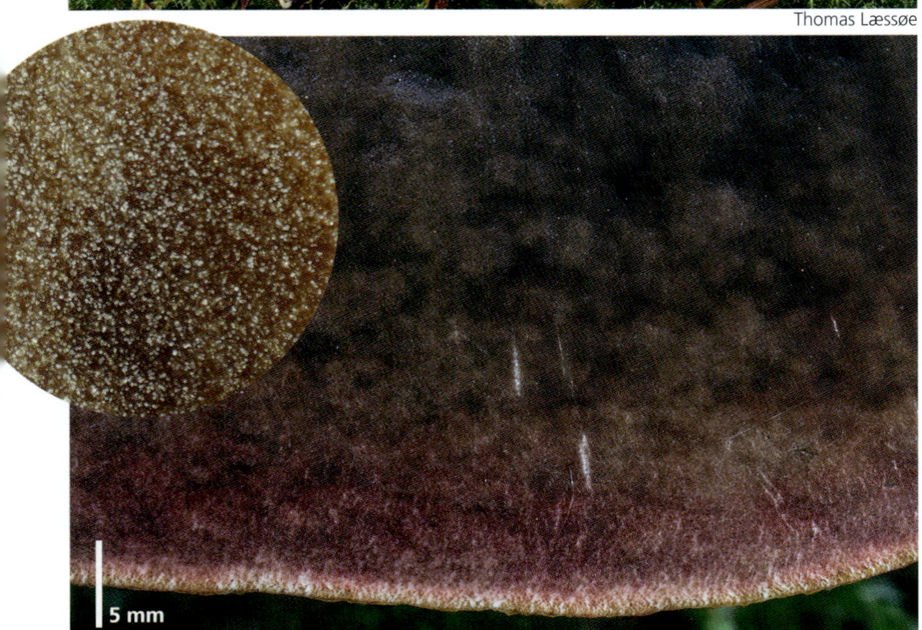

• ***Xerocomellus pruinatus*** is a small to medium-sized bolete with a pruinose to finely felty, at first purple-brown to red-brown, later grey-brown cap with a very narrow reddish margin, and pale yellow, rather coarse pores. The stem is yellow at the apex and covered below by reddish dots and fibres. There is usually no evident bluing, but the flesh may turn somewhat blue where the stem is joined to the cap flesh and at the stem base. The spores measure 13–14.5 × 4.5–5.5 µm and are very finely striate when seen in Melzer's reagent. Occurs predominantly with *Fagus*.

Can easily be confused with young specimens of *X. chrysenteron* ◁◁ but that species always has a distinctly felty cap surface when seen with a hand lens; *X. pruinatus* may, at most, develop a very thin felt in old specimens. *Xerocomellus chrysenteron* also lacks a reddish cap margin and tends to peak much earlier in the season. It is very difficult to use the ornament on the spores as a separating character as the striations are very hard to see using a normal compound microscope. Species in the genus *Hortiboletus* ◁◁ may, when young, display pruinose cap surfaces but have different colours and normally orange grains in the flesh at the stem base.

Widespread and common, but absent from the boreal zone; August–November, peaking late in the season.

• ***Pseudoboletus parasiticus*** is a small, thick-fleshed, finely felty, ± cracking bolete with a cinnamon-buff, honey-yellow to brownish-olive cap and stem, and pale yellow to yellowish-brown pores. The stem ± tapers towards the base and has an unevenly roughened surface. Surfaces and flesh either do not stain blue or only stain faintly blue. The ± fusiform spores measure 12–17.5 × 4–5.5 µm. Parasitizes *Scleroderma citrinum* (page 1232), mostly in damp *Betula* forests, but does not appear to cause obvious harm to the fruitbodies of the host. It has also been reported as occurring on other species of *Scleroderma*.

Distinctive and unlikely to be misidentified. A closely related species, *P. astraeicola* ⋉, parasitizes species of *Astraeus* (page 1255) and is known from, *e.g.,* Japan.

Widespread, occasional to rather rare and virtually absent from the hemiboreal–boreal zones; July–November.

The details concerning this species' parasitism are not well understood. It is striking how site-faithful it is, and frutibodies will appear from the same *Scleroderma* mycelium year after year, while neighbouring mycelia are free of frutibodies. This would indicate that the parasitism is carried out at the level of the mycelium, possibly involving the mycorrhizae of the *Betula/Scleroderma* association.

Thomas Læssøe

Jens Maarbjerg

• ***Imleria badia*** is a medium-sized bolete with an at first finely felty, but later greasy-slimy, greyish-brown to orange-brown cap. The stem is fibrillose-marbled and pale orange-brown. The fine, pale yellow pores stain blue after bruising, whereas the flesh hardly stains. The ± fusiform spores measure 12–15 × 4–5 µm. Occurs predominantly with conifers on poor soils, but also with *Fagus* on mor soils.

The most similar species is *Lanmaoa fragrans* ⋉, which is somewhat more robust, mostly clustered and has tapering stems; it occurs with deciduous trees in more southern parts of Europe. *Xerocomus* ◁◁, *Xerocomellus* ◁◁ and *Hortiboletus* ◁◁ generally have thinner stems, larger pores and do not have caps that become sticky with age and in wet weather. Although *Suillus* ▷▷ species are mainly sticky-slimy, the slime layer also runs down the stem and many species have a slimy ring. Due to the absence of red colours, *Imleria badia* may recall a species of *Boletus* but these have reticulate stems and the pores never stain blue.

Widespread and common; June–December.

Aureoboletus gentilis is a rather small bolete with sour-tasting cap slime, a pale vinaceous-brown to pale brownish-red cap and bright yellow pores and stem apex. No parts stain blue. Cheilo- and pleurocystidia have yellow contents; no clamps. Spores ± fusiform, 11–16 × 4.5–6 µm. Occurs with deciduous trees in warm habitats on clay-rich soils, mostly with *Fagus*.

The sour slime is unique. *Suillus granulatus* ▷▷ is rather similar but it occurs with *Pinus*, is paler and mild-tasting. Some species of *Xerocomus* ◁◁ and *Xerocomellus* ◁◁ can look similar but are never slimy.

Widespread, but rare and local, absent from the boreal zone; June–October.

Gyrodon lividus is a small to medium-sized bolete with a very greasy, pale olive-grey, ± unevenly coloured cap with an inrolled margin. The yellow, rather coarse, deeply decurrent tubes are well connected to the cap flesh (unlike in typical boletes) and stain blue. The pale, ± smooth stem becomes brownish towards the base and stains reddish after bruising. The cap flesh stains somewhat blue, whilst the stem flesh becomes orange-red. The ellipsoid, pale brown spores measure 4.5–6 × 3–4 µm. Mycorrhizal with *Alnus*, mainly in fairly dry forests.

Other species of *Suillus* ▷▷ with decurrent tubes hardly stain blue and occur with conifers.

Widespread, rather common in parts of the boreal zone, rarer in the south; July–November.

Suillus is characterized by a slimy universal veil. In the few species that have scaly caps, *e.g. S. variegatus* ▷▷, this can be difficult to observe, but in most species it is very obvious. The slimy veil is best detected on the lower part of the stem and often leaves a slimy ring zone two-thirds of the way up. Blue stains are rare in this genus. All species have smooth spores.

Mycorrhizal with conifers, starting on this and the next page spread with those that have *Larix* as their host.

• ***Suillus grevillei*** is a medium-sized *Suillus* with a slimy, smooth, yellow-brown to brownish-orange cap and pale yellow pores. The yellow to somewhat orange-brown stem is slimy and has a slimy ring zone. The spores measure 8–11 × 3–4 μm. Occurs with *Larix* on various soil types.

There are two varieties: var. *grevillei* (images on this page); and the darker var. *clintonianus* ▷. This is the only *Suillus* species associated with *Larix* that has a completely smooth cap. *Suillus viscidus* ▷ has an olive-grey, ± scaly cap. *Suillus luteus* ▷▷ and *S. flavidus* ▷▷ have smooth caps and stem rings but both occur with *Pinus*.

Widespread and common; May–November.

• ***Suillus grevillei* var. *clintonianus*** is a dark brownish-orange to red-brown variety of *Suillus grevilei*. The pores are pale yellow, and, as in the main variety, the stem has a slimy ring zone. It has incrusted pigment in the cap cuticle. The spores measure 8.5–12 × 3.5–4.5 µm. Occurs with *Larix*.

Recalls a very dark version of *S. grevillei* var. *grevillei* ◁. Could also be confused with *S. luteus* ▷▷ but that species occurs with *Pinus* and has a different, less red, stem colour.

Widespread but rather rare; mainly August–October.

• ***Suillus viscidus*** is a medium-sized, very slimy *Suillus* with an olive-grey, somewhat scaly marbled cap and olive-grey, rather coarse pores. The stem has a grey-brown, slimy coating and an indistinct ring zone. The spores measure 10–13 × 4–5.5 µm. Occurs with *Larix*, typically on ± disturbed mull soils in gardens and parks, more rarely in woodlands.

A very distinctive species and unlikely to be misidentified. It is also known as *Fuscoboletinus aeruginascens* and *F. viscidus*. A variety (or possibly species) with yellower colours is known from the southern parts of temperate Europe, and is typically named *S. bresadolae* var. *flavogriseus* ⋉ or *S. nueschii*.

Widepread, occasional to rather rare, absent from the boreal zone; May–November.

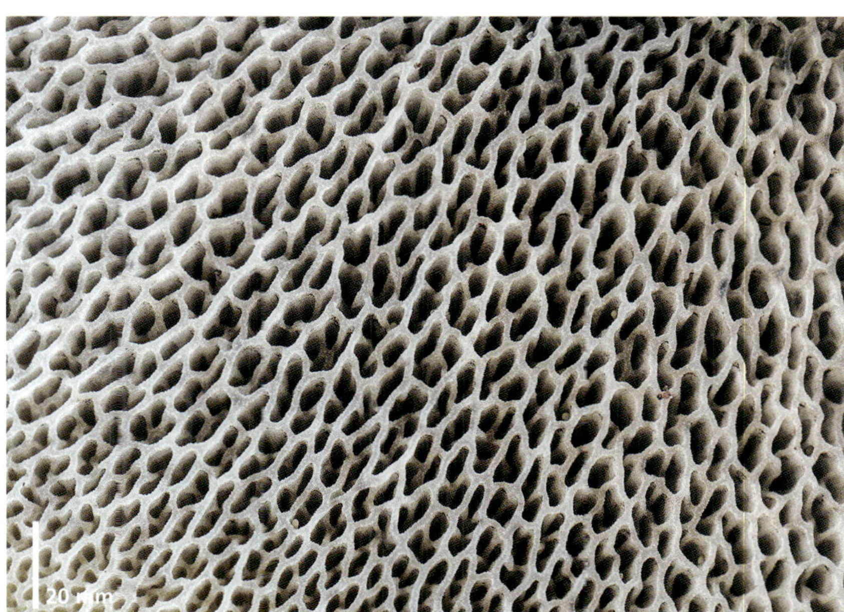

Gianluigi Boerio

Suillus tridentinus is a rather large *Suillus* with a ± yellow-brown cap that has darker scales along the margin, coarse, apricot-orange pores and somewhat decurrent tubes. The stem is massive and has a slimy ring zone. The spores are ± fusiform and measure 9–13 × 4–6 µm. Occurs in *Larix* plantations but also in natural stands in the Alps.

Suillus grevillei ◁◁ has small, pale yellow pores. *Suillus cavipes* ▷ has large, pale yellow pores and a hollow stem. The very rare *S. amabilis* ▷▷ has sordid pale yellow pores and occurs with *Pseudotsuga*.

Widespread, but very rare in temperate Europe; late summer–autumn.

Thomas Læssøe

Suillus asiaticus is a small to medium-sized, scaly *Suillus* with a carmine-red cap, decurrent tubes and large, pale yellow pores. The stem is massive, ± carmine red and has a greasy ring zone. The spores measure 11–12 × 4.5–5 µm. Occurs with *Larix*.

There are no other European species of *Suillus* with a carmine-red cap.

Occurs in the taiga from north-eastern Finland, through northern Russia to the Pacific coast, but has recently also been reported from France; summer–autumn.

Suillus cavipes is a rather large, soft-fleshed *Suillus* with an orange-brown, rarely yellow or whitish, greasy, scaly cap, large, pale yellow pores and distinctly decurrent tubes. The stem is hollow with a woolly ring zone. The spores measure 7–10 × 3.5–4 µm. Occurs with *Larix*, mostly on sandy soils.

Also known as *Boletinus cavipes*. The hollow stem is characteristic. There are two forms: *S. cavipes f. cavipes* (top two images) is reddish-brown; and *S. cavipes f. aureus* ▽ is orange-yellow. *Suillus tridentinus* ▷▷ is also scaly and has decurrent tubes but has a massive stem and yellow-brown to apricot-orange colours on both cap and pores. *Suillus ampliporus* ⋉ is a North American sister-species.

Widespread and locally common, very rare in the boreal zone and in the UK; June–November.

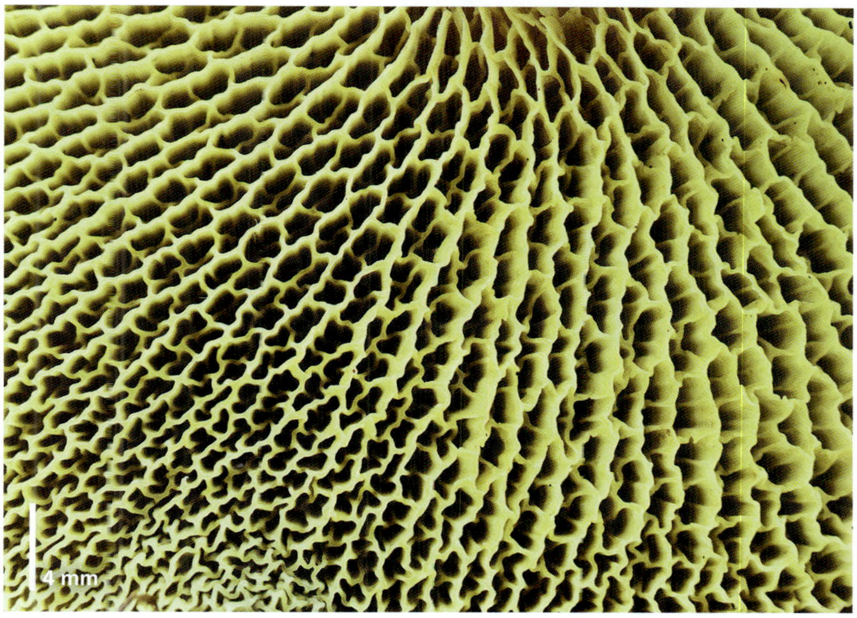

Suillus cavipes* f. *aureus is an orange-yellow form of *S. cavipes*. Apart from the colour it completely resembles *S. cavipes* f. *cavipes* △ (the main variety) and even the spores are the same. It can also occur in a paler, creamy-white form, but only rarely. Occurs with *Larix*, often co-occuring with *S. cavipes f. cavipes*.

Distributed as the main variety, but rarer; mainly July–October.

• ***Suillus luteus*** is a rather large, very slimy *Suillus* with a smooth, red-brown to orange-brown cap and fine, pale yellow pores. The tubes are not decurrent. The stem has a large, membranous, slimy ring, below which it is partly tinted the same colour as the cap. The spores measure 7–10 × 3–3.5 µm. Occurs with two-needle *Pinus* species on mostly poor soils.

Suillus grevillei ◁◁ also has a very evident ring zone but the cap is usually more orange-brown and it always occurs with *Larix*. *Suillus granulatus* ▷ and *S. collinitus* ▷ have ± similar cap colours and hosts but lack a ring on the stem.

Widespread and common; May–November.

• ***Suillus granulatus*** is a medium-sized *Suillus* with a viscid-slimy, smooth, orange-brown to cinnamon-brown cap and fine, pale yellow, mostly guttulating pores. The stem is yellow, lacks a ring and has a grainy-glandular surface, especially on the upper part and when young and fresh. The spores measure 8–10 × 2.5–3.5 µm. Occurs with two-needled *Pinus* species, mostly on calcareous soils, including in parks and in churchyards.

Suillus collinitus ▽ is very similar but has a somewhat wider cap and a rose tinge to the stem base. *Suillus luteus* ◁ has slightly darker cap colours and an distinct ring on the stem. *Suillus weaverae* ⋉ is a North American sister-species.

Widespread, occasional to locally rather common; May–November.

• ***Suillus collinitus*** is a rather large *Suillus* with a smooth, ± cinnamon-brown cap and pale yellow, rather fine tubes and pores. The stem is ± yellow, lacks a ring but has a grainy-glandular top and pinkish mycelium at the base. The spores measure 8.5–9.5 × 3.5–4 µm. Occurs with two-needled *Pinus* species, mostly on disturbed, ± calcareous soils.

The very similar, but slightly paler and smaller *S. granulatus* △ does not have a pinkish stem base. Suillus luteus ◁ has a stem ring.

A southern species that reaches as far north as southern Scandinavia, rare to occasional; August–November.

Jan Vesterholt

Suillus bovinus is a rather small *Suillus* with a convex, viscid, smooth, pale brown cap, rather large, straw-yellow pores and slightly decurrent tubes of uneven length. The stem is pale and lacks a ring zone. The spores measure 8–10 × 3–4 µm. Occurs with two-needled *Pinus* on poor, sandy soils, including in dune plantations.

Other ringless *Suillus* species have finer pores. *Suillus flavidus* ▽ occurs in similar places but has a conical cap and an evident ring zone on the stem. *Suillus bovinus* is associated with *Gomphidius roseus* (page 599).

Widespread and common; July–November.

Suillus flavidus is a rather small *Suillus* with a viscid, smooth, conical cap, an evident, slimy ring zone, rather pale buff to straw-yellow colours and rather coarse, irregular pores. The spores measure 8–10 × 3.5–4.5 µm. Occurs with two-needled *Pinus* on poor, often damp soils.

This species belongs to a difficult complex of closely related species. It may recall a pale *S. grevillei* ◁◁, but that species occurs with *Larix*. *Suillus bovinus* △ has ± the same colours but no ring zone.

Widespread, rare in the nemoral zones, but increases in abundance towards the north; July–November.

• ***Suillus variegatus*** is a medium-sized to large, greasy to almost dry, minutely scaly *Suillus* with clay-buff to grey-olive cap and pores, and usually somewhat bluing flesh and tubes. Lacks a ring. It has a metallic smell. The spores measure 7.5–11 × 3–4 µm. Occurs with two-needled *Pinus* species, mostly on sandy soils.

The pinkish-buff to grey-olive pores and ringless stem are good characters. *Suillus plorans* ▷ is very similar but occurs with five-needled *Pinus* species.

Widespread and common but slightly less frequent in southern parts of temperate Europe; June–November.

• ***Suillus plorans*** is a medium-sized, greasy, minutely scaly *Suillus* with clay-buff to grey-olive cap and pores and no stem ring. The spores measure 8.5–12 × 3.5–4.5 µm. Occurs with five-needled *Pinus* species.

Suillus variegatus ◁ is very similar, but occurs with two-needled *Pinus* species.

Found mainly in the Alps, but with some scattered records from arboreta or churchyards farther north; August–October.

Yves Deneyer

Suillus placidus is a medium-sized *Suillus* with a viscid, smooth, whitish to pale buff cap and white to pale yellow pores. The ringless stem has a ± red dotted-speckled surface. The spores measure 7.5–9 × 3–3.5 µm. Occurs with five-needled *Pinus* species (*Pinus strobus* and *P. cembra*).

Suillus granulatus ◁◁ is darker, only has dots on upper part of the stem and occurs with two-needled *Pinus* species.

Known from scattered plantations in northern parts of the nemoral zone, where it can be quite site faithful, and from the Alps; June–October.

Flemming Rune

Suillus amabilis is a rather large *Suillus* with a greasy, minutely scaly, red-brown to orange-brown cap, pale yellow, rather coarse pores and somewhat decurrent tubes. The stem is massive and has a woolly ring zone. The spores measure 8–12 × 3.5–5.5 µm. Mycorrhizal with *Pseudotsuga* in plantations.

The similar *S. cavipes* ◁◁ has a hollow stem and occurs with *Larix*. *Suillus lakei* is a synonym of *S. amabilis*.

This species originates in North America and was introduced to Europe with tree seedlings, very rarely recorded; late summer–autumn.

Fred Stevens

Rubroboletus is a genus of large, bulky, moderately bluing, poisonous boletes with fine ± orange pores.

† ***Rubroboletus satanas*** is a large to very large, firm-fleshed *Rubroboletus* with fine orange pores, a pale grey cap. The stem has a yellow to rose reticulum on the upper part and becomes ± rose and smooth towards the base. The flesh is pale yellow and moderately bluing. Mature specimens smell nauseous, recalling chicken manure, strongest from the cap surface. The hyphae in the stem base are amyloid. The spores are ± fusiform and measure 11–13.5 × 5–6 µm. Occurs with deciduous trees (*Fagus*, *Quercus*, *Tilia*) in warm habitats, typically on clay-rich or calcareous soils on coastal slopes.

Rubroboletus legaliae ▷ and *R. rhodoxanthus* ▷▷ are similar but have rose colours on the cap and lack the nauseous smell. Species of *Suillellus* ▷▷, *Neoboletus* ▷▷ and *Imperator* ▷▷ have an immediate blackish-blue reaction in the flesh and on the surfaces of fresh specimens.

Rather rare and southern, extending northwards to southern nemoral parts of Scandinavia; mainly July–October.

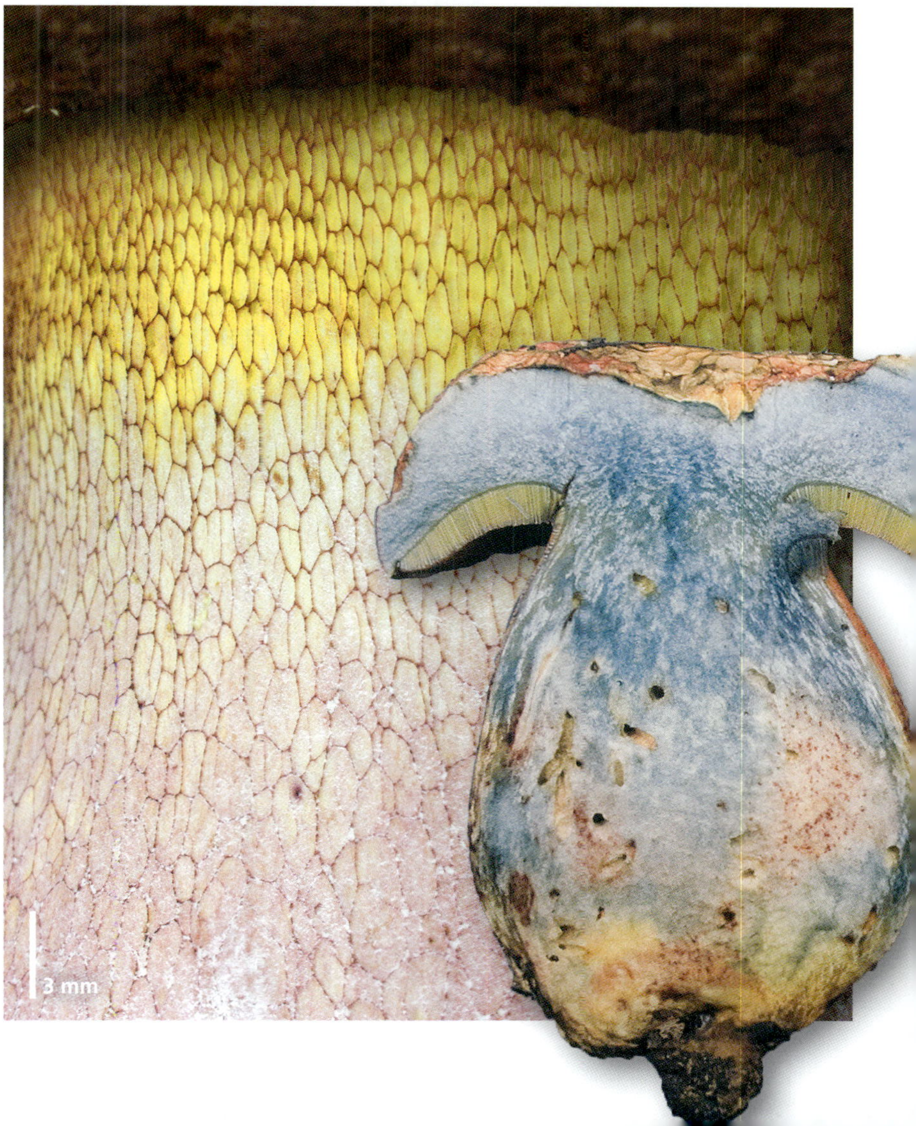

† ***Rubroboletus legaliae*** is a large *Rubroboletus* with orange pores (may be yellow in very young specimens) and ± pale rose tinges at the cap margin. The stem has an orange to carmine reticulum. The stem flesh is pale yellow and stains moderately blue along ± its full length. The basal hyphae have amyloid walls. The spores measure 11.5–13 × 5–6 µm. Occurs with *Fagus* and *Quercus* in warm sites, mostly open parkland with older trees.

Rubroboletus satanas ◁ is rather similar but has a completely pale cap that lacks rose colours, and an unpleasant, nauseous smell. *Rubroboletus rhodoxanthus* ▷▷ has non-staining stem flesh. The very similar *R. rubrosanguineus* ⋉ has stem flesh that is bluing along its entire length, but grows in montane *Abies* forest. *Rubroboletus pulchrotinctus* ⋉ occurs farther south in Europe; it is just as poisonous and has bright rose colours on the cap and yellow pores.

A rare southern species extending northwards to southern nemoral parts of Scandinavia; July–October.

Michal Mikšík

Patrice Tanchaud

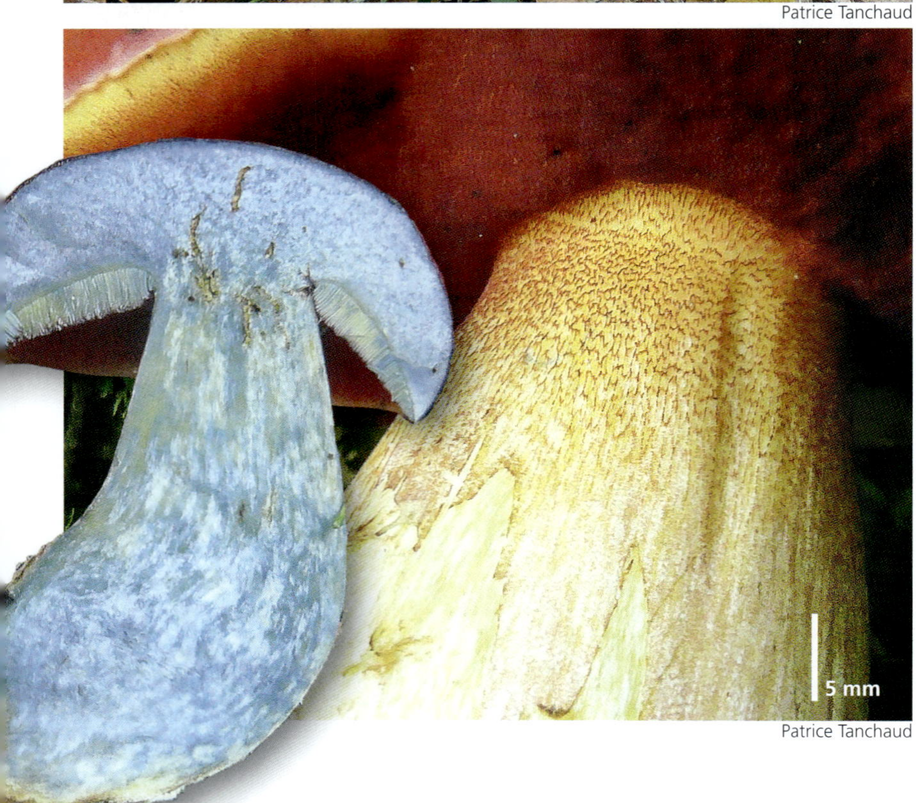

Patrice Tanchaud

† ***Rubroboletus rhodoxanthus*** is a large *Rubroboletus* with orange pores and a ± pale rose cap. The stem has an orange reticulum and yellowish flesh. Stains moderately blue in the cap flesh but not in the stem. The basal stem hyphae have amyloid walls. Spores 10–14 × 4–5.5 µm. With deciduous trees in warm forests and parks.

The stem flesh in *R. legaliae* ◁◁ stains entirely blue; *R. satanas* ◁◁ has a pale cap that lacks rose colours and a nauseous smell; *R. rubrosanguineus* ⋉ occurs with *Abies* and the stem stains entirely blue.

A rare southern species that reaches the Netherlands and south-eastern Sweden; late summer–autumn.

Rubroboletus dupainii is a medium-sized *Rubroboletus* with orange pores and a ± red, shiny cap. The stem has predominantly small, orange scales, but may also have a rather narrow zone with a fine reticulum at the top. The flesh stains moderately blue and a more monochrome pale blue towards the stem base. The spores measure 11–12 × 4.5–5.5 µm. Occurs with *Quercus* in warm habitats.

The stem scales recall the dots/scales on the stem surface of *Neoboletus erythropus* ▷▷, but that species has a brown cap. The other *Rubroboletus* species all have a prominent reticulum on the stem and never have a shiny red cap surface.

A southern species, which only reaches as far north as southern Germany, rare; late summer–autumn.

Gianluigi Boerio/Jens H. Petersen

†***Imperator rhodopurpureus*** is a large, extremely strongly blue-staining *Imperator* with a pale grey, yellow, vinaceous-brown to purple-flamed cap and varying yellow and orange colours on the stem surface and reticulum. The pores vary from reddish orange to pure yellow (see page 778). The spores measure 12–13.5 × 5–5.5 µm. Occurs with deciduous trees (mostly *Quercus* and *Fagus*) in warm habitats.

Fruitbodies with yellow pores may be confused with *I. torosus* ◁◁. *Suillellus* ▷▷ and *Neoboletus* ▷▷ stain blue almost as strongly but have ± brown caps. The southern *I. luteocupreus* ⋉ has distinctly yellow colours on the cap. *Neoboletus erythropus* f. *pseudosulphureus* ◁◁ may recall yellow forms of *I. rhodopurpureus*, but lacks a reticulate stem. *Cyanoboletus pulverulentus* ◁◁ also lacks a reticulum. Pale-capped species of *Rubroboletus* ◁ stain much less strongly.

A rare southern species that reaches as far north as southern nemoral Scandinavia; mainly July–October.

Jan Vesterholt

Suillellus is characterized by fleshy fruitbodies with fine orange to red pores, strong and fast blue staining on all parts, and wine-red flesh in the stem base. When old, dry or attacked by moulds, the blue reaction may be partly absent. Somewhat poisonous as raw.

•• ***Suillellus luridus*** is a rather large to large, strongly bluing bolete with a honey-yellow, cinnamon-brown to yellow-brown, dry cap, sometimes with purple-red 'flames', and orange-red or very rarely yellow pores on a red surface (in section a red line is seen above the tubes). The stem has an orange-red reticulum almost to the base. The flesh in the base reacts blue with iodine reagents. The spores measure 11–15 × 5.5–7 µm, Q 2.1–2.4. Occurs with deciduous trees on clay soils, mostly in open forests, parks, *etc.*

The slender *S. mendax* ▽ only has a reticulum on the upper part of the stem and has red spots below. It is mostly more carmine-red to purple-red on the cap, pores and stem and should have wider spores (Qav. 2.6–2.8). Species of *Rubroboletus* ◁◁ stain less strongly blue and have paler caps. Species of *Imperator* ◁◁ stain very blue, including on the cap surface, but mostly have yellower pores and paler caps.

Widespread and common, but ± absent from the boreal zone; May–October, peaking early in the season.

Michal Mikšík

•• ***Suillellus mendax*** is similar to a rather purple version of *S. luridus* △, and in section a red line can be seen above the tubes. The reticulum quickly turns into dots down the stem. The spores measure 12.5–15.5 × 4.5–5.5 µm, Qav. 2.6–2.8. Occurs with deciduous trees in warm habitats.

Almost inseparable from *S. luridus* that usually is more orange and has a more pronounced reticulum; it should also have narrower spores. *Suillellus comptus* ⋉ has a reticulum as in *S. mendax* and spores as in *S. luridus*, but usually lacks a red zone between the tubes and cap flesh – a southern species.

Distribution and frequency not clear but widespread; summer–autumn.

Thomas Læssøe

•• ***Suillellus queletii*** is a medium-sized, strongly bluing *Suillellus* with a cinnamon-brown, red-brown to brownish-orange, ± felty cap and apricot-orange to brownish-orange, rarely more yellow pores. When kept, the whole fruitbody may discolor wine-red. The stem is finely dotted and ± tapering towards the base. The flesh in the base is wine-red. The spores measure 10–14 × 5.5–7 µm, Q 1.9–2.3. Occurs with deciduous trees on mostly clay-rich soils in warm habitats.

Neoboletus xanthopus and *N. erythropus* ▷▷ have similar, but more distinctly punctate stems, usually ± scarlet pores and do not normally have wine-red flesh at the stem base; they also have a higher Q value (2.7–3.3). Other, similar *Suillellus* species have ± reticulate stems. *Cyanoboletus pulverulentus* ◁◁ is strongly blue staining and smooth-stemmed, but has yellow pores and lacks wine-red flesh in the stem base.

Widespread and locally common, rare in the hemiboreal zone and absent from the boreal zone; June–October.

Jan Vesterholt/Jens H. Petersen

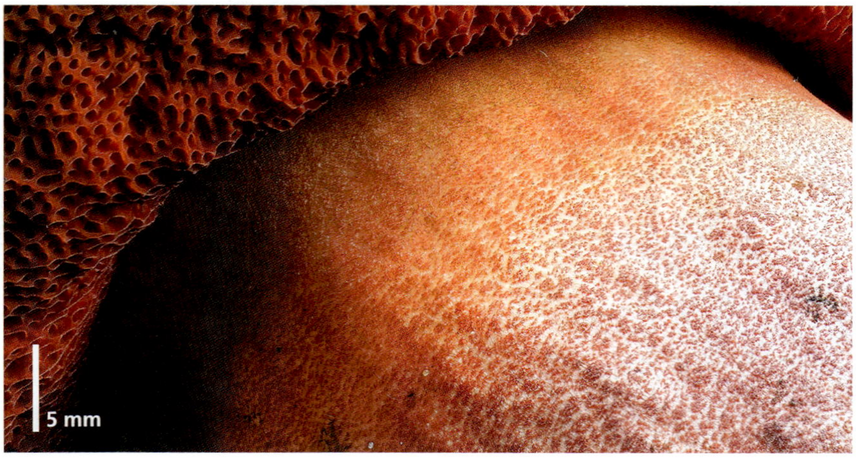

Neoboletus is recognized by a strong, blue stain, absence of a reticulum, mostly yellow, blue-staining flesh in the stem base (it may rarely be more wine-red) and a high spore Q. The pores are usually orange-red. The correct genus name might be *Sutorius*.

•• ***Neoboletus xanthopus*** is a large *Neoboletus* with a yellow-brown, almost smooth cap and with yellowish to orange colours on the stem, which is very fine-dotted on the upper part (hand lens required!). The spores measure 13.5–16 × 4–5.5 µm, Q 2.8–3.3. Known from clay soils with *e.g. Carpinus* and *Quercus* in warm habitats, including open, old parks.

Neoboletus erythropus ▷ has a ± felty cap surface and coarser dots on the stem, which may be dominated by orange-red colours; it also usually occurs on mor soils and it is not thermophilic. *Suillellus queletii* ◁◁ has more brownish-orange pores and a lower spore Q. *Cyanoboletus pulverulentus* ◁◁ has yellow tubes.

A recently described species that is known from central Europe and also recently confirmed as being present in southern nemoral Scandinavia; summer–autumn.

•• ***Neoboletus erythropus*** is a rather large, strongly blue-staining *Neoboletus* with an orange-brown to red-brown, finely felty cap and ± scarlet, fine pores. The stem is covered with small orange-red to scarlet dots or scales. The flesh is yellow, including in the stem base. The spores measure 13–17 × 4–6 µm, Q 2.7–3.1. Occurs with deciduous and coniferous trees, mostly on mor soils.

Neoboletus xanthopus ◁ has a smoother, yellow-brown cap and a more finely-dotted stem, often dominated by yellow colours. *Suillellus queletii* ◁◁ may also have tiny red dots on the stem, but the pores are more brownish-orange, the cap colour is more variable, the spores have a lower Q value and the flesh in the stem base is wine-red. *Suillellus luridus* ◁◁ and *S. mendax* ◁◁ both have a well-defined reticulum on the stem and a distinct red line separating the tubes from the cap flesh. *Neoboletus erythropus* may occur in yellower forms and can sometimes be completely sulphur-yellow (see *Neoboletus erythropus* f. *pseudosulphureus* ◁◁).

Widespread and common; May–December, peaking early in the season.

Chalciporus piperatus is a small bolete with a greasy, pale brown to vinaceous-brown cap and coarse, brownish-orange to coral pores. The stem is relatively long and the basal flesh is strikingly yellow. The taste is peppery. The smooth, ± fusiform spores measure 8–11 × 4–5 µm. Occurs predominantly with conifers and typically together with *Amanita muscaria* (page 363), whose mycelia it apparently parasitizes.

Chalciporus amarellus ⋉ and *Rubinoboletus rubinus* ▽ have a similarly yellow flesh in the stem base, but somewhat more rose to carmine-red colours on the pores and stem. *Rubinoboletus rubinus* also tastes mild rather than peppery, and has unusually pale spores.

Widespread and common; June–November.

Jaroslav Malý

Rubinoboletus rubinus is a small bolete with a ± pale brown cap and rather coarse, rose to carmine-red pores. The stem has rose to carmine-red colours and the basal flesh is yellow. The taste is mild. The rather pale spores measure 5.5–8.5 × 4–5.5 µm. Occurs mostly with *Quercus* in warm habitats.

May be confused with *Chalciporus piperatus* △ and *C. amarellus* ⋉ but these species taste hot and peppery and have more orange-red colours on the pores and stem.

Widespread in Central Europe and is also found in England, uncommon; summer–autumn.

Porphyrellus porphyrosporus is a medium-sized, dull grey-brown to greyish-brown bolete with large, cinnamon-brown pores and a smooth stem with similar colours. Fresh specimens stain sordid red and bluish when bruised or sectioned. The smooth, ± fusiform spores measure 12–19 × 5–7.5 µm. Occurs predominantly with *Fagus*, but may also occur with other deciduous trees and with *Abies*.

A unique and easily identifiable species in temperate Europe.

Widespread, occasional, absent from the boreal zone; July–November.

Strobilomyces strobilaceus is a rather large to large bolete with a coarsely scaly, dark grey cap and large, grey pores. The cap margin has coarse, triangular scales. The grey stem is very firm (almost woody), floccose-scaly and has a woolly ring. The flesh stains orange-red when cut. The ± spherical, blackish spores have a coarse, reticulate ornament and measure 10–12 × 8.5–11 µm. Occurs with deciduous trees, mostly on clay soils, typically with *Fagus*.

Several similar species occur In other parts of the world, but in temperate Europe this species is unique and unmistakable.

Widespread, occasional, locally common, absent from the boreal zone and rare in the hemiboreal zone; July–November.

Fungi of Temperate Europe

Fungi of Temperate Europe is translated from *Nordeuropas svampe* by Thomas Læssøe and Jens H. Petersen.

Published by Princeton University Press
41 William Street, Princeton, New Jersey 08540
99 Banbury Road, Oxford OX2 6JX

press.princeton.edu

ISBN (two-volume set): 978-0-691-18037-3.

Project management, photo work, composition and graphic design: Jens H. Petersen.
Prepress: Jens H. Petersen and Narayana Press, Gylling, Denmark.
Editor: Axel Kielland.
Print: Grafotisak, GPS Group.

Printed and bound in Bosnia & Herzegovina.

FSC www.fsc.org MIX Supporting responsible forestry FSC® C110418

10 9 8 7 6 5

Colour coding used within the identification wheels

other similar fungi

the group contains fungi living in mutualistic symbiosis (*e.g.* lichens and ectomycorrhizal fungi, p. 17)

the group contains parasitic fungi (fungi that weaken or kill their host, p. 16)

the group contains brown rotters, p 14 (white rotters are shown on a neutral background)

Edibility symbols – only important edible and seriously or deadly poisonous species are marked.

- ● acceptable edible after proper cooking
- ●● high quality edible after proper cooking
- †? presumed seriously poisonous
- † seriously poisonous
- †† deadly poisonous

References to other species

△ = found above on the same page
▽ = found below on the same page
◁ & ▷ = found on the opposite page
◁◁ = covered earlier in the chapter
▷▷ = covered later in the chapter
⋉ = not treated further